T0331394

An Introduction
to Quantum Transport
in Semiconductors

An Introduction to Quantum Transport in Semiconductors

David K. Ferry

PAN STANFORD PUBLISHING

Published by

Pan Stanford Publishing Pte. Ltd.
Penthouse Level, Suntec Tower 3
8 Temasek Boulevard
Singapore 038988

Email: editorial@panstanford.com
Web: www.panstanford.com

British Library Cataloguing-in-Publication Data
A catalogue record for this book is available from the British Library.

An Introduction to Quantum Transport in Semiconductors
Copyright © 2018 by Pan Stanford Publishing Pte. Ltd.
All rights reserved. This book, or parts thereof, may not be reproduced in any form or by any means, electronic or mechanical, including photocopying, recording or any information storage and retrieval system now known or to be invented, without written permission from the publisher.

For photocopying of material in this volume, please pay a copying fee through the Copyright Clearance Center, Inc., 222 Rosewood Drive, Danvers, MA 01923, USA. In this case permission to photocopy is not required from the publisher.

ISBN 978-981-4745-86-4 (Hardcover)
ISBN 978-1-315-20622-6 (eBook)

Printed in the USA

Contents

Preface

The density of transistors in integrated circuits has grown exponentially since the first circuit was created. This growth has been dubbed Moore's Law. Now, why should this be of interest to the engineer or scientist who wants to study the role of quantum mechanics and quantum transport in today's world? Well, if you think about the dimensions that are intrinsic to an individual transistor in modern integrated circuits, about 5–20 nm, then it is clear that these are really quantum mechanical devices. In fact, we live in a world in which basically all of our modern microelectronics have become quantum objects, ranging from these transistors to the world of lasers and light-emitting diodes. It is also not an accident that this world is created from semiconductor materials, because these materials provide a canvas upon which we can paint our quantum devices as we wish. Of course, silicon is the dominant material since it is the base for the integrated circuits. But, optical devices are created from a wide range of semiconducting materials in order to cover the wide spectrum of light that is desired; from the ultraviolet to the far infrared.

I have had the good fortune to be an observer, and occasional contributor, to this ever-increasing world of microelectronics. I have followed the progress from the very first transistor radio to today's massive computing machines which live on a chip of about 1 cm^2. Over these years, I have become involved in the study of quantum devices and the attempts to try to write down the relevant theoretical expressions and find their solutions. As an educator, this led to many attempts to devise a course in which to teach these complicated (both then and now) quantum approaches to device physics. As with most people, the effort began with Kadanoff and Baym's excellent but small book on Green's functions. It became easier when Steve Goodnick and I undertook to write the book *Transport in Nanostructures*, which appeared in 1997. But, neither this book, nor its later second edition, was a proper textbook, and it contained far too much material to contemplate a one semester course on the topic. Nevertheless, we pressed forward with its use as a text several times in the intervening years. As age has crept not so slowly upon me, it became evident that it was time to try to put down my vision of a textbook on the topic. I guess it became evident that it was going to be now or never, and so I undertook to create this textbook (and I have to thank Stanford Chong for pushing me to do this). There are, of course, many other textbooks on Green's functions, but not so many that each one of them can treat all of the approaches to quantum transport. According to me, a more thorough coverage is essential. Despite the glorious claims of its practioners, nonequilibrium Green's functions are not the entire answer to the problem, and this is becoming evident as we experimentally probe more and more into questions of quantum coherence in real systems.

As evidenced by this book, I have finished the task with this version. I am sure that no author has ever finished a science text without immediately (or at least within a few minutes of seeing the published book) being worried that they have missed important points or should have said it differently. I know from my other books that, in looking back at them (which is often with the textbooks), I wonder what I was thinking when I wrote certain passages, especially as there are better ways to express something, which also crop up in retrospect. Nevertheless, I hope that this book will serve as a good reference for others as well as myself. It is designed to be more than a one semester course, so that the teacher can pick and choose among the topics and still have enough to fill a semester. It is not a first-year graduate course, as the student should have a good background in quantum mechanics itself. Typically, the prior attempts to put the course together have suggested that the student be "a serious-minded doctoral student," a phrase my own professor used to describe a one semester course out of the old 1100+ page Morse and Feshbach. The field has a lot of mathematical detail, but sometimes the simpler aspects have been blurred by confusing presentations. I don't know if I can claim that I have overcome this, but I have tried. Hopefully, the readers will find this book easier to use than some others.

I have benefitted from the interaction with a great many very bright people over the years, who have pushed me forward in learning about quantum transport. To begin with, there were John Barker, Gerry Iafrate, Hal Grubin, Carlo Jacoboni, Antti-Pekka Jauho, and Richard Akis, who remain friends to this day, in spite of my inherent grumpy nature. In addition, I have learned with and from Wolf Porod,

Walter Pötz, Jean-Jacques Niez, Jacques Zimmermann, Al Kriman, Bob Grondin, Steve Goodnick, Chris Ringhofer, Yukihiko Takagaki, Kazuo Yano, Paolo Bordone, Mixi Nedjalkov, Anna Grincwajg, Roland Brunner, and Max Fischetti, as they passed through my group or were collaborators at Arizona State University. Then, there were my bright doctoral students who worked on quantum theory and simulations: Tampachen Kunjunny, Bob Reich, Paolo Lugli, Umberto Ravaioli, Norman "Mo" Kluksdahl, Rita Bertoncini, Jing-Rong Zhou, Selim Günçer, Toshishige Yamada, Dragica Vasileska, Nick Holmberg, Lucian Shifren, Irena Knezevic, Matthew Gilbert, Gil Speyer, Aron Cummings, and Bobo Liu.

In addition, I have had the good fortune to collaborate with a number of excellent experimentalists, particularly John Bird, but also over the years with Yuichi Ochiai, Koji Ishibashi, and Nobuyuki Aoki in Japan. Then, there are my doctoral students who labored on the quantum device experiments: Jun Ma, David Pivin, Kevin Connolly, Neil Deutscher, Carlo da Cunha, and Adam Burke. These are long lists, both here and in the previous paragraph, but the present work is really the result of their work. Of course, I have to thank my long persevering wife, who puts up with my shenanigans, and without whom I probably wouldn't have amounted to much.

David K. Ferry

Fall 2017

Chapter 1

Introduction

The transport of carriers, electrons and holes, in semiconductors has been of interest for quite some time. It certainly became a subject of central interest when the inventors of the transistor were trying to understand the properties of the carriers in this new device [1]. But almost immediately, there was interest in the behavior of the carriers at high electric fields, in efforts to understand the breakdown of the oxides in use at that time [2]. Of course, there was increased interest in the materials important to the new semiconductor devices, such as silicon [3]. By understanding the transport properties of the carriers, one could certainly understand more about the physics governing the interactions between the carriers and their environment—the surfaces, the phonons, and so on. Over the decades since, we have found that the careful modeling of transport and the semiconductor devices has contributed to the ability to push the technology to ever smaller physical sizes. Today, the critical length in a modern tri-gate transistor is approaching the distance between the individual atoms of the underlying semiconductor. Indeed, we have seen the fabrication of a device in which the active region consists of a single phosphorus atom [4]! If the atoms of the semiconductor are held together by quantum mechanical forces, then it is quite likely that we will need to describe the transport in such small transistors via

An Introduction to Quantum Transport in Semiconductors
David K. Ferry
Copyright © 2018 Pan Stanford Publishing Pte. Ltd.
ISBN 978-981-4745-86-4 (Hardcover), 978-1-315-20622-6 (eBook)
www.panstanford.com

a fully quantum mechanical approach (and, indeed, this was done to gain understanding of the physics within the single-atom transistor).

Thus, it is clear that more detailed modeling of the quantum contributions in modern semiconductor devices is required. These contributions appear in many forms: (1) changes in the statistical thermodynamics within the devices themselves as well as in its connection and interaction with the external world, (2) new critical length scales, (3) an enlarged role for ballistic transport and quantum interference, and (4) new sources of fluctuations, which will affect device performance. Indeed, many of these effects have already been studied at low temperatures where the quantum effects appear more readily in such devices [5].

A fair question to ask at this point is why are not quantum effects seen in today's very small devices? In fact, quantum effects are an integral part of the design of today's devices, but they are not seen in the observed output characteristics for one good reason. Most of the important quantum effects are in a direction normal to that in which the current flows. But this does not diminish their importance. For example, strain is a common part of every device in a modern microprocessor. This strain is used to distort the band structure and improve the mobility of the electrons and holes. So controlled introduction of quantum modifications has been a part of the fabrication of devices for more than a decade. And there has been an ongoing effort to design and create simulation tools for the semiconductor world, which incorporate the quantum effects in the very base of the physics included within the tool. On the other hand, many people have studied quantum transport (and written books on the subject) in metals for quite a long time. But semiconductors are not metals. The differences are large and significant. So while one would like to extrapolate from what is known in metals, this can be taken only so far. What we would like to do here is to examine what approaches work for semiconductors and to try to learn from the many places where studies have been done for these materials and the resulting devices. In the following sections, we will try to describe what the key features are that differentiate quantum transport from the classical transport world that has been used so successfully in semiconductors and semiconductor devices.

1.1 Life Off the Shell

One of the hallmarks of classical physics is the strong connection between energy and momentum. For a free particle, it is clear that the relationship $E = p^2/2m$ is a fundamental relationship. Even when we move to the quantum mechanics of the energy bands in semiconductors, we still hold to the tenet that the energy is well defined at each and every value of momentum, in this case the crystal momentum. But then we have to remind ourselves that the normal calculation of energy bands is done for the so-called empty lattice, in which we seek the allowed energy levels for a single electron [6]. This is the reason that density functional theory calculations for the energy bands of semiconductors are notorious for not getting the energy gap (between the conduction and valence bands) correct. The usual approach is to use a linear density approximation for the exchange and correlation energies, but without really correcting the single-electron approximation. What has to be done is to account for the interactions among the electrons, usually through what is known as a *GW* approximation, where *G* is a Green's function and *W* is an interaction self-energy. This lowers the entire valence band through the cooperative interactions and provides a correction to the energy gap. How this is performed is not the subject we want to address here, but the phrase "self-energy" is a topic we want to begin to discuss, especially as it is connected with going beyond the simple relationship $E = p^2/2m$. In three-dimensional momentum space, this latter relationship defines a spherical shell of the given energy, and all states with this energy lie on this shell. In any computation, this connection between energy and momentum is introduced by the delta function

$$\delta(E - p^2/2m). \tag{1.1}$$

When we move to the quantum world, however, this constraining relationship is no longer either valid or required. As we will see later, energy and momentum are separate variables, although they do have a connection, just one not so strict as Eq. (1.1).

The relationship (1.1) is a sort of resonance, and we can actually see how it is modified with a classical analog. Let us consider an *R*, *L*, *C* circuit from undergraduate circuit theory. It does not matter whether this is a parallel or a series circuit in principle, although the exact relationships will differ. So let us consider the series case. This circuit has a resonant frequency in the absence of the resistance, which is given by

$$\omega_0 = \frac{1}{\sqrt{LC}}.$$ (1.2)

However, when the resistance is present, the oscillation of the circuit is damped exponentially in time with the time constant $2L/R$, and the frequency of the oscillation is shifted to

$$\omega = \omega_0 \sqrt{1 - \frac{R^2}{4L^2\omega_0^2}}.$$ (1.3)

The resistance leads to a quality factor $Q = \omega_0 L/R$, which tells us how good the resonant circuit is at its task. But if the resistor has too large a value, then the second term in the bracket can be larger than 1, and the circuit will no longer oscillate. In order to connect to our energy and momentum relationship, we note that the resistance moves the frequency away from the shell value of ω_0.

Two things happened when the resistance was added to the resonant circuit. First, damping appeared, and second, the frequency was shifted by the non-infinite Q of the circuit. In the quantum world, the resistor represents interactions that take the system from a one-electron picture to what is called the many-body picture. These interactions can be among the many electrons that are present, or among the electrons and the lattice vibrations, or among the electrons and the impurities in the system, and so on. But the interactions lead to the same two effects. However, we define them as a single term, the *self-energy*. In the world of Schrödinger wave mechanics, the time variation for an energy eigenstate is the simple exponential

$$e^{i\omega t} = e^{iEt/\hbar},$$ (1.4)

where we have used the Planck relationship between frequency and energy. In the absence of interactions, the energy is given by the momentum according to Eq. (1.1), but when the interactions are present, the energy is shifted to

$$E = \frac{p^2}{2m} - \Sigma(p).$$ (1.5)

The last term in Eq. (1.5) is the self-energy. The real part of the self-energy corresponds to a shift downward of the energy (frequency shift). The imaginary part of the self-energy provides a damping of the state in time, which is the analog of the resistive damping of the resonant circuit. So in situations when the self-energy is nonzero, states that do not lie on the energy shell can be important in the transport. The impact of these states is cleverly described as off-shell effects, thereby obfuscating their role to anyone who is not one of the cognoscenti. But how do we account for these off-shell effects? The answer lies in Eq. (1.1).

Let us consider the two parabolic energy bands shown in Fig. 1.1. In panel (a), we have the free electron energy band, which is a crisp single line representing the exactness of Eq. (1.1). On the other hand, in panel (b), we have a fuzzy line that has more or less the same shape but does not represent the crisp behavior expected from Eq. (1.1). This is because the fuzzy curve in panel (b) is in the presence of the self-energy. So there is no single shell, but a range of momentum values that can have the same energy. To describe this, we define a new quantity, which we will call the *spectral density*, that describes the relationship between the energy and the momentum. For the crisp energy band of panel (a), we have

$$A(E, k) = \delta(E - E_k), \tag{1.6}$$

where we have introduced the quantity

$$E_k = \frac{\hbar^2 k^2}{2m} = \frac{p^2}{2m}, \tag{1.7}$$

with the wave momentum $\hbar k$. Of course, in a semiconductor, the mass should be the effective mass corresponding to the band of interest. When we include the self-energy, the delta function broadens into a Lorentzian lineshape as

$$A(E,k) = \frac{2\mathrm{Im}\{\Sigma(p)\}}{\left[E - E_k - \mathrm{Re}\{\Sigma(p)\}\right]^2 + \left[\mathrm{Im}\{\Sigma(p)\}\right]^2}, \tag{1.8}$$

which appears to be far more formidable. Yet this is just the Fourier transform of the damped, shifted oscillation represented by Eq. (1.4) with the two parts of the self-energy stated explicitly. If one takes a horizontal cut through the fuzzy line of panel (b), the shape should be this Lorentzian rather than the delta function of panel (a).

Of course, the integral under both functions should be unity, as the broadening is not producing or destroying any states.

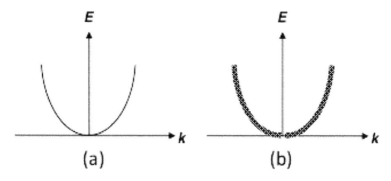

(a) (b)

Figure 1.1 (a) The crisp single-electron energy band that arises in the non-interacting regime. (b) The fuzzy energy band that arises when interactions are taken into account to create the self-energy.

Now it has to be pointed out that not all formulations of quantum transport make clear the role of this new spectral density. Indeed, in many formulations, it is sometimes obscure as to whether or not the fuzziness actually occurs, or is being considered. Nevertheless, a key distinction between quantum and classical transport is just this existence of the spread of states off the energy shell.

1.2 Schrödinger Equation

Waves came to the quantum mechanics of particles primarily from Prince Louis de Broglie. He postulated in his doctoral thesis that all matter had wave-like properties and demonstrated that Bohr's model of the atom could be explained by requiring each quantized orbit to have an integer number of wavelengths in its circumference [7]. It was almost immediately recognized by Walter Elsasser that this wave nature could explain recent experiments on electron scattering from metals as diffraction of the waves [8, 9]. Now it was clear that not only did light have particle properties, but solid matter had wave properties. Finally, in late 1925 and inspired by de Broglie's work, Schrödinger developed his famous wave equation and showed that this differential equation approach yielded the correct solutions for Bohr's atomic model, but also produced equivalent results to the matrix mechanics developed slightly earlier [10]. Today, the Schrödinger wave equation is the most commonly taught approach to quantum mechanics but is also the most important starting place for any treatment of quantum transport.

In one dimension, the time-dependent Schrödinger equation can be written quite simply as, where we also introduce the Hamiltonian operator $H(x)$,

$$i\hbar \frac{\partial \psi(x,t)}{\partial t} = H(x)\psi(x,t) = -\frac{\hbar^2}{2m}\frac{\partial^2 \psi(x,t)}{\partial x^2} + V(x)\psi(x,t). \quad (1.9)$$

The equation is simply a statement about the conservation of energy, as the left-hand side is the total energy, while the right-hand side is the sum of the kinetic energy and the potential energy. In all terms, these energies are acting upon the wavefunction itself to produce the time and space variations that the equation suggests. While we have written it in only one spatial dimension, the first term on the right-hand side is clearly extendable to the full Laplacian operator for three dimensions.

The Schrödinger equation (1.9) is a partial differential equation in position and time. Often such equations are solvable by separation of variables, and this is also the case here. To do so, we introduce a product form for the wavefunction as $\psi(x, t) = \chi(t)\varphi(x)$, from which we can obtain the new form of the equation as

$$\frac{i\hbar}{\chi(t)}\frac{\partial \chi(t)}{\partial t} = -\frac{\hbar^2}{2m\varphi(x)}\frac{\partial^2 \varphi(x)}{\partial x^2} + V(x). \quad (1.10)$$

The left-hand side is a function of time alone, while the right-hand side is a function of position alone. This equality is only possible if both sides of the equation are equal to a constant, that we relate to the energy. Thus, we can write the time variation as

$$\chi(t) = e^{-iEt/\hbar}, \quad (1.11)$$

while the time-independent Schrödinger equation is given by

$$-\frac{\hbar^2}{2m}\frac{\partial^2 \varphi(x)}{\partial x^2} + V(x)\varphi(x) = E\varphi(x). \quad (1.12)$$

In this last formulation, we recognize that the time-independent equation is an energy eigenvalue equation. That is, the important constant is the energy. But if we are in a confined system, there will be a number of eigenstates, each with its own eigenenergy. That is, let us suppose that we are in a confined geometry, in which there are a number of possible eigenstates, which we denote by $\varphi_i(x)$, and for which we write the total wavefunction as

$$\varphi(x) = \sum_i c_i \varphi_i(x). \tag{1.13}$$

Here, c_i are expansion coefficients, and the basis functions are required to satisfy orthonormality via

$$\int_R \phi_j^*(x)\varphi_i(x)dx = \delta_{ij}, \tag{1.14}$$

where the last term is the Kronecker delta function and R is the region in which the wavefunction have support. In order for the total wavefunction to be normalized, we now have to have

$$\int_R \varphi^*(x)\varphi(x)dx = \sum_i |c_i|^2 = 1. \tag{1.15}$$

These properties will be important later. If we reinsert the time variation, then the total wavefunction can be written as

$$\psi(x,t) = \sum_i c_i \varphi_i(x)e^{-iE_i t/\hbar}, \tag{1.16}$$

where E_i are the corresponding eigenenergies for each of the basis wavefunctions.

Let us now consider the example of the free electron. Then there is no potential energy in the system, although a constant background potential would serve merely to shift the energy axis, so that we can consider the *kinetic* energy is measured from this potential. Then Eq. (1.12) can be rewritten as

$$\frac{\partial^2 \varphi(x)}{\partial x^2} + k^2 \varphi(x) = 0 \quad, \quad k^2 = \frac{2mE}{\hbar^2}. \tag{1.17}$$

This is a simple harmonic equation for which the solutions may be written as

$$\varphi(x) = Ae^{ikx} + Be^{-ikx}. \tag{1.18}$$

Once more, we can reinsert the time variation from Eq. (1.11) to give

$$\varphi(x) = Ae^{ikx-i\omega t} + Be^{-ikx-i\omega t}, \ E = \hbar\omega. \tag{1.19}$$

Thus, the free particle can be represented by a plane, which propagates either in the positive x-direction or in the negative x-direction. It is important to note that these plane waves have uniform probability throughout space, so that they are certainly not localized in any one position. Hence, if we are interested in localizing the particle that is represented by this wave, we have to construct a wave packet composed of a great number of waves of slightly differing momenta. One such quite useful wave packet is of the form, often called a coherent wave packet in optics,

$$\varphi(x) = \frac{1}{(2\pi)^{1/4}\sqrt{\sigma}} e^{-(x-x_0)^2/4\sigma^2 - ik_0 x}. \tag{1.20}$$

This is clearly a Gaussian packet localized at x_0, which is a moving coordinate, and with a momentum $\hbar k_0$. Again, this is a very useful construct, which will be useful to us in the later chapters.

1.3 On the Velocity and Potentials

It is useful at this point to diverge a bit from the Schrödinger equation with a variant that gives some insight into some of the parts of the equation. This approach was first put forward by Madelung [11] and is now connected in more detail with Bohm [12]. In the previous section, it became clear that the wavefunction possesses both a magnitude and a phase, as it is a complex quantity. In what we can call the hydrodynamic approach, we express the wavefunction explicitly in terms of these two quantities, so that we can examine each of them in some detail. Hence, we write the wavefunction as

$$\psi(x, t) = A(x, t)e^{iS(x, t)/\hbar}. \tag{1.21}$$

The quantity $S(x,t)$ is known as the action, which is the time integral of the energy. If we insert this into the Schrödinger equation (1.9) and separate the resulting real and imaginary parts, which must both be satisfied, we can write these as

$$\frac{\partial S}{\partial t} + \frac{1}{2m}\left(\frac{\partial S}{\partial x}\right)^2 + V - \frac{\hbar^2}{2mA}\frac{\partial^2 A}{\partial x^2} = 0 \tag{1.22}$$

and

$$\frac{\partial A}{\partial t} + \frac{A}{2m}\frac{\partial^2 S}{\partial x^2} + \frac{1}{m}\frac{\partial S}{\partial x}\frac{\partial A}{\partial x} = 0. \tag{1.23}$$

In these equations, there is only a single term that includes Planck's constant. Hence, if we consider an approach to the classical limit when Planck's constant is set to zero, we should obtain the classical equations that appear in some forms of hydrodynamics. For example, if we multiply Eq. (1.23) by the factor A and rearrange the terms, we obtain

$$\frac{\partial A^2}{\partial t} + \frac{\partial}{\partial x}\left(\frac{A^2}{m}\frac{\partial S}{\partial x}\right) = 0.$$ (1.24)

The factor of A^2 is obviously related to the quantity $|\psi|^2$, which is the probability as a function of position. If we connect the gradient of the action with the momentum, the second term is the divergence of the probability current. Thus, the quantity in the parentheses in Eq. (1.24) is the product of the probability and the velocity; the velocity always lies in the phase of the wavefunction. With these connections, we can now return to Eq. (1.22) to note that the terms, from left to right, are the total energy, the kinetic energy, the potential energy, and a new quantity, which is identified as the *quantum potential*. In device simulation, the wavefunction is often connected to the density of carriers (e.g., the probability of carriers being at a position, normalized to the total number of carriers in the device), so that the quantum potential can be written as

$$V_Q = -\frac{\hbar^2}{2m\sqrt{n}}\frac{\partial^2 \sqrt{n}}{\partial x^2}.$$ (1.25)

This quantum potential was used by Philippidis et al. [13] to illustrate how particles subject to this potential can show the interference effects associated with the two-slit experiment in waves.

A slightly different version of this potential was found by Wigner [14] and is often referred to as the density-gradient potential, which is used as a quantum correction in device simulation [15]. Wigner's form is expressed as

$$V_W = -\frac{\hbar^2}{8m}\frac{\partial^2 \ln(n)}{\partial x^2}.$$ (1.26)

The two forms are quite similar and can be connected as

$$V_Q = V_W + \frac{\hbar^2}{8mn}\frac{\partial \ln(n)}{\partial x}.$$ (1.27)

While the above is instructive, it is not complete. The deconstruction of the Schrödinger equation via Eq. (1.21) is not complete for a quantum picture unless we add the constraint that the action is cyclic; e.g., the action must satisfy [16, 17]

$$S = S \bmod h, \quad \int_C dr \cdot \nabla S = nh,$$ (1.28)

where n is an integer. In normal hydrodynamic terms, this corresponds to quantization of the velocity circulation, which is related to the vorticity via Stoke's theorem. Generally, in classical hydrodynamics, $n = 0$, although situations are known where this is not the case, for example in shocks, boundaries, or scattering processes. In quantum cases, n is certainly not zero. Indeed, vortices are known in many quantum processes [18], particularly in open quantum dots where the carrier flow can be described in terms of open or closed orbits that separate regions of trapped vortex flow [19]. Of course, the semi-classical equivalent to Eq. (1.28) is the Einstein–Brillouin–Keller quantization condition [20–22], and when S is reduced by $\hbar$, the integral is an important element of the Gutzwiller trace formula [23]. In quantum mechanics, we can replace the gradient of the action with the expectation value of the momentum, and

$$\gamma_n = -\frac{1}{\hbar} \oint_C \langle \mathbf{p}(\mathbf{x}) \rangle \cdot d\mathbf{x} = i \oint_C \langle \psi(\mathbf{x}) | \nabla \psi(\mathbf{x}) \rangle \cdot d\mathbf{x}$$ (1.29)

is the geometric phase accumulated by the wavefunction around the closed trajectory, which is also known as the Berry phase [24]. When the momentum is generalized to include the coupling to a vector potential, this phase becomes the well-known Aharonov–Bohm phase [25].

The importance of the vortices lies in an interpretation of the quantum flow of the wavefunction, and one may pursue a study of the topology of this flow [18]. Around the points where the velocity, or ∇S, vanishes, the local flow in the absence of a magnetic field is hyperbolic, and these points act as saddle points. For non-vanishing magnetic field, there are also elliptical orbits around points that act as the centers for these orbits and thus appear as vortices. A second class of vortex flow arises even when there is no magnetic field. In this latter case, $\nabla \times \nabla S = 0$ everywhere except at the points where $A = 0$. Since the amplitude is considered to be a positive semi-definite function, the phase has to shift by π as one passes through these nodes to account for the change of sign of the wavefunction. Hence, at these points, the phase and the velocity are ill-defined, which is a pure quantum effect. The vortex flow around the various centers is governed by the quantum potential Eq. (1.25).

We can illustrate the manner in which the interfering trajectories, or paths, and the onset of vortices appear through the consideration of a relatively simple example. Let us consider a simple GaAlAs/GaAs heterojunction, which is modulation doped. The conduction band discontinuity at the interface creates a triangular potential well on the GaAs side of the interface. When the dopants are placed in the AlGaAs, the electrons can leave the dopants and drop into this potential well, creating a quasi-two-dimensional electron gas whose motion normal to the interface is quantized. Hence, they are free to move in the two dimensions along the plane of the interface. When this material is patterned into a mesoscopic device, the lateral dimension, which is constrained (the longitudinal dimension is option for current flow) and a set of transverse modes are created in the confinement potential (we discuss this further in the last section of the chapter). Each of these modes has a different confinement energy, so that the latter creates a ladder of discrete energy levels. The current in each of these transverse modes provides one channel through the device, and the carrier density observable for multiple channels is shown in Fig. 1.2a, for a Fermi energy of 13.5 meV. This corresponds to a two-dimensional electron density of about 3.8×10^{11} cm^{-2}. One can see the peaks and valleys of the electron density running from left to right in the figure, corresponding to the variations in the local probability density due to the multiple modes present. Now, however, let us introduce a random potential that arises from disorder in the sample. This disorder can come from defects, the impurities in the AlGaAs, or a number of other sources. But this random potential creates a varying potential landscape, which induces random variations in the electron density, as shown in Fig. 1.2b. This latter is a three-dimensional view so that the peaks and valleys may be more easily observed. Now it is clear how various paths through the sample can wind around the "hills" via the "valleys" and lead to wave interference effects. Such interference effects give rise to the phase variations through the geometrical phase. These, in turn, can be significantly modified by small changes in the Fermi energy and can appear as sizable fluctuations in the value of the conductance through the sample.

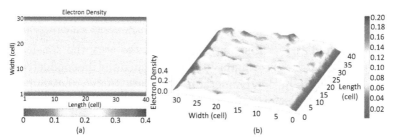

Figure 1.2 (a) Electron density in a two-dimensional channel of finite width, with no disorder in the local potential. (b) Local density variation when a random potential is added. The current flow direction is from lower left to upper right. In both cases, the density is indicated by the color bar in units of 10^{12} cm^{-2}, although the density is lower in (b) as the random potential localizes lower energy modes.

1.4 Single-Atom Transistor

It should be obvious from the aforementioned discussion that there is a need for quantum transport in order to understand some transport properties of semiconductors. In this section, we want to reinforce this view with a discussion of what may be the ultimate small semiconductor device, one in which the "channel" region is composed of the states of a single impurity atom. Let us start with a bare (100) surface of Si. When this is exposed to the environment in a high vacuum chamber, the interaction of the electrons in the surface dangling bonds leads to a reconstruction of the surface layer into the 2×1 configuration, which means that the surface unit cell is two bulk unit cells in one direction and a single bulk unit cell in the other. The structure creates rows of dimerized dangling bonds through the movement of the surface atoms, and this lowers the overall energy of the surface [26]. Even so, there remain dangling bonds, but these can be effectively passivated by treatment with hydrogen, with each dangling bond attaching to a single hydrogen atom. Now one can begin to build structures on the surface by recognizing that each individual hydrogen atom can be affected by using a scanning tunneling microscope to break the individual bond [27]. Hence, the lithographic process consists of removing the hydrogen atoms at a single site, or a group of sites to define a region. Then the resulting dangling bonds are exposed to other molecules, such as phosphine, which will bond and then allow the phosphorous to move into the Si lattice with subsequent processing [28]. When the phosphorus is incorporated at each dangling bond of a large area, the density is sufficiently large that an impurity band is formed and the conductivity remains high even at low temperature. Similarly, if only a single bond is broken and phosphorus is placed there, then only a single isolated atom is contained in the structure. In any case, the surface is secured by the growth of a thin layer of Si over this structure. This structure can then be made into a single-atom transistor [4]. In Fig. 1.3, we show the basic structure of the lithographic definition of the device. It may be seen in panel (a) that heavily doped areas define the source and drain, as well as two gates symmetrically placed to either side of the current direction. Then, there is a single atom at the center that acts as a "quantum dot" to which electrons can move from the source or drain (or depart to these terminals). The local dot potential is controlled by the gates. A blowup of the active region is shown in panel (b) of the figure, where it may be seen just how close the quantum dot is to the source and drain. In Fig. 1.4, we show an image of the local potential that is computed for the device. Here, it may be seen that there is a potential well at the site of the single atom, which forms the quantum dot. The electrons can then charge the bound states within this well, and the structure of the well and its bound states can be determined by fitting a simulation to the experimental data.

The simulation approach uses a tight-binding atomistic formulation to get the band structure for the phosphorus-doped electrodes, and then a self-consistent solution to the local potential is obtained for various gate voltages [29], without the single-atom dot present. Then these are updated with the dot in place, and the local potentials are determined. This method also gives rise to the potential profile shown in Fig. 1.4. The channel modulation by the gates is simulated self-consistently in the Thomas–Fermi approximation. Overall, this is a multilevel modeling exercise, which allows a very accurate comparison of the predicted charging energies with those observed experimentally. Indeed, an observable change in the charging energy can be found if the central atom is moved by as little as a single atomic position. This is a clear example of how coupling the quantum transport simulation to experiment lays a very strong foundation upon which to base a thorough understanding of the physics at work in the detailed device.

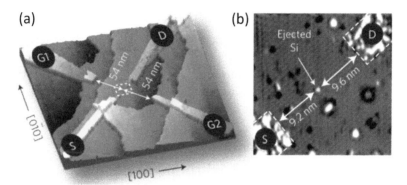

Figure 1.3 (a) A perspective STM image of the single-atom transistor. The hydrogen-desorbed regions defining the source, drain, and two gates appear raised due to the increased tunneling to these regions. (b) Close-up of the inner region of the transistor showing the single atomic site where the phosphorus quantum dot will be located. Reprinted with permission from Macmillan Publishers Ltd: *Nature Nanotechnology*, Ref. [4], copyright 2012.

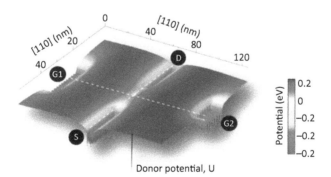

Figure 1.4 False-color plot of the self-consistently determined potential within the single-atom transistor. The deep potential well at the central atom can be seen at the bottom of the figure. Reprinted with permission from Macmillan Publishers Ltd: *Nature Nanotechnology*, Ref. [4], copyright 2012.

This particular device, with the single-atom quantum dot in the center, is what is known as a single-electron tunneling transistor

[30]. There has been significant work over the past few decades in these devices, and we will return to them in Section 6.6.

It seems to be clear that today's semiconductor devices are truly small, whether it be the single-atom transistor or a modern industrial tri-gate field-effect transistor with an effective gate length of only 14 nm. If we are to understand the full nature of the transport within the device in any detail, which is necessary to begin to understand the physical limits and approaches to optimization, then the transport will need to be considered with the full nature of the quantum behavior inherent in the device. That is, it seems to no longer be adequate to avoid dealing with the details of quantum transport.

1.5   Discretizing the Schrödinger Equation

In many of the transport approaches that will be discussed in the following chapters, a numerical technique will be used to evaluate the results via computer. Hence, it is useful to develop the discretization of the Schrödinger equation here, once and for all. For this purpose, we will do this in two dimensions to illustrate the approach, and to make it useful for later chapters. We begin with the Schrödinger equation in two dimensions, which we take here to the x- and y-axes. For simplicity, we take the x-axis as aligned along the length of the semiconductor strip in the current flow direction and the y-axis in the width direction. Thus, we may write the equation as

$$-\frac{\hbar^2}{2m^*}\left(\frac{\partial^2}{\partial x^2}+\frac{\partial^2}{\partial y^2}\right)\psi(x,y)+V(x,y)\psi(x,y)=E\psi(x,y). \quad (1.30)$$

In this latter equation, we have introduced the effective mass m^* and have assumed that this is isotropic in the two dimensions. In later work, where we have multiple conduction band valleys, this approximation will have to be revisited to account for the anisotropic nature of the effective mass in spheroidal bands. In order to discretize this equation, we need to introduce a grid in the two dimensions; e.g., we will evaluate the wavefunction and the potential at a set of points, which form a two-dimensional grid, as shown in Fig. 1.5. The grid shown has another simplification in that the spacing of the grid points in both the x-direction and the *y*-direction is taken to be the same. This is not required and, in fact, is often not done in real device simulations. Here, we do it to simplify the approach for clarity.

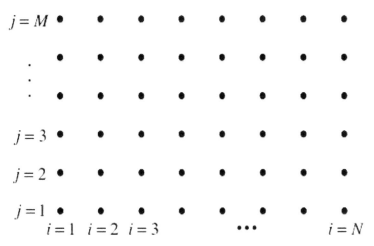

Figure 1.5 A typical two-dimensional grid with equal spacing in the two directions.

Now, around any arbitrary point (i,j), we can expand the wavefunction, taking the grid spacing to be a, as

$$\psi(i+1,j)=\psi(i,j)+a\frac{\partial\psi(i,j)}{\partial x}+\frac{a^2}{2}\frac{\partial^2\psi(i,j)}{\partial x^2}+\dots$$

$$\psi(i-1,j)=\psi(i,j)-a\frac{\partial\psi(i,j)}{\partial x}+\frac{a^2}{2}\frac{\partial^2\psi(i,j)}{\partial x^2}+\dots$$

$$\psi(i,j+1)=\psi(i,j)+a\frac{\partial\psi(i,j)}{\partial y}+\frac{a^2}{2}\frac{\partial^2\psi(i,j)}{\partial y^2}+\dots \qquad (1.31)$$

$$\psi(i,j-1)=\psi(i,j)-a\frac{\partial\psi(i,j)}{\partial y}+\frac{a^2}{2}\frac{\partial^2\psi(i,j)}{\partial y^2}+\dots.$$

With this expansion, we can write the second derivative term in Eq. (1.30) as

$$\frac{\partial^2\psi(i,j)}{\partial x^2}+\frac{\partial^2\psi(i,j)}{\partial y^2}$$

$$\sim\frac{\psi(i+1,j)+\psi(i-1,j)+\psi(i,j+1)+\psi(i,j-1)-4\psi(i,j)}{a^2} \qquad (1.32)$$

which is known as the five-point template. Other versions can be adapted by a more refined set of partial derivatives, such as including the second partial with respect to x and y, but this is seldom done in practice. The Schrödinger equation can now be rewritten as

$$\left[4t - E + V(i,j)\right]\psi(i,j) = t\left[\psi(i+1,j) + \psi(i-1,j)\right.$$
$$\left. + \psi(i,j+1) + \psi(i,j-1)\right], \tag{1.33}$$

where

$$t = \frac{\hbar^2}{2m^*a^2} \tag{1.34}$$

is known as the *hopping integral*. This quantity gives the coupling between the nearest neighbor sites on the grid. In general, the vertical set of sites, for a constant value of i, is known as a *slice*, while the horizontal set of sites, for a constant value of j, is known as a *row*. For example, if we use a grid spacing of 5 nm, for GaAs where the effective mass is $0.067m_0$, the hopping energy is 22.8 meV.

An interesting aspect of the discretization is that it introduces an artificial band structure. This may be seen in Fig. 1.6, where we plot the eigenenergies for a single transverse slice of the two-dimensional grid. Here, we used a grid spacing of 5 nm and there were 65 grid points in the width. The 65 grid points mean that we will have 65 energy values, and the state corresponds to the ordering of the various energies. The material is assumed to be GaAs. Near the bottom of the energy band, the energy varies nearly parabolically, but deviation sets in just above the quarter point. The total width of the energy range is $4t$, or 90.9 meV in this case. Hence, the message is that one needs to keep the energy in the lower quarter of the band if good connection to the proper results for GaAs is to be obtained.

We can go a little bit further by defining a vector wavefunction for each slice. For example, for slice i, we can write this as

$$\overline{\psi}_i = \begin{bmatrix} \psi(i,1) \\ \psi(i,2) \\ \psi(i,3) \\ \cdots \\ \psi(i,M) \end{bmatrix}. \tag{1.35}$$

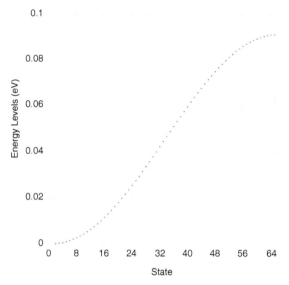

Figure 1.6 The eigenenergies for a single transverse slice of the two-dimensional grid with $a = 5$ nm and 65 grid points.

With this nomenclature, we can rewrite Eq. (1.33) as

$$tU\bar{\psi}_{i+1} = H\bar{\psi}_i - tU\bar{\psi}_{i-1}, \tag{1.36}$$

where U is a unit matrix and H is the Hamiltonian

$$H = \begin{bmatrix} 4t - E - V(i,1) & t & 0 & \cdots & 0 \\ t & 4t - E - V(i,2) & t & \cdots & 0 \\ 0 & t & 4t - E - V(i,3) & \cdots & \cdots \\ \cdots & \cdots & \cdots & \cdots & t \\ 0 & 0 & \cdots & t & 4t - E - V(i,M) \end{bmatrix}. \tag{1.37}$$

This creates an interactive approach in which the coupling from one slice to the next acts as a perturbative connection. This is often rewritten in the matrix form

$$\begin{bmatrix} \bar{\psi}_{i+1} \\ \bar{\psi}_i \end{bmatrix} = \begin{bmatrix} t^{-1}H_i & -U \\ U & 0 \end{bmatrix} \begin{bmatrix} \bar{\psi}_i \\ \bar{\psi}_{i-1} \end{bmatrix}. \tag{1.38}$$

The propagation is often made by matching the wavefunction and its derivative at each interface between two slices, but this is an unstable approach, as is that of Eq. (1.38). The instability arises when a Fermi energy is imposed in the contacts, evanescent modes arise for states whose energy is above the Fermi energy. These modes have both decaying and growing exponentials (with real arguments), and this leads to an exponential growth of some amplitudes, which is unphysical. As we will see later, one normally transforms the iteration into a variant of the scattering matrix, or uses recursive Green's functions, both of which will be discussed in later chapters.

Problems

1. Consider a normal MOS structure on Si at room temperature. Using the classical approach, in which the density decreases from the surface as $n(x) = n(0)\exp(-eV(x)/k_BT)$, to determine the effective width of the inversion layer, compute the width for densities of 5×10^{11}, 1×10^{12}, 3×10^{12}, 5×10^{12}, 1×10^{13}, all in units of cm^{-2}. Assuming an acceptor density of 5×10^{17}, compute the effective electric field in the oxide. Plot both the effective thickness and the effective electric field as a function of the inversion density on a single plot, using a logarithmic axis for the density. (In the equation, the potential is assumed to be zero at the surface and rises to its maximum value in the bulk.)

2. Go to nanoHUB.org and launch the "bound states calculation lab." Using the data for the Si MOS structure above, and using the heavy mass for the Si inversion layer, enter the effective electric fields computed in the previous problem. Using the triangular geometry to approximate the inversion layer, determine the effective thickness of each of the first four sub-bands as a function of the inversion density corresponding to the effective electric field entered in the program.

3. In a quasi-two-dimensional system, one can calculate the non-interacting density of states by consideration of the size of the Fermi sphere for a given density, and show that this density of states for a single sub-band is $m^*/\pi\hbar^2$. Thus, when the system is confined in the third dimension, a new sub-band arises at each of the quantized (in z-direction) energies. One may think of the sub-band as arising from integration in momentum (over E_k) over the delta function characterizing the spectral

density (1.6). Show that, if one replaces this delta function with the Lorentzian line of Eq. (1.8), there is a smooth onset of the density of states instead of a sharp step function. Ignore the real part of the self-energy and take the imaginary part from a mobility of 2000 cm^2/Vs in GaAs.

4. Consider the expansion of the wavefunction into an arbitrary basis set, which is complete and orthonormal. Consider an operator A, which satisfies an eigenvalue equation with this basis set. Show that this operator has zero uncertainty.

5. Consider the MOSFET in Problems 1 and 2. Assume the potential in the semiconductor is approximately a linear potential, so that the electrons are confined in a triangular potential well. Use the Airy function solutions for this potential well to find the wavefunction of the lowest energy level for the case of an inversion density of 5×10^{12} cm^{-2} and a doping density of 5×10^{17} cm^{-3} (these define the electric field for the potential). From Eq. (1.25), plot the quantum potential in the direction normal to the interface.

6. Consider that a random potential exists in a quasi-two-dimensional semiconductor layer. This random potential has a peak-to-peak amplitude of 10 $k_B T$ at 4.2 K and a zero average value. Using the normal two-dimensional statistics and density of states, find the variation of the local density, and the peak-to-peak amplitude of the density variation for the lowest sub-band.

7. Let us consider a single-electron tunneling structure composed of two capacitors, of equal value C, in series and connected to a bias voltage V_{sd}. Between the two capacitors, there is a small quantum dot upon which charge of ne can accumulate (with $e < 0$ for electrons). Compute the energy stored in the system (hint: do not forget the energy provided by the bias supply). Then, compute the change in the energy as a single electron is added to, or removed from, the quantum dot and find the conditions for the Coulomb blockade.

8. Consider a one-dimensional confined system in GaAs, in which the confinement distance is 1 micron in length. Using a 5 nm grid, construct the one-dimensional discretized Schrödinger equation. Solve for the energy values and find the lowest five

wavefunctions and show that they agree well with the exact wavefunctions for an infinite potential well of this size.

References

1. W. Shockley, *Electrons and Holes in Semiconductors* (Van Nostrand, Princeton, 1950).

2. H. Fröhlich and F. Seitz, *Phys. Rev.* **79**, 526 (1950).

3. E. J. Ryder, *Phys. Rev.* **90**, 766 (1953).

4. M. Fuechsle, J. A. Miwa, S. Mahapatra, H. Ryu, S. Lee, O. Warschkow, L. C. L. Hollenberg, G. Klimeck, and M. Y. Simmons, *Nat. Nanotechnol.* **7**, 242 (2012).

5. D. K. Ferry, *Transport in Semiconductor Mesoscopic Devices (IOP Publishing, Bristol, 2015)*.

6. J. Kohanoff, *Electronic Structure Calculations for Solids and Molecules* (Cambridge University Press, Cambridge, 2006).

7. L.-V. de Broglie, *Ann. Physique* **3**(10), 22 (1925).

8. W. M. Elsasser, *Naturwissenschaften* **16**, 720 (1925).

9. C. Davisson and C. H. Kunsman, *Phys. Rev.* **22**, 242 (1923).

10. E. Schrödinger, *Ann. Phys.* **79**, 361, 489 (1926).

11. E. Madelung, *Naturwissenschaften* **14**, 1004 (1926).

12. D. Bohm, *Phys. Rev.* **85**, 166, 180 (1952).

13. C. Philippidis, C. Dewdney, and B. J. Hiley, Il, *Nuovo Cim.* **52B**, 15 (1979).

14. E. Wigner, *Phys. Rev.* **40**, 749 (1932).

15. J.-R. Zhou and D. K. Ferry, *IEEE Trans. Electron Dev.* **39**, 473 (1992).

16. J. R. Barker, *Semicond. Sci. Technol.* **9**, 911 (1994).

17. J. R. Barker and D. K. Ferry, *Semicond. Sci. Technol.* **13**, A135 (1998).

18. J. R. Barker, *VLSI Design* **13**, 237 (2001).

19. J. R. Barker, R. Akis, and D. K. Ferry, *Superlattices Microstruct.* **27**, 319 (2000).

20. A. Einstein, *Verh. Deutsch. Phys. Ges.* **19**, 82 (1917).

21. L. Brillouin, *J. Phys. Radium* **7**, 353 (1926).

22. J. B. Keller, *Ann. Phys.* (New York) **4**, 180 (1958).

23. M. C. Gutzwiller, *Chaos in Classical and Quantum Mechanics* (Springer, Berlin, 1990)

24. M. V. Berry, *Proc. Roy. Soc. London A* **392**, 45 (1984).

25. Y. Aharonov and D. Bohm, *Phys. Rev.* **115**, 485 (1959); **123**, 1511 (1961).

26. D. K. Ferry, *Semiconductors* (Macmillan, New York, 1991) Sec. 5.10.2.

27. J. W. Lyding, T. C. Shen, J. S. Hubacek, J. R. Tucker, and G. C. Abeln, *Appl. Phys. Lett.* **64**, 2010 (1994).

28. B. Weber, S. Mahapatra, H. Ryu, S. Lee, A. Fuhrer, T. C. G. Reusch, D. L. Thompson, W. C. T. Lee, G. Klimeck, L. C. L. Hollenberg, and M. Y. Simmons, *Science* **335**, 64 (2012).

29. H. Ryu, S. Lee, M. Fuechsle, J. A. Miwa, S. Mahapatra, L. C. L. Hollenberg, M. Y. Simmons, and G. Klimeck, *Small* **11**, 374 (2015).

30. D. K. Ferry, S. M. Goodnick, and J. P. Bird, *Transport in Nanostructures*, 2nd Ed. (Cambridge University Press, Cambridge, 2009) Sec. 6.1.

Chapter 2

Approaches to Quantum Transport

There are many approaches to computing the quantum transport properties of carriers in semiconductors. Nearly all of these start with the Schrödinger equation, which we discussed and discretized in Chapter 1. The form of this equation is both reversible and dissipation free, while its use in actual devices is normally modified to incorporate the presence of dissipation. We will illustrate this in Section 2.4.4 with the recursive version of an approach to computing the transport with a technique based on the Schrödinger equation. Others, however, have often discussed transport in cases in which the dissipation was ignored—so-called ballistic transport cases. For this approach, it must be assumed that the length of the region being simulated is considerably smaller than any characteristic dissipation length, such as the mean free path or the inelastic scattering length (often related to the energy relaxation length). Several such characteristic lengths are important in transport, particularly quantum transport.

One of the most important lengths deals with processes that can break phase coherence of the wavefunction. Processes that can break phase coherence are not necessarily energy relaxing, although nearly all energy relaxation processes also break phase coherence. For example, phase-breaking processes, which do not relax energy, can arise from elastic interactions that are sufficiently strong that they introduce localization. In general, we distinguish between

An Introduction to Quantum Transport in Semiconductors
David K. Ferry
Copyright © 2018 Pan Stanford Publishing Pte. Ltd.
ISBN 978-981-4745-86-4 (Hardcover), 978-1-315-20622-6 (eBook)
www.panstanford.com

different situations by the discussion of appropriate *lengths*. Hence, we can define a phase-breaking length as

$$l_\varphi = \sqrt{D\tau_\varphi} \qquad (2.1)$$

where D is the diffusion constant related to the mobility as

$$D = \frac{\mu k_B T}{e} \qquad (2.2)$$

for non-degenerate carrier statistics. In the case in which the statistics are degenerate, there is a correction to Eq. (2.3), which involves the ratio of a pair of Fermi–Dirac integrals whose specific order depends on the dimensionality of the system. In addition to this important length, we also have various lengths that are associated with the scattering processes. One of these is the mean free path

$$l = v\tau_m \qquad (2.3)$$

where v is usually either the thermal velocity or the Fermi velocity and τ_m is the mean free time. Note that this is not the actual scattering time τ, but is usually thought of as the momentum relaxation time. The second is the inelastic length

$$l_{in} = v\tau_{in} \qquad (2.4)$$

Many other lengths have been introduced for various situations, and these will be discussed as the need arises.

The basic concepts of transport in mesoscopic quantum systems in the presence of localized scatterers, or tunneling barriers, can be traced to Landauer [1]. From many approaches, it is also clear that the onset of inelastic scattering will suppress many of the quantum effects that are of interest, primarily through broadening of the density of states as discussed in the last chapter, but also broadening the important nature of the quantum states themselves. It has often been assumed that the Landauer formula is only applicable in cases of ballistic transport, but this is just not true. The Landauer approach can be applied to nearly any situation and with nearly any scattering process as long as it is done in a smart manner. That is, the Landauer approach is quite fundamental as its extensions to the use of the distribution functions in the contact regions is all that is required. We will discuss this in the next few sections.

From the wavefunctions that are given by the Schrödinger equation, one can generate a quite general quantum statistical quantity, whether this is the density matrix or Green's function. These are related to each other and also lead to the Wigner function, which is the phase–space version of them. Green's function can also be extended to a nonequilibrium form, which is necessary when the system is far from equilibrium. We will discuss the connection among these different functions in the last section of the chapter.

2.1 Modes and the Landauer Formula

To develop the method of utilizing the discretized Schrödinger equation developed in Chapter 1, we will assume that the transport is confined to a two-dimensional system as described there. The actual waveguide is created by a confining potential in the experimental situation, often by the use of gates that impose an electrostatic potential upon the overall two-dimensional system. Then, the Schrödinger equation can be written as

$$-\frac{\hbar^2}{2m}\left(\frac{\partial^2}{\partial x^2}+\frac{\partial^2}{\partial y^2}\right)\psi(x,y)+V(x,y)\psi(x,y)=E\psi(x,y) \qquad (2.5)$$

with

$$V(x,y) = V_c(x,y) + V_a(x,y) \qquad (2.6)$$

defining the confinement potential V_c as well as the applied potential V_a. When we use a fixed grid width, the solution actually assumes an infinite confinement potential at the edges of the grid. If one wants a softer confinement potential, then grid points must be included within the region of finite potential. This is a minor complication that is easily included in practice. The applied potential can be any potential describing impurities, bias, scattering through an imaginary potential, etc. Before using the fully discretized form of Chapter 1, let us discuss basically the nature of the solutions.

The general solution to the wavefunction within the waveguide region, which is defined by the confinement potential, may be written as a product form in terms of various modes as

$$\psi(x,y)=\sum_n \varphi_n(x)\chi_n(y), \qquad (2.7)$$

where, in general for only the infinite potentials at the edges of the grid,

$$\chi_n(y) = \sqrt{\frac{2}{W}} \sin\left(\frac{n\pi y}{W}\right). \tag{2.8}$$

If a soft wall potential is present, then the wavefunction will assume another form, but this one is sufficient for our present purposes. The longitudinal modes are described, in general, by a combination of forward and backward waves as

$$\varphi_n(x) = a_n e^{\gamma_n x} + b_n e^{-\gamma_n x}, \tag{2.9}$$

where γ_n is the propagation constant. If the mode is a propagating mode, which usually corresponds to an eigenenergy below the Fermi energy, then this propagation constant is imaginary and describes the wave nature of the mode. If, on the other hand, the mode is evanescent, which usually corresponds to an eigenenergy above the Fermi energy, then the propagation constant is real. It is these evanescent modes that create the problem with simple matching of modes at interfaces in the waveguide region. One solution to Eq. (2.9) is always growing in either the forward or backward direction, and this leads to an instability in which the amplitude of the wave grows without limit. It would be simple to just ignore the evanescent modes, but they must be included in the matching problem as they will correspond to reactive elements at the boundaries, just as in microwave waveguides. They can also represent tunneling transfer across short regions.

We can see the nature of the modes by referring to Fig. 2.1. Here we plot the kinetic energy of each of the lowest few transverse modes. This kinetic energy is given by the wave number k_n ($\gamma_n = ik_n$) for the modes where the curves lie below the Fermi energy. This wave number is for propagation in the x-direction, with quantization in the y-direction. Here, we show four sub-bands below the Fermi level, and two of the many sub-bands that have quantized energies above the Fermi level. Thus, there will be four channels flowing through the waveguide.

As remarked earlier, Rolf Landauer presented an approach to transport, and the calculation of conductance, that was dramatically different from the microscopic kinetic theory based on the Boltzmann equation that had been utilized previously (and is still heavily utilized in macroscopic conductors) [1, 2]. He suggested that one could compute the conductance of low dimensional systems simply by computing the transmission of a mode from an input reservoir to a similar mode in an output reservoir. The transmission probability from one mode to the other was then very similar to the computation of a tunneling probability, except that there was no requirement that the process be one of tunneling. The only real constraint was that of lateral confinement so that the two reservoirs could be discussed in terms of their transverse modes (2.8). The key property of the two reservoirs is that they are in equilibrium with any applied potentials. That is, the electrons in the reservoirs are to be described by their intrinsic Fermi–Dirac distributions with any applied potentials appearing only as a shift of the relative energies (which would shift one Fermi level relative to the other). While he originally considered that the transport was ballistic, this is not required. Rather, the requirement is that we can assign a definitive mode to the electron when it is in either of the two reservoirs, which means that if scattering is present, it must be described specifically as a transfer of the electron from one internal mode to another within the active region and not in the reservoir. If we consider a potential applied to the right (output) reservoir relative to the left (input) reservoir, then the right reservoir emits carriers into the active region with energies up to the local Fermi level plus the applied bias, $E_F + eV$ (note that the energy eV will be negative for a positive voltage). The left reservoir emits electrons into the active region with energies only up to E_F. In the following discussion, we will assume that the applied voltage is quite small, although this also is not a stringent requirement.

For simplicity, we make the initial assumption that no scattering takes place within the active region, so that we can make a definitive association between the energy of the carriers and their direction of propagation. That is, electrons injected into the active region from the left reservoir have a positive momentum and travel from left to right. On the other hand, electrons injected into the active region from the right reservoir have a negative momentum and thus travel from the right to the left. Those with positive momentum come from the left reservoir and travel to the right, while those with negative momentum come from the right reservoir and travel to the left.

Because of the applied bias, there are more electrons traveling to the right than are traveling to the left and this gives us a net current through the constriction. The imbalance between left-going and right-going electrons means that the constriction is in a condition of nonequilibrium, which is required to support a net current.

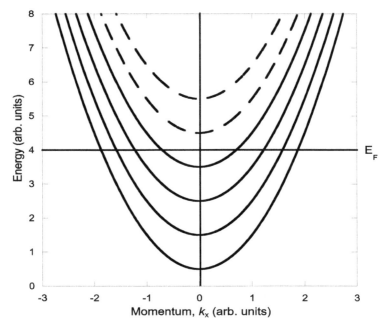

Figure 2.1 Dispersion relation for a quantum waveguide, as described in the text.

In order to compute the current through the active region, we need to first evaluate the charge that is occupied in each transverse mode, or *channel*, in the energy range between E_F and $E_F + eV$. As we remarked earlier, if the applied bias is small, then we can estimate the excess charge as

$$\delta Q = e \cdot eV \cdot \frac{1}{2}\rho_{1d}(E_F) = e^2 V \sqrt{\frac{m^*}{2\pi^2 \hbar^2 E_F}}, \qquad (2.10)$$

where ρ_{1d} is the one-dimensional density of states for the semiconductor with effective mass m^*. It should be noted that we used only one-half of the density of states since we are only interested in those electrons that are traveling to the right (those with positive momentum). Also, because of our definition of the density of states, Eq. (2.10) is the charge per unit length. In order to compute the current, we need only multiply Eq. (2.10) by the group velocity of the carriers at the Fermi energy, which is

$$v_F = \frac{1}{\hbar} \frac{dE}{dk_x}\bigg|_{E=E_F} = \frac{\hbar k_F}{m^*} = \sqrt{\frac{2E_F}{m^*}} .$$ (2.11)

and this leads to the current

$$I = \delta Q \cdot v_F = \frac{2e^2}{h} V .$$ (2.12)

A very interesting effect occurred in the development of the current, and that is the energy has dropped out of the equation. Hence, the current is completely independent of the energy, and this is true for each quasi-one-dimensional channel that flows through the constriction. The result Eq. (2.12) has no dependence on the channel index; the current is identical in each channel. This result has been called the equipartition of the current; e.g., the current is divided equally among the available channels. Since the current in each channel is equal, the total current carried in the constriction is simply the number of such channels that are occupied and the current per channel. Hence,

$$I_{total} = NI = N\frac{2e^2}{h} V .$$ (2.13)

From this last expression, we obtain the conductance through the constriction as

$$G = \frac{I_{total}}{V} = \frac{2e^2}{h} N .$$ (2.14)

and this last expression is usually called the Landauer formula. Thus, the conductance of the quantum wire (our constriction) is quantized in units of $2e^2/h$ (~77.28 µS) with a resulting magnitude that depends only on the number of occupied channels. However, this value is obtained for spin degeneracy; e.g., the factor of 2 arises for the two spin degenerate channels. In the case of some materials, such as Si, there can also be a valley degeneracy, usually 2. This will double the conductance given in Eq. (2.14). These can be confirmed by various experimental techniques that can split the degeneracies.

We note from Eq. (2.14) that, as one raises the Fermi energy, the conductance increases through a set of steps. These steps in the conductance occur as individual channels are occupied or emptied (depending on a rising or decreasing potential). As each new channel is occupied, the conductance jumps upward by $2e^2/h$, and this increase is the same for each and every channel. To clearly see the steps in the conductance requires high mobility materials so that the transport is almost ballistic. As we will see later, the presence of significant scattering will wash out the steps by introducing transitions in which electrons jump from one channel to another and this will upset the balance in the channels.

The conductance steps were observed experimentally in what are known as quantum point contacts (QPCs), which are narrow channels defined in a quasi-two-dimensional semiconductor by a set of gates [3, 4]. In Fig. 2.2, we show the inverse of the conductance steps, plateaus in the resistance of a QPC [5]. As shown in panel (a), the structure is a GaAs/AlGaAs heterostructure, with the quasi-two-dimensional electron gas existing at the interface between the GaAs substrate and the first subsequent GaAlAs layer. The doping is located away from this interface, but electrons fall into the quantum well at the interface by the process of modulation doping. The gates are placed on the top surface, and negative voltage depletes the regions under the gates of electrons. Panel (b) shows the top-down view, with the narrow channel existing between the two surface gates. Finally, panel (c) shows the resistance of the channel as negative gate bias is applied. Here, the resistance is given by the inverse of Eq. (2.14) as

$$R = \frac{h}{2e^2 N}.$$
(2.15)

The first plateau that is observed is at about −0.8 V on the gates and has a resistance of 12.9 kΩ and corresponds to a single spin degenerate mode in the QPC. For a slightly more positive gate voltage, the second plateau at about half this resistance appears and is the onset of the second mode. One can then follow the sequence of more and more modes flowing through the QPC as the gate voltage is made less negative.

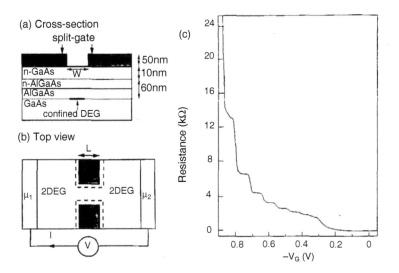

Figure 2.2 (a) Cross section of a typical GaAs/AlGaAs heterostructure with the two gates placed on the surface. (b) Top-down view of the device showing the positioning of the two gates. The device used had an opening of 0.2 micron for the channel and a value of L = 0.3 micron. (c) Resistance of the channel as a function of the top gate voltage. The actual channel is some 100 nm below the surface of the heterostructure. Reprinted with permission from Ref. [5], Copyright 1994, IOP Publishing. All rights reserved.

2.2 Scattering Matrix Approach

As discussed earlier, the normal approach to matching the wavefunction and its derivative across interfaces is generally unstable when a large number of such interfaces are present. However, stability can often be restored by modifying the recursion to one based on the scattering matrix. The scattering matrix has long been a staple of microwave systems and entered quantum mechanics through the Lippmann–Schwinger equation [7]. The strength of this approach lies in the fact that modal solutions to the Lippmann–Schwinger equation maintain their orthogonality through the scattering process [7]. In mesoscopic structures and nanostructures, the connection to microwave theory becomes stronger as the transport becomes dominated by the modes introduced by the lateral confinement. Even in the non-recursion mode, the use of scattering states provides a viable method of building up an orthogonal ensemble, even with weighting by, e.g., a Fermi–Dirac distribution [8]. This is ultimately usable to provide the initial distribution for a semiconductor device [9]. In the recursive form, however, it reaches a new level of viability when coupled with the modal structure in small devices [10]. This approach allows one to determine the transmission probability from one excited mode to a second mode in the output end of the structure and, therefore, to use the well-known Landauer formula [1, 2].

The procedure generally begins with the structure of Eq. (1.38) from Chapter 1. The wavefunction of each transverse slice is given in the *site* representation. That is, it is evaluated at each grid point in the lateral slice. As there are different modes, there will be different wavefunctions whose values on the sites vary significantly. However, the procedure must be initiated, and this is done in the so-called *contact* layer. In this layer, it is assumed that equilibrium conditions exist, and for the present, we consider that the system is at a zero temperature state (we will add nonzero temperature and Fermi–Dirac statistics in a later section). As there are forward and reverse modes and eigenvalues, we have to first find these modes and eigenvalues in the contact layer. This is done through the equation [11]

$$\det[\mathbf{T}_0 - \lambda\mathbf{I}] = 0, \tag{2.16}$$

where $\mathbf{T}_0$ is a matrix whose rank is twice that of the Hamiltonian in Eq. (1.38), and is given as

$$\mathbf{T}_0 = \begin{bmatrix} \mathbf{U}_+ & \mathbf{U}_- \\ \lambda_+\mathbf{U}_+ & \lambda_-\mathbf{U}_- \end{bmatrix}. \tag{2.17}$$

Here, $\mathbf{U}_\pm$ are matrices whose rank is equal to that of the Hamiltonian in Eq. (1.38) and which represent the modes of the structure. The positive subscript is for modes propagating from this contact to the right (and to the other contact), while the negative one is for modes going in the opposite direction. Each column of $\mathbf{U}$ is a wavefunction for a particular mode with the elements representing the value on each grid point of the slice. In essence, these matrices represent the needed mode-to-site transformation matrices. Similarly, the $\lambda_\pm$ are the diagonal eigenvalue matrices whose elements are the values of the propagation constants that appear in, e.g., Eq. (2.9). Associated with this zero slice is a value of the Fermi energy, so that modes whose eigenvalues are below the Fermi energy have propagating wavevectors, while those whose eigenvalues are above the Fermi energy are evanescent modes.

To continue to the scattering matrix, we introduce two matrices $\mathbf{C}_1^s$ and $\mathbf{C}_2^s$, which represent the amplitudes of the forward and backward modes, respectively, at slice s. We will work in the recursion procedure with these matrices. Now, the scattering matrix recursion relation may be expressed as [12]

$$\begin{bmatrix} \mathbf{C}_1^{s+1} & \mathbf{C}_2^{s+1} \\ \mathbf{0} & \mathbf{I} \end{bmatrix} = \begin{bmatrix} \mathbf{0} & \mathbf{I} \\ -\mathbf{I} & t^{-1}(\mathbf{H}_{0,s} - E\mathbf{I}) \end{bmatrix} \begin{bmatrix} \mathbf{C}_1^s & \mathbf{C}_2^s \\ \mathbf{0} & \mathbf{I} \end{bmatrix} \begin{bmatrix} \mathbf{I} & \mathbf{0} \\ \mathbf{P}_{1,s} & \mathbf{P}_{2,s} \end{bmatrix}, \quad (2.18)$$

where $\mathbf{H}_{0,s}$ is the slice Hamiltonian (1.37) (with the values for the local potentials that exist on that slice). The matrices $\mathbf{P}_1$ and $\mathbf{P}_2$ are unknowns but are required to satisfy this recursion equation. Hence, they may be found by expanding the matrix products as

$$\begin{aligned} \mathbf{C}_1^{s+1} &= \mathbf{P}_{1,s} = -\mathbf{P}_{2,s}\mathbf{T}_{21,s}\mathbf{C}_1^s \\ \mathbf{C}_2^{s+1} &= \mathbf{P}_{2,s} = (\mathbf{T}_{21,s}\mathbf{C}_2^s + \mathbf{T}_{22,s})^{-1}. \end{aligned} \quad (2.19)$$

Here, the $\mathbf{T}$ matrices refer to the various entries in the first block matrix on the right-hand side of Eq. (2.18). Hence, $\mathbf{T}_{22}$ refers to the Hamiltonian and energy matrix. The iteration is started by the initial conditions $\mathbf{C}_{1,0} = \mathbf{I}$ and $\mathbf{C}_{2,0} = \mathbf{0}$. Once the last slice is reached, the final transmission matrix is found using the mode-to-site transformation as

$$t = -(\mathbf{U}_+ \, \lambda_+)^{-1}[\mathbf{C}_{1,N+1} - \mathbf{U}_+(\mathbf{U}_+ \, \lambda_+)^{-1}]^{-1}. \quad (2.20)$$

The numerical stability of this method in large part stems from the fact that the iteration implied by Eqs. (2.18) and (2.19) involves products of matrices with *inverted* matrices. Taking such products tends to cancel out most of the troublesome exponential factors.

Once the transmission is determined, we can also find the wavefunctions in the interior of the structure by back propagating from the output contact. At the last slice, it can be shown that the wavefunction is just

$$\psi_N = \mathbf{P}_{2,N}. \quad (2.21)$$

Of course, this wavefunction is a matrix whose columns are the mode wavefunctions, and in turn whose elements are the values on each grid point. Propagating backward, we do the iteration

$$\psi_s = \mathbf{P}_{1,s} + \mathbf{P}_{2,s}\psi_{s+1}, \qquad (2.22)$$

and the density at each grid point can be found from a sum over the various mode contributions at that site, as

$$n(x,y) = n(i,j) = \sum_{k=1}^{M} |\psi_k(i,j)|^2, \qquad (2.23)$$

where k is the mode index (there are as many modes as there are transverse grid points). Because each mode may have a different velocity associated with it, Landauer's formula must be modified to account for this as

$$G = \frac{2e^2}{h} \sum_{n,m} \frac{v_n}{v_m} |t_{nm}|^2, \qquad (2.24)$$

where t_{nm} is the transmission from mode n in the input contact to mode m in the output contact. The summation in Eq. (2.24) is carried out only over the propagating modes, and the velocities are, of course, those in the longitudinal direction of the device.

In Fig. 2.3, we plot the conductance through a QPC of length 100 nm and width 200 nm, but with hard wall confining potentials. In the experiment, the Schottky barrier depletion gates usually produce a soft wall potential that is very harmonic oscillator-like. Here, however, we use the sharp barriers of the hard wall potential for convenience to illustrate the effect. The simulation is done at very low temperature, and one may notice that there are slight peaks and oscillations in the conductance. These arise from so-called "over the barrier resonances" in the tunneling/transmission of new modes. We may recall from introductory quantum mechanics that the tunnel wave does not have unity transmission when the energy is exactly at the top of the barrier, but displays a continued climb to unity then decreases and oscillates. That is the effect that is occurring here. But this will be washed out as the temperature increases and leads to level broadening or there is dissipation in the system due to scattering.

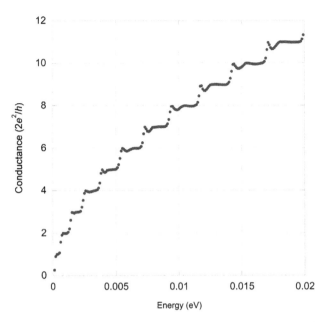

Figure 2.3   The conductance through a hard wall QPC with a width of 200 nm and a length of 100 nm, as computed with the techniques of Section 2.2.

2.3   Including Magnetic Field

The magnetic field may be included in the solutions to the Schrödinger equation once one decides upon an appropriate gauge for the vector potential. There are various gauges that are commonly used, but we almost always work in the overall Coulomb gauge (or electrostatic gauge), where the divergence of the vector potential is set to zero. This then allows us to uniquely connect the magnetic field to the vector potential through

$$\mathbf{B} = \nabla \times \mathbf{A} . \tag{2.25}$$

When we study the magnetic field effect in various mesoscopic devices, we usually make two further choices for the gauge. These arise from the manner in which we can force the magnetic field and the vector potential to satisfy Eq. (2.25). One of these is the *Landau* gauge

$$\mathbf{A} = Bx\mathbf{a}_y , \tag{2.26}$$

where the last quantity is a unit vector in the direction transverse to the current flow direction in the approach of the last section. The other choice often used is the *symmetric* gauge

$$\mathbf{A} = \frac{1}{2}(-By\mathbf{a}_x + Bx\mathbf{a}_y).$$ (2.27)

In quantum mechanics, it is quite general to make the wavefunction gauge invariant, especially if we want to talk about the wavefunction as a field. Normally, we invoke gauge invariance through the condition that we create a new function Λ, which we use to change the vector potential through

$$\mathbf{A} \rightarrow \mathbf{A} + \nabla\Lambda.$$ (2.28)

This now also requires that the scalar potential be modified, and the wavefunction is modified according to

$$\psi(\mathbf{r}) \rightarrow e^{ie\Lambda/h}\psi(\mathbf{r}).$$ (2.29)

If we make a connection between this gauge change and the vector potential itself, we then arrive at Peierls' substitution, which changes the momentum via the vector potential as

$$\mathbf{p} \rightarrow \mathbf{p} - e\mathbf{A}.$$ (2.30)

Now, as we move around each small square of the discretized grid in Fig. 1.5, we couple phase to the wavefunction from the presence of a magnetic field normal to the two-dimensional plane. In the Landau gauge, the amount of phase change is given by

$$\delta\varphi = \frac{e}{h}\oint \mathbf{A}\cdot d\mathbf{l} = \frac{e}{h}\iint B_z dx dy = 2\pi\frac{\delta\Phi}{\Phi_0},$$ (2.31)

where $\delta\Phi$ is the flux coupled through that square and $\Phi_0 = h/e$ is the flux quantum. This is accomplished by adding a phase factor to the hopping energy t, in which the phase is given by

$$t \rightarrow te^{i\varphi}, \quad \varphi = \frac{eBa^2 j}{\hbar},$$ (2.32)

where a is the grid spacing (uniform in this case) and j is the transverse row index. It should be noted that this is an almost universal approach, which can be applied to any numerical simulation scheme, including the recursive Green's functions we will encounter in Section 3.3. One point that is worth noting is that the simulations are often more stable if the number of rows is odd, and the j index above is taken to be zero at the center line of the simulation space. Hence, this would give positive and negative values for the vector potential, although this has no effect on the resulting magnetic field or the overall results of the simulation.

2.4 Simple Semiconductor Devices: The MOSFET

There have been many suggestions for different full quantum methods to model ultra-small semiconductor devices. However, in most of these approaches, the length and the depth are modeled rigorously, while the third dimension (the channel width) is usually included through the assumption that there is no interesting physics in this dimension (lateral homogeneity). Even when a mode structure is assumed, it usually is assumed that the mode does not change shape as it propagates from the source of the device to the drain of the device. Other simulation proposals have simply assumed that only one sub-band in the orthogonal direction is occupied, therefore making higher-dimensional transport considerations unnecessary. These are generally not valid assumptions, especially as we approach devices whose width is comparable to the channel length, both of which may be less than 10 nm in the not-too-distant future.

It is important to consider all the modes that may be excited in the source (or drain) region, as these may be responsible for some of the interesting physics that we wish to capture. In the source, the modes that are excited are three dimensional in nature, even in a thin SOI device. These modes are then propagated from the source to the channel, and the coupling among the various modes will be dependent on the details of the total confining potential at each point along the channel. Moreover, as the doping and the Fermi level in short-channel MOSFETs increase, we can no longer assume that there is only one occupied sub-band. In an effort to provide a more complete simulation method, we have developed a full three-dimensional quantum simulation, based on the use of recursive scattering matrices, as described in the previous section [13, 14].

2.4.1 Device Structure

The device under consideration is an Si nanowire MOSFET structure with a cross-sectional area of 6×9 nm. The x- and z-directions correspond to the length and the height of the device, respectively, while the y-direction is the width of the structure. The source and the drain contact regions are 10 nm in length and 18 nm in width but have the same height as the active channel. In an actual device, the length of the source and the drain of a MOSFET would be much longer, but this length captures the important energy relaxation length. We implement open boundary conditions at the ends of the structure and on the sidewalls. The gate is applied only to the top surface of the nanowire and has a length of 11 nm, which overlaps onto the source and drain. The actual channel length of the device used in these simulations is 9 nm. The channel itself is 9 nm in width, so that the Si layer is a wide–narrow–wide structure. The entire structure is on a 10 nm buried oxide (BOX) layer. The gate oxide is taken to be 2 nm thick.

An important point relates to the crystal orientation of the device. As is normal, we assume that the device is fabricated on a [100] surface of the Si crystal, and we then orient the channel so that the current will flow along the <100> direction. This direction is chosen so that the principal axes of the conduction band valleys line up with the coordinate axes. By this, we mean that the <010> direction lines up along the y-direction and the <001> direction lines up with the z-direction, and the six equivalent ellipsoids are oriented along the Cartesian coordinate axes. This is important so that the resulting quantization will split these ellipsoids into three pairs. Moreover, the choice of axes is most useful as the resulting Hamiltonian matrix will be diagonal. In contrast, if we had chosen the <110> direction to lie along the channel, the six ellipsoids would have split into a two-fold pair (those normal to the [100] plane) and a four-fold pair, but the Hamiltonian would not be diagonal since the current axis makes an angle with each ellipsoid of the four-fold pair. Using our orientation complicates the wavefunction, as we will see, but allows for simplicity in terms of the amount of memory needed to store the Hamiltonian and to construct the various scattering matrices (as well as the amount of computational time that is required).

2.4.2 Wavefunction Construction

The total wavefunction is composed of three major parts, one for each of the three sets of valleys. That is, we can write the wavefunction as a vector

$$\psi_T = \begin{bmatrix} \psi^{(x)} \\ \psi^{(y)} \\ \psi^{(z)} \end{bmatrix},$$

(2.33)

where the superscript refers to the momentum space direction of the long axis of the ellipsoidal constant energy surfaces; e.g., the principal axis of the ellipsoid. Each wavefunction defined in this manner accounts for two ellipsoids of the six total. So the x-axis accounts for the pair of <100> ellipsoids. Each of these three component wavefunctions is a complicated wavefunction on its own. Consider the Schrödinger equation for one of these sets of valleys (i corresponds to x, y, or z valleys):

$$-\frac{\hbar^2}{2}\left[\frac{1}{m_x}\frac{\partial^2}{\partial x^2} + \frac{1}{m_y}\frac{\partial^2}{\partial y^2} + \frac{1}{m_z}\frac{\partial^2}{\partial z^2}\right]\psi^{(i)} + V(x,y,z)\psi^{(i)} = E\psi^{(i)}.$$

(2.34)

Here, it is assumed that the mass is constant, in order to simplify the equations (for nonparabolic bands, the reciprocal mass enters between the partial derivatives). We have labeled the mass corresponding to the principal coordinate axes, and these take on the values of m_L and m_T as appropriate. We then implement this on a finite difference grid with uniform spacing a, as discussed earlier. This leads to another complication, which is that the hopping energy is now directionally dependent, as we have

$$t_x = \frac{\hbar^2}{2m_x a^2}, \quad t_y = \frac{\hbar^2}{2m_y a^2}, \quad t_z = \frac{\hbar^2}{2m_z a^2}.$$

(2.35)

Each hopping energy corresponds with a specific direction in the silicon crystal. The fact that we are now dealing with three sets of hopping energies is quite important.

The next important point is that we are now dealing with a three-dimensional structure. For this, we will develop the method in terms of slices. We begin first by noting that the transverse plane has $N_y \times N_z$ grid points. Normally, this would produce a second-rank tensor (matrix) for the wavefunction, as in the previous sections. However, we can re-order the coefficients into a $N_y N_z \times 1$ first-rank tensor (vector), so that the propagation is handled by a simpler matrix multiplication. Since the smaller dimension is the z-direction, we use N_z for the expansion and write the vector wavefunction as

$$\psi^{(i)} = \begin{bmatrix} \psi_{1,N_y}^{(i)} \\ \psi_{1,N_y}^{(i)} \\ \dots \\ \psi_{1,N_y}^{(i)} \end{bmatrix}. \tag{2.36}$$

Now, this vector can be incorporated into the discretized Schrödinger equation as

$$H^{(i)}\psi_s^{(i)} - T_x^{(i)}\psi_{s-1}^{(i)} - T_x^{(i)}\psi_{s+1}^{(i)} = EI\psi_s^{(i)}, \tag{2.37}$$

where I is the unit matrix as before. The new slice Hamiltonian is a block tri-diagonal matrix of dimension $N_y N_z \times N_y N_z$ because of the stacking of the two-dimensional planes in this three-dimensional structure. This Hamiltonian may be written as

$$H^{(i)} = \begin{bmatrix} \mathbf{H}_0^{(i)}(\mathbf{r}) & \mathbf{t}_z^{(i)} & \dots & 0 \\ \mathbf{t}_z^{(i)} & \mathbf{H}_0^{(i)}(\mathbf{r}) & \dots & 0 \\ \dots & \dots & \dots & \mathbf{t}_z^{(i)} \\ 0 & 0 & \mathbf{t}_z^{(i)} & \mathbf{H}_0^{(i)}(\mathbf{r}) \end{bmatrix}, \tag{2.37}$$

where each of the blocks is given as

$$\mathbf{H}_0^{(i)}(\mathbf{r}) = \begin{bmatrix} V(s,1,\eta)+W & t_y^{(i)} & \dots & 0 \\ t_y^{(i)} & V(s,2,\eta)+W & \dots & 0 \\ \dots & \dots & \dots & t_y^{(i)} \\ 0 & 0 & t_y^{(i)} & V(s,N_y,\eta)+W \end{bmatrix}, \tag{2.38}$$

with $W = 2(t_x^{(i)} + t_y^{(i)} + t_z^{(i)})$. Similarly, the hopping matrix is also of this large dimension and may be written as the diagonal

$$\mathbf{T}_x^{(i)} = \begin{bmatrix} t_x^{(i)}I & 0 & \cdots & 0 \\ 0 & t_x^{(i)}I & \cdots & 0 \\ \cdots & \cdots & \cdots & 0 \\ 0 & 0 & 0 & t_x^{(i)}I \end{bmatrix}. \tag{2.39}$$

With this setup of the matrices, the general procedure follows that laid out in the previous recursive scattering matrix approach of Section 2.2. The only new problem is the non-scalar effective mass, and the much larger size of the matrices due to the three-dimensional approach to the simulation.

2.4.3 The Landauer Formula Again

Before proceeding, we wish to develop a more general form of the Landauer formula that can be utilized for cases in which there is a nonzero temperature. Quite generally, in a quasi-one-dimensional conductor, we can write the current flowing from the left reservoir (or contact) to the right reservoir as

$$I_{LR} = 2e \int dE \rho_1(E) v(E) T(E) f(E, E_{FL}). \tag{2.40}$$

Here, the transmission is a sum over the propagating modes as

$$T(E) = \sum_i T_i(E), \tag{2.41}$$

and Eq. (2.40) can be rewritten as

$$J_{LR} = \frac{2e}{h} \sum_i \int dE T_i(E) f(E, E_{FL}). \tag{2.42}$$

Similarly, we may write the current flowing from the right reservoir to the left reservoir as

$$J_{RL} = \frac{2e}{h} \sum_i \int dE T_i(E) f(E, E_{FR}). \tag{2.43}$$

If we apply a voltage bias to the right reservoir relative to the left reservoir, we can connect the two Fermi levels as

$$E_{FR} = E_{FL} + eV = E_F + eV, \tag{2.44}$$

and the net current can be written as

$$J = \frac{2e}{h} \sum_i \int dE T_i(E) \left[f(E, E_F) - f(E, E_F + eV) \right]. \tag{2.45}$$

The above equation can be compared to the equivalent standard equation for a simple MOSFET, which may be written as

$$
\begin{aligned}
I &= \frac{W \mu C_{ox}}{L} \left(V_G - V_T - \frac{1}{2} V_D \right) V_D \\
&= \frac{W \mu C_{ox}}{2L} \left[(V_G - V_T)^2 - (V_G - V_T - V_D)^2 \right].
\end{aligned} \tag{2.46}
$$

Here, W and L are the gate width and length, μ is the carrier mobility, C_{ox} is the gate oxide specific capacitance per unit area, V_G and V_D are the biases applied to the gate and drain, and V_T is the threshold voltage at which mobile charge begins to accumulate in the inversion layer. It is important to note that this latter equation has a similar form as Eq. (2.45) in that there is a term for current that enters from the source (left reservoir) and a counter-flowing current that enters from the drain (right reservoir). If the applied bias is very small, and we want only the small-signal conductance, then we need only calculate the simple transmission from our numerical technique. However, if we are going to apply large bias voltages and need to solve Poisson's equation self-consistently along with the quantum transport, then we need to carefully determine the contributions to the charge density that come from both of these currents. This must be done at each and every grid point in order to have the correct density to include in Poisson's equation. Hence, the procedure is iterated as we first solve Poisson's equation for the local potential and then the quantum transport equation for the transmission and the local density. It is important that we account for both the left-derived density and the right-derived density. Even though the current through the MOSFET saturates when the right-derived term goes to zero, one must still determine the density contributed by this term as it affects the local potential within the device. Thus, we must use our quantum transport approach twice, once from each end in order to determine both the left-derived density and the right-derived density. This is true whatever methodology is used to determine the quantum transport.

Second, the presence of the Fermi–Dirac distributions is also important. This is because, at nonzero temperature, the "constant" $\mathbf{C}_{1,0}$ is no longer the unit matrix. Instead, the initial value for each transverse eigenmode is taken from the value of the Fermi–Dirac distribution at the eigenenergy of that mode. This introduces not only the temperature dependence, but also the additional modifications of the transmission that can occur as the local potential changes with bias and self-consistent density.

In Fig. 2.4, the drain current through the Si nanowire device discussed earlier is plotted as a function of the gate voltage [14]. Here, the threshold voltage is about –0.5 V, as can be determined by extrapolating the linear rise of current back to zero. In this case, the drain voltage is relatively small. The saturation in the current is not the normal saturation but arises from the fact that only a few modes are excited in the source. Here, the acceptor density in the nanowire itself is 2×10^{18} cm^{-3}. A simple calculation says that there is, on average, only a single impurity in the channel, so its position is quite important in determining the current. The three curves in the figure are different implementations of the doping in the channel and the source and drain. The randomness in the doping and the impurity positions leads to the variation in current that is seen.

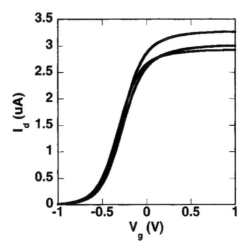

Figure 2.4   The current flowing through the small Si nanowire transistor. Here, the potential is found self-consistently with the charge density for the device. Reprinted with permission from Ref. [14], Copyright 2004, AIP Publishing LLC.

2.4.4   Incorporating Scattering

Many authors have stated that real scattering could not be incorporated in the Landauer formula. But, Landauer's original paper had "scattering" in the title [1]. Indeed, so long as the transmission is computed *completely* through the structure, from one contact to another, and from one mode to another, this approach will work. We need only assure that the method is density conserving. The problem is in making sure that the total transmission is computed. Indeed, using a relaxation time will not work, since it does not conserve probability density. Rather, the correct approach must insure a gain–loss formalism. What this means is that for each out-scattering process, there must be a corresponding in-scattering process treated within the formalism. Only with this complete treatment will probability density be conserved throughout the simulation regime. We have found that this can be achieved with a self-energy formulation, so long as the actual self-energy matrix is transformed into the site representation used in the recursive scattering matrix [15].

Generally, scattering is quite weak in semiconductors. This is why they have high mobilities. Moreover, the dominant scattering is quite often impurity scattering, and this is treated in real space via the actual self-consistent potential [16, 17]. While this approach was originally developed for classical simulations, it readily carries over to quantum simulations since the potential is a local one-point function and is readily incorporated into the Hamiltonian in the recursive simulation.

Scattering is treated via the self-energy, which we introduced in Chapter 1. The self-energy Σ has both real and imaginary parts, with the latter representing the dissipative interactions. The real part contributes an energy shift. In semiconductors, the scattering is weak and is traditionally treated by first-order time-dependent perturbation theory, which yields the common Fermi golden rule for scattering rates. With such weak scattering, the real part of the self-energy can generally be ignored for the phonon interactions, and the part that arises from the carrier–carrier interactions is incorporated into the solutions to Poisson's equation by a local-density approximation, which approximately accounts for the Hartree–Fock corrections. Now, because we are going to use the modes that exist in our nanoribbon, we have to consider that the "particle" may shift position during the scattering process, so write the self-energy as a two-point function in the form

$$\Sigma(\mathbf{r}_1, \mathbf{r}_2).$$ (2.47)

In the case in which we are using transverse modes in the quantum wire of the transistor, this may be rewritten as

$$\Sigma(i, j; i', j'; x_1, x_2).$$ (2.48)

In this notation, the scattering accounts for transitions from transverse mode (i, j) at longitudinal position x_1 to transverse mode (i', j') at longitudinal position x_2. To facilitate the following, let us introduce center-of-mass coordinates in the longitudinal direction with

$$X = \frac{x_1 + x_2}{2}, \quad \xi = x_1 - x_2.$$ (2.49)

We then Fourier transform on the difference variable to give

$$\Sigma(i, j; i', j'; X, k) = \frac{1}{2\pi} \int d\xi e^{i\xi k} \Sigma(i, j; i', j'; X, \xi).$$ (2.50)

The center-of-mass position X remains in the problem since the mode structure may change as one moves along the channel, particularly if the device is gated. At this point, the left-hand side of Eq. (2.50) is the self-energy computed by the normal scattering rates, as was done in quantum wells and quantum wires previously [18]. However, these previous calculations usually used the Fermi golden rule, which is an evaluation of the bare self-energy. In many-body approaches, one normally does not use the energy-conserving delta function, which is the central part of the Fermi golden rule. Rather, this function is broadened into the *spectral density*, through the use of the self-consistent Born approximation. In this way, off-shell effects are taken into account through the broadening of the relationship between momentum and energy. In semiconductors, however, we have already noted that the scattering is weak. It has been pointed out that these off-shell corrections are only important in fast processes where we are interested in femtosecond response, and their neglect introduces only slight errors for times large compared to the collision duration [19]. Moreover, the broadening of the delta function will not be apparent when we reverse the

Fourier transform of Eq. (2.50), as the area under the spectral density remains normalized to unity. Since our recursion is in the site representation, rather than in a mode representation, we have to reverse the Fourier transform in Eq. (2.50) to get the *x*-axis variation, and then do a mode-to-site unitary transformation to get the self-energy in the form necessary for the recursion. This is the subject of the rest of this section. Hence, we begin by seeking the imaginary part of the self-energy, which is related to the scattering rate via

$$\text{Im}\{\Sigma(i, j; i', j'; X, k)\} = \hbar \left(\frac{1}{\tau}\right)\bigg|_{i,j}^{i',j'}. \tag{2.51}$$

It is this latter scattering rate that we calculate using the Fermi golden rule. This result will be a function of the *x*-directed momentum (which is related, in turn, to the energy of the carrier) in the quantum wire. Finally, this scattering rate must be converted to the site representation with a unitary transformation developed above in Eq. (2.17). These can be either the forward or the reverse transformation matrices, as they should be the same, and

$$\Gamma = \text{Im}\{\Sigma\} = U^+ \left(\frac{\hbar}{\tau}\right)\bigg|_{i,j}^{i',j'} U. \tag{2.52}$$

The actual scattering rates for Si devices in this formulation have been calculated earlier [20], and we present only the results here. For the acoustic phonons, we have

$$\frac{1}{\tau}\bigg|_{i,j}^{i',j'} = \frac{m_x^* D_{\text{ac}}^2 k_B T}{4\hbar^3 \rho v_s^2} \sqrt{\frac{\pi}{2}} I_{i,j}^{i',j'}, \tag{2.53}$$

where

$$I_{i,j}^{i',j'} = \frac{1}{4\pi^2} \int\int dy \, dz \left[\varphi_{i,j}^*(y,z)\varphi_{i',j'}(y,z)\right]^2 \tag{2.54}$$

is the overlap integral between the two transverse wavefunctions for the lateral modes. In Eq. (2.53), D_{ac} is the deformation potential, ρ is the mass density, and v_s is the longitudinal sound velocity in Si.

Scattering between the equivalent valleys of Si is carried out by high-energy optical modes, usually with momentum wavevectors that are relatively large. There are two types of these phonons: the

f phonons between valleys on different axes in momentum space and g phonons between the two valleys along the same momentum axis [21, 22]. There are several phonons that can contribute to these processes, but in the lowest sub-band of the quantized Si nanowire, we usually only need to consider the g phonons. The difference from the acoustic modes is that we can no longer ignore the energy of these phonons, and their distribution cannot be taken to be in the equipartition limit. As a result, we have to treat the emission and absorption processes separately, but the results are similar. In general, the absorption process is simpler, but we have to consider the energy separation of the initial and final modes via the quantity

$$\Delta_{i,j}^{i',j'} = E_{0,i,j} - E_{0,i',j'} + \hbar\omega_0,$$
(2.55)

where E_0 is the energy of the lowest sub-band (and the added energy for the particular mode) and the last term is the phonon energy. In the absorption of the phonon, this energy difference creates no problem, and the scattering rate can be computed as

$$\left.\frac{1}{\tau_{ab}}\right|_{i,j}^{i',j'} = \frac{m_x^* D_{op}^2 N_q}{8\pi^2 \rho\hbar^2 \omega_0}\sqrt{\frac{\pi}{2}} I_{i,j}^{i',j'}.$$
(2.56)

In this equation, D_{op} is the optical deformation potential and N_q is the Bose–Einstein distribution evaluated at the energy of the optical phonon. In the emission case, one has to worry about the energy separation Eq. (2.55), and the size and sign of this quantity give rise to different results. Nevertheless, the emission term can be evaluated in a straightforward manner, with the result

$$\left.\frac{1}{\tau_{em}}\right|_{i,j}^{i',j'} = \frac{m_x^* D_{op}^2 (N_q+1)}{8\pi^2 \rho\hbar^2 \omega_0} I_{i,j}^{i',j'} \cdot \begin{cases} 1, & \Delta_{i,j}^{i',j'} < 0, \\ \dfrac{2}{(2\pi)^{3/2}}, & \Delta_{i,j}^{i',j'} = 0, \\ \dfrac{\pi^{3/2}}{4}, & \Delta_{i,j}^{i',j'} > 0. \end{cases}$$
(2.57)

These scattering rates have been used in a quantum calculation of a large Si nanowire of ~20 nm width and height, and the density dependence of the mobility has been found to fit that of bulk material [20]. Hence, there is pretty good confidence in the use of these scattering rates.

2.4.5 Ballistic to Diffusive Crossover

In general, the transport in small transistors is often claimed to be ballistic, but this is seldom the case in Si devices, and this has been demonstrated in small transistors [23]. Here we discuss the presence of ballistic transport and the transition to diffusive behavior. We use the aforementioned simulation techniques and consider a small square cross-sectional device with a larger source and drain, so that the dominant resistance arises from the nanowire. We then vary the length of the nanowire section and study the resistance. When the transport is ballistic, the resistance of the channel will be given by the inverse of the Landauer formula as

$$R = \frac{h}{2e^2 N},$$ (2.58)

where N is the number of transverse modes in the cross section of the nanowire. We note that this resistance is independent of the length of the nanowire section, as the resistance exists in the two contact regions. On the other hand, when scattering dominates the transport, we expect the resistance to be

$$R = \frac{L}{\sigma A} = \frac{L}{ne\mu A},$$ (2.59)

where n is the density, μ is the mobility, L is the length of the nanowire, and A is the cross-sectional area of the nanowire. In Fig. 2.5, we show the resistance of the nanowire section as a function of its length for two different temperatures for a cross section of 6.5×6.5 nm^2. At a lattice temperature of 100 K, the resistance is independent of the length of the nanowire, at least up to 8 nm. But at 300 K, the phonons are sufficiently excited that the resistance is almost linear in the length of the nanowire, indicating diffusive transport. So for this nanowire, the ballistic to diffusive transport at 300 K occurs somewhere near or below 2 nm. Now, we might expect the mobility in an MOS structure to be of the order of 300 cm^2/Vs or less. For this mobility and transport along the (100) direction, this would give a momentum relaxation time of 32 fs, or an apparent mean free path of about 7 nm. However, the scattering rate is much faster than the momentum relaxation time, perhaps by as much as a factor of 4, and this lowers the real mean free path to just under 2 nm, which is what is observed. For comparison, in Fig. 2.6, we

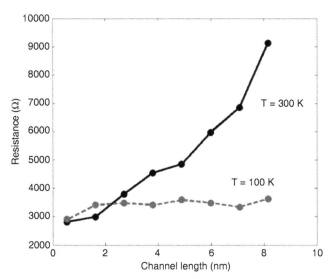

Figure 2.5 Resistance versus the nanowire length for a device with a width of 6.5 nm at the indicated temperatures. Reprinted with permission from Ref. [15], Copyright 2006, IOP publishing. All rights reserved.

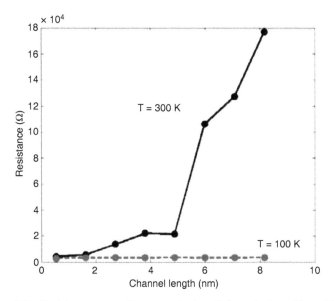

Figure 2.6 Resistance versus the nanowire length for a device with a width of 4.3 nm at the indicated temperatures. Reprinted with permission from Ref. [15], Copyright 2006, IOP publishing. All rights reserved.

show the resistance at the same two temperatures for a nanowire cross section of 4.3 × 4.3 nm². In this case, the smaller cross section produces more quantization of the phonon modes, making them less effective. Now it appears that the ballistic to diffusive crossover at 300 K is about 5 nm. However, the overall resistance is much higher, and there is likely only a single mode occupied in the ballistic regime, judging from the size of the resistane.

2.5 Density Matrix and Its Brethren

In this section, we want to give a broad introduction to the various forms of quantum distributions that have found use in quantum transport. To be sure, each of these forms arises from the wavefunction that we have introduced in this chapter, and each of these will be treated in more detail in later chapters. Here, however, we want to show how they all fit together and derive from the wavefunction. Generally, one can solve the Schrödinger equation by assuming an expansion of the wavefunction in a suitable basis set so that each basis function is an energy eigenfunction according to

$$H\varphi_n = E_n\varphi_n \qquad (2.60)$$

in which E_n is the energy level corresponding to the particular basis function. Then, the total wavefunction can be written as

$$\psi(\mathbf{r}, t) = \sum_n c_n \varphi_n(\mathbf{r}) e^{-iE_n t/\hbar} . \qquad (2.61)$$

If we now multiply this equation with the complex conjugate of another basis function, say $\varphi_m^*(\mathbf{r})$, and integrate over the volume, we can determine the coefficients of the expansion as

$$\int d\mathbf{r} \varphi_m^*(\mathbf{r})\psi(\mathbf{r}, t) = \sum_n c_n \delta_{nm} e^{-iE_n t/\hbar} = c_m e^{-iE_m t/\hbar} , \qquad (2.62)$$

and the coefficient can be inserted into Eq. (2.61), using different time and position, to give

$$\psi(\mathbf{r}, t) = \sum_n \int d\mathbf{r}' \varphi_n^*(\mathbf{r}')\varphi_n(\mathbf{r})\psi(\mathbf{r}', t_0) e^{-iE_n(t - t_0)/\hbar} \qquad (2.63)$$

Now this equation can be rewritten as

$$\psi(\mathbf{r}, t) = \int d\mathbf{r}' K(\mathbf{r}, t; \mathbf{r}', t_0)\psi(\mathbf{r}', t_0) \qquad (2.64)$$

which defines the generalized propagator or kernel K. This quantity is a function of two positions and two times. It is often called a Green's function, but we will define this more efficiently below.

The kernel describes the general propagation of any initial wavefunction at time t_0 to any arbitrary time t (which is normally $>t_0$, but not always). There are a number of methods of evaluating it, either by differential equations or by integral equations known as path integrals [24]. In general, the usefulness of the present form (2.64) lies in the existence of the entire set of basis functions, which are characteristic of the problem at hand. As a result, the kernel generally represents an admixture of states, which differs from the above formulation in which we write it as

$$K(\mathbf{r}, t; \mathbf{r}', t_0) = \sum_{n,m} c_{nm} \varphi_m^*(\mathbf{r}, t) \varphi_n(\mathbf{r}', t_0). \tag{2.65}$$

The indication of Eq. (2.65) is that the basis functions are now time dependent, but this could arise merely through the exponential functions of the energy eigenvalues for these functions.

In a great many situations, the separation into two times is not necessary, as the processes that lead to these separate times are incredibly fast. For the general situation, one may then reduce the kernel to a single time variable, and we then have the density matrix

$$\rho(\mathbf{r}, \mathbf{r}', t) = \sum_{n,m} c_{nm} \varphi_m^*(\mathbf{r}, t) \varphi_n(\mathbf{r}', t) \equiv \psi^*(\mathbf{r}, t) \psi(\mathbf{r}', t). \tag{2.66}$$

There are many different definitions of the density matrix, which differ only in the fine detail. One can also introduce the imaginary time $t \to -i\hbar\beta$, where $\beta = 1/k_B T$ is the inverse temperature. Then, the density matrix can be written in the thermal equilibrium form

$$\rho(\mathbf{r}, \mathbf{r}', t) = \sum_n e^{-\beta E} \varphi_n^*(\mathbf{r}, t) \varphi_n(\mathbf{r}', t). \tag{2.67}$$

2.5.1 Liouville and Bloch Equations

The temporal differential equation for the density matrix can best be developed by the standard approach for the time variation of any operator, which involves its commutator with the Hamiltonian

$$i\hbar \frac{\partial \rho}{\partial t} = [H, \rho] \tag{2.68}$$

and this leads to the form

$$i\hbar\frac{\partial\rho}{\partial t}=\left[-\frac{\hbar^2}{2m^*}\left(\frac{\partial^2}{\partial \mathbf{r}^2}-\frac{\partial^2}{\partial \mathbf{r'}^2}\right)+V(\mathbf{r})-V(\mathbf{r'})\right]\rho(\mathbf{r},\mathbf{r'},t).\qquad(2.69)$$

which is known as the *Liouville equation*. While no dissipation is indicated here, it is quite easy to incorporate a dissipative term [25–27].

A different form of the equation arises when we introduce the imaginary time used for Eq. (2.67). If we use the same definition in Eq. (2.69), we arrive at

$$-\frac{\partial\rho}{\partial\beta}=\left[-\frac{\hbar^2}{2m^*}\left(\frac{\partial^2}{\partial \mathbf{r}^2}-\frac{\partial^2}{\partial \mathbf{r'}^2}\right)+V(\mathbf{r})-V(\mathbf{r'})\right]\rho(\mathbf{r},\mathbf{r'}),\qquad(2.70)$$

which is known as the *Bloch equation*. In a sense, the Bloch equation is a quasi-steady-state, or quasi-equilibrium, form in which the time variation is either nonexistent or sufficiently slow as to not be important in the form of the statistical density matrix. It is important to note that the Bloch equation possesses an adjoint equation, which arises from the anti-commutator form as

$$-\frac{\partial\rho}{\partial\beta}=\left[-\frac{\hbar^2}{2m^*}\left(\frac{\partial^2}{\partial \mathbf{r}^2}+\frac{\partial^2}{\partial \mathbf{r'}^2}\right)+V(\mathbf{r})+V(\mathbf{r'})\right]\rho(\mathbf{r},\mathbf{r'}).\qquad(2.71)$$

The importance of the adjoint equation is that it allows one to find the equilibrium form of the density matrix, and this can be used as the initial condition for the time evolution of Eq. (2.69).

2.5.2 Wigner Functions

One problem that is always raised with respect to the density matrix is that it is a function only of time and the position variables. In classical physics, one has the Boltzmann equation, which is a function of position, momentum, and time, so that one can generate a phase–space formulation of the transport problem. None of the quantum functions we have discussed so far has this convenient phase–space formulation. But it is often very convenient to describe the system with a phase–space formulation. And, in fact, such a formulation exists in which we first introduce the center-of-mass and difference coordinates through

$$\mathbf{R} = \frac{\mathbf{r} + \mathbf{r}'}{2}, \quad \mathbf{s} = \mathbf{r} - \mathbf{r}'. \tag{2.72}$$

We can now introduce the phase–space Wigner function as the Fourier transform on the difference coordinate as [28]

$$f_W = \frac{1}{h^3} \int d\mathbf{s} \rho(\mathbf{R}, \mathbf{s}, t) e^{i\mathbf{p}\cdot\mathbf{s}/\hbar} = \frac{1}{h^3} \int d\mathbf{s} \rho(\mathbf{R} + \frac{\mathbf{s}}{2}, \mathbf{R} - \frac{\mathbf{s}}{2}, t) e^{i\mathbf{p}\cdot\mathbf{s}/\hbar}. \tag{2.73}$$

If we incorporate the coordinate transformations of Eq. (2.72) into Eq. (2.69), we arrive at a new equation of motion for the density matrix as

$$i\hbar \frac{\partial \rho}{\partial t} = \left[-\frac{\hbar^2}{m^*} \frac{\partial^2}{\partial \mathbf{R} \partial \mathbf{s}} + V\left(\mathbf{R} + \frac{\mathbf{s}}{2}\right) - V\left(\mathbf{R} - \frac{\mathbf{s}}{2}\right) \right] \rho(\mathbf{R}, \mathbf{s}, t). \tag{2.74}$$

The transform in Eq. (2.73), although introduced by Wigner, is often called the Weyl transform [29–32]. When this transformation is applied to the above equation, we get a new equation of motion for the Wigner function as

$$\frac{\partial f_W}{\partial t} + \frac{\mathbf{p}}{m^*} \cdot \frac{\partial f_W}{\partial \mathbf{R}} - \frac{1}{i\hbar} \left[V\left(\mathbf{R} + \frac{i\hbar}{2} \frac{\partial}{\partial \mathbf{p}}\right) - V\left(\mathbf{R} - \frac{i\hbar}{2} \frac{\partial}{\partial \mathbf{p}}\right) \right] f_W = 0. \tag{2.75}$$

In the absence of any dissipative processes, this can be rewritten in a more useful form as

$$\frac{\partial f_W}{\partial t} + \frac{\mathbf{p}}{m^*} \cdot \frac{\partial f_W}{\partial \mathbf{R}} - \frac{1}{h^3} \int d\mathbf{P} W(\mathbf{R}, \mathbf{P}) f_W(\mathbf{R}, \mathbf{p} + \mathbf{P}) = 0, \tag{2.76}$$

where

$$W(\mathbf{R}, \mathbf{P}) = \int d\mathbf{s} \sin\left(\frac{\mathbf{P} \cdot \mathbf{s}}{\hbar}\right) \left[V\left(\mathbf{R} + \frac{\mathbf{s}}{2}\right) - V\left(\mathbf{R} - \frac{\mathbf{s}}{2}\right) \right]. \tag{2.77}$$

The use of the Wigner function is particularly important in scattering problems [33] and clearly shows the transition to the classical world.

In general, the Wigner function is not a positive definite function. This is a consequence of the uncertainty relationship. On the other hand, if we integrate the Wigner function over either position or momentum, we will get a positive definite function. In fact the negative excursions of the function exist over phase–space regions whose volume is $\hbar^3$, so that smoothing the function over a volume corresponding to the uncertainty principle will produce a positive definite function.

2.5.3 Green's Functions

In both the density matrix and the Wigner function, only a single time variable appears in the problem, as it was assumed that the two wavefunctions were to be evaluated at the same time. In essence, these two approaches build in correlations in space but do not consider that there may be correlations in time. On the other hand, the kernel Eq. (2.65) possesses both times, and this will lead to the concept of Green's functions [34]. In general, we separate the kernel into a term that is forward in time $(t > t_0)$ and a term that is backward in time $(t < t_0)$, by changing the order of the two wavefunctions. Then we combine the two, which produces a commutator relationship, and define the *retarded* Green's function as

$$G_r(\mathbf{r}, \mathbf{r}'; t, t_0) = -i\Theta(t - t_0)\langle\{\psi(\mathbf{r}, t), \psi^\dagger(\mathbf{r}', t_0)\}\rangle. \qquad (2.78)$$

In this equation, we have included several new notations. First, $\Theta(x)$ is the Heaviside step function, which is 1 for $x > 0$ and zero otherwise. Then the angle brackets have been added to indicate an ensemble average, which corresponds to a sum over the appropriate basis states, while the curly brackets denote an *anti*-commutator relation where the normal negative sign is replaced with a positive sign, which is the normal case for fermions. By the same token, we can arrive at the retarded Green's function as

$$G_a(\mathbf{r}, \mathbf{r}'; t, t_0) = i\Theta(t_0 - t)\langle\{\psi^\dagger(\mathbf{r}', t_0), \psi(\mathbf{r}, t)\}\rangle. \qquad (2.79)$$

Note here that the argument of the Heaviside function has been reversed, as has the order of the two wavefunctions. With these two definitions, one can write the kernel as

$$\langle K(\mathbf{r}, \mathbf{r}'; t, t_0)\rangle = i\left[G_r(\mathbf{r}, \mathbf{r}'; t, t_0) - G_a(\mathbf{r}, \mathbf{r}'; t, t_0)\right]. \qquad (2.80)$$

Now these two Green's functions do not exhaust the possibilities, and these are normally described as the equilibrium Green's functions. There are certainly other Green's functions with which we will deal in later chapters. We will see in the next chapter that the kernel in Eq. (2.80) is, in fact, quite close to the spectral density that we introduced in Eq. (1.8), and we will discuss this in the next chapter.

Finding Green's functions from the Schrödinger equation, or from the Liouville equation, is not difficult for simple Hamiltonians, as for any quantum mechanical problem. For complicated Hamiltonians, such as in the case of many-body interactions or electron–phonon interactions, it is not so simple, and one normally resorts to perturbation theory. However, perturbation theory is not without its own problems, in that the terms may be difficult to evaluate or the series may not converge. We will deal with these problems in later chapters.

2.6   Beyond Landauer

In earlier sections, we discussed the transport through a nanostructure through the use of the scattering matrix and the wavefunction itself. This approach is very like the scattering of light through various structures or the scattering of electromagnetic waves through complex waveguide circuits. But it provides the dc conductance of the system, which is fine. But one can do more. In steady-state, or dc, transport, the circuit must satisfy Kirchhoff's voltage and current laws. However, in the transient condition, this is not required, as deviations from these continuity laws lead to the accumulation or depletion of charge at various points in the circuit, a result that satisfies Gauss' law in electromagnetic fields. Indeed, in single electron circuits, the charging of isolated quantum dots clearly obeys Gauss' law and not necessarily Kirchhoff's law [35].

Büttiker began to address ac conductance in these quantum circuits with a discussion of noise in them [36]. He introduced the idea that particles in the incoming states could be created and annihilated by a set of operators $a^\dagger$ and a, respectively. Similarly, one can define a set of operators that create and annihilate a particle in the outgoing (or reflected) states $b^\dagger$ and b, respectively. Then the scattering matrix provides a unique unitary transformation between the a operators and the b operators. We used this property in the development of Section 2.2. Let us define the scattering matrix $\mathbf{s}_{\alpha\beta}$ as the element of the scattering matrix that describes a wave entering the system from mode β of the input lead and is scattered back into mode α of this lead. Büttiker then determined that the equilibrium noise, equivalently the noise conductance, is determined by the quantity

$$\frac{e^2}{h} Tr\{\mathbf{s}_{\alpha\beta}^\dagger \mathbf{s}_{\alpha\beta}\}. \tag{2.81}$$

These bilinear terms are identical to the coefficients determined in Section 2.2 to describe the transport coefficients of the dc current. In a sense, this is a form of the Nyquist relation, or equivalently the Kubo formula, in which the fluctuations (noise, arising from a current–current correlation function) are related to the conductance (or current) through the structure.

The connection between the ac response (of the noise) and the dc conductance is interesting. Once we determine the conductance and the scattering matrix, the ac-transport properties are sensitive to the *phases* of the scattering matrix elements [37]. In the latter paper, it is shown that the conductance formula Eq. (2.24) can be extended to the frequency-dependent conductance via the formula [38]

$$g_{\alpha\beta}^{(m)}(\omega) = \frac{e^2}{h} \int dE \cdot Tr\left\{ \mathbf{I}_\alpha^{(m)}(E)\delta_{\alpha\beta} - \mathbf{s}_{\alpha\beta}^{(m)\dagger}(E)\mathbf{s}_{\alpha\beta}^{(m)}(E + \hbar\omega) \right\}$$
$$\times \frac{f_\beta^{(m)}(E) - f_\beta^{(m)}(E + \hbar\omega)}{\hbar\omega}, \tag{2.82}$$

where $\mathbf{I}$ is an identity matrix for lead m and mode α, and $f_\beta^{(m)}(E)$ is the Fermi–Dirac distribution appropriate for lead m and mode β, which would reflect the sub-band energy for this mode and the Fermi energy for this lead. In general, this conductance leads to a complex admittance, in which the imaginary part can be related to the effective capacitance (charge storage) of the circuit [39].

The fact that the ac transport is dispersive suggests that the corresponding time dependence is related to a propagation time for the mode through the nanostructure. Indeed, this time delay is referred to as the Wigner–Smith [40, 41] delay time, which may be expressed as [42]

$$Q = -i\hbar \mathbf{s}^{-1/2} \frac{\partial \mathbf{s}}{\partial E} \mathbf{s}^{-1/2}, \tag{2.83}$$

which is a symmetrized form. A more common form is given as [43]

$$Q = i\hbar \frac{\partial \mathbf{s}^\dagger}{\partial E} \mathbf{s}. \tag{2.84}$$

An important point is that these matrices are reduced over the full scattering matrix as they only include those modes that are propagating through the system. That is, they ignore the evanescent modes whose sub-band energies lie above the Fermi energy in that lead. Another form in which this delay time arises in a single-channel (or mode) conductor is just the derivative of the phase of the scattering matrix element of that mode [44], as

$$\tau_d = \hbar \frac{\partial \varphi(E)}{\partial E}. \tag{2.85}$$

The delay time can be an important approach to gain more information about the properties of a nanostructure. In the conductance through a structure such as an open quantum dot, all transmitted modes contribute equally to the conductance. But if one is interested, for example, in isolating individual modes to determine their respective trajectory through the structure, then the delay time can be quite useful [45]. In particular, a subspace of modes, each of which enters through a single lead and also exits through a single lead, creates a noiseless subspace. These modes can then be separated by a delay time analysis. Rotter et al. [45] carried out this procedure for a square cavity of edge $4d$, where d is the width of the leads that are attached to the cavity. These leads are offset from one another with the entrance lead at the top and the exit lead at the bottom, as can be seen in Fig. 2.7. The modes of interest enter with a Fermi momentum of $75.5\pi/d$, so many modes propagate in the leads and enter the cavity. In Fig. 2.7, two back-scattering states are shown and two propagating states are shown, which have complex multiple scattering off the walls before exiting the output lead [46]. The delay time is used to separate these structures as each trajectory obviously spends a different amount of time within the cavity. The different states each correspond to a different initial condition in the input lead from which the wave is injected. With this high an initial momentum, the wave is nearly classical, so that the results are quite similar to a classical orbit simulation. It may be seen that one of the back-scattering modes enters almost parallel to the cavity wall and is back-scattered from the right wall of the cavity, while the second enters at an angle and encounters the lower left corner, which acts like a corner reflector mirror. The transmitted modes that are shown encounter many reflections from the walls before finally exiting.

More information on the calculations can be found in Rotter et al. [45].

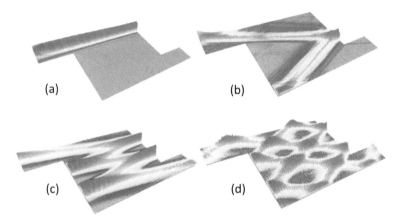

Figure 2.7 (a,b) Two back-scattered orbits in which the incident wave is reflected from within the cavity. (c,d) Two transmitted trajectories that undergo multiple scattering within the cavity before exiting. Reprinted from Refs. [45] and [46] with permission from Stefan Rotter, Technical University of Vienna.

2.7 Bohm Trajectories

In treating the quantum transport of carriers through small semiconductor devices, it is often desired to find an approach that gives the important aspects of the quantum effects without all the mathematical detail that is necessary for some of the more fundamental approaches. Of course, the scattering matrix approach of Section 2.2 allows this, but one would like to have something closer to the ubiquitous ensemble Monte Carlo approach. One such approach, for example, relies upon an effective mass solution to the Schrödinger equation coupled to a Monte Carlo introduction of scattering. The problem with this approach is that one needs to have the particle trajectories to fully employ the Monte Carlo procedure, and these do not come from the normal Schrödinger equation. However, these trajectories can be found from the hydrodynamic variant of the Schrödinger equation that has been developed by Madelung and Bohm [47, 48]. This approach has been applied to a study of a resonant-tunneling diode by Oriols et al. [49]. The use of Bohm trajectories has also been applied to semiconductor nanostructures such as quantum wires, and QPCs [50–52]. In this section, we will develop this approach to quantum transport [53].

It is important to note that the hydrodynamic approach is developed from the Schrödinger equation itself, so is another interpretation of the wavefunction theory for quantum mechanics. We begin by defining the wavefunction to be defined in terms of a real amplitude A and real phase $S/\hbar$, where S is the action (integral of the energy over time). Thus, we write the wavefunction as

$$\psi(\mathbf{x}, t) = A(\mathbf{x}, t)e^{iS(\mathbf{x}, t)/\hbar}. \tag{2.86}$$

Now we insert this wavefunction into the time-dependent Schrödinger equation (2.5) (but not limited to just two dimensions). This leads to the complex equation

$$-A\frac{\partial S}{\partial t} + i\hbar\frac{\partial A}{\partial t} = \frac{A}{2m^*}(\nabla S)^2 - \frac{i\hbar A}{2m^*}\nabla^2 S$$
$$-\frac{i\hbar}{m^*}\nabla S \cdot \nabla A - \frac{\hbar^2}{2m^*}\nabla^2 A + V(\mathbf{x})A. \tag{2.87}$$

For this equation to be valid, it is required that both the real parts and the imaginary parts balance separately, so that this is really a set of two equations, which are given as

$$\frac{\partial S}{\partial t} + \frac{1}{2m^*}(\nabla S)^2 + V - \frac{\hbar^2}{2m^* A}\nabla^2 A = 0 \tag{2.88}$$

and

$$\frac{\partial A}{\partial t} + \frac{A}{2m^*}\nabla^2 S + \frac{1}{m^*}\nabla S \cdot \nabla A = 0. \tag{2.89}$$

Equation (2.89) can be rearranged by multiplying by A, which leads to

$$\frac{\partial A^2}{\partial t} + \nabla \cdot \left(\frac{A^2}{m^*}\nabla S\right) = 0. \tag{2.90}$$

The factor A^2 is obviously the magnitude squared of the wavefunction itself and, therefore, relates to the probability density that the wavefunction represents. Dimensionally, the quantity in the parentheses is the product of the probability and the velocity of the wavefunction, so that

$$\mathbf{p} = \nabla S, \quad \mathbf{v} = \frac{1}{m^*} \nabla S. \tag{2.91}$$

There is another constraint on this momentum and action that is not apparent from this present situation. There is a requirement for the quantum system that

$$\oint_C \mathbf{p} \cdot d\mathbf{l} = 2\pi\hbar n, \tag{2.92}$$

where n is an integer and C is a closed contour. This is just a quantization condition, and normally the contour is an extremal orbit that the closed trajectory may take on what is called the invariant torus of the system. But this contour can be almost any closed path. Now this integral appears in many guises. For example, the Einstein, Brillouin [54], and Keller [55] form is often called EBK quantization. In a more modern version, the right-hand side of Eq. (2.92) is modified, such as by the addition of a factor of 1/2 in Wentzel–Kramer–Brillouin (WKB) approximations, and more generally it is written as

$$\oint_C \mathbf{p} \cdot d\mathbf{l} = 2\pi\hbar \left(n + \frac{1}{4}\beta \right), \tag{2.93}$$

where β is the Morse or Maslov index [56, 57]. In general, this index relates to the number of *turning points* that the trajectory makes. But all this index really does is to shift the energy levels that arise from this quantization. If we now apply Stoke's theorem, we can rewrite Eq. (2.92) as

$$\frac{1}{m^*} \int_W (\nabla \times \mathbf{v}) \cdot \hat{\mathbf{n}} dW = 2\pi n, \tag{2.94}$$

where W is the surface area enclosed by the contour C and $\hat{\mathbf{n}}$ is the surface normal unit vector. The quantity in parentheses in Eq. (2.94) is the vorticity of the hydrodynamic "fluid" and this equation requires it to be quantized. In the integral Eq. (2.92), there is also the important concept that passing around the contour may introduce an additional phase to the wavefunction, often called the Berry phase or geometric phase [58]. Examples of this arise in, for example, the Aharonov–Bohm effect [18].

Now let us return to Eq. (2.88). In this equation, there is only one term that involves Planck's constant and this would vanish if we let this constant go to zero. We have already remarked that the action is the time integral of the energy, so that the first term in Eq. (2.88) is obviously the energy, while the third term is the potential energy. From Eq. (2.91), we recognize the second term as the kinetic energy, so that Eq. (2.88) can be recognized as just the condition of conservation of energy. But the last term is new and has not been recognized as an energy, which it must be. This last term is usually referred to as the Bohm potential, given as

$$U_B = -\frac{\hbar^2}{2m*A}\nabla^2 A,\tag{2.95}$$

and is a true quantum potential. There are several variants in the literature, including one introduced earlier by Wigner [28]. This quantum potential is important to the trajectory approach, as we can define the acceleration of the trajectory from the total potential as

$$\frac{d\mathbf{v}}{dt} = -\frac{1}{m*}\nabla\left[V(\mathbf{x},t)+U_B(\mathbf{x},t)\right].\tag{2.96}$$

This directly allows us to develop a time-resolved approach to electron transport using the quantum trajectories [59]. Once the Schrödinger equation is solved at each instant of time, the quantum potential can be computed along with the self-consistent potential energy (subject to any charge movement or rearrangement), and then the directional velocities of the various trajectories determined. These velocities now allow one to create the motion of the trajectories during the next time increment.

In Fig. 2.8a, we illustrate the motion of a Gaussian wave packet, indicated at the lower left of the figure, through a double barrier, resonant-tunneling diode [49]. Here the barriers are each 2 nm thick and have an energy height of 0.3 eV. The enclosed quantum well is 18 nm thick. It is assumed that the material is GaAs and the barriers are AlGaAs. The Gaussian wave packet is a coherent wave packet with a momentum corresponding to a central energy of 0.16 eV and has a spatial dispersion of 10 nm. No bias is considered as being applied to the structure. In Fig. 2.8b, the resulting phase–space distribution function for a bias of 0.39 V. Note the tunneling ridge in the lower right (arrow) that corresponds to the ballistically tunneling electron. It can be shown that the Bohm trajectory approach agrees very well with the use of the Schrödinger equation directly [52] and with the use of the Wigner phase–space distribution function [60].

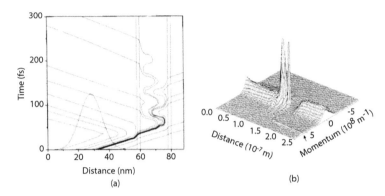

Figure 2.8 (a) Bohm trajectories for a Gaussian wave packet tunneling through a double barrier resonant-tunneling diode. Details are given in the text. (b) Phase–space distribution for such a diode with full density under a bias of 0.39 V. Reprinted with permission from Ref. [49], Copyright 1998, AIP Publishing LLC.

In Fig. 2.9, the results for a simple asymmetric QPC are shown, with the trajectories overlaid on the quantum potential itself. In the top panel, the Schrödinger equation was solved assuming hard walls for the boundaries and without considering a self-consistent rearrangement of the charge in response to the barrier potential. This was then used to find the trajectories and the quantum potential, both of which are shown in the top panel. The Fermi energy for the propagating modes was 18.84 meV, which allowed five modes to propagate through the structure. Note that here, as well as in Fig. 2.8, none of the trajectories cross, which is required since the velocity field is uniquely determined, and only one trajectory can pass through any point [61, 62]. One can also see that the trajectories are bent around the constriction, a consequence of electron diffraction through the slit. In the lower panel of the figure, we now consider results in which the electron density is computed self-consistently with the potential. This would lead to a self-energy that causes more modes to propagate, so the Fermi energy has been lowered to 5.46 meV to maintain the same number of propagating modes. Now the trajectories are laid over the sum of the quantum and self-consistent potentials. Use of the self-consistent potential leads to some variations of the trajectories between the two panes.

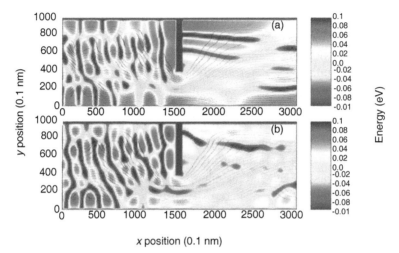

x position (0.1 nm)

Figure 2.9 Bohm trajectories for a quantum point contact in GaAs. (a) The background shading corresponds to the quantum potential. The constriction is the vertical blue bar at x = 150 nm. The electrostatic potential is taken to be zero. (b) Now the electrostatic potential and the density are computed self-consistently to account for the effect of the barrier on the density. The background shading is now the sum of the electrostatic potential and the quantum potential. Reprinted from Ref. [52], Copyright 2000, with permission from Elsevier.

Problems

1. When a voltage (V_{sd}) is applied across a quantum point contact, it is typical to assume that the quasi-Fermi level on one side of the barrier is raised by αeV_{sd} while that on the other drops by $(1 - \alpha)eV_{sd}$, where α is a phenomenological parameter that, in a truly symmetrical structure, should be equal to ½. If we consider a device in which only the lowest sub-band contributes to transport, then the current flow through the QPC may be written as:

$$I_{sd} = \frac{2e}{h}\left[\int_L T(E)dE \ - \int_R T(E)dE \right],$$

where $T(E)$ is the energy-dependent transmission coefficient of the lowest sub-band and L and R denote the left and right reservoirs, respectively. If we assume low temperatures, we can treat the transmission as a step function, $T(E) = \theta(E - E_1)$, where E_1 is the threshold energy for the lowest sub-band.

2. At nonzero temperature, the conductance through a nanodevice is determined by weighting the transmission at each energy by the value of the (negative of the) derivative of the Fermi–Dirac distribution. This leads to the fact that the conductance takes place near the Fermi level. Compute the derivative of the Fermi–Dirac function and evaluate the full-width at half-maximum for this function.

3. A Poisson solver that may be used to calculate the energy bands of different heterostructures may be downloaded at: http://www.nd.edu/~gsnider/. Now consider a hetero-structure comprising (starting from the top layer): 5 nm thick GaAs cap layer; 40 nm of undoped $Al_xGa_{1-x}As$ ($x = 0.33$); 10 nm of $Al_xGa_{1-x}As$ ($x = 0.33$) doped with Si at a concentration of 1.5×10^{18} cm^{-3}; 20 nm of undoped $Al_xGa_{1-x}As$ ($x = 0.33$), 100 nm of undoped GaAs; and a GaAs substrate with an unintentional p-type doping of 5×10^{14} cm^{-3}.

(a) The Poisson solver is a self-consistent program that computes the band structure by solving two important equations simultaneously. What are these equations?

(b) Plot the calculated energy bands and ground-state electron wavefunction for the heterostructure at 1 K. You may assume full ionization of the donors.

(c) Explain *quantitatively* the reasons for the different energy variations in each of the layers of the heterostructure.

(d) What is the number of two-dimensional sub-bands that are occupied at this temperature? Explain this result by using the value of the electron density determined from the program.

(e) Determine the minimum doping density that may be used to realize a two-dimensional electron gas in the heterostructure. Plot the energy bands for this doping condition and the ground-state electron wavefunction.

4. A rectangular quantum wire with cross section 15×15 nm^2 is realized in GaAs.
   (a) Write an expression for the sub-band threshold energies of the quantum wire.
   (b) Write a general expression for the electron dispersion in the quantum wire.
   (c) List the quantized energies and their degeneracies (neglecting spin) for the first 15 energetically distinct sub-bands in the form of a table.
   (d) Plot the density of states of the wire over an energy range that includes the first 15 sub-band energies.

5. In Problem 2, we solved for the density of states in a rectangular GaAs quantum wire with cross section 15×15 nm^2. For the same wire:
   (a) Write an expression for the electron density (per unit length) as a function of the Fermi energy of the wire.
   (b) Plot the variation of the electron density as a function of energy for a range that corresponds to filling the first five distinct energy levels, taking proper account of level degeneracies.

6. Consider a free electron in a magnetic field. Using Peierls' substitution and the Landau gauge, show that the electron satisfies a Hamiltonian that has the form of a harmonic oscillator.

7. Consider a GaAs quantum dot of diameter d. By making suitable assumptions, plot the variation of the charging energy and the characteristic level spacing as a function of diameter, indicating the value of d for which these energy scales show a crossover in their relative magnitude. (Note: you need to determine the capacitance of the dot as a function of its diameter, assuming that its thickness is negligible—this is the capacitance of a dielectric disc.)

8. Using the description of the recursive scattering matrix from the chapter, develop a computer code that solves for the transmission through a QPC. Assume the QPC exists in a GaAlAs/GaAs heterostructure, which is modulation doped with 10^{12} cm^{-2} Si atoms set 10 nm from the interface. Assume that the QPC is defined by etched trenches so that it is 30 nm

wide and 30 nm long. You can determine the actual carrier concentration in the 2DEG at the interface from the doping and the spacer layer thickness. Assume that you are at 10 mK, and plot the transmission through the QPC as the magnetic field is varied from 0 to 2 T.

References

1. R. Landauer, *IBM J. Res. Dev.*, **1**, 223 (1957).

2. R. Landauer, *Philos. Mag.*, **21**, 863 (1970).

3. D. A. Wharam, T. J. Thornton, R. Newbury, M. Pepper, H. Ahmed, J. E. Frost, D. G. Hasko, D. C. Peacock, D. A. Ritchie, and G. A. C. Jones, *J. Phys. C*, **21**, L325 (1988).

4. B. J. van Wees, H. van Houten, C. W. J. Beenakker, J. G. Williamson, L. P. Kouwenhouven, D. van der Marel, and C. T. Foxon, *Phys. Rev. Lett.*, **60**, 848 (1988).

5. R. P. Taylor, *Nanotechnology*, **5**, 183 (1994).

6. R. E. Collin, *Field Theory of Guided Waves* (McGraw-Hill, New York, 1960).

7. E. Merzbacher, *Quantum Mechanics*, 2nd Ed. (Wiley, New York, 1970).

8. A. M. Kriman, N. C. Kluksdahl, and D. K. Ferry, *Phys. Rev. B*, **36**, 5953 (1987).

9. N. C. Kluksdahl, A. M. Kriman, and D. K. Ferry, *Phys. Rev. B*, **39**, 7720 (1989).

10. T. Ando, *Phys. Rev. B*, **44**, 8017 (1991).

11. R. Akis, J. P. Bird, D. Vasileska, D. K. Ferry, A. P. S. de Moura, and Y.-C. Lai, "On the influence of resonant states on ballistic transport in open quantum dots," in *Electron Transport in Quantum Dots*, edited by J. P. Bird (Kluwer Academic Publishers, Boston, 2003), pp. 209–276.

12. T. Usuki, M. Saito, M. Takatsu, R. Kiehl, and N. Yokoyama, *Phys. Rev. B*, **52**, 8244 (1995).

13. M. J. Gilbert, R. Akis, and D. K. Ferry, *J. Comp. Electron.*, **2**, 329 (2003).

14. M. J. Gilbert and D. K. Ferry, *J. Appl. Phys.*, **95**, 7954 (2004).

15. R. Akis, M. Gilbert, and D. K. Ferry, *J. Phys. Conf. Ser.*, **38**, 87 (2006).

16. R. P. Joshi and D. K. Ferry, *Phys. Rev. B*, **41**, 9734 (1991).

17. W. J. Gross, D. Vasileska, and D. K. Ferry, *IEEE Trans. Electron. Dev.*, **20**, 563 (1999).

18. D. K. Ferry, S. M. Goodnick, and J. P. Bird, *Transport in Nanostructures*, 2nd Ed. (Cambridge University Press, Cambridge, UK, 2009).

19. D. K. Ferry, A. M. Kriman, and H. Hida, *Phys. Rev. Lett.*, **67**, 633 (1991).

20. M. J. Gilbert, R. Akis, and D. K. Ferry, *J. Appl. Phys.*, **98**, 094303 (2005).

21. D. Long, *Phys. Rev.*, **120**, 2024 (1960).

22. D. K. Ferry, *Phys. Rev. B*, **12**, 2361 (1975).

23. A. Svizhenko and A. P. Anantram, *IEEE Trans. Electron Dev.*, **50**, 1459 (2003).

24. R. P. Feynman and A. R. Hibbs, *Quantum Mechanics and Path Integrals* (McGraw-Hill, New York, 1965).

25. W. Kohn and J. M. Luttinger, *Phys. Rev.*, **108**, 590 (1957).

26. J. L. Siegel and P. N. Argyres, *Phys. Rev.*, **178**, 1016 (1969).

27. P. N. Argyres and J. L. Siegel, *Phys. Rev. Lett.*, **31**, 1397 (1973).

28. E. P. Wigner, *Phys. Rev.*, **40**, 749 (1932).

29. J. E. Moyal, *Proc. Cambridge Philos. Soc.*, **45**, 99 (1949).

30. T. B. Smith, *J. Phys. A*, **11**, 2179 (1978).

31. A. Janusis, A. Streklas, and K. Vlachos, *Physica*, **107A**, 587 (1981).

32. A. Royer, *Phys. Rev. A*, **43**, 44 (1991).

33. E. A. Remler, *Ann. Phys.*, **95**, 455 (1975).

34. A. L. Fetter and J. D. Walecka, *Quantum Theory of Many-Particle Systems* (McGraw-Hill, New York, 1971).

35. J. Gabelli, G. Fève, J.-M. Berroir, B. Plaçais, A. Cavanna, B. Ettiene, Y. Jin, and D. C. Glattli, *Science*, **313**, 499 (2006).

36. M. Büttiker, *Phys. Rev. B*, **46**, 12485 (1992).

37. A. Prêtre, H. Thomas, and M. Büttiker, *Phys. Rev. B*, **54**, 8130 (1996).

38. M. Büttiker, A. Prêtre, and H. Thomas, *Phys. Rev. Lett.*, **70**, 4114 (1993).

39. M. Büttiker, H. Thomas, and A. Prêtre, *Phys. Lett. A*, **180**, 364 (1993).

40. E. P. Wigner, *Phys. Rev.*, **98**, 145 (1955).

41. F. T. Smith, *Phys. Rev.*, **118**, 349 (1960).

42. P. W. Brouwer, K. M. Frahm, and C. W. J. Beenakker, *Phys. Rev. Lett.*, 78, 4737 (1997).

43. L. Reichl, *The Transition to Chaos*, 2nd Ed. (Springer, New York, 2004).

44. Z. Ringel, Y. Imry, and O. Entin-Wohlman, *Phys. Rev. B*, **78**, 165304 (2008).

45. S. Rotter, P. Ambichl, and F. Libisch, *Phys. Rev. Lett.*, **106**, 120602 (2011).

46. http://concord.itp.tuwien.ac.at/~florian/prl_bundle/

47. E. Madelung, *Z. Phys.*, **40**, 322 (1926).

48. D. Bohm, *Phys. Rev.*, **85**, 166 (1952).

49. X. Oriols, J. J. Garcia-Garcia, F. Martin, J. Suñé, T. Gonález, J. Mateos, and D. Pardo, *Appl. Phys. Lett.*, **72**, 806 (1998).

50. H. Wu and D. W. L. Sprung, *Phys. Lett. A*, **196**, 229 (1994).

51. L. Lundberg, E. Sjöqvist, and K. F. Berggren, *J. Phys.: Condens. Matter*, **10**, 5583 (1998).

52. L. Shifren, R. Akis, and D. K. Ferry, *Phys. Lett. A*, **274**, 75 (2000).

53. X. Oriols and J. Mompart, *Applied Bohmian Transport: From Nanoscale Systems to Cosmology* (Pan Stanford, Singapore, 2012).

54. L. Brillouin, *J. Phys. Radium*, **7**, 353 (1926).

55. J. B. Keller, *Ann. Phys.* (New York), **4**, 180 (1958).

56. H.-J. Stockmann, *Quantum Chaos: An Introduction* (Cambridge University Press, Cambridge, 1999).

57. W. H. Miller, *J. Chem. Phys.*, **63**, 995 (1975).

58. M. V. Berry, *Proc. Roy. Soc. Lond. A*, **392**, 45 (1984).

59. G. Albareda, D. Marian, A. Benali, S. Yaro, N. Zanghi, and X. Oriols, *J. Comput. Electron.*, **12**, 405 (2013).

60. E. Colomés, Z. Zhan, and X. Oriols, *J. Comput. Electron.*, **14**, 894 (2015).

61. D. Bohm and B. J. Hiley, *The Undivided Universe* (Routledge, New York, 1993).

62. P. R. Holland, *The Quantum Theory of Motion* (Cambridge University Press, Cambridge, 1997).

Chapter 3

Equilibrium Green's Functions

We briefly introduced Green's function in the last chapter. In defining the different quantities, we also alluded to the use of the anti-commutator relationship. Now, this turns out to be a very important concept for electrons that are required not only to satisfy Fermi–Dirac statistics but also to possess the anti-symmetry expected for such fermions. Hence, we need to distinguish between bosons and fermions. This is not only important for Green's functions but for most advanced quantum transport, as we will encounter Green's functions for both fermions (the electrons) and bosons (the phonons, or lattice vibrations). While the Green's functions we will deal with in this chapter are called equilibrium Green's functions, they are not strictly confined by this name. Perhaps more properly, we should refer to them as just being near to equilibrium and at low temperature.

In normal undergraduate courses on quantum mechanics, one normally deals only with commuting operators, such as the raising and lowering operators (or creation and annihilation operators, as they are often called) for the harmonic oscillator. These were boson operators, as the harmonic oscillator is often used for applications in which the particles are bosons. For these operators, the commutators could be written simply as

$$[a, a] = [a^\dagger, a^\dagger] = 0, \quad [a, a^\dagger] = 1, \tag{3.1}$$

An Introduction to Quantum Transport in Semiconductors
David K. Ferry
Copyright © 2018 Pan Stanford Publishing Pte. Ltd.
ISBN 978-981-4745-86-4 (Hardcover), 978-1-315-20622-6 (eBook)
www.panstanford.com

where the superscript dagger indicates the adjoint operator. On the other hand, electrons are fermions. When a fermion state, which may be defined by its position and spin, is full, a second fermion cannot be created in this state. This is the Pauli exclusion principle. Hence, we require that (and we now use the letter c to indicate fermion operators)

$$c^\dagger |1\rangle = 0, \quad c|0\rangle = 0, \tag{3.2}$$

where the last expression indicates that the empty state cannot have its occupancy set to zero as it is already zero. The latter term was the same for bosons, and it allowed us to determine the wavefunction in the harmonic oscillator. However, the former term is new for fermions. These two operations lead to the existence of an anti-commutator relationship for the fermion operators, which may be expressed as

$$\left\{c, c^\dagger\right\} = cc^+ + c^+ c = 1. \tag{3.3}$$

If the state is empty, only the first term works to create a particle in the empty state and then destroy this particle, returning it to the empty state. If the state is full, then the second term first destroys this particle and then recreates a particle in the state. In either case, the result leaves the state unchanged, thus the unity of the anti-commutator. Now, here we have used the curly brackets to indicate the anti-commutator, but other notations are commonly used, such as the square brackets with a subscript "+" attached to them.

Another important point is the use of field operators, in which we turn the wavefunction itself into an operator. Normally, we have no trouble referring to the wavefunction as a field, since mathematically this means only that the wavefunction has a defined value at every point in space. But, if we start with a basis set, as in the previous chapters, then we can use our operators to create excitations into the electron basis set and write the field operator for the wavefunction as

$$\widehat{\psi}(x,t) = \sum_i \alpha_i c \varphi_i(x,t)$$

$$\widehat{\psi}^\dagger(x,t) = \sum_i \alpha_i c^\dagger \varphi_i^*(x,t). \tag{3.4}$$

That is, for each basis function of interest, there is an operator which creates or destroys an excitation in that state, with an amplitude of probability of a_j. Generally, we do not write out the full form of the field operators. For electrons, these operators must satisfy the same commutation relationships as the fermion operators themselves, so that

$$\{\psi(1),\psi(2)\}=\{\psi^\dagger(1),\psi^\dagger(2)\}=0$$

$$\{\psi(1),\psi^\dagger(2)\}=\delta(1-2).$$

(3.5)

In these expressions, we have introduced another short-hand notation, in that we define the arguments as 1= x_1,t_1, 2 = x_2,t_2, so that the commutator requires the field operators to represent the same position and time, basically meaning that they are required to represent the same particle.

In the use of the basis sets, we recall that these have properties of generalized Fourier series. One of these important properties is that of closure, in which

$$\sum_j |j\rangle\langle j|=1.$$

(3.6)

This quantity is also often called the resolution of unity, and we can use it to get the time rate of change of the field operator in the appropriate basis set expansion. We can write the general expression as

$$i\hbar\frac{\partial \psi_m}{\partial t}=[H,\psi_m]=\left[\sum_j |j\rangle\langle j|H\sum_k |k\rangle\langle k|,\psi_m\right].$$

(3.7)

We can now rearrange the terms by recognizing the sums can be rewritten as

$$\sum_{j,k}|j\rangle\langle j|H|k\rangle=\sum_{j,k}\psi_j\langle j|H|k\rangle\psi_k^\dagger,$$

(3.8)

and

$$i\hbar\frac{\partial \psi_m}{\partial t}=[H,\psi_m]=\sum_{j,k}\langle j|H|k\rangle\left[\psi_j\psi_k^\dagger,\psi_m\right]$$

$$=\sum_{j,k}\langle j|H|k\rangle\psi_j\{\psi_k^\dagger,\psi_m\}=\sum_j\langle j|H|m\rangle\psi_j.$$

(3.9)

In the step from the third to fourth terms, the anti-commutator of ψ_j and ψ_m has introduced a minus sign, which converts the commutator into an anti-commutator. Now, the variation of one basis state field operator depends on the matrix elements between all of the basis set states and is a simple expansion. As previously, any physical results are independent of the basis states themselves.

One last point about the field operators and anti-symmetry lies in the way in which the states are filled. For example, we can define a many-boson state as

$$|n_1, n_2, \ldots\rangle = \prod_j \frac{\left(\psi_j^\dagger\right)^{n_j}}{\sqrt{n_j!}} |0\rangle, \tag{3.10}$$

where each n_j can be 0, 1, 2, ..., since there is no limit to the population in each boson state. In this equation, the 0 state upon which the operator act is termed the empty or vacuum state. On the other hand, a many-body fermion state is defined as

$$|n_1, n_2, \ldots\rangle = \ldots\left(\psi_r^\dagger\right)^{n_r} \ldots\left(\psi_2^\dagger\right)^{n_2}\left(\psi_1^\dagger\right)^{n_1} |0\rangle, \tag{3.11}$$

where, now, each n_j can only be 0 or 1. We have indicated here the normal ordering in that the first operator that works upon the vacuum state is placed to the right, and sequential operations then are appended, working toward the left. This leads to an important point. Suppose we first create an electron in state 3 and then an electron in state 1. This is denoted by

$$\psi_1^\dagger \psi_3^\dagger |0\rangle = |101\rangle. \tag{3.12}$$

But, what if we reversed the order of the particle generation, writing instead

$$\psi_3^\dagger \psi_1^\dagger |0\rangle = -\psi_1^\dagger \psi_3^\dagger |0\rangle = -|101\rangle. \tag{3.13}$$

In the second term, the anti-commutator has been used to create the normal ordering of the creation operators, and the minus sign carries through to the result. This minus sign is a characteristic of the anti-symmetry of the fermions.

3.1 Role of Propagator

In the previous chapter, we briefly introduced the kernel or propagator for the solutions to the Schrödinger equation. Replacing the t_0 we had there with an arbitrary t' we can write the wavefunction as

$$\psi(\mathbf{x},t) = \int d\mathbf{x}' K(\mathbf{x},\mathbf{x}';t,t') \psi(\mathbf{x}',t'), \tag{3.14}$$

where

$$K(\mathbf{x},\mathbf{x}';t,t') = \sum_n \varphi_n^*(\mathbf{x}')\varphi_n(\mathbf{x})e^{-iE_n(t-t')/\hbar}. \tag{3.15}$$

Now, it would be very tempting to write Green's function directly as the kernel, or propagator. But, this would be a mistake (in terms of common usage). When we want to use the field operators, we want to write the retarded and advanced Green's functions as commutator products, as [1]

$$G_r(\mathbf{x},\mathbf{x}';t,t') = -i\vartheta(t-t')\left\langle\left\{\widehat{\psi}(\mathbf{x},t),\widehat{\psi}^\dagger(\mathbf{x}',t)\right\}\right\rangle \tag{3.16a}$$

$$G_a(\mathbf{x},\mathbf{x}';t,t') = i\vartheta(t'-t)\left\langle\left\{\psi^\dagger(\mathbf{x}',t'),\psi(\mathbf{x},t)\right\}\right\rangle. \tag{3.16b}$$

For the retarded Green's function, we need to first create an excitation from the ground state (or vacuum state) and then destroy this excitation at a later time. The Green's function tells us how this excitation propagates. To see this, let us expand the right-hand side of (3.16a) as

$$\left\langle\left\{\widehat{\psi}(\mathbf{x},t),\widehat{\psi}^\dagger(\mathbf{x}',t')\right\}\right\rangle = \langle 0|\widehat{\psi}(\mathbf{x},t)\widehat{\psi}^\dagger(\mathbf{x}',t')|0\rangle + \langle 0|\widehat{\psi}^\dagger(\mathbf{x}',t')\widehat{\psi}(\mathbf{x},t)|0\rangle. \tag{3.17}$$

Now, in the second term, the $\widehat{\psi}$ operating on the $|0\rangle$ state produces zero. But, the adjoint creation operator working on this state in the first term produces an electron at $\mathbf{x}'$. This electron is then annihilated at a later time t at position $\mathbf{x}$. The two positions and times cannot be the same, since that would lead to self-annihilation of the excitation. This is given by the anti-commutator

$$\left\{\widehat{\psi}(\mathbf{x},t),\widehat{\psi}^\dagger(\mathbf{x}',t')\right\} = \delta(\mathbf{x}-\mathbf{x}')\delta(t-t'). \tag{3.18}$$

To see how we formulate these two functions, let us begin with the general non-interacting ground-state Green's function, which must satisfy two forms of the Schrödinger equation, in terms of the different sets of variables. These equations are

$$\left[i\hbar\frac{\partial}{\partial t}-H_0(\mathbf{x})-V(\mathbf{x})\right]G_0(\mathbf{x},\mathbf{x}';t,t')=\hbar\delta(\mathbf{x}-\mathbf{x}')\delta(t-t') \quad (3.19a)$$

$$\left[-i\hbar\frac{\partial}{\partial t'}-H_0(\mathbf{x}')-V(\mathbf{x}')\right]G_0(\mathbf{x},\mathbf{x}';t,t')=\hbar\delta(\mathbf{x}-\mathbf{x}')\delta(t-t').$$
$$(3.19b)$$

If we Fourier transform in both position (the Fourier variable is $\mathbf{k}$) and time (the Fourier variable is ω) and assume that the Green's function is a function of only the difference in the two positions and the two times, this Fourier version becomes

$$G_0(\mathbf{k},\omega)=\frac{\hbar}{\hbar\omega-E(\mathbf{k})}, \quad (3.20)$$

where we note that the bare Hamiltonian H_0 produces the energy eigenvalue, and we have assumed that the potential term is zero. Neglecting the potential means that we are looking at a free propagating Green's function for a homogeneous system with no boundaries. Now, if we want to invert this Fourier transform, we can recover the retarded and advanced Green's functions by introducing a convergence factor into Eq. (3.20). Hence, we write the retarded and advanced bare (or ground state) Green's functions as

$$G^0_{r,a}(\mathbf{k},\omega)=\frac{\hbar}{\hbar\omega-E(\mathbf{k})\pm i\eta}. \quad (3.21)$$

For example, let us do the inverse transform in time, so that we have

$$G^0_{r,a}(\mathbf{k},t)=\int\frac{d\omega}{2\pi}G^0_{r,a}(\mathbf{k},\omega)e^{-i\omega t}. \quad (3.22)$$

In general, the integration proceeds along the real frequency axis. But, as for any integration involving a complex variable, we have to decide how to close the contour to obtain a full contour integration. Consider the two possible closures for the contour, as indicated in Fig. 3.1. On the contour, we can rewrite the frequency as

$$\omega=Ae^{i\theta}=A\cos\theta+iA\sin\theta, \quad (3.23)$$

and the exponential term in Eq. (3.22) becomes

$$e^{-i\omega t}=e^{-iA\cos\theta t}e^{A\sin\theta t}. \quad (3.23)$$

For stability, we require that the argument of the last exponential be less than zero, so that the integral converges as we let A go to infinity. So for $t > 0$, we require that the sine term be negative, which means that we have to close in the lower half plane (the green curve in Fig. 3.1). On the other hand, for $t < 0$, we require the sine term to be positive, which means that we have to close in the upper half plane (the blue curve in Fig. 3.1). The importance of this closure lies in the convergence factor that was added to Eq. (3.21). We require that the pole of this latter equation lie within the closed contour for there to be a solution to the integral. Hence, when we close in the lower half plane, for positive time, we take the upper sign in Eq. (3.21), and this defines the retarded Green's function as

$$G_r^0(\mathbf{k},\omega) = \frac{\hbar}{\hbar\omega - E(\mathbf{k}) + i\eta}. \qquad (3.24)$$

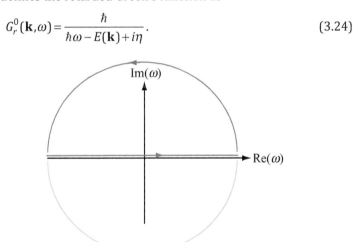

Figure 3.1 When doing the Fourier inversion operation of Eq. (3.22), the integration is done along the real frequency axis (red line). However, one must close the contour in either the upper half plane or the lower half plane, which leads to two different Green's functions.

Similarly, when we close in the upper half plane, we require the pole to lie in this upper half plane, and this requires the lower, negative sign in Eq. (3.21). This leads us to the advanced Green's function as

$$G_a^0(\mathbf{k},\omega) = \frac{\hbar}{\hbar\omega - E(\mathbf{k}) - i\eta}. \qquad (3.25)$$

These two definitions now give us an important result, which relates the two functions via the property

$$G_r^0(\mathbf{k},\omega) = \left[G_a^0(\mathbf{k},\omega) \right]^* . \tag{3.26}$$

This property is readily apparent from their respective equations but carries forward beyond just the bare Green's functions. Indeed, when we replace later the convergence factor by the full self-energy, this property will continue to be obeyed, since the imaginary parts of the self-energy play the convergence roles in the full Green's functions.

3.2 Spectral Density

The integral (3.22) leads to a number of interesting and important properties that we associate with Green's functions. First, there is an important property of complex functions such as Eqs. (3.24) and (3.25). When η is vanishingly small, as it is for these two functions where it serves only as a convergence factor, we can write the functions as

$$\frac{1}{x \pm i\eta} = P\frac{1}{x} \mp i\pi\delta(x), \tag{3.27}$$

where P denotes the principal part of the function (and the resulting integral). We recognize the delta function as giving us the on-shell relationship between the energy $\hbar\omega$ and the momentum-dependent kinetic energy $E(\mathbf{k})$. Just as in Chapter 1, this delta function assures us that the energy is given by the classical value. This relationship tells us that the delta function arises from the imaginary part of the Green's function. In the present case, we get a delta function because we have no interactions in the system as yet. When these arise, then our discussion in Chapter 1 tells us that the delta function will be broadened into a spectral function and accounts for the off-shell contributions to the integrals. Consequently, this will give us the interacting Green's functions, which we will encounter in the next chapter. Hence, for our non-interacting Green's functions, we can write

$$2\pi\delta(\hbar\omega - E(\mathbf{k})) = -\text{Im}\left\{ G_r^0(\mathbf{k},\omega) - G_a^0(\mathbf{k},\omega) \right\}, \tag{3.28}$$

and the corresponding spectral density for the interacting system will be (for historical reasons, we incorporate the 2π into the definition of the spectral density)

$$A(\mathbf{k},\omega) = -\text{Im}\left\{G_r(\mathbf{k},\omega) - G_a(\mathbf{k},\omega)\right\} = i\left\{G_r(\mathbf{k},\omega) - G_a(\mathbf{k},\omega)\right\}. \quad (3.29)$$

In condensed matter physics, an important quantity is the density of states, which is often written as

$$\rho(E) = \sum_k \delta(E - E(\mathbf{k})), \quad (3.30)$$

where the states are assumed to be of plane wave form. We can extend this to the spectral density if we account for the actual basis states in the definition, and

$$A(E) = 2\pi \sum_k |k\rangle\langle k| \delta(E - E(\mathbf{k})). \quad (3.31)$$

Using the relationships developed above, we can rewrite this as

$$A(\hbar\omega) = 2i \sum_k \frac{|k\rangle\langle k|}{\hbar\omega - E(\mathbf{k}) + i\eta} = 2i \sum_k |k\rangle\langle k| G_r^0(\mathbf{k},\omega). \quad (3.32)$$

In a sense, the wavefunctions retain the position information (we recognize the form of Eq. (3.31) as introducing the resolution of unity once again), and some useful normalizations of the spectral function can be found by rewriting Eq. (3.31) as [2]

$$A(\mathbf{x},\mathbf{x}',E) = 2\pi \sum_k \widehat{\psi}(\mathbf{x})\widehat{\psi}_k^\dagger(\mathbf{x}')\delta(E - E(\mathbf{k})). \quad (3.33)$$

If we now integrate this over the energy, we find

$$\int \frac{dE}{2\pi} A(\mathbf{x},\mathbf{x}',E) = \sum_k \widehat{\psi}_k(\mathbf{x})\widehat{\psi}_k^\dagger(\mathbf{x}') = \delta(\mathbf{x}-\mathbf{x}'), \quad (3.34)$$

which is another way of writing the completeness of the basis set. If we now integrate the diagonal form over the position, we get

$$\int d\mathbf{x} A(\mathbf{x},\mathbf{x},E) = 2\pi \sum_k \delta(E - E(\mathbf{k})) = 2\pi\rho(E), \quad (3.35)$$

which is the density of states per unit energy defined in Eq. (3.30).

While these normalizations are useful, the key element is the spectral density (3.29), which exactly determines the broadening of the off-shell effects that arise as a result of the interactions within the many-body system. We will begin to study these interactions in the next chapter, where they will be connected to the self-energy as well. First, however, we want to discuss the use of Green's functions in a recursion approach to simulate quantum systems in the ballistic limit (no interactions), just as we did for the recursive scattering matrix in the last chapter.

3.3 Recursive Green's Function

In the last chapter, we discussed the simulation of quantum transport through a nanowire section with the recursive scattering matrix. A second approach is to use Green's functions, derived from the Schrödinger equation, in a recursive approach [3–5]. As with the recursive scattering matrix, we have used this Green's function technique successfully in the past [6]. With the recursive Green's function, the contact area is assumed to be a semi-infinite metallic wire of a given width. There is no requirement that the two contacts have the same width, as we use the Landauer formula again, and Green's functions give us the transmission of carriers from one mode in the source to another mode in the drain. Thus, the modes in the source and drain contacts do not have to be the same as the modes in the nanowire. The modes and eigenvalues are computed for this contact for a given Fermi energy. To begin, the nanowire region, as well as the source and drain regions, is discretized on a finite difference grid with grid spacing a. The width of the contact is given as $W = Ma$. Thus, the transverse eigenfunctions in the contacts are those of an infinite potential well, given as

$$\varphi_r(y) = \sqrt{\frac{2}{W}} \sin\left(\frac{r\pi y}{W}\right), \quad r = 1,2,3,...,M \cdot$$ (3.36)

We have taken the transverse direction as the y-axis and will take the longitudinal direction as the x-axis. This wavefunction is evaluated at each grid point j, which of course runs from 0 to M, in the transverse direction to evaluate the wavefunction at the sites in the zero slice. Once we are given the Fermi energy, then we can compute the longitudinal wave number from the knowledge of whether or not the mode energy is greater or less than the Fermi energy. We first compute the quantity

$$\xi = 2 - \cos\left(\frac{r\pi y}{M+1}\right) - \frac{E_F}{2t},$$ (3.37)

where

$$t = \frac{\hbar^2}{2m^* a^2}$$ (3.38)

is the hopping energy in the Schrödinger equation. If $|\xi| \leq 1$, then the wave is a propagating mode, and we can write the longitudinal wavefunction amplitude in the contact as

$$\varphi_r = \sqrt{1-\xi^2} \quad , \quad x \leq 0 , \tag{3.39}$$

and the wavevector is given as $k = i\xi/a$. This wavefunction needs to be properly normalized, of course, but represents the assumption that the contact leads to the vanishing of the wavefunction at $x = 0$ in the absence of the active channel. If the mode energy is greater than the Fermi energy, we are dealing with an evanescent wave, and Eq. (3.39) is replaced by

$$\lambda = \frac{1}{a}\left[\xi - \sqrt{\xi^2 - 1}\right]. \tag{3.40}$$

The recursion begins by first establishing a self-energy correction, which will appear in the Green's function for the first slice but is based on the left contact and the connection of the contact to the first slice. That is, we define the left contact self-energy by

$$\Sigma_L = H_{10}G_{00}H_{01}, \quad \Gamma_L = 2\text{Im}\{\Sigma_L\}, \tag{3.41}$$

where

$$G_0(0) = \left(E_F I - H_{0,0} + i\eta I\right)^{-1}, \tag{3.42}$$

and a small damping factor has been added to assure convergence of the matrix inversion process. The use of a positive damping factor insures that we have the retarded, causal Green's function in our calculations. In the above equations, I is a unit matrix and H_{01} is the hopping energy times a unit matrix, although this is modified in the presence of a magnetic field. The quantity $H_{0,0}$ is the bare slice Hamiltonian for the zero slice of the contact, the last slice before the nanowire. This Hamiltonian is given by

$$H_0(j) = \begin{bmatrix} V_{M,j}-4t & t & 0 & 0 & 0 \\ t & \cdots & \cdots & \cdots & \cdots \\ \cdots & \cdots & \cdots & \cdots & \cdots \\ 0 & \cdots & \cdots & V_{2,j}-4t & t \\ 0 & \cdots & \cdots & t & V_{1,j}-4t \end{bmatrix}, \tag{3.43}$$

but perhaps without the potential terms. Generally, one assumes that the potential applied to the source or drain is vanishingly small and can be neglected in the slice Hamiltonians for these contacts. However, various potentials such as QPCs may be added within the nanowire region, and so will appear in the slice Hamiltonian. Now, Eq. (3.42) is used to determine the self-energy coupling between the contact and the channel, and the actual zeroith slice Green's function is computed from

$$\mathbf{G}(0,0) = \left(E_F \mathbf{I} - \mathbf{H}_0(0) + \Sigma_L + i\eta \mathbf{I} \right)^{-1},\tag{3.44}$$

which is just Eq. (3.42) modified to include the self-energy term. Note that we have changed notation here, using the slice arguments in Green's functions to differentiate from the initialization Green's function of Eqs. (3.41) and (3.42).

Now we propagate across the active channel with a recursion that computes first the slice Green's function and then the connecting Green's functions. For slice $j \geq 1$, we can compute the slice Green's function as

$$\mathbf{G}(j,j) = \left(E_F \mathbf{I} - \mathbf{H}_0(j) + \mathbf{H}_{j,j-1}\mathbf{G}(j-1,j-1)\mathbf{H}_{j-1,j} \right)^{-1}.\tag{3.45}$$

In this expression, the slice Hamiltonian is just Eq. (3.43) and includes the local site potential at each grid point, while the two nonlocal Green's functions are just the diagonal hopping terms involving t described previously. These are equal to one another except when we add a magnetic field, which will be discussed below. In addition to Eq. (3.45), we also construct the two propagating Green's functions

$$\mathbf{G}(0,j) = \mathbf{G}(0,j-1)\mathbf{H}_{j-1,j}\mathbf{G}(j,j),\tag{3.46}$$

and

$$\mathbf{G}(j,0) = \mathbf{G}(j,j)\mathbf{H}_{j,j-1}\mathbf{G}(j-1,0).\tag{3.47}$$

At the right contact, we have to connect to the second lead, which we assume occurs for $j = L$. Then, the right contact Green's functions become

$$\mathbf{G}(L+1,L+1) = \left(E_F \mathbf{I} - \mathbf{H}_0(L+1) + \Sigma_R - \mathbf{H}_{L,L+1}\mathbf{G}(L,L)\mathbf{H}_{L+1,L} \right)^{-1}$$
$$\mathbf{G}(0,L+1) = \mathbf{G}(0,L)\mathbf{H}_{L,L+1}\mathbf{G}(L+1,L+1)$$
$$\mathbf{G}(L+1,0) = \mathbf{G}(L+1,L+1)\mathbf{H}_{L+1,L}\mathbf{G}(L,0).$$
$$\tag{3.48}$$

Here, the right self-energy has been computed in exactly the same manner as in Eq. (3.41), but with the right contact Hamiltonian. This approach allows one to actually use leads that have different sizes, neither of which actually needs to be the same as the active strip. The transmission through the total system is now found from (there are several variants of this formula available in the literature, for example [7])

$$T = Tr\{\Gamma_L \mathbf{G}(0, L+1)\Gamma_R \mathbf{G}(L+1, 0)\}. \tag{3.49}$$

A more extensive derivation of the approach is given in Ref. [8]. A magnetic field can be easily added to the simulation via Peierls' substitution and Landau gauge discussed in the last section.

As in the last chapter, the recursive Green's function approach discussed here is well suited to vanishingly small bias applied between the two reservoirs. Temperature can be included by computing the transmission for a range of energies around the Fermi energy, and then weighting each transmission by the appropriate value of the Fermi–Dirac distribution. But this is an approximation that cannot be extended to large values of the applied bias. Nor can the recursive Green's function approach discussed here be used in this latter case. The approach here is based on equilibrium Green's functions. Even weighting each value of transmission as mentioned is an approximation, as reasonable temperatures really require use of the thermal, or Matsubara, Green's functions [1]. And, if we go to larger values of bias applied between the two reservoirs, then the density is driven out of equilibrium and one has to go to the more complicated nonequilibrium, or real-time Green's functions, which can also yield the nonequilibrium distribution function within the device [9–11]. We will encounter these more complicated Green's functions in Chapters 9 and 10.

3.4 Propagation through Nanostructures and Quantum Dots

In this section, we will consider the application of the recursive Green's function technique to a few structures to illustrate the type of information that can be obtained. We first consider a single tunneling barrier placed across a nanowire. The barrier is assumed to have infinitely steep walls and a potential height of U. In addition, the length of the barrier (in the transport direction) is L and the width of the nanowire is W. Everything is normalized through the ratio to the Fermi wavelength λ_F. This makes the solutions independent of the choice of material and the electron concentration by burying the effective mass and doping into this wavelength. The device itself is shown in the inset to Fig. 3.2 [6]. A magnetic field will be applied normal to the nanowire strip itself, and the solution technique, with a limited size of the grid, considers there to be an infinite confining potential at each side of the nanowire. The total transmission probabilities for the conductance through the barrier are plotted in the main panel of Fig. 3.2 as a function of the barrier height U. Several curves are shown for different lengths of the

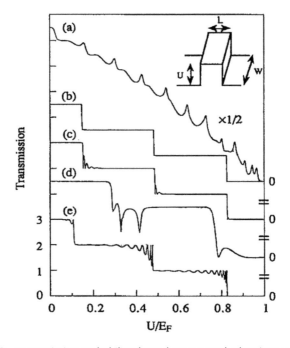

Figure 3.2 Transmission probability through a rectangular barrier as a function of the height U. The length of the barrier is L/λ_F = (a) 3.0, (b) 3.0, (c) 2.0, (d) 1.0, and (e) 6.1. The width of the wire is W/λ_F = 6.1 for (a)–(d) and (e) 2.0. The magnetic field strength is chosen such that $\hbar\omega_c/E_F$ = 0 for (a) and 0.35 for (b)–(e). The inset illustrates the nature of the barrier. Reprinted with permission from Ref. [6], Copyright 1993, American Physical Society.

barrier. Curve (a) is for no magnetic field, while the remaining curves are for a magnetic field such that $\hbar\omega_c/E_F = 0.35$, where $\omega_c = eB/m^*$ is the cyclotron frequency in the material of interest. In the absence of the magnetic field (curve (a)), a resonant structure due to multiple (intramode) scattering, in which this scattering is from the abrupt potential steps, dominates the transmission and basically wipes out the conductance steps as various modes are propagated. The overall drop in conductance as one raises the potential height is caused by the depletion of the modes, which have energies above the barrier height. When a magnetic field is applied (curve (b)) for the same length, this resonant structure disappears. For a narrow wire (curve (e)), other oscillatory structure appears, which are the normal "over the barrier" resonances with the barrier length that are seen in tunneling problems, but these resonances are very sensitive to the length of the barrier, as is common in resonant situations.

The second example we consider is an anti-dot localized in the center of a nanowire strip. By anti-dot, we mean a circular symmetric barrier through which no propagation is allowed. However, with a magnetic field present, the transport is carried by edge states, which arise from the Landau levels [12]. The device itself is shown in Fig. 3.3 [13]. The anti-dot has a diameter of D and is spaced from either side of the nanowire by the distance D. As before, both of these quantities will be normalized to the Fermi wavelength λ_F. The basic idea of the simulation is that the openings between the anti-dot and the edge of the nanowire will be sufficiently small that not all of the edge states can fit through. Thus, the edges states that lie further from the edge of the nanowire, those with higher Landau level indices, will be reflected, as shown in Fig. 3.3. In addition, an edge state can be trapped such that it propagates around the anti-dot [14–16]. This leads to a periodic oscillation superimposed upon the slowly varying background, which is thought to be due to the Aharonov–Bohm effect [12]. In order to affect the edge states, a large diameter of the anti-dot relative to the wire width is required. In addition, there is a resonant-tunneling process that can couple the propagating edge states to those trapped around the anti-dot and, in addition, can couple the reflected edge states to those trapped around the anti-dot. The coupling from the propagating edge states will produce resonant reflection, as the amplitude is transferred into the oppositely directed edge states on the opposite side of the nanowire.

By the same token, the coupling from the reflected edge states gives rise to a forward scattering and thus a resonant transmission past the anti-dot. These resonant processes are indicated by the dashed and dotted lines in Fig. 3.3.

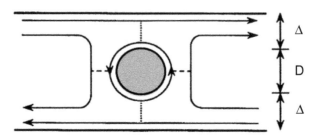

Figure 3.3 Illustration of an anti-dot (grey area) inserted within a nanowire of width $W = 2\Delta + D$. The illustrated edge states lie at the Fermi energy, and forward tunneling and backward tunneling are illustrated by the dashed and dotted lines, respectively. Reprinted with permission from Ref. [13], Copyright 1993, American Physical Society.

The conductance of the nanowire of Fig. 3.3 is plotted in Fig. 3.4 as a function of the magnetic field, normalized as $\hbar\omega_c/E_F$. The width of the openings, Δ, has been chosen so that only the lowest sub-band is occupied in each opening at $B = 0$. With no magnetic field, the sample appears to have two QPCs, one on either side, each of which propagates a single mode through the structure. As the magnetic field is increased, this changes over to the edge state picture where only the single edge state goes through and the conductance drops to $2e^2/h$. Once the magnetic field becomes sufficiently large that $r_c < \Delta/2$, with $r_c = \hbar k_F/eB$, the radius of the cyclotron orbit, the magnetic field effect dominates the quantization effects. In between these two effects, conductance exhibits a sinusoidal oscillation due to resonances within the QPC. Once the magnetic field effect dominates the transport, the peaks and dips are signs of the resonant transmission and reflections between the various edge states as discussed above.

Conductance fluctuations have been observed in semiconductors for many decades (while they are also seen in mesoscopic metals, our focus here will be on semiconductors due to their much larger mobilities and coherence properties). The first observations seem to have been in MOSFETs at low temperatures [17, 18]. They have subsequently been seen in many materials. In experimental studies, variations of the gate voltage (corresponding to sweeping the Fermi energy in the electron gas) led to fluctuations whose r.m.s. amplitude could be scaled with the ratio of the measurement length to the coherence length, with a value of e^2/h when this ratio was unity [19]. Following these early papers, theorists began to establish the source of the fluctuations as a form of quantum interference within a random potential landscape that existed in these mesoscopic structures. Here, it was assumed that an electron trajectory could pass around a potential hill, at the Fermi surface, and create interference effects between the two paths. The random nature of the potential landscape would then lead to random fluctuations rather than periodic Aharonov–Bohm oscillations. These fluctuations can be efficiently simulated with the use of the recursive Green's function.

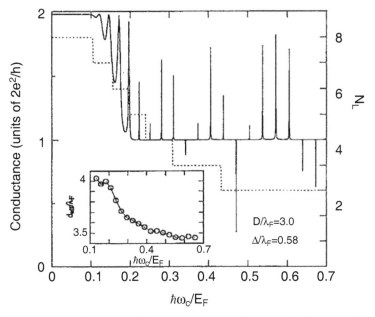

Figure 3.4 Conductance of the nanowire as a function of the normalized magnetic field $\hbar\omega_c/E_F$. Only the lowest mode is occupied at the Fermi energy in the constrictions. The dotted line represents the number of occupied Landau levels in the wire away from the anti-dot potential. The effective loop diameter corresponding to the period is plotted in the inset. Reprinted with permission from Ref. [13], Copyright 1993, American Physical Society.

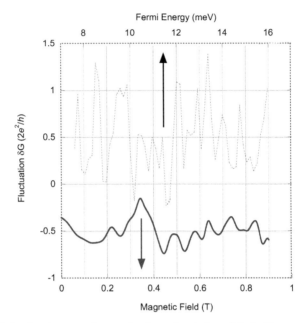

Figure 3.5 Conductance fluctuation for a simulation of GaAs. The upper curve has been offset upward by 0.5, while the lower curve has been offset downward by the same amount to provide clarity. The Fermi energy for the magnetic field sweep is 15.2 meV, and the random potential amplitude is W = 45.6 meV.

The most common method of creating the random potential is to use the finite difference grid, such as shown in Fig. 1.5, and then to use the Anderson model potential [20]. For the potential, we assume that the potential at each grid point is a random value taken from the range

$$-\frac{W}{2} \leq V(x,y) \leq \frac{W}{2} \tag{3.50}$$

with uniform probability. This potential value is added to the diagonal on-site terms of the Hamiltonian. Then the conductance (transmission) is computed as the Fermi energy or a magnetic field applied normal to the plane is varied [21]. The fluctuations will ride on a varying background conductance, which can be removed by standard signal processing techniques to isolate the fluctuation part of the conductance. Typical curves for a GaAs sample are shown in Fig. 3.5. In this figure, the curves have been offset for clarity, but one can see that the fluctuation with magnetic field is much smaller than that with the Fermi energy, a result that has been found in graphene [22]. Both curves in the figure are for a potential amplitude of $W = 45.6$ meV. From curves such as this, we can determine the r.m.s. amplitude of the conductance fluctuation, and this is plotted in Fig. 3.6 as a function of the amplitude of the random potential. Several points are important in this figure. First, the amplitude of the fluctuations goes to zero as the amplitude of the random potential goes to zero. That is, there are no fluctuations if there is no random potential. Second, the amplitude of the fluctuations increases with the random potential and then saturates beyond $W \sim 0.5t$, where t is the hopping energy (22.8 meV). The saturation value can be estimated given the observation that, in the saturation regime, the fluctuation involves only the switching on and off of a single transverse mode in the structure. This mode has a conductance of $2e^2/h$ according to the Landauer formula, so that the r.m.s. value is simply

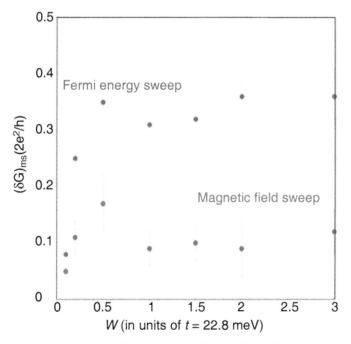

Figure 3.6 Dependence of the r.m.s. amplitude of the fluctuations on the strength of the random potential. Here, W is measured in units of the hopping energy, which is determined by the grid size and the material.

$$\delta G = \frac{1}{2\sqrt{2}} \frac{2e^2}{h} \sim 0.354 \frac{2e^2}{h} \sim 0.707 \frac{e^2}{h}. \tag{3.51}$$

This value is seen in most experiments, although it is reduced by the fact that the phase coherence length is usually much smaller than the sample size. In the simulations, the phase coherence length is exactly the sample size. As remarked, the general trend that shows that the fluctuation for varying magnetic field is about a factor of 3 smaller than that for the Fermi energy variation has been seen in graphene [22], and this precludes any possibility that one can assign ergodicity to the fluctuations. Ergodicity would result if all possible variations in the sample led to the same amplitude of fluctuations, and this is just not the case as can be seen in the figure.

3.5 Electron–Electron Interaction

In the introduction to this chapter, we discussed the creation and annihilation operators for the electrons. These operators satisfy the anti-commutation relationship (3.3) rather than the commutation relationship appropriate for bosonic operators. The importance of these anti-commuting operators is that they assure us that a given state can have only a single electron in it, thus satisfying the Pauli exclusion principle. These operators can be used to support further studies of the electron gas itself. For example, we can write the total number of electrons as a spatial integral over the density as

$$\int \rho(\mathbf{x}) d^3\mathbf{x} = \int \psi^\dagger(\mathbf{x}) \psi(\mathbf{x}) d^3\mathbf{x}. \tag{3.52}$$

Each of these wavefunctions can be expanded in momentum (or energy) field operators using Eq. (3.4) (we assume the various α are unity) to give

$$\int \rho(\mathbf{x}) d^3\mathbf{x} = \int \sum_k c_k^\dagger \varphi_k^* \sum_l c_l \varphi_l d^3\mathbf{x}$$
$$= \sum_{k,l} c_k^\dagger c_l \int \varphi_k^* \varphi_l d^3\mathbf{x} = \sum_k c_k^\dagger c_k, \tag{3.53}$$

where we have used the orthonormality of the basis functions. It must be noted, however, that this result is for a homogeneous electron gas. Otherwise we have trouble with the Fourier transformation that is indicated in the expansion of the wavefunctions. But we notice that this final form is the same as that for the number operator in the harmonic oscillator, a bosonic system. To some extent, this similarity is desired, but it also means that we can write the kinetic energy in the electron gas as

$$H_0 = \sum_k E_k c_k^+ c_k , \tag{3.54}$$

which is just a sum of the energy of the occupied states.

When the electrons interact with each other, they are subject to the Coulomb interaction between individual electrons. We can write this new energy term as

$$V = \frac{1}{2} \int d^3\mathbf{x} V(\mathbf{x} - \mathbf{x}') \rho(\mathbf{x}) \rho(\mathbf{x}') . \tag{3.55}$$

The factor of 1/2 arises from double counting of the electrons in the integration and the subsequent summations. We can now introduce the operators for the two densities, but we have to be very careful about this. The reason is that when this energy operator operates on the vacuum (empty) state, it must produce zero and not induce any excitations into the vacuum state. This is a reflection of the fact that one cannot have a Coulomb interaction among electrons if there are no electrons in the system. Hence, when we introduce the number operators, we must move the annihilation operators to the right-hand side of the expression so that they are the first to operate on the vacuum (or any other) state. In addition, we wish to have the primed coordinate operators surrounding those of the unprimed coordinates. This is called *normal ordering* of the operators. This form is achieved by using the anti-commutation relationships as follows:

$$N[\rho(\mathbf{x})\rho(\mathbf{x}')] = N[\psi^\dagger(\mathbf{x})\psi(\mathbf{x})\psi^\dagger(\mathbf{x}')\psi(\mathbf{x}')]$$
$$= -N[\psi^\dagger(\mathbf{x})\psi^\dagger(\mathbf{x}')\psi(\mathbf{x})\psi(\mathbf{x}')] \tag{3.55}$$
$$= \psi^\dagger(\mathbf{x}')\psi^\dagger(\mathbf{x})\psi(\mathbf{x})\psi(\mathbf{x}'),$$

where N indicates the normal ordering operator. We could go further and add a spin index to each of the wavefunctions, and then sum over the spin indices for the total potential energy, but we will ignore this for the present time. Now, we want to put this into the same form as the kinetic energy in Eq. (3.54), so will have to go over to the Fourier space and introduce the operators, as

$$\psi(\mathbf{x}) = \int \frac{d^3\mathbf{k}_1}{(2\pi)^3} c_k e^{i\mathbf{k}_1 \cdot \mathbf{x}}$$

$$\psi^\dagger(\mathbf{x}) = \int \frac{d^3\mathbf{k}_2}{(2\pi)^3} c_k^\dagger e^{-i\mathbf{k}_2 \cdot \mathbf{x}} \tag{3.56}$$

$$V(\mathbf{x} - \mathbf{x}') = \int \frac{d^3\mathbf{q}}{(2\pi)^3} V_q e^{i\mathbf{q} \cdot (\mathbf{x} - \mathbf{x}')}.$$

Similar Fourier transforms must be introduced for the primed coordinates, and for these we use $\mathbf{k}_3$ and $\mathbf{k}_4$. Then the entire messy integral that results in Fourier space for Eq. (3.55) can be written as

$$\hat{V} = \frac{1}{2} \int \frac{d\mathbf{k}_1}{(2\pi)^3} \int \frac{d\mathbf{k}_2}{(2\pi)^3} \int \frac{d\mathbf{k}_3}{(2\pi)^3} \int \frac{d\mathbf{k}_4}{(2\pi)^3} V_q c_{k_4}^\dagger c_{k_2}^\dagger c_{k_1} c_{k_3} I, \tag{3.57}$$

where

$$I = \int d^3\mathbf{x} \int d^3\mathbf{x}' e^{i(\mathbf{k}_1 - \mathbf{k}_2 + \mathbf{q}) \cdot \mathbf{x}} e^{i(\mathbf{k}_3 - \mathbf{k}_4 - \mathbf{q}) \cdot \mathbf{x}'}$$

$$= (2\pi)^6 \delta(\mathbf{k}_1 - \mathbf{k}_2 + \mathbf{q}) \delta(\mathbf{k}_3 - \mathbf{k}_4 - \mathbf{q}). \tag{3.58}$$

With the delta functions, we can now rewrite Eq. (3.57) by summing over the second and fourth momentum vectors (but will then replace the third momentum vector by the index 2, as these are dummy indices) to obtain

$$\hat{V} = \frac{1}{2} \int \frac{d\mathbf{k}_1}{(2\pi)^3} \int \frac{d\mathbf{k}_2}{(2\pi)^3} V_q c_{k_2 - q}^\dagger c_{k_1 + q}^\dagger c_{k_1} c_{k_2}. \tag{3.59}$$

The important aspect of this latter equation is that the interaction energy involves the transfer of momentum from particle 2 to particle 1 during the process. The nature of this transfer is by the emission of a low-energy photon by particle 2 and its absorption by particle 1. That is, the Coulomb interaction is mediated by the transfer of photons—massless particles that can carry energy and momentum from one electron to another. This is verified by remembering that the electromagnetic field arises from the Coulomb potential of the particles, and electromagnetic waves are the transfer of photons.

Messy integrals, such as that in Eq. (3.59), are often hard to manage due to the massive bookkeeping needed to track all of the

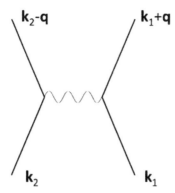

Figure 3.7 The Feynman diagram for the direct Coulomb interaction between two electrons, denoted by k_1 and k_2. This interaction accounts for the transfer of momentum q from particle 2 to particle 1.

indices and coordinates. While Eq. (3.59) is not too messy, we will encounter much messier integrals as we progress through this book (and even later in this chapter). Consequently, one would like to have a short-hand method to keep track of the interaction terms. A very powerful form of such a short-hand was introduced by Feynman [23, 24]. We show the Feynman diagram for the direct interaction that can be inferred from Eq. (3.59) in Fig. 3.7. In the interaction, no particular direction of time has been indicated. Normally, for such an interaction, time is considered to progress from left to right, although it might be more logical to have it flow upward in this particular case. If we were to do this, we should add an arrow to the squiggly line corresponding to the transfer of momentum from left to right. Feynman diagrams have become a very commonplace usage in many-body physics. By comparing the diagram to the integral, we note that the vertex on the left is the place where the two lines corresponding to particle 2 meet the Coulomb line. The vertex on the right corresponds to where the two particle lines meet the Coulomb line. At each vertex, there is an integral over the momentum of the corresponding particle and a factor of $1/(2\pi)^3$ for each of these integrals. The Coulomb line corresponds to adding the factor of the interaction potential V_q to the result. So one can easily write down the proper integral that corresponds to the diagram, although we will see some additional factors when we get to perturbation theory. The use of such diagrams has been given the name *diagrammatica*, and books have appeared on the use of these diagrams [25, 26]. Finally, the result (3.57) allows us to write down the total energy of the interacting electron gas as

$$H = \sum_{k,\sigma} \frac{\hbar^2 k^2}{2m^*} c_{k,\sigma}^\dagger c_{k,\sigma} + \frac{1}{2} \sum_{k,k',q,\sigma,\sigma'} \frac{e^2}{\varepsilon_s q^2} c_{k'-q,\sigma'}^\dagger c_{k+q,\sigma}^\dagger c_{k,\sigma} c_{k',\sigma'} , \quad (3.60)$$

where we have added the spin indices σ and σ' and indicated a summation over these extra indices.

3.5.1 Hartree Approximation

Now if we could determine the interaction energy from all of the electrons, then we would have the exact solution. In fact, that is a very difficult problem except for very few electrons. Thus, a number of approaches have been made and we will describe some of them here. To do this, however, it is better to work with the equation of motion for Green's function itself (3.19). To this, we now add the interaction energy described above, and this equation becomes

$$\left[i\hbar \frac{\partial}{\partial t} + \frac{\hbar^2}{2m^*} \nabla_x^2 - V(x) \right] G(x,x';t,t') = \hbar \delta(t-t')\delta(x-x')$$

(3.60)

$$- \int d^3x'' \psi^\dagger(x',t')\psi^\dagger(x'',t^-) \frac{e^2}{4\pi\varepsilon_s |x-x''|} \psi(x'',t)\psi(x,t).$$

The potential term in the square brackets includes all of the external potentials, plus any that arise from the impurities, phonons, or any other such disturbance that affects the Green's function (or the wavefunction) locally. The last term is the electron–electron interaction, and the form is somewhat more complicated than treated earlier, primarily due to the fact that we are facing an equation for Green's function rather than just the total energy. The difficulty with this last term is that the combination of wavefunctions within the integral composes a two-particle Green's function and is not a simple product of two single-particle Green's functions. We define this two-particle Green's function as the time-ordered product of the field operators as

$$G_2(x'',t;x'',t^-;x,t;x',t') = -\left\langle T\left[\psi^\dagger(x',t')\psi^\dagger(x'',t^-)\psi(x'',t)\psi(x,t) \right] \right\rangle.$$

(3.61)

Here T is the time-ordering operator, which moves the earliest operations to the right-hand side of the sequence, while obeying the anti-commutator relationships. The superscript minus sign on the second time indicates that this occurs just before the time t. This will put this creation operator on the far right, without changing the sign as there are two anti-commutator moves involved.

Now we have to decide how we wish to group the creation and annihilation operators, if we want to simplify this two-particle Green's function. One of the simplest approaches is to put the two which are at $\mathbf{x}''$ together. Then, we can write Eq. (3.61) as

$$G_2(\mathbf{x}'',t;\mathbf{x}'',t^-;\mathbf{x},t;\mathbf{x}',t') = i\left\langle T\left[\psi^\dagger(\mathbf{x}'',t^-)\psi(\mathbf{x}'',t)\right]\right\rangle$$

$$\times i\left\langle T\left[\psi^\dagger(\mathbf{x}',t')\psi(\mathbf{x},t)\right]\right\rangle$$

$$= G(\mathbf{x}'',\mathbf{x}'';t,t^-)G(\mathbf{x},\mathbf{x}';t,t'). \qquad (3.62)$$

When this is done, the first Green's function will be totally integrated by the interaction as it appears in Eq. (3.6), while the second Green's function is just the one for which the differential equation has been written. Hence, we can rewrite the differential equation (3.60) as

$$\left[i\hbar\frac{\partial}{\partial t} + \frac{\hbar^2}{2m^*}\nabla_x^2 - V(\mathbf{x}) - \int d^3\mathbf{x}'' \frac{e^2\rho(\mathbf{x}'',t)}{4\pi\varepsilon_s |\mathbf{x}-\mathbf{x}''|}\right]G(\mathbf{x},\mathbf{x}';t,t')$$

$$= \hbar\delta(t-t')\delta(\mathbf{x}-\mathbf{x}'). \qquad (3.63)$$

This approximation is termed the Hartree approximation and is what is known as a mean field approximation. That is, the average charge density, sitting within the integral in the square brackets, produces a change in the external potential, known as screening. In fact, the integral itself is one of the Lienard–Weichert potentials, which may be found in any good book on electromagnetic theory. Hence, we recognize that this potential is actually found by the solution to Poisson's equation subject to the boundary conditions defined by the external potential. Diagrammatically, the integral looks like Fig. 3.8. One notes that the Green's function line closes upon itself, which is a result of the beginning and ending positions being identical. In momentum space, this means that a single momentum can be identified for Green's function, but the momentum carried by the photon line is identically zero. This is the result that the Hartree screening potential, in this approximation, is a static potential, which is uniform as a result of the average (mean) density throughout the system.

Figure 3.8 The diagram for the Hartree integral in Eq. (3.63). The circle represents Green's function, or the density, since the beginning and ending positions are identical.

3.5.2 Exchange Interaction

There is a second possible grouping of the creation and annihilation operators that can arise in Eq. (3.61), and this leads to

$$G_2(\mathbf{x}'',t;\mathbf{x}'',t^-;\mathbf{x},t;\mathbf{x}',t') = -i\left\langle T\left[\psi^\dagger(\mathbf{x}',t')\psi(\mathbf{x}'',t)\right]\right\rangle$$
$$\times i\left\langle T\left[\psi^\dagger(\mathbf{x}'',t^-)\psi(\mathbf{x},t)\right]\right\rangle$$
$$= -G(\mathbf{x},\mathbf{x}'';t,t)G(\mathbf{x}'',\mathbf{x}';t^-,t'). \quad (3.64)$$

Here, the wavefunctions at $\mathbf{x}$ and $\mathbf{x}''$ have been interchanged, so that one thinks about the electrons at these two positions being exchanged. Indeed, this term is called the exchange interaction, and it makes the differential equation for Green's function very difficult to handle. Neither of these two Green's functions in Eq. (3.64) is the one for which the differential equation has been written. Thus, this term must remain as a forcing function on the right-hand side of the equation. This exchange is indicated by the first Green's function, which makes the positional move at a single instance of time, following which it moves on to the later time from the new position. The approximation in which both of the two terms are kept is termed the *Hartree–Fock approximation*. We will see in the next chapter that these two terms are the leading terms in a perturbation expansion of the interaction term, and many more terms can be generated.

To analyze this exchange term in a simple manner, let us rewrite the differential equation (3.60) for a particular basis function instead of Green's function, as

$$\left[i\hbar \frac{\partial}{\partial t} + \frac{\hbar^2}{2m^*} \nabla_x^2 - V(\mathbf{x}) - \int d^3\mathbf{x}'' \frac{e^2 \rho(\mathbf{x}'',t)}{4\pi\varepsilon_s |\mathbf{x}-\mathbf{x}''|} \right] \psi_n(\mathbf{x},t)$$

$$= -\sum_{j\neq n} \int d^3\mathbf{x}'' \frac{e^2}{4\pi\varepsilon_s |\mathbf{x}-\mathbf{x}''|} \psi_j^\dagger(\mathbf{x}'',t)\psi_n(\mathbf{x}'',t^-)\psi_j(\mathbf{x},t). \quad (3.65)$$

The source of the problem is that the particular basis function in the last integral is at the wrong position due to this exchange interaction. Thus, we shall finesse this last term by introducing the proper basis function through writing the last term as

$$I_{HF} = -\sum_{j\neq n} \int d^3\mathbf{x}'' \frac{e^2}{4\pi\varepsilon_s |\mathbf{x}-\mathbf{x}''|} \psi_j^\dagger(\mathbf{x}'',t)\psi_n(\mathbf{x}'',t^-)\psi_j(\mathbf{x},t)$$

$$= -\sum_{j\neq n} \int d^3\mathbf{x}'' \frac{e^2}{4\pi\varepsilon_s |\mathbf{x}-\mathbf{x}''|} \frac{\psi_j^\dagger(\mathbf{x}'',t)\psi_n(\mathbf{x}'',t^-)\psi_j(\mathbf{x},t)}{\psi_n(\mathbf{x},t)} \psi_n(\mathbf{x},t)$$

$$= -\frac{e}{4\pi\varepsilon_s} \int d^3\mathbf{x}'' \frac{\rho(\mathbf{x},\mathbf{x}'')}{|\mathbf{x}-\mathbf{x}''|} \psi_n(\mathbf{x},t).$$

$$(3.66)$$

In this last expression, we have introduced the two-particle density. This density has an important property, in that

$$\int d^3\mathbf{x}'' \rho(\mathbf{x},\mathbf{x}'') = e \sum_{j\neq n} \int d^3\mathbf{x}'' \frac{\psi_j^\dagger(\mathbf{x}'')\psi_n(\mathbf{x}'')\psi_j(\mathbf{x})}{\psi_n(\mathbf{x})}$$

$$= e \sum_{j\neq n} \frac{\psi_j(\mathbf{x})}{\psi_n(\mathbf{x})} \delta_{jn} = e.$$

$$(3.67)$$

Hence, the total charge in the two-particle density is just a single electron, and this term represents the correlation among the electrons that is important in the exchange interaction. In addition, evaluating this at a single position leads to

$$\rho(\mathbf{x},\mathbf{x}) = e \sum_{j \ne n} \psi_j^\dagger(\mathbf{x})\psi_j(\mathbf{x}) < e, \tag{3.68}$$

where the last inequality arises because of the missing state n.

From these arguments, one can conceive that a particular electron, of a given spin, is surrounded by an effective positive charge as this electron pushes the other electrons away from it. This creates what is known as the *exchange hole* around this electron due to its repulsive force on other electrons of the same spin. This is a direct result of the Pauli exclusion principle. Now let us go from the positional basis functions to plane waves in the Fourier representation, so that instead of an index for the basis function, each function will have a different momentum vector. Hence, we can write the two-particle density in this form as

$$\begin{aligned}
\rho(\mathbf{x},\mathbf{x}'') &= e \sum_{\mathbf{k}'} \frac{\psi_{\mathbf{k}'}^\dagger(\mathbf{x}'')\psi_{\mathbf{k}}(\mathbf{x}'')\psi_{\mathbf{k}'}(\mathbf{x})}{\psi_{\mathbf{k}}(\mathbf{x})} \\
&= e \sum_{\mathbf{k}'} e^{-i\mathbf{k}'\bullet\mathbf{x}''} e^{i\mathbf{k}\bullet\mathbf{x}''} e^{-i\mathbf{k}\bullet\mathbf{x}} e^{i\mathbf{k}'\bullet\mathbf{x}} \\
&= e \sum_{\mathbf{k}'} e^{-i(\mathbf{k}'-\mathbf{k})\bullet(\mathbf{x}-\mathbf{x}'')}.
\end{aligned} \tag{3.69}$$

The integral term in Eq. (3.66) can now be written as

$$I = -\frac{e^2}{4\pi\varepsilon_s} \int d^3\mathbf{x}'' \sum_{\mathbf{k}'} \frac{e^{i(\mathbf{k}'-\mathbf{k})\bullet(\mathbf{x}-\mathbf{x}'')}}{|\mathbf{x}-\mathbf{x}''|} = -\frac{e^2}{\varepsilon_s} \sum_{\mathbf{k}'} \frac{1}{|\mathbf{k}'-\mathbf{k}|^2}, \tag{3.70}$$

where we have performed the three-dimensional Fourier transform on the relative position of the Coulomb potential. The sum can be converted to an integral and evaluated as

$$\begin{aligned}
\sum_{\mathbf{k}'} \frac{1}{|\mathbf{k}-\mathbf{k}'|^2} &= \frac{1}{8\pi^3} \int_0^{2\pi} d\varphi \int_0^\pi \sin\vartheta d\vartheta \int_0^{k_F} \frac{k'^2 dk'}{k^2 + k'^2 - 2kk'\cos\vartheta} \\
&= -\frac{1}{4\pi^2} \int_0^{k_F} \frac{k' dk'}{k} \left[\ln\left(k^2 + k'^2 - 2kk'\cos\vartheta\right) \right]_0^{2\pi} \\
&= -\frac{1}{4\pi^2 k} \int_0^{k_F} k' dk' \ln\left(\frac{k'+k}{k'-k}\right) \\
&= -\frac{k_F}{4\pi^2} \left(1 + \frac{k_F^2 - k^2}{2kk_F} \ln\left|\frac{k_F + k}{k_F + k}\right| \right).
\end{aligned} \tag{3.71}$$

Thus, the exchange energy renormalizes the kinetic energy of the electron, as this term adds to that energy in Eq. (3.65). Ignoring the screened external potential, the energy is now

$$E(k) = \frac{\hbar^2 k^2}{2m^*} - \frac{e^2 k_F}{4\pi^2 \varepsilon_s}\left(1 + \frac{k_F^2 - k^2}{2kk_F}\ln\left|\frac{k_F + k}{k_F - k}\right|\right). \tag{3.72}$$

The leading term in the large parentheses is a downward shift of the energy, a renormalization of the energy due to the contribution of the many-body interaction to the self-energy of the electron gas as a whole. The second term is a k-dependent term, which is generally believed to be a renormalization of the effective mass. If the semiconductor has a relatively small electron (or hole) concentration, then the Fermi wavevector becomes imaginary as the Fermi energy lies below the conduction band (or above the valence band) edge. In this case, the exchange energy is negligible and can be ignored. At low temperatures or high doping densities, however, this is seldom the case, and this energy can be quite significant.

Problems

1. Write a computer program that computes the recursive Green's function for a two-dimensional system and determines the transmission through a mesoscopic device of 0.4 micron transverse dimension. Set the grid step in the code to 5 nm, and the longitudinal dimension to 101 nodes. Use the Landauer formula to convert the transmission into conductance. (a) Get the code running on your favorite computational engine. (b) Create an output file containing the eigenvalues in the transverse direction (in the absence of the magnetic) for one of the early slices. Plot these eigenvalues (energy versus number). (c) Set the code to carry out a sweep of the Fermi energy from 5 meV to 20 meV (in 0.2 meV steps). Plot the transmission as a function of the Fermi energy. (d) Set the code to carry out a sweep of the magnetic field from 0 to 0.5 T (in 1 mT steps), for a Fermi energy of 14 meV. Plot the transmission as a function of the magnetic field.

2. Using the mesoscopic structure in Problem 1, apply a harmonic oscillator potential of the form

$$V(x,y)=V_0 + \frac{m^*\omega_0^2}{2}\left(y^2 - x^2\right)$$

where y is the transverse direction and x is the longitudinal direction. Set V_0 to 5 meV, and adjust the angular frequency of the potential so that the opening is 100 nm at a Fermi energy of 14 meV. (a) Set the code to carry out a sweep of the Fermi energy from 5 meV to 20 meV (in 0.2 meV steps). Plot the transmission as a function of the Fermi energy. (b) Set the code to carry out a sweep of the magnetic field from 0 to 0.5 T (in 1 mT steps), for a Fermi energy of 14 meV. Plot the transmission as a function of the magnetic field.

3. Using the mesoscopic structure in Problem 1, apply a potential of $4t$, where t is the hopping energy in the form of a circular barrier of radius 100 nm, cited in the center of the structure (midway in both x- and y-directions). (a) Set the code to carry out a sweep of the Fermi energy from 5 meV to 20 meV (in 0.2 meV steps). Plot the transmission as a function of the Fermi energy. (b) Set the code to carry out a sweep of the magnetic field from 0 to 0.5 T (in 1 mT steps), for a Fermi energy of 14 meV. Plot the transmission as a function of the magnetic field. (c) Repeat for a radius of 50 nm.

4. Using the mesoscopic structure in Problem 1, apply a potential that is sinusoidal in both x and y. The period of the sinusoid so that the potential is zero at both ends of the grid and at both sides of the grid, with four periods in the width and five periods in the length. The peak amplitude of the potential is to be 3 meV. (a) Set the code to carry out a sweep of the Fermi energy from 1 meV to 20 meV (in 0.2 meV steps). Plot the transmission as a function of the Fermi energy. (b) Set the code to carry out a sweep of the magnetic field from 0 to 0.5 T (in 1 mT steps), for a Fermi energy of 14 meV. (c) Repeat for a peak amplitude of the potential of 8 meV.

5. Consider a slab of GaAs at a temperature of 8 mK, where it is certainly degenerate. If it is doped to 10^{18} cm^{-3}, it is certainly not frozen out (the carriers sit well into the conduction band). (a) Determine the Fermi energy and free electron concentration if the donor ionization energy is 5 meV. (b) Determine the Hartree–Fock energy as a function of the

momentum for an energy of one-half the Fermi energy to one of twice the Fermi energy.

References

1. A. L. Fetter and J. D. Walecka, *Quantum Theory of Many-Particle Systems* (McGraw-Hill, New York, 1971).

2. J. Rammer, *Quantum Transport Theory* (Perseus, Reading, MA, 1998).

3. P. A. Lee and D. S. Fisher, *Phys. Rev. Lett.*, **47**, 882 (1981).

4. D. J. Thouless and S. Kirkpatrick, *J. Phys. C*, **14**, 235 (1981).

5. A. MacKinnon, *Z. Phys. B*, **59**, 385 (1985).

6. Y. Takagaki and D. K. Ferry, *Phys. Rev. B*, **47**, 9913 (1993).

7. Y. Meir and N. S. Wingreen, *Phys. Rev. Lett.*, **68**, 2512 (1992).

8. K. Shepard, *Phys. Rev. B*, **44**, 9088 (1991).

9. R. Lake and S. Datta, *Phys. Rev. B*, **45**, 6670 (1992).

10. R. Lake, G. Klimeck, and S. Datta, *Phys. Rev. B*, **47**, 6427 (1993).

11. D. K. Ferry, S. M. Goodnick, and J. P. Bird, *Transport in Nanostructures*, 2nd Ed. (Cambridge University Press, Cambridge, 2009) Sec. 9.4.

12. D. K. Ferry, *Transport in Semiconductor Mesoscopic Devices* (Institute of Physics Publishing, Bristol, UK, 2015).

13. Y. Takagaki and D. K. Ferry, *Phys. Rev. B*, **48**, 8152 (1993).

14. C. G. Smith, M. Pepper, R. Newbury, H. Ahmed, D. G. Hasko, D. C. Peacock, J. E. F. Frost, D. A. Ritchie, G. A. C. Jones, and G. Hill, *J. Phys.: Condens. Matter*, **2**, 3405 (1990).

15. D. Weiss, M. L. Roukes, A. Menschig, P. Grambow, K. von Klitzing, and G. Weiman, *Phys. Rev. Lett.*, **66**, 2790 (1991).

16. D. K. Ferry, *Prog. Quantum Electron.* **16**, 251 (1992).

17. W. E. Howard and F. F. Fang, *Solid State Electron.*, **8**, 82 (1965).

18. A. B. Fowler, A. Hartstein, and R. A. Webb, *Phys. Rev. Lett.*, **48**, 196 (1982).

19. W. J. Skocpol, *Physica Scripta*, **T19**, 95 (1987).

20. P. W. Anderson, *Phys. Rev.*, **109**, 1492 (1958).

21. B. Liu, R. Akis, and D. K. Ferry, *J. Phys.: Condens. Matter*, **25**, 395802 (2013), and references therein.

22. G. Bohra, R. Somphonsone, N. Aoki, Y. Ochiai, R. Akis, D. K. Ferry, and J. P. Bird, *Phys. Rev. B*, **86**, 161405 (2012).

23. R. P. Feynman, *Phys. Rev.*, **76**, 749, 769 (1949).

24. A. L. Fetter and J. D. Walecka, *Quantum Theory of Many-Particle Systems* (McGraw-Hill, New York, 1971).

25. R. D. Mattuck, *A Guide to Feynman Diagrams in the Many-Body Problem*, 2nd Ed. (McGraw-Hill, New York, 1974).

26. M. Veltman, *Diagrammatica: The Path to Feynman Diagrams* (Cambridge University Press, Cambridge, 1994).

Chapter 4

Interaction Representation

At the end of the last chapter, we considered the electron–electron interaction. Here, we found two terms: the Hartree energy and the exchange energy. These work fine if we can solve for the energy exactly. However, as was remarked, this can only be done for a few electron systems. For more meaningful systems, we need to develop an approximation technique for the interacting Green's functions. The standard approach to this is via the interaction representation, which gives us a prescription for developing the perturbation series. As one might infer from our previous discussion, this will lead to multiple, messy integrals. Fortunately, the Feynman diagrams give us an elegant and pictorial method of looking at the various terms in the perturbation series. As a result, they also give us a much more intuitive method of examining the various terms.

It is important to note, however, that it is basically impossible to include all possible perturbative terms. The number of these terms is just too massive. In addition, one must be sure that the perturbation series actually converges, although it is not always possible to do so. This presents a problem in some cases, where it is necessary to push ahead while not being confident that the approach actually converges to the correct, or even to any, answer. In practice, one normally only keeps certain sets of terms, for example those based on the two terms we have already found—the Hartree and exchange terms. The decision as to which sets of terms one retains is based on

An Introduction to Quantum Transport in Semiconductors
David K. Ferry
Copyright © 2018 Pan Stanford Publishing Pte. Ltd.
ISBN 978-981-4745-86-4 (Hardcover), 978-1-315-20622-6 (eBook)
www.panstanford.com

intuition, and on the fact that, if the series is to converge, the lower-order terms are larger than higher-order terms. Hence, one normally proceeds on the basis of faith that the most important terms are the ones that are kept. This can be a self-fulfilling prophecy, but it can also result in misleading results. So care must be taken as one moves ahead with this perturbative approach.

We generally begin directly with the Schrödinger equation itself, in which we keep the Hamiltonian as an operator, and write as

$$\psi(\mathbf{x}, t) = e^{-iHt/\hbar} \psi(\mathbf{x}, 0). \tag{4.1}$$

Here we assume that the operators appearing in the Hamiltonian are not explicit functions of time (although this will change when we get to the electron–phonon interaction). The result (4.1) is often referred to as the Schrödinger representation, or picture, in which the time variation is placed with the wavefunction. An alternative view is the Heisenberg representation, or picture, in which the operators are written as

$$i\hbar \frac{\partial A(t)}{\partial t} = [A(t), H] \quad \Rightarrow \quad A(t) = e^{iHt/\hbar} A(0) e^{-iHt/\hbar}. \tag{4.2}$$

In the interaction representation, or picture, we combine these by assuming that the Hamiltonian has a base form H_0 and a perturbing potential V. Then, Eq. (4.2) is modified to be

$$A(t) = e^{iH_0 t/\hbar} A(0) e^{-iH_0 t/\hbar}, \tag{4.3}$$

and the wavefunction is written as

$$\psi(\mathbf{x}, t) = e^{iH_0 t/\hbar} e^{-iHt/\hbar} \psi(\mathbf{x}, 0). \tag{4.4}$$

One must be very careful as this point, because in general these two exponentials cannot be easily combined. Generally, this is handled through the Baker–Hausdorf formula [1]

$$e^A e^B = e^{A+B} e^{-[A, B]/2}. \tag{4.5}$$

With the interaction representation, we can write matrix elements for two arbitrary basis states as

$$\left\langle \widehat{\psi}_1^\dagger(t) A(t) \widehat{\psi}_2(t) \right\rangle = \left\langle \widehat{\psi}_1^\dagger(0) e^{iHt/\hbar} e^{-iH_0 t/\hbar} \left(e^{iH_0 t/\hbar} A e^{-iH_0 t/\hbar} \right) \right.$$

$$\left. e^{iH_0 t/\hbar} e^{-iHt/\hbar} \widehat{\psi}_2(0) \right\rangle$$

$$= \left\langle \widehat{\psi}_1^\dagger(0) e^{iHt/\hbar} A e^{-iHt/\hbar} \widehat{\psi}_2(0) \right\rangle, \tag{4.6}$$

which gives us the correct time dependence. That is, we can now assign the exponentials either to the wavefunctions for the Schrödinger picture or to the operator for the Heisenberg picture.

We recognize that the perturbing potential is the term that will upset this nice balance, and we need to generate an approach for handling this term. To begin, we can write the time derivative of the field operator as

$$\frac{\partial \widehat{\psi}(t)}{\partial t} = \frac{i}{\hbar} e^{iH_0 t/\hbar} (H_0 - H) e^{-iHt/\hbar} \widehat{\psi}(0)$$

$$= -\frac{i}{\hbar} e^{iH_0 t/\hbar} V(e^{-iH_0 t/\hbar} e^{iH_0 t/\hbar}) e^{-iHt/\hbar} \widehat{\psi}(0) \qquad (4.7)$$

$$= -\frac{i}{\hbar} V(t) \widehat{\psi}(t).$$

The two exponentials that precede the initial time wavefunction can be defined as an evolution operator, which we give the short-hand definition as

$$U(t) = e^{iH_0 t/\hbar} e^{-iHt/\hbar}, \qquad (4.8)$$

which we recognize as a unitary operator. In addition, it has the initial value $U(0) = 1$. If we put this definition into Eq. (4.7), then we find that

$$\frac{\partial U(t)}{\partial t} = \frac{i}{\hbar} e^{iH_0 t/\hbar} (H_0 - H) e^{-iHt/\hbar}$$

$$= -\frac{i}{\hbar} e^{iH_0 t/\hbar} V(e^{-iH_0 t/\hbar} e^{iH_0 t/\hbar}) e^{-iHt/\hbar} \qquad (4.9)$$

$$= -\frac{i}{\hbar} V(t) U(t),$$

which can be formally integrated to yield

$$U(t) = U(0) - \frac{i}{\hbar} \int_0^t dt' V(t') U(t'). \qquad (4.10)$$

The entire perturbative approach is based on an iterative solution to this last equation. For the lowest-order term, we just use the initial value and proceed as

$$U_0(t) = 1$$

$$U_1(t) = 1 - \frac{i}{\hbar} \int_0^t dt' V(t')$$

$$U_2(t) = 1 - \frac{i}{\hbar} \int_0^t dt' V(t') + \left(\frac{i}{\hbar}\right)^2 \int_0^t dt_1 \int_0^{t_1} dt_2 V(t_1)V(t_2). \qquad (4.11)$$

In the last equation, it is always required that $t_2 < t_1$.

In the sense that the first potential occurs after the second, the two potentials are time ordered. Let us explore this a little further. We define the time-ordering operator T such that

$$T[V(t_1)V(t_2)] = \theta(t_1 - t_2)V(t_1)V(t_2) + \theta(t_2 - t_1)V(t_2)V(t_1). \quad (4.12)$$

Now, the integral over this quantity becomes

$$\int_0^t dt_1 \int_0^t dt_2 T[V(t_1)V(t_2)] = \int_0^t dt_1 \int_0^{t_1} dt_2 V(t_1)V(t_2) + \int_0^t dt_2 \int_0^{t_2} dt_1 V(t_2)V(t_1)$$

$$= 2\int_0^t dt_1 \int_0^{t_1} dt_2 V(t_1)V(t_2), \qquad (4.13)$$

where in the last term we have interchanged the two dummy variables. Thus, the limits of the second integral can be changed to the final time by introducing the time-ordering operator in Eq. (4.11). Then, it can be observed that the additional numerical factor makes the summation into a representation of the exponential function, and we can write our evolution operator as

$$U(t) = T\left[\exp\left\{-\frac{i}{\hbar}\int_0^t dt' V(t')\right\}\right]. \qquad (4.14)$$

Thus, we have the evolution of our system from $t = 0$ to an arbitrary time. Suppose we do not want to start at $t = 0$, but at some more convenient time, say t'. Then, we can introduce the adjoint of the evolution operator as

$$\widehat{\psi}(t) = U(t)\widehat{\psi}(0) = U(t)U^\dagger(t')\widehat{\psi}(t') \equiv S(t,t')\widehat{\psi}(t'), \qquad (4.15)$$

where S is called the S-matrix or, often, the scattering matrix. By natural evolution, we can introduce the operator for this matrix as

$$S(t,t') = T\left[\exp\left\{-\frac{i}{\hbar}\int_{t'}^t dt'' V(t'')\right\}\right]. \qquad (4.16)$$

Obviously, this differs from Eq. (4.14) only by the lower limit of the integral.

4.1 Green's Function Perturbation

It is very difficult to begin at $t = 0$ with the perturbation turned on immediately. This is because we do not know the state of the system at this time, since it must be in some sort of equilibrium, which may not be the normal ground state of the system. That is, the presence of the perturbation will lead to a state at $t = 0$, which could well be different from that in which the perturbing potential is not present. Hence, we make an assumption that, in the infinite past, the perturbation was not present and the system was in its unexcited ground state, which is denoted as φ_0 or $|0\rangle$. Now, at $t = 0$, we are not sure just what the state of the system really is, so we will denote it by the undefined state $|\ \rangle$; that is, the empty bra indicates that we really do not know what the state is. However, with the scattering operator, we can say some things about this state. For example, we do know that

$$\widehat{\psi}(0) = S(0, -\infty)|0\rangle = S(0, -\infty)\varphi_0$$
$$\widehat{\psi}(t) = S(t, 0)\widehat{\psi}(0) = S(t, -\infty)|0\rangle \qquad (4.17)$$
$$\widehat{\psi}(\infty) = S(\infty, 0)\widehat{\psi}(0) = S(\infty, -\infty)|0\rangle.$$

The last of these equations implies that the system has evolved from its ground state to some new state in the infinite future, which has this form, and hence we can assume that we may define an exponential operator as

$$\varphi_0 e^{iL} = \widehat{\psi}(\infty) = S(\infty, -\infty)\varphi_0, \qquad (4.18)$$

for which the exponential operator may be defined as

$$e^{iL} = \langle 0|S(\infty, -\infty)|0\rangle, \qquad (4.19)$$

that is as the expectation of the scattering operator on the ground state from the infinite past to the infinite future. This will be useful to us going forward.

With these preliminaries, we can now proceed to define the Green's function at $t = 0$, even when we do not know the exact state of the system. This general definition becomes

$$G(\mathbf{x}, \mathbf{x}'; t, t') = -i \langle \; |\psi(\mathbf{x}, t)\psi^\dagger(\mathbf{x}', t')| \; \rangle. \qquad (4.20)$$

The difficulty with the state at $t = 0$ can be eliminated by using the assumption that the wavefunction at this initial time is related to that at a general time as

$$
\begin{aligned}
\psi(\mathbf{x}, t) &= e^{iHt/\hbar} \psi(\mathbf{x}, 0) e^{-iHt/\hbar} \\
&= e^{iHt/\hbar} e^{-iH_0 t/\hbar} \widehat{\psi}(\mathbf{x}, t) e^{iH_0 t/\hbar} e^{-iHt/\hbar} \\
&= S(0, t) \widehat{\psi}(\mathbf{x}, t) S(t, 0).
\end{aligned}
\qquad (4.21)
$$

Hence, the wavefunction is given by the field operator and its evolution from the infinite past, the ground state. Thus, we can now rewrite Green's function as

$$
\begin{aligned}
G(\mathbf{x}, \mathbf{x}'; t, t') = &-i\theta(t - t')\langle 0 | S(-\infty, t)\psi(\mathbf{x}, t)S(t, t')\psi^\dagger(\mathbf{x}', t')S(t', -\infty) | 0\rangle \\
&+i\theta(t' - t)\langle 0 | S(-\infty, t')\psi^\dagger(\mathbf{x}', t')S(t', t)\psi(\mathbf{x}, t)S(t, -\infty) | 0\rangle.
\end{aligned}
\qquad (4.22)
$$

Now, we can use our useful operator developed in the previous paragraph to write

$$\langle 0 | S(-\infty, 0) = e^{-iL} \langle 0 | S(-\infty, \infty) S(-\infty, 0) = \frac{\langle 0 | S(\infty, 0)}{\langle 0 | S(-\infty, \infty) | 0\rangle}. \qquad (4.23)$$

The useful thing about introducing the denominator term is that this term provides continuous normalization of the state as the wavefunction evolves through time. With the introduction of this important normalization, the Green's function can be written as

$$
\begin{aligned}
G(\mathbf{x}, \mathbf{x}'; t, t') = &-i\theta(t - t')\frac{\langle 0 | S(\infty, t)\psi(\mathbf{x}, t)S(t, t')\psi^\dagger(\mathbf{x}', t')S(t', -\infty) | 0\rangle}{\langle 0 | S(\infty, -\infty) | 0\rangle} \\
&+i\theta(t' - t)\frac{\langle 0 | S(\infty, t')\psi^\dagger(\mathbf{x}', t')S(t', t)\psi(\mathbf{x}, t)S(t, -\infty) | 0\rangle}{\langle 0 | S(\infty, -\infty) | 0\rangle}.
\end{aligned}
\qquad (4.24)
$$

We can refine this further by the introduction of the time-ordering operator, which also allows us to recombine the two terms,

$$G(\mathbf{x}, \mathbf{x}'; t, t') = -i \frac{\langle 0 | T[\psi(\mathbf{x}, t)\psi^\dagger(\mathbf{x}', t')S(\infty, -\infty)] | 0 \rangle}{\langle 0 | S(\infty, -\infty) | 0 \rangle}. \quad (4.25)$$

In the future, we will eliminate the square brackets, and it will be important to remember that the time order operator acts upon all terms to its right.

As the first example, let us consider the simple case termed the "empty band." For this, the annihilation operator acting upon the ground state produces zero, or

$$\hat{\psi}(\mathbf{x}, t) | 0 \rangle = 0. \quad (4.26)$$

As a result, the Green's function must consist only of the retarded time ordering, where the creation operator is to the right, and occurs earlier, and

$$G(\mathbf{x}, \mathbf{x}'; t, t') = -i\theta(t - t')\langle \psi(\mathbf{x}, t)\psi^\dagger(\mathbf{x}', t') \rangle, \quad (4.27)$$

where the averaging is over the ground state, and the proper normalization must be incorporated. At this point, it is useful to introduce the ground-state Green's function where there is no perturbation and the normalization is always maintained without any effort (no denominator). We may write this ground state in general as

$$G^0(\mathbf{x}, \mathbf{x}'; t, t') = -i\langle 0 | T\hat{\psi}(\mathbf{x}, t)\hat{\psi}^\dagger(\mathbf{x}', t') | 0 \rangle, \quad (4.28)$$

and in the empty band this becomes

$$G^0(\mathbf{x}, \mathbf{x}'; t, t') = -i\theta(t - t')e^{iE(t-t')/\hbar} \langle 0 | \hat{\psi}(\mathbf{x}, t)\hat{\psi}^\dagger(\mathbf{x}', t') | 0 \rangle \quad (4.29)$$
$$= -i\theta(t - t')e^{iE(t-t')/\hbar}.$$

In the last term, the expectation value is evaluated by using the commutator relationship, so that the operator term obeys Eq. (4.26) and the delta function gives unity under the integral over the ground-state wavefunctions. This last expression can be Fourier transformed in the time difference to give

$$G^0(\mathbf{k}, \omega) = \frac{\hbar}{\hbar\omega - E(k) + i\eta}, \quad (4.30)$$

which, of course, is just the retarded ground–state Green's function (3.21).

Now let us consider the degenerate electron gas in which all states up to the Fermi energy are full and all states above this energy are empty. This is sometimes referred to as the Fermi sea. Thus, in Fourier space, we can write the electron operator, using the Fermi–Dirac distribution as

$$\langle 0|\widehat{\psi}^{\dagger}(\mathbf{k})\widehat{\psi}(\mathbf{k})|0\rangle = \theta(k_F - k) = \theta(\xi_k)$$

$$= \lim_{\beta \to \infty} \frac{1}{e^{\beta\xi_k}+1} = n_F(\xi_k), \tag{4.31}$$

where $\xi_k = E(k) - \mu$, $\mu = E_F$, and $\beta = 1/k_BT$ and the temperature has been set to zero. Here, n_F is the number occupancy in each state, so is just 0 or 1, the latter occurring if the state lies below, or at, the Fermi energy. Similarly, for holes, we have the relationship

$$\langle 0|\widehat{\psi}(\mathbf{k})\widehat{\psi}^{\dagger}(\mathbf{k})|0\rangle = \theta(k - k_F). \tag{4.32}$$

The degenerate Fermi gas is homogeneous in position, so it is best to work with the Fourier space, as the Green's function will be a function of the difference in the two positions, which is easily Fourier transformed into momentum. With these quantities, we can write the ground-state Green's function as

$$G^0(\mathbf{k};t,t') = \langle 0|T\psi(\mathbf{k},t)\psi^{\dagger}(\mathbf{k},t')|0\rangle$$

$$= \left[-i\theta(t-t')\theta(\xi_k) + i\theta(t'-t)\theta(-\xi_k)\right]e^{-i\xi_k(t-t')/\hbar}. \tag{4.33}$$

Fourier transforming the time difference variable now leads to the Green's functions we had in Eqs. (3.21) and (3.22), but with the existence of the Fermi energy as

$$G^0(\mathbf{k},\omega) = \frac{\hbar\theta(\xi_k)}{\hbar\omega - \xi_k + i\eta} + \frac{\hbar\theta(-\xi_k)}{\hbar\omega - \xi_k - i\eta}. \tag{4.34}$$

The two terms represent the retarded and advanced Green's functions, respectively, and we can form the spectral density from these as

$$A(\mathbf{k},\omega) = -\left[G_r^0(\mathbf{k},\omega) - G_a^0(\mathbf{k},\omega)\right] = -2\text{Im}\{G_r^0(\mathbf{k},\omega)\}$$

$$= \frac{\hbar\eta\left[\theta(\xi_k) + \theta(-\xi_k)\right]}{(\hbar\omega - \xi_k)^2 + \eta^2} = \frac{\hbar\eta}{(\hbar\omega - \xi_k)^2 + \eta^2}. \tag{4.35}$$

We can check the proper normalization of the spectral density by integrating over the frequency as

$$\int_{-\infty}^{\infty} \frac{d\omega}{2\pi} A(\mathbf{k}, \omega) = \frac{1}{\pi} \int_{-\infty}^{\infty} d\omega \frac{\hbar\eta}{(\hbar\omega - \xi_k)^2 + \eta^2} = \frac{\eta}{\pi} \int_{-\infty}^{\infty} \frac{dx}{x^2 + \eta^2}$$

$$= \frac{1}{\pi} \int_{-\infty}^{\infty} \frac{du}{u^2 + 1} = \frac{1}{\pi} \tan(u)\Big|_{-\infty}^{\infty} = 1.$$

(4.36)

This important relationship between the spectral density and the retarded (or advanced) Green's function, along with the conservation of the spectral density, will both remain valid in the presence of interactions. Indeed, even as Green's function evolves from the ground state, these relationships will hold. But we remember that we are working with small deviations from the true ground state.

4.2 Electron–Electron Interaction Again

We now want to return to the electron–electron interaction to see how we go beyond the ground state (non-interacting) situation. If we use Eqs. (4.16) and (4.25), we can write the Green's function with the presence of the scattering operator as

$$G(\mathbf{k}, t - t') = \sum_{n=0}^{\infty} \frac{1}{n!} \left(\frac{-1}{\hbar} \right)^n \int dt_1 \ldots \int dt_n$$

$$\times \frac{\langle 0 | T \hat{\psi}(\mathbf{k}, t) V(t_1) \ldots V(t_n) \hat{\psi}^\dagger(\mathbf{k}, t') | 0 \rangle}{\langle 0 | S(\infty, -\infty) | 0 \rangle}.$$

(4.37)

All the integrals in this last equation run from t' to t, and we now have to include the denominator to assure that we retain the normalization that is required. The first term of the series ($n = 0$) just gives the ground-state Green's function, so it is all the other terms that lead to the interacting Green's function. We can now see how this creates the perturbation series and the need that exists to assure ourselves that this series actually converges. Hence, we normally have to assume that the perturbing potential V is actually smaller than the relevant energies in the problem. The next term in the series creates the first perturbation term, with

$$V(t_1) = \frac{e^2}{2\varepsilon_s} \sum_{\mathbf{k}'',\mathbf{k}',\mathbf{q}} \frac{1}{q^2} \hat{\psi}^\dagger(\mathbf{k}''+\mathbf{q},t_1)\hat{\psi}^\dagger(\mathbf{k}'-\mathbf{q},t_1)\psi(\mathbf{k}',t_1)\psi(\mathbf{k}'',t_1),$$

$$(4.38)$$

and, with the two operators in the expectation term of Eq. (4.37), this term in the expansion now has six wavefunction operators. The problem of this term boils down to evaluating the expectation value

$$\langle 0|T\hat{\psi}(\mathbf{k},t)\hat{\psi}^\dagger(\mathbf{k}''+\mathbf{q},t_1)\hat{\psi}^\dagger(\mathbf{k}'-\mathbf{q},t_1)\psi(\mathbf{k}',t_1)\psi(\mathbf{k}'',t_1)\hat{\psi}^\dagger(\mathbf{k},t')|0\rangle.$$

$$(4.39)$$

Even this first term, with six wavefunction operators, can be really messy, and one wonders just how to approach this task. The answer lies with Wick's theorem [2, 3].

Wick showed how to simplify an earlier evaluation due to Dyson in a manner that allowed any type of field to be included. One important step along the way is to recognize that

$$\langle 0|T\hat{\psi}(\mathbf{k}',t_1)\hat{\psi}^\dagger(\mathbf{k},t')|0\rangle \sim \delta(\mathbf{k}'-\mathbf{k}),$$

$$(4.40)$$

so that any pair of operators in this form will introduce a conservation of momentum expression. We can use this to match one creation and one annihilation operator, until we have used all six operators. The expression is then reduced to three pairings. However, we have to include all possible pairings, which goes as 3! or six total terms. Hence, we can rewrite Eq. (4.39) as

$$\begin{aligned}
(4.39) = &\langle 0|T\hat{\psi}^\dagger(\mathbf{k}''+\mathbf{q},t_1)\hat{\psi}(\mathbf{k},t)\hat{\psi}^\dagger(\mathbf{k}'-\mathbf{q},t_1)\psi(\mathbf{k}',t_1)\hat{\psi}^\dagger(\mathbf{k},t')\psi(\mathbf{k}'',t_1)|0\rangle\\
&-\langle 0|T\hat{\psi}^\dagger(\mathbf{k}''+\mathbf{q},t_1)\hat{\psi}(\mathbf{k},t)\hat{\psi}^\dagger(\mathbf{k}'-\mathbf{q},t_1)\psi(\mathbf{k}'',t_1)\hat{\psi}^\dagger(\mathbf{k},t')\psi(\mathbf{k}',t_1)|0\rangle\\
&-\langle 0|T\hat{\psi}^\dagger(\mathbf{k}''+\mathbf{q},t_1)\psi(\mathbf{k}'',t_1)\hat{\psi}^\dagger(\mathbf{k}'-\mathbf{q},t_1)\psi(\mathbf{k}',t_1)\hat{\psi}^\dagger(\mathbf{k},t')\hat{\psi}(\mathbf{k},t)|0\rangle\\
&-\langle 0|T\hat{\psi}^\dagger(\mathbf{k}''+\mathbf{q},t_1)\psi(\mathbf{k}',t_1)\hat{\psi}^\dagger(\mathbf{k}'-\mathbf{q},t_1)\hat{\psi}(\mathbf{k},t)\hat{\psi}^\dagger(\mathbf{k},t')\psi(\mathbf{k}'',t_1)|0\rangle\\
&+\langle 0|T\hat{\psi}^\dagger(\mathbf{k}''+\mathbf{q},t_1)\psi(\mathbf{k}'',t_1)\hat{\psi}^\dagger(\mathbf{k}'-\mathbf{q},t_1)\hat{\psi}(\mathbf{k},t)\hat{\psi}^\dagger(\mathbf{k},t')\psi(\mathbf{k}',t_1)|0\rangle\\
&+\langle 0|T\hat{\psi}^\dagger(\mathbf{k}''+\mathbf{q},t_1)\psi(\mathbf{k}',t_1)\hat{\psi}^\dagger(\mathbf{k}'-\mathbf{q},t_1)\psi(\mathbf{k}'',t_1)\hat{\psi}^\dagger(\mathbf{k},t')\hat{\psi}(\mathbf{k},t)|0\rangle
\end{aligned}$$

$$(4.41)$$

The various signs correspond to the exchanges of terms that must follow the anti-commutator relationships. Each of these six pairings is now separated into three separate expectation values as was done for the Hartree approximation in the last chapter. However, the Green's functions have the order of the operators reversed with the creation operator to the right, as it has to work on the empty ground state. And we need to add the ubiquitous factor of $(-i)$ to each of these Green's functions. The three order reversals give us a minus sign, and the latter factors lead to an overall factor of i, so that we can write the terms as

$$
\begin{aligned}
(4.39) = &-iG^0(\mathbf{k}, t - t_1)\delta_{\mathbf{k}, \mathbf{k}'' + \mathbf{q}}G^0(\mathbf{k}', 0)\delta_{\mathbf{k}', \mathbf{k}' - \mathbf{q}}G^0(\mathbf{k}, t_1 - t')\delta_{\mathbf{k}, \mathbf{k}''} \\
&+iG^0(\mathbf{k}, t - t_1)\delta_{\mathbf{k}, \mathbf{k}'' + \mathbf{q}}G^0(\mathbf{k}'', 0)\delta_{\mathbf{k}'', \mathbf{k}' - \mathbf{q}}G^0(\mathbf{k}, t_1 - t')\delta_{\mathbf{k}, \mathbf{k}'} \\
&+iG^0(\mathbf{k}'', 0)\delta_{\mathbf{k}'', \mathbf{k}'' + \mathbf{q}}G^0(\mathbf{k}', 0)\delta_{\mathbf{k}', \mathbf{k}' - \mathbf{q}}G^0(\mathbf{k}, t - t') \\
&+iG^0(\mathbf{k}', 0\delta_{\mathbf{k}', \mathbf{k}'' + \mathbf{q}}G^0(\mathbf{k}, t - t_1)\delta_{\mathbf{k}, \mathbf{k}' - \mathbf{q}}G^0(\mathbf{k}, t_1 - t')\delta_{\mathbf{k}, \mathbf{k}''} \\
&-iG^0(\mathbf{k}'', 0)\delta_{\mathbf{k}'', \mathbf{k}'' + \mathbf{q}}G^0(\mathbf{k}, t - t_1)\delta_{\mathbf{k}, \mathbf{k}' - \mathbf{q}}G^0(\mathbf{k}, t_1 - t')\delta_{\mathbf{k}, \mathbf{k}'} \\
&-iG^0(\mathbf{k}', 0)\delta_{\mathbf{k}', \mathbf{k}'' + \mathbf{q}}G^0(\mathbf{k}'', 0)\delta_{\mathbf{k}'', \mathbf{k}' - \mathbf{q}}G^0(\mathbf{k}, t - t').
\end{aligned}
$$

$$(4.42)$$

As with the approach of the last chapter, it is visually much easier to understand these terms pictorially with the Feynman diagrams. These six terms are shown in Fig. 4.1, in the same order in which they are written in Eq. (4.42). (We should remark that time progresses along the horizontal axis in these figures.) Since the momentum wavevector on the isolated loop in the first and fifth terms is a dummy variable, these two terms are the same and so the first term is multiplied by a factor of 2. Similarly, in the second and fourth terms, the value of q can be easily reversed so that these two terms are the same. The first term is just the Hartree term from the last chapter, while the second term is the exchange interaction for which we did not diagram in that discussion. It is the third and sixth terms that are new and could give us some trouble. These two terms are called disconnected diagrams, since they are separated into two parts that have no real physical connection.

Before continuing, let us look at the normalization term in the denominator of Eq. (4.37). Since this term does not have the extra pair of wavefunction operators that appear in the numerator, the first correction term now has only four wavefunction operators, as

$$\langle 0|T\hat{\psi}^\dagger(\mathbf{k}''+\mathbf{q},t_1)\hat{\psi}^\dagger(\mathbf{k}'-\mathbf{q},t_1)\psi(\mathbf{k}',t_1)\psi(\mathbf{k}'',t_1)|0\rangle. \quad (4.43)$$

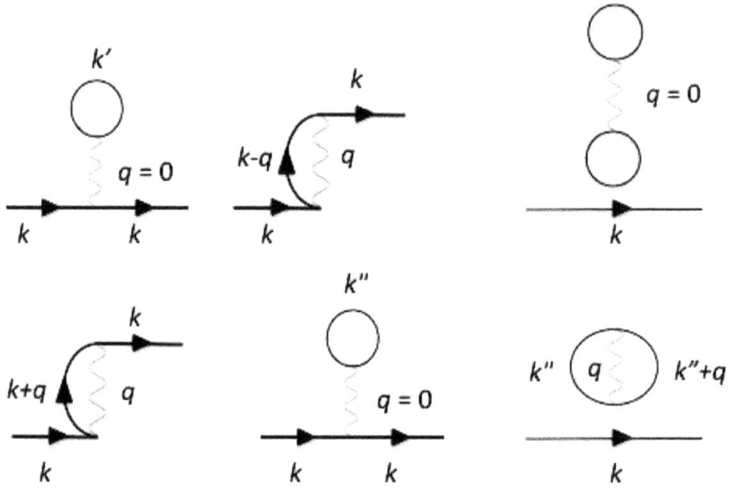

Figure 4.1 The six diagrams that arise from Eq. (4.42). The diagrams, across the top left to right and then across the bottom left to right, correspond to the six combinations of Green's functions in the equation.

We can proceed exactly in the same manner as we did for the numerator term. However, now there are only two options for the pairing of the operators, so we will have only two terms, and the various reversal signs and the complex terms all combine for a single minus sign change. Hence, we can write this term as the Green's functions

$$G^0(\mathbf{k}',0)\delta_{\mathbf{k}',\mathbf{k}''+\mathbf{q}}G^0(\mathbf{k}'',0)\delta_{\mathbf{k}'',\mathbf{k}'-\mathbf{q}}$$
$$-G^0(\mathbf{k}',0)\delta_{\mathbf{k}',\mathbf{k}'+\mathbf{q}}G^0(\mathbf{k}'',0)\delta_{\mathbf{k}'',\mathbf{k}''-\mathbf{q}}. \quad (4.44)$$

If we take the numerator term and keep the zero-order term (1) plus just the third and sixth diagrams from Fig. 4.1, it gives exactly the same two terms in the series that we get from the zero and first terms of the denominator. That is, the Green's function expression in Eq. (4.44) is exactly the same two Feynman diagrams as the two bubbles in the third and sixth terms of Fig. 4.1. This is an important result, because it tells us that the normalization terms of the denominator will cancel the disconnected diagram at the first order. In fact, we could show with a great deal of effort that the normalization term in the denominator will cancel the disconnected diagrams to all orders of the perturbation series. Hence, the first-order interaction that remains, after normalization, is just the Hartree and exchange diagrams, with a factor of two multiplier. In addition, there is a factor of $-i/\hbar$ that arises from the pre-factors in Eq. (4.37). Hence, we can write the interacting Green's function, at first order, as

$$G(\mathbf{k}, t-t') = G^0(\mathbf{k}, t-t') - \frac{1}{\hbar}\Lambda(\mathbf{k}, t-t') + \dots ,\tag{4.45}$$

where Λ is shown diagrammatically in Fig. 4.2.

Each of the Green's functions that appear in the Feynman diagrams depends only on the difference of the two times, so that we easily Fourier transform to frequency as well. Hence, we can now write Eq. (4.45) as

$$G(\mathbf{k}, \omega) = G^0(\mathbf{k}, \omega)\left\{ 1 - \frac{1}{\hbar}\sum_{\mathbf{k}'}\lim_{q\to 0}\frac{e^2}{\varepsilon_s q^2}G^0(\mathbf{k}, \omega)G^0(\mathbf{k}', 0) \right.$$

$$\left. + \frac{1}{\hbar}\sum_{\mathbf{q}}\frac{e^2}{\varepsilon_s q^2}G^0(\mathbf{k}, \omega)G^0(\mathbf{k}+\mathbf{q}, 0)\right\}.\tag{4.46}$$

Note that, in this Fourier transformed form, the bare Green's function factors out of each term, even though a remaining term still sits in each term of the series.

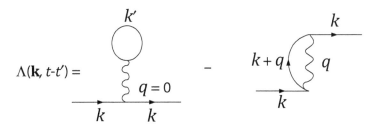

Figure 4.2 The diagrammatic expression of the first-order interaction term for the interacting Green's function. The two terms are the Hartree term and the exchange term, from left to right.

The second term in the expansion (4.37) contains 10 wavefunction operators. Even taking out the common diagrams that appear multiple times and removing the disconnected diagrams, there remain 10 diagrams that should be considered. We will not go through the mathematical details, but the diagrams are important to us, because they lead us through the full expansion of the interacting Green's function. These diagrams appear in Fig. 4.3. Starting with the top row, the first two terms are simply the two diagrams for the Hartree and exchange interactions repeated to second order. The third term and the fourth (second row at left) are the cross product of these two terms, in the opposite orders. The last two terms in the second and third rows are complicated higher-order terms that we will tend to ignore. The fact is these are very difficult terms to evaluate and are certainly beyond the simple approach we want to take in the present treatment. By ignoring them, we are making an approximation that the terms we are keeping are in fact the important terms. Of course, we do not know this for sure, without going to the trouble of evaluating the "difficult" terms to see if they are important, but since no one really keeps them, our approximation may be good enough. However, the little triangle of two Green's functions and the interaction line that sits at the center bottom of the last diagram on the second row will reappear later as a *vertex correction*. In essence, it corresponds to the Coulomb potential that appears in the first-order Hartree term being corrected due to screening. Now, the first two terms on the third row are corrections to the Green's function that appears in the bubble of the first-order Hartree term. So if we were to employ the full interacting Green's function for the bubble, and the screened Coulomb interaction, in the first-order Hartree term, we would have three of these other diagrams included in the calculation. We will come back to this discussion later.

Now we can write out the corrections to the bare Green's function that lead to the full interacting Green's function diagrammatically and we do this in Fig. 4.4. Here, we have kept the three terms on the first row and the first term from the second row of Fig. 4.3 and then rearranged them. In addition, we have factored the last Green's function from each row for the first-order diagrams to include within the two brackets. What appears in each bracket is the leading terms for the interacting Green's function itself, which means that this last Green's function can be replaced in the normal first-order Hartree terms by the full interacting Green's function, which we denote by the double line. This is depicted in Fig. 4.5. In addition, the interaction line and Green's function go together to create the self-energy Σ^0, as we have indicated in the second and third lines of Fig. 4.5. This greatly simplifies the overall diagrammatic picture that we have developed, as we can recollect a great number of the terms into the full interacting Green's function, which is a great savings in the pictorial view.

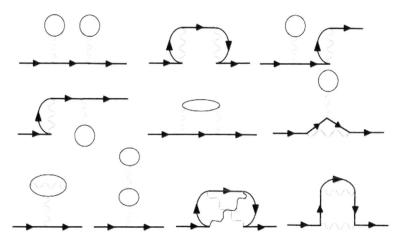

Figure 4.3 The 10 distinct diagrams that arise in the second-order electron–electron interaction. These diagrams are discussed in the text.

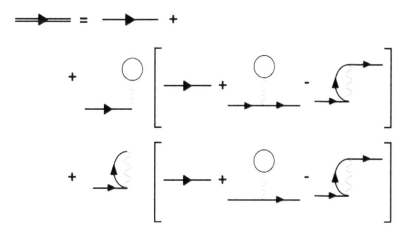

Figure 4.4 The expansion for the interacting Green's function in which we have used the first four diagrams from Fig. 4.3 to show how the first-order Hartree and exchange terms are modified to include the full Green's function.

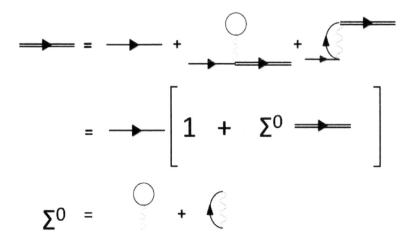

Figure 4.5 Reassembling the interacting Green's function in terms of itself in the expansion and introducing the bare self-energy, defined in the last line.

Now we remarked in connection with Fig. 4.3 that the first two terms on the third row modified the Green's function in the bubble (and the interaction Green's function in the exchange term). These are just the first corrections, and more will appear if we go to higher orders in the interaction. But these are sufficient to tell us that there are additional terms that will modify the Green's functions in the self-energy. These modifications are such that the bare self-energy Σ^0 will evolve into the full interacting Σ. Hence, we can now write the diagrammatic equation in the second line of Fig. 4.5 as

$$G = G^0[1 + \Sigma G] = G^0 + G^0 \Sigma G . \tag{4.47}$$

This last equation is known as Dyson's equation and represents a resummation of the diagrams and perturbation theory. This is more easily expressed by taking the last term to the left-hand side and rewriting everything as

$$G(\mathbf{k}, \omega) = \frac{G^0(\mathbf{k}, \omega)}{1 - G^0(\mathbf{k}, \omega)\Sigma(\mathbf{k}, \omega)} = \frac{1}{[G^0(\mathbf{k}, \omega)]^{-1} - \Sigma(\mathbf{k}, \omega)}, \tag{4.48}$$

which is the more commonly recognized form of Dyson's equation. One advantage of working in the Fourier-transformed world is that the products of all these diagrams are simple products. In real space, we would have many convolution integrals involved in evaluating the various terms.

4.3   Dielectric Function

The idea of a screening process around a Coulomb potential was mentioned above, and we want to now investigate this further. Let us consider a spin-independent Coulomb potential such as might arise from the electron–electron interaction (or later we will encounter it again with charged impurities), which is given as $V(\mathbf{x}, \mathbf{x}') = V(\mathbf{x} - \mathbf{x}')$. In the semi-classical approach used in semiconductor physics, this potential produces a fluctuation in the electron density, which we saw as the exchange hole around the electron in the last chapter. The density fluctuation in turn leads to a modification of the potential itself, as one must examine the closed response of the potential and the electrons in a self-consistent manner. One normally sees the potential written with an exponential screening term described either by the Debye length or the Fermi–Thomas screening length. In this section, we would like to examine just how this arises and will discuss it with our Green's functions to study the fluctuations and their correlations. The interaction potential was given in Eq. (4.38), but let us rewrite it in position space as

$$
\begin{aligned}
\left\langle \hat{V} \right\rangle &= \frac{1}{2} \int d^3\mathbf{x} \int d^3\mathbf{x}' V(\mathbf{x}-\mathbf{x}') \left\langle \hat{\psi}^\dagger(\mathbf{x})\hat{\psi}^\dagger(\mathbf{x}')\hat{\psi}(\mathbf{x}')\hat{\psi}(\mathbf{x}) \right\rangle \\
&= \frac{1}{2} \int d^3\mathbf{x} \int d^3\mathbf{x}' V(\mathbf{x}-\mathbf{x}') \left[\left\langle \hat{n}(\mathbf{x})n(\mathbf{x}') \right\rangle - \delta(\mathbf{x}-\mathbf{x})\left\langle \hat{n}(\mathbf{x}) \right\rangle \right],
\end{aligned}
\tag{4.49}
$$

where the term in the square brackets defines the correlation function of the density fluctuations. The latter are defined by

$$
\delta n(\mathbf{x}) = \hat{n}(\mathbf{x}) - \left\langle \hat{n}(\mathbf{x}) \right\rangle.
\tag{4.50}
$$

With this definition, we can define the time-ordered correlation function

$$
iD(\mathbf{x}, \mathbf{x}'; t, t) = \frac{\left\langle \; |\delta n(\mathbf{x}, t)\delta n(\mathbf{x}', t)| \; \right\rangle}{\left\langle 0|S(\infty, -\infty)|0 \right\rangle},
\tag{4.51}
$$

where once again we truly do not quite know the state of the system at time t. On the other hand, we can define the non-interacting correlation function from the true ground state as

$$
iD^0(\mathbf{x}, \mathbf{x}'; t, t) = \left\langle 0|\delta n(\mathbf{x})\delta n(\mathbf{x}')|0 \right\rangle.
\tag{4.52}
$$

Now it can be shown that this is the term we really need to deal with and the energy can be shown to be

$$E = \langle 0|H_0 + \hat{V}|0 \rangle + E_{corr}, \tag{4.53}$$

where the last term is the correlation energy and involves the difference between D and D^0. Generally, it is known that this term is much smaller than the normal terms arising from the Hartree and exchange interactions, so that we need only deal with the ground-state evaluation of the correlation function.

In general, we know that $(\delta n)^2 = \langle n^2 \rangle - \langle n \rangle^2$, so that we can write the density correlation function for the ground-state average as

$$iD^0(\mathbf{x}, \mathbf{x}'; t) = \langle 0|T\hat{\psi}^\dagger(\mathbf{x}, t)\hat{\psi}(\mathbf{x}, t)\hat{\psi}^\dagger(\mathbf{x}', t)\hat{\psi}(\mathbf{x}', t)|0 \rangle$$

$$- \langle 0|\hat{\psi}^\dagger(\mathbf{x}, t)\hat{\psi}(\mathbf{x}, t)|0 \rangle \langle 0|\hat{\psi}^\dagger(\mathbf{x}', t)\hat{\psi}(\mathbf{x}', t)|0 \rangle \tag{4.54}$$

$$= 2G^0(\mathbf{x}, \mathbf{x}'; t, t)G^0(\mathbf{x}', \mathbf{x}; t, t)$$

$$\equiv i\hbar\Pi^0(\mathbf{x}, \mathbf{x}'; t).$$

Now there is a lot going on in this equation. First, the leading two-particle Green's function gives rise to the product of two Green's functions, each with its own spin, minus another pair of Green's functions in which the two wavefunctions may have different spins. This last pair of Green's functions cancels with the last term, leaving the two Green's functions shown on line 2. Second, the factor of 2 comes from summing over the allowed spins for the two Green's functions. Finally, the last line introduces the bare *polarization* function, or bare polarization bubble term. This bubble is composed of the two Green's function lines written in terms of the bare Green's functions. If we incorporate higher-order terms, by which the bare Green's functions become the full interacting Green's functions, then we arrive at the full polarization function. In Fourier transform space, Eq. (4.54) can be written as

$$D^0(\mathbf{q}, \omega) = -iG^0(\mathbf{k}, \omega)G^0(\mathbf{k} + \mathbf{q}, \omega) = \hbar\Pi^0(\mathbf{q}, \omega). \tag{4.55}$$

A word of caution is needed here. The correlation function in Eq. (4.54) is the product of two equal time Green's functions *at the same instance of time*. In fact, we require a convolution integral in frequency in Eq. (4.55), which we will actually perform just below. But for the moment, we leave them in the present form in order to formulate the resummation in the next few lines. The inherent lack of a dependence of either D or Π on the particle momentum wavevector tells us that the polarization formula includes an integration over this momentum wavevector. In this transformed space, we can now write down the general interaction for the total effective potential as

$$V(\mathbf{q}, \omega) = V^0(\mathbf{q}, \omega) + V^0(\mathbf{q}, \omega)\Pi(\mathbf{q}, \omega)V^0(\mathbf{q}, \omega) + \dots$$
$$= [1 + V^0(\mathbf{q}, \omega)\Pi(\mathbf{q}, \omega) + \dots]V^0(\mathbf{q}, \omega) \quad\quad (4.56)$$
$$= \frac{V^0(\mathbf{q}, \omega)}{1 - V^0(\mathbf{q}, \omega)\Pi(\mathbf{q}, \omega)},$$

where V^0 is the bare Coulomb potential. The last line is recognized as the Lindhard dielectric function as the denominator is the effective relative dielectric constant

$$\varepsilon_s(\mathbf{q}, \omega) = \varepsilon_0[1 - V^0(\mathbf{q})\Pi(\mathbf{q}, \omega)], \quad\quad (4.57)$$

where we have recognized that the bare Coulomb potential is a static potential, and the free space permittivity comes from this potential. In Fig. 4.6, we show the polarization bubble in both real space and in the Fourier-transformed form.

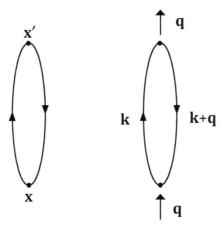

Figure 4.6 The diagrams for the polarization bubble in real space (left) and in Fourier transform space (right). The dots have been added to indicate the vertices where other Green's functions might be attached. The integration over x′ in the real space formulation goes into an integration over **k** in the momentum space formulation.

To proceed to the normal dielectric function, we need to perform the convolution integral mentioned above [2]. First, we write the full form of the polarization as

$$\Pi(\mathbf{q},\omega)=\frac{-2i}{\hbar}\int\frac{d^3k}{(2\pi)^3}\int\frac{d\omega'}{2\pi}\frac{\hbar}{\hbar\omega'-E(\mathbf{k})\pm i\eta}\frac{\hbar}{\hbar(\omega'-\omega)-E(\mathbf{k}+\mathbf{q})\pm i\eta},$$
(4.58)

where the factor of 2 comes from the sum over spins. The plus/minus signs reflect that we do not really know whether to use the retarded or advanced Green's functions. Hence, we have to consider all possibilities. If we use either retarded only or advanced only, the two poles lie on the same side of the real axis for ω' and the contour can be closed in the opposite half plane, yielding zero. Hence, we need to use one retarded function and one advanced function, which is reflected in Fig. 4.6 by the arrows flowing in opposite directions. Then, we get two integrals, one of which flows from each pole (which are in opposite half-planes and the integral has to close in opposite planes, which will lead to a minus sign between the two terms. Then, the residue theorem leads to the result

$$\Pi(\mathbf{q})=\frac{2}{\hbar}\int\frac{d^3k}{(2\pi)^3}\left[\frac{\hbar\theta(|\mathbf{k}+\mathbf{q}|-k_F)\theta(k_F-k)}{\hbar\omega+E(\mathbf{k})-E(\mathbf{k}+\mathbf{q})+i2\eta}\right.$$
$$\left.-\frac{\hbar\theta(k_F-|\mathbf{k}+\mathbf{q}|)\theta(k-k_F)}{\hbar\omega+E(\mathbf{k}+\mathbf{q})-E(\mathbf{k})-i2\eta}\right],$$
(4.59)

where the two θ functions remind us that the two energies must lie on opposite sides of the Fermi energy (for a degenerate electron gas). We can replace the θ functions with the densities of the individual states. This leads to

$$\Pi(\mathbf{q},\omega)=2\int\frac{d^3k}{(2\pi)^3}\left[\frac{n_\mathbf{k}-n_{\mathbf{k}+\mathbf{q}}}{-\hbar\omega+E(\mathbf{k})-E(\mathbf{k}+\mathbf{q})+i\eta}\right]$$
$$=-2\int\frac{d^3k}{(2\pi)^3}\left[\frac{f(\mathbf{k})-f(\mathbf{k}+\mathbf{q})}{E(\mathbf{k}+\mathbf{q})-E(\mathbf{k})+\hbar\omega-i\eta}\right],$$
(4.60)

where we have also introduced the distribution functions themselves into the last equation. This last form is that commonly found in discussions of the Lindhard dielectric function. Let us now consider some standard approximations to this function.

4.3.1 Optical Dielectric Constant

In this approximation, we consider optical excitations in which the photon energy corresponds to the average energies of the conduction and valence bands. These energies are typically near the centers of the two bands, so we are talking about photon energies somewhat above 10 eV. Thus, we can make an approximation that

$$E(\mathbf{k}+\mathbf{q})-E(\mathbf{k}) \sim E_{\text{Gav}} \gg \hbar\omega, \tag{4.61}$$

which means we are looking at probing the system with photons smaller than this average energy gap (photons with energy << 10 eV). To proceed, we change the momentum wavevector in the second term to $\mathbf{k} \to \mathbf{k} - \mathbf{q}$, which allows us to factor out the distribution function. The two energy terms are then combined to give, incorporating the additional factors from Eq. (4.57),

$$\varepsilon_s = \varepsilon_0 + \frac{e^2}{q^2}\int\frac{d^3\mathbf{k}}{4\pi^3}f(\mathbf{k})\frac{2E(\mathbf{k})-E(\mathbf{k}+\mathbf{q})-E(\mathbf{k}-\mathbf{q})}{[E(\mathbf{k}+\mathbf{q})-E(\mathbf{k})+\hbar\omega-i\eta][E(\mathbf{k})-E(\mathbf{k}-\mathbf{q})+\hbar\omega-i\eta]}. \tag{4.62}$$

The denominator terms just become, according to Eq. (4.61), $-(E_{\text{Gav}})^2$, since the first transition is downward, while the second is upward. We can write the numerator as

$$2E(\mathbf{k})-E(\mathbf{k}+\mathbf{q})-E(\mathbf{k}-\mathbf{q}) \sim -q^2\frac{\partial^2 E}{\partial k^2} = -\frac{\hbar^2 q^2}{m_0}, \tag{4.63}$$

which uses the free electron mass, as the effective mass approximation is not valid for such large excitations. There is no k dependence left in the integral, other than the distribution function, so this integral just becomes the number of bonding electrons, or the total number of carriers in the valence band per unit volume. Then, these terms can be rewritten as

$$\varepsilon_s = \varepsilon_0\left[1+\left(\frac{\hbar\omega_P}{E_{\text{Gav}}}\right)^2\right], \tag{4.64}$$

where

$$\omega_P^2 = \frac{n_{\text{val}}e^2}{m_0\varepsilon_0} \tag{4.65}$$

is the square of the valence plasma frequency. In Si, this plasma frequency corresponds to an energy of 16.6 eV, while in GaAs, it is 15.6 eV [4]. Equation (4.64) is often called the Penn dielectric function. We must remember that this is the optical dielectric constant. Materials such as GaAs, which lacks an inversion frequency, will contain a polarization contribution from the effective charges on the individual Ga and As atoms, and this polarization will contribute to the low-frequency dielectric constant. In most cases, the transition between the high-frequency dielectric constant (4.64) and the low-frequency value occurs in the far infrared, typically near a photon energy of 30 meV or so, for semiconductors [4]. This is reflected in the fact that the zone center TO phonon lies below the LO phonon in energy. The TO mode can couple to the photons, and this produces a resonance behavior in the dielectric function, which is referred to as the *reststrahlen* region. Below this resonance, the low-frequency dielectric constant is valid.

4.3.2 Plasmon–Pole Approximation

In this approximation, we are looking at high (optical) frequency effects, where the photon energy is still larger than either of the energies in the denominator terms (and smaller than their differences). Thus, the denominator in Eq. (4.62) becomes $(\hbar\omega)^2$ and the numerator is still evaluated with Eq. (4.63), although the sum is only over the free carriers and uses the effective mass. Thus, we have

$$\varepsilon(\omega) = \varepsilon_s + \frac{e^2}{q^2} \int \frac{d^3\mathbf{k}}{4\pi^3} f_{fc}(\mathbf{k}) \left(-\frac{\hbar^2 q^2}{m^*} \right) \frac{1}{\hbar^2 \omega^2} = \varepsilon_s \left[1 - \frac{\omega_p^2}{\omega^2} \right], \quad (4.66)$$

where

$$\omega_p^2 = \frac{ne^2}{m^* \varepsilon_s} \quad (4.67)$$

defines the free carrier plasma frequency and n is the free carrier density.

4.3.3 Static Screening

For static screening, we take the small-momentum, low-frequency limit of Eq. (4.60). In this approach, we ignore the frequency and the convergence factor as both being small. Then the numerator and the denominator are expanded as

$$f(\mathbf{k}) - f(\mathbf{k} + \mathbf{q}) \sim -\mathbf{q} \cdot \frac{\partial f}{\partial \mathbf{k}} = -\left(\mathbf{q} \cdot \frac{\partial E}{\partial \mathbf{k}}\right) \frac{\partial f}{\partial E} \tag{4.68}$$

and

$$E(\mathbf{k} + \mathbf{q}) - E(\mathbf{k}) \sim \mathbf{q} \cdot \frac{\partial E}{\partial \mathbf{k}}. \tag{4.69}$$

Now we are left with the integral, which can be evaluated for a Maxwell–Boltzmann distribution as

$$2\sum_{\mathbf{k}} \frac{d^3 \mathbf{k}}{(2\pi)^3} \left(-\frac{\partial f}{\partial E}\right) \sim 2\beta \sum_{\mathbf{k}} \frac{d^3 \mathbf{k}}{(2\pi)^3} f(\mathbf{k}) = \frac{n}{k_B T}. \tag{4.70}$$

The dielectric function then becomes

$$\varepsilon(0) = \varepsilon_s \left(1 + \frac{q_D^2}{q^2}\right), \tag{4.71}$$

where

$$q_D^2 = \frac{n e^2}{\varepsilon_s k_B T} \tag{4.72}$$

is the square of the Debye–Thomas screening wavevector (inverse of the screening length). This is for the non-degenerate case, where the approximation in Eq. (4.70) is valid.

For the degenerate case, we can go a little deeper into the problem and consider the case for a finite momentum transfer in the screening. The low-frequency approximation will be retained here, but the desire is to evaluate carefully the summation over the free carriers that is involved in the summation (4.60), and to remove the approximation $q \ll k$. Thus, the dielectric function may be rewritten as

$$\varepsilon(\mathbf{q}, 0) = \varepsilon_s + \frac{2e^2}{q^2} \int \frac{d^3 \mathbf{k}}{(2\pi)^3} f(\mathbf{k}) \left[\frac{1}{E(\mathbf{k} + \mathbf{q}) - E(\mathbf{k})} - \frac{1}{E(\mathbf{k}) - E(\mathbf{k} - \mathbf{q})}\right]. \tag{4.73}$$

We can expand the two energy denominators as

$$E(\mathbf{k} \pm \mathbf{q}) - E(\mathbf{k}) = \frac{\hbar^2 q^2}{2m^*} \pm \frac{\hbar^2 k q}{m^*} \cos \vartheta, \tag{4.74}$$

where the mass is the effective mass in the semiconductor of interest. The integration will be carried out for three dimensions but is readily extendible to lower dimensionality. The integration over the azimuthal angle is straightforward, and this leaves us with

$$
\varepsilon(q) = \varepsilon_s + \frac{m^* e^2}{q^2 \hbar^2 \pi^2} \int k^2 dk \sin \vartheta d\vartheta f(k)
$$

$$
\times \left[\frac{1}{q^2 + 2kq\cos\vartheta} - \frac{1}{q^2 - 2kq\cos\vartheta} \right] \tag{4.75}
$$

$$
= \varepsilon_s + \frac{m^* e^2}{2q^2 \hbar^2 \pi^2} \int f(k) k \ln\left|\frac{k+2q}{k-2q}\right| dk.
$$

The form of the argument of the logarithm arises from having factored $\hbar^2 q / 2m^*$ out of each term in the numerator and the denominator, so the magnitude sign is required to assure that the argument is positive definite. To proceed further, the following normalized variables are now introduced:

$$
\xi^2 = \frac{\hbar^2 q^2}{8m^* k_B T}, \quad x^2 = \frac{\hbar^2 k^2}{2m^* k_B T}, \quad \mu = \frac{E_F}{k_B T}. \tag{4.76}
$$

It may be noted that the temperature here is that of the distribution function. By incorporating these normalizing factors into the expression (4.75), we obtain [5]

$$
\varepsilon(q) = \varepsilon_s \left[1 + \frac{q_D^2}{q^2} F(\xi, \mu) \right], \tag{4.77}
$$

where

$$
F(\xi, \mu) = \frac{1}{2\xi F_{1/2}(\mu)} \int_0^\infty \frac{x}{1 + e^{x^2 - \mu}} \ln\left|\frac{x+\xi}{x-\xi}\right| dx \tag{4.78}
$$

and

$$
F_{1/2}(\mu) = \frac{2}{\sqrt{\pi}} \int_0^\infty \frac{x^{1/2} dx}{1 + e^{x^2 - \mu}} \tag{4.79}
$$

is a standard Fermi integral (and is equal to 1 in the non-degenerate case).

In the case of a non-degenerate semiconductor (μ large and negative), $F(\xi, \mu)$ becomes Dawson's integral and is a tabulated function. In either case, however, best results are obtained from numerical simulation. In Fig. 4.7, the overall behavior of this function is plotted, and the topmost curve (μ = −30) corresponds to the non-degenerate limit. The behavior is not very dramatic. As $q \to 0$ (i.e., $\xi \to 0$), $F(\xi,\mu) \to 1$ for non-degenerate material, and the usual Debye screening behavior is recovered. On the other hand, as $q \to \infty$, $F(\xi,\mu) \to 0$, and the screening is broken up completely. Thus, for high-momentum transfer in the scattering process, the scattering potential is completely descreened $q \gg q_D$. The upshot of this is that the nonlinearity in Coulomb scattering appears as a scattering cross section that depends on itself through the momentum transfer $\hbar q$ (and hence through the scattering angle ϑ). $F(\xi,\mu)$ has decreased to a value of one-half its maximum already at a value of $\xi = 1.07$, for which $q = 1.75q_T$, where $q_T = mv_T / \hbar$ is the thermal wavevector, corresponding to the thermal velocity of a carrier. For GaAs at room temperature, $q_T = 2.61 \times 10^6$ cm^{-1}. This value is only 3.5 times larger than the Debye wavevector at a carrier density of 10^{17} cm^{-3} and becomes smaller than the Debye screening wavevector at higher carrier densities. Although screening is not totally eliminated, it is greatly reduced, and this can lead to more effective single-particle scattering than expected.

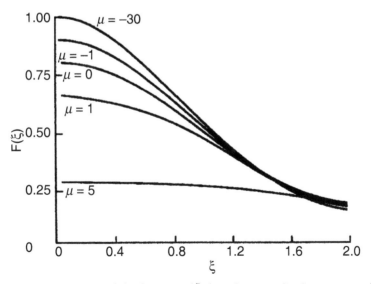

Figure 4.7 Variation of the function $F(\xi,\mu)$ as the normalized momentum ξ and Fermi energy μ are varied. Reprinted from Ref. [5], Copyright 2015, with permission from Elsevier.

On the other hand, as the carrier density is increased, the material becomes degenerate, and the Maxwellian approximation cannot be used. In the case of the Fermi–Dirac distribution above, the integrals are more complicated but still readily evaluated, as evident in Fig. 4.7. At small values of q, the degeneracy lowers the screening wavevector as one moves to the Fermi–Thomas screening regime. But for larger values of q, it is clear that the screening is strongly reduced at all wavevectors as the carrier density is increased, but the variation with the wavevector is also strongly reduced. The equivalent results for two dimensions have been evaluated by Ando et al. [6].

4.4 Impurity Scattering

There is another source of Coulomb interaction in semiconductors, and this is from the ionized impurities that exist in the material. These charge centers also interact with the electrons via the Coulomb potential. This interaction is a form of scattering that contributes to the resistivity (or conductivity, as the case may be). The impurity has an associated Coulomb potential that is long range in nature. This long-range interaction is usually cut off by assuming that it is screened by the electrons (through the electron–electron interaction, as discussed above). The contribution of the impurity potential to the energy is given by

$$V_{I,total} = \int d^3x \sum_j V_i(\mathbf{x} - \mathbf{x}_j)\bar{n}(\mathbf{x})$$

$$= \frac{1}{\Omega}\sum_{q,j} V_i(\mathbf{q})e^{i\mathbf{q}\cdot\mathbf{x}}\sum_{\mathbf{k}',\mathbf{k}} \widehat{\psi}^\dagger_{s'}(\mathbf{k}')\widehat{\psi}_s(\mathbf{k})\delta_{\mathbf{k}',\mathbf{k}-\mathbf{q}}\delta_{ss'}. \tag{4.80}$$

In the second line, we inserted the Fourier transform version and *assumed* that the spin of the electron will be conserved in the interaction. In some cases, the spin is not conserved and is commonly flipped by magnetic impurities. But this is beyond the introductory level that we want to maintain, so the spin conservation indicated in Eq. (4.80) will be maintained. Here Ω is the volume over which the wavefunctions are normalized. Usually, the wavefunctions (taken here to be field operators) reduce to the creation and annihilation operators for plane-wave states. Although the primary role may well be the scattering of the electron (or hole) from one plane-wave state to another, there is also a recoil of the impurity itself, which can ultimately couple into local modes of the lattice. This latter complication will not be considered here, although it can be a source of short-wavelength phonons that can cause impurity-dominated intervalley scattering. Generating the perturbation series usually relies upon the S-matrix expansion of the unitary operators, just as in the case of electron–electron scattering, and the perturbing potential (4.80) will enter the expansion (4.16). Let us see how this interaction modifies the Green's functions.

The expansion of the S-matrix in the scattering operator leads to an infinite series of terms, which change the equilibrium state. The higher-order terms are usually broken up by the use of Wick's theorem, and this leads to a diagram expansion. The formation of the Green's functions in the S-matrix expansion is accompanied by an averaging process over the equilibrium state (this is actually coupled to an average over the impurity configuration as well, which is discussed below). The equilibrium state must be renormalized in this process, and this causes a cancellation of all disconnected diagrams as previously. The result is an expansion in only the connected diagrams. Typical diagrams for the impurity scattering are shown in Fig. 4.8a. One still must carry out an averaging process over the position of the impurities, since the end result (at least in macroscopic samples) should not depend on the unique distribution of these impurities. The impurity averaging may be understood by noting the summation over the impurity positions that appear in the exponential factors in the second line of Eq. (4.80). Terms like those in Fig. 4.8a involve the average

$$\left\langle \sum_{\mathbf{q}} e^{i\mathbf{q}\circ\mathbf{x}} \right\rangle \rightarrow \delta(\mathbf{q}), \tag{4.81}$$

where the angular brackets denote the average over impurity positions. If the number of impurities is large, then the averaging of these positions places the important contribution of the potential from the impurities as that of a regular array, which can be thought of as creating a super-lattice. The vectors $\mathbf{x}_j$ are then the basis vectors for this lattice, and the summation is over all such vectors. In short, the summation then represents the closure of a complete set, which yields the delta function shown on the right side of the arrow

[2]. Thus, only the series of events in which the impurity imparts zero momentum to the electrons is allowed in the scattering process after the impurity averaging. Now, consider the double scattering processes shown in Fig. 4.8b, in which a single impurity interacts to second order with the propagating electron. Now there are two momenta imparted by the impurity, and the averaging in Eq. (4.80) becomes

$$\left\langle \sum_q e^{i(q+q')\cdot x} \right\rangle \rightarrow \delta(q+q'), \qquad (4.82)$$

so that one arrives at $q' = -q$. Thus, the interaction matrix element contained in the resultant expansion term from Eq. (4.80) for the second interaction is the complex conjugate of that for the first interaction, and the overall scattering process is proportional to the magnitude squared of the matrix element. This is the obvious result expected if one had started with the Fermi golden rule rather than Green's functions. Impurities can also interact with three lines and four lines, and more, as obvious extensions of the two situations shown in Fig. 4.8. The terms from Eq. (4.81) produce only an unimportant shift in the energy that arises from the presence of the impurities in the real crystal lattice, and the second-order interaction is the dominant scattering process. In general, the impurity interaction is sufficiently weak that all terms with more than two coupled impurity lines usually can be ignored (at least in semiconductors in the absence of significant impurity-induced disorder).

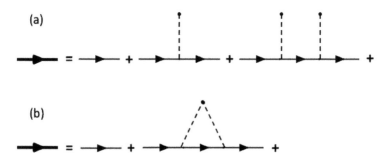

Figure 4.8 (a) Typical impurity scattering single interactions, which will not contribute to scattering after impurity averaging. (b) Second-order interactions that will contribute to scattering after impurity averaging.

It may be noticed first that, after the impurity averaging, we always have $\mathbf{k'} = \mathbf{k}$. Thus, as described above, each of the Green's functions, for which the spatial variation is in the *difference* of the two coordinates, is described by a single momentum state. The remaining set of diagrams in Fig. 4.8b corresponds to the same series that was encountered in the electron–electron interaction, and we can, therefore, do the resummation of that series by writing the interacting Green's function as

$$G_{r,a}(\mathbf{k}, \omega) = G_{r,a}^0(\mathbf{k}, \omega) + G_{r,a}^0(\mathbf{k}, \omega)\Sigma_{r,a}(\mathbf{k}, \omega)G_{r,a}(\mathbf{k}, \omega), \quad (4.83)$$

which, of course, is just Eq. (4.47) and can be rewritten as Dyson's equation. The self-energy is Σ and its expansion is shown in Fig. 4.9. Note that the internal Green's functions are the full Green's functions. In many cases, where the number of impurities is small, the scattering is weak, and the self-energy can be approximated by keeping only the first term and using the bare Green's function. Another, slightly better, approach is to keep only the lowest-order diagram but to use the full Green's function, solving the overall problem by iteration. This is often called the *self-consistent Born approximation*. We note that the real part of the self-energy can be considered a correction to the single-particle energy $E(\mathbf{k})$. This represents the *dressing* of (or change in) the energy due to the interaction with the impurities, as discussed in Chapter 1. This dressing can cause a general overall momentum-dependent shift in the energy, which also causes a change in the effective mass of the particle. On the other hand, the imaginary part of the self-energy represents the dissipative interaction that was included in an ad hoc manner by the insertion of the parameter η. Clearly the sign of the imaginary part of the self-energy is important to determine whether Eq. (4.83) represents the retarded or the advanced Green's function and this sign comes from the appropriate version of the self-energy.

$$\Sigma(\mathbf{k}, \omega) = \quad \text{} \quad +$$

Figure 4.9 The leading term in the self-energy contribution from impurity scattering. The Green's function is the full interacting Green's function, but is often approximated as the bare form, since the scattering is weak in semiconductors.

Let us now proceed to compute the scattering for the simplest case. This means that the task is really to evaluate the self-energy. However, if the full Green's function is retained in the latter term, then the process must be iterated. Here, we retain only the lowest-order term in the self-energy and treat the included Green's function with the bare equilibrium Green's function. We need to sum over the momentum variable contained in the impurity interaction. For this, we will keep only the first term in Fig. 4.9, so that we assume the impurity scattering is weak. Then,

$$\Sigma_{r,a}(\mathbf{k}, \omega) = \frac{N_i}{\Omega \hbar^2} \sum_{\mathbf{q}} |V_i(\mathbf{q})|^2 \, G_{r,a}^0(\mathbf{k} - \mathbf{q}, \omega), \tag{4.84}$$

where the additional factor $1/\hbar^2$ comes from the prefactor of the S-matrix terms and assures us that Eq. (4.83) retains the proper dimensionality. The factor N_i is the total number of impurities in the entire semiconductor, but together with Ω yields the average density of impurities. The impurity average has replaced the scattering by the various assortment of impurities with a single interaction, the averaged interaction, and this is multiplied by the number of impurities to arrive at the total scattering strength. At this point, it is pertinent to note that we expect the impurity potential to be reasonably screened by the free carriers, which means that the potential is very short range. It is easiest to assume a δ-function potential (in real space), which means that the Fourier-transformed potential is independent of momentum, or $V(\mathbf{q}) = V_0$. This is equivalent to assuming that the screening wavevector is much larger than any scattering wavevector of interest, or that $V_0 \sim e^2/\varepsilon_s q_s^2$. However, we will retain the form shown in Eq. (4.84) since we will need to produce a higher-order correction that modifies this term.

We are primarily interested in the imaginary parts of the self-energy, and this can be obtained by recognizing that η is small, so that we retain only the imaginary parts of the bare Green's function. The real parts of the self-energy are no more than a shift of the energy scale, and we expect this to be quite small in the weak scattering limit. Then, we take the limit of small η and ignore the principal part of the resulting expansion (since it leads to the real part of the self-energy, which we have decided to ignore), and

$$\lim_{\eta \to 0} \frac{\hbar}{\hbar \omega - E(\mathbf{k} - \mathbf{q}) \pm i\eta} \to i\pi\hbar\delta(\hbar\omega - E). \tag{4.85}$$

We now convert the summation into an integration in **k**-space, so that

$$\sum_q \rightarrow \frac{1}{4\pi} \int_0^{2\pi} d\phi \int_0^{\pi} \sin\vartheta d\vartheta \int \rho(E_q) dE_q = \frac{\hbar^2 q^2}{2m^*}, \tag{4.86}$$

under the assumption that there is no angular variation in the potential term, which could occur if q were larger than the screening wavevector. The first two integrals represent an integration over the solid angle portions of the overall three-dimensional integral. If the integrand is independent of these angular variables, then these integrals yield unity. It may be noted that the impurity scattering conserves energy, so that the final state energy is the same as the initial energy, or $E(\mathbf{k} - \mathbf{q}) = E(\mathbf{k})$. Thus, the energy integral can be evaluated as well, and the angular integral is at most an angular averaging of the scattering potential. This allows us to evaluate the self-energies, with $n_i = N_i/\Omega$ (Ω is the volume of the crystal) the impurity density, as

$$\Sigma_{r,a}(\mathbf{k}, \omega) = \pm \frac{i\pi n_i}{\hbar} \int_0^{2\pi} \int_0^{\pi} \frac{\sin\vartheta d\vartheta d\phi}{4\pi} |V(\mathbf{q})|^2 = \pm i \frac{1}{2\tau(\omega)}. \tag{4.86}$$

With the approximations used, the self-energies are independent of the momentum and are only functions of the energy $\hbar\omega$.

4.5 Conductivity

In the classical case and certainly before many-body theories were fully developed, the theory of electrical conduction was based on the Boltzmann transport equation for a one-particle distribution function. Certainly this approach still is heavily used to study transport in semiconductors [7], particularly in the response of semiconductor systems in which the full quantum response is not necessary. The essential assumption of this theory is the Markovian behavior of the scattering processes, that is, each scattering process is fully completed and independent of any other process. Coherence of the wavefunction is fully destroyed in each collision. On the other hand, here we are talking about possible coherence between a great many scattering events. Certainly, replacements for the Boltzmann equation exist for the non-Markovian world, but a more general approach is that of the Kubo formula, in which the conductance is computed from the current–current correlation function [8]. This linear response behavior describes transport with only an assumption that the current is linear in the applied electric field. The result, the Kubo formula, is that the current itself (or an ensemble averaged current) is given by a correlation function as

$$\langle \mathbf{J}(\mathbf{x},t) \rangle = \langle \mathbf{j}(\mathbf{x},t) \rangle + \int dt' \int d^3 \mathbf{x}' \tilde{\mathbf{G}}(\mathbf{x}-\mathbf{x}', t-t') \cdot \mathbf{A}(\mathbf{x}',t'), \quad (4.87)$$

where $\tilde{\mathbf{G}}$ is the conductance tensor and $\mathbf{A}$ is the vector potential. In the diffusive limit, we can ignore the ballistic response (which leads to plasma oscillation effects in the dielectric function), and the conductance is given by the current–current correlation function

$$\tilde{\mathbf{G}}(\mathbf{x}-\mathbf{x}', t-t') = -i\theta(t-t')\langle [\mathbf{j}(\mathbf{x},t), \mathbf{j}(\mathbf{x}',t')] \rangle . \quad (4.88)$$

If we recognize the current operator as

$$
\begin{aligned}
\mathbf{j}(\mathbf{x},t) &= -i\frac{e\hbar}{2m^*}\left\{ \widehat{\psi}^\dagger(\mathbf{x},t)\nabla\widehat{\psi}(\mathbf{x},t) - \left[\nabla\widehat{\psi}^\dagger(\mathbf{x},t)\right]\widehat{\psi}(\mathbf{x},t) \right\} \\
&\rightarrow \frac{e\hbar\mathbf{k}}{m^*}\widehat{\psi}^\dagger(\mathbf{x},t)\widehat{\psi}(\mathbf{x},t),
\end{aligned}
\quad (4.89)
$$

we can write the Fourier-transformed conductivity as

$$\sigma_{\alpha\beta} = \frac{e^2\hbar^2}{m^{*2}}\sum_{\mathbf{k},\mathbf{k}'} k_\alpha k'_\beta G_2(\mathbf{k},t;\mathbf{k},t;\mathbf{k}',t';\mathbf{k}',t')\delta(E_\mathbf{k} - E_\mathrm{F}), \quad (4.90)$$

where the subscripts define the coordinate axes, and we encounter once more the two-particle Green's function. The last delta function assures that we measure the conductivity at the Fermi surface. The latter form is arrived at rather simply by Fourier-transforming Eq. (4.87) in both space and time, noting that the commutator in Eq. (4.88) will lead to four terms, and then by using Eq. (4.89) to replace each current term. The four field operators combine to define the *two-particle Green's function.*

While the above has been written in momentum space and in time, we can examine the behavior somewhat differently in real space. The two-particle Green's function was defined previously in Eq. (3.61). Under the conditions for the above derivation (homogeneous conductance and the *zero-frequency* static conductivity), we may rewrite this in Fourier transform form as the product of two time functions, giving a convolution integral in Fourier space; since we are interested in the static result, only the frequency integral survives. The momentum integration is already contained in Eq. (4.90) and so does not appear. This result is then

$$G_2(\mathbf{k}, \mathbf{k}; \mathbf{k}', \mathbf{k}') = \int \frac{d\omega}{2\pi} G_r(\mathbf{k}, \omega) G_a(\mathbf{k}', \omega),$$

(4.91)

which is another bubble diagram. But this bubble is different from the one that appeared in Fig. 4.6. In this latter case, the connectors (the two arrows indicating the momentum $\mathbf{q}$) were in the exchanged momentum coordinate so that the summation was over the electron momentum $\mathbf{k}$. Here, however, the connectors are the vectors $\mathbf{k}$ and $\mathbf{k}'$ that precede the two-particle Green's function in Eq. (4.91), so that the integration is over the primed coordinate, or alternatively over the exchanged momentum $\mathbf{q}$. Moreover, this bubble is going to be horizontal as the times at the beginning and end are different (in Fig. 4.6, they were the same). This bubble is often called the conductivity bubble.

It is important to reiterate that the result (4.91) is a d.c. result and the factor ω in the integral is not the applied frequency, but the energy $E/\hbar$. The result is actually the d.c. (not the a.c.) conductivity, as has been stated several times. The interpretation of this term is that an electron is excited across the Fermi energy, which creates an electron–hole pair in momentum state $\mathbf{k}'$. The creation is assumed to be accomplished by a non-momentum-carrying process, such as a photon (whose momentum is considerably smaller than that of the electron and/or hole). This pair then propagates, scattering from various processes such as the charge impurities, to state $\mathbf{k}$, where it recombines, again giving up the excess energy to a momentum-less particle of some type. This is then a particle–hole propagator, since the electron excited above the Fermi energy is called a quasi-particle with its characteristic energy measured from the Fermi energy itself and is represented by the retarded Green's function. Similarly, the hole is a quasi-particle existing below the Fermi energy, and its energy is measured downward from the Fermi energy and is represented by the advanced Green's function.

We can now evaluate the conductivity, where we will seek only the diagonal conductivity under the assumption that the current is parallel to the electric field. Hence, we take $\alpha = \beta$, and noting that the two Green's functions have to be at the same momentum as each involves only a single momentum and they come together at the end of the bubble as shown in Fig. 4.10, and write Eq. (4.91) as

$$\sigma_{xx} = \frac{e^2\hbar^4}{m^{*2}}\sum_k k_x^2 \int \frac{d\omega}{2\pi} \frac{1}{\hbar\omega - E(\mathbf{k}) - i\hbar/2\tau} \frac{\delta(E(\mathbf{k}) - E_F)}{\hbar\omega - E(\mathbf{k}) + i\hbar/2\tau}, \quad (4.92)$$

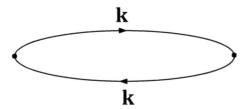

Figure 4.10 The lowest-order bubble for the conductivity. As shown in the text, this leads to the Drude conductivity.

where we have replaced the convergence factor with the imaginary part of the self-energy (4.86). Either the frequency integral or the energy integral can be evaluated quickly by residues, with the other being replaced by the use of the delta function at the Fermi surface. Then, using $k_x^2 = k^2/3$ and $\rho(E_F)E_F = dn/2$, where $\rho(E)$ is the density of states (per unit volume), the conductivity is finally found to be

$$\sigma = \frac{e^2\hbar^4}{3m^{*2}}\sum_k k^2 \frac{\tau}{\hbar^2}\delta(E(\mathbf{k}) - E_F) = \frac{ne^2\tau}{m^*}. \quad (4.93)$$

where

$$n = \frac{1}{\pi^2}\left(\frac{2m^*E_F}{\hbar^2}\right)^{3/2} \quad (4.94)$$

at low temperatures for a strongly degenerate electron gas. This is the normal low-frequency result of the Drude formula, and it is the usual conductivity one arrives at in transport theory. It must be noticed, however, that most semi-classical treatments of the impurity mobility include a factor $(1 - \cos\theta)$, which is missing in this formulation, so that τ is a scattering time and not a relaxation time. It also must be noted that this result, which leads to a part of the normal Drude conductivity, is a result in which the two Green's functions are evaluated in isolation from one another, and there is no interaction between the two. If we are to recover the angular variation that appears semi-classically, we shall have to consider higher-order terms in which the two Green's functions interact with each other.

The higher-order corrections that are most important are those in which the impurities connect the two Green's functions that appear in Eq. (4.91). The typical types of new diagrams that appear are shown in Fig. 4.11. The diagrams shown all have four interaction lines from either a single or a pair of impurities. To handle this complex situation, we will have to expand the two-particle Green's function. (As an aside, we remind ourselves that the frequencies in the two-particle Green's function are most often seen as $\omega_\pm = \omega \pm \omega_a/2$, where ω_a is the applied a.c. frequency at which the conductivity is evaluated. Here, however, we are interested in the d.c. conductivity so that $\omega_a = 0$ and we can ignore this complication.) In general, the two-particle Green's function is frequency dependent as shown, although the one we need for the static conductivity does not have any frequency dependence. We define the kernel of the conductivity (for the static conductivity) in the isotropic limit as [9, 10]

$$\sigma = \frac{e^2\hbar^2}{3m^{*2}}\int\frac{d\omega}{2\pi}F(\omega) , \qquad (4.95)$$

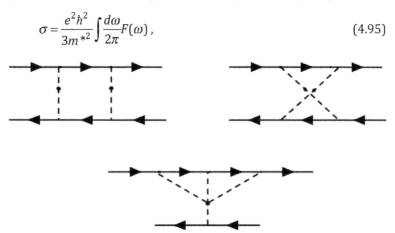

Figure 4.11 Some typical fourth-order (in the impurity interaction) terms that span the two Green's function lines. The top two involve two impurities, while the bottom one is a higher-order interaction arising from a single impurity.

where

$$F(\omega) = \sum_{\mathbf{k},\mathbf{k}'} \mathbf{k} \cdot \mathbf{k}' G_2(\mathbf{k},\mathbf{k}',\omega)$$

$$= \sum_{\mathbf{k}} G_r(\mathbf{k},\omega) G_a(\mathbf{k},\omega) \left\{ k^2 + \sum_{\mathbf{k}'} \mathbf{k} \cdot \mathbf{k}' \Lambda(\mathbf{k},\mathbf{k}',\omega) G_r(\mathbf{k}',\omega) G_a(\mathbf{k}',\omega) \right\}.$$

$$(4.96)$$

In going from the tensor form of the conductivity to the form here, where the two wavevectors form a dot product, we note the previous choice of k_x^2 is replaced by k^2, which leads to the factor of 3 in the denominator. The first term in the curly brackets leads to the results obtained in the previous paragraph. The new term is the second one in the curly brackets. Again, the various terms of Fig. 4.10 can be used. But the ladder terms shown as the upper diagram can be summed nicely. It is apparent that the dot product in the second term is going to give rise to the angular term we need. The last pair of Green's functions in the second term in curly brackets is an approximation to the two-particle Green's function. The particular form, illustrated in Fig. 4.11, gives the leading terms. With the full two-particle Green's function in Eq. (4.96), the result is known as the Bethe–Salpeter equation. Here, the quantity Λ is referred to as the irreducible scattering vertex, which we shall approximate by the ladder terms (because the diagram in the upper left of Fig. 4.10 looks like the lower rungs of a ladder lying on its side).

The key contributions to the scattering vertex arise from the second-order interaction, which spans the two Green's functions. The impurity lines do not transport energy (the scattering from impurities is elastic), so the upper and lower Green's functions are at the same energy throughout; for the static case, the two frequencies are the same. We keep only the lowest-order correction in these ladder diagrams, which is the iterated diagrams such as that of the upper left diagram in Fig. 4.10. The kernel for the single scattering pair that spans the upper and lower Green's function is given by

$$\Gamma_0 = \frac{n_i}{\hbar^2} |V(\mathbf{k}-\mathbf{k}')|^2, \tag{4.97}$$

where the two momenta are those on either side of the scattering line. We want to form a series for the set of iterated interactions and have the two integrations over $\mathbf{k}$ and $\mathbf{k}'$. The scattering kernel

(4.97) can be rewritten in terms of the scattering wavevector $\mathbf{q} = \mathbf{k} - \mathbf{k}'$, but this will make the second set of Green's functions depend on $\mathbf{k}' + \mathbf{q}$, and the integration is then taken over $\mathbf{q}$. However, since energy is conserved in the interaction, the argument of the potential is a function only of $\mathbf{k}$ and a scattering angle, which is not part of the integration over the last Green's functions. This means that the kernel can be treated separately from the last pair of Green's functions. We can write the series of ladder diagrams as

$$\Lambda = \Gamma_0 + \Gamma_0 \Pi \Gamma_0 + \Gamma_0 \Pi \Gamma_0 \Pi \Gamma_0 + \ldots = \frac{\Gamma_0}{1 - \Pi \Gamma_0}. \qquad (4.98)$$

The set of Green's function integrals for Π can be integrated to give

$$\Pi(\omega) = \int \rho(E)dE \frac{\hbar}{\hbar\omega - E + i\hbar/2\tau} \frac{\hbar}{\hbar\omega - E - i\hbar/2\tau} = -2\pi\hbar\tau\rho(\hbar\omega). (4.99)$$

The last pair of Green's functions in the second term of Eq. (4.96) just adds another factor of Π to the numerator of Eq. (4.98), if we take the kerm in $\mathbf{k}'$ out of the integral. What remains from the dot product $\mathbf{k} \cdot \mathbf{k}' = k^2 \cos\vartheta$, where ϑ is the scattering angle. Using these last equations and properties, the kernel can be rewritten as

$$F(\omega) = \sum_{\mathbf{k}} k^2 G_r(\mathbf{k}, \omega)G_a(\mathbf{k}, \omega)\left\{1 + \cos\vartheta \frac{\Gamma_0\Pi}{1 - \Gamma_0\Pi}\right\}. \qquad (4.100)$$

Generally, one makes the assumption here that the product $\Pi\Lambda_0$ is large compared to unity, so that one arrives at an additional term involving just the cosine of the scattering angle with its sign changed. Carrying out the frequency integration in Eq. (4.93) leads to the result

$$\sigma = \frac{e^2\hbar^2}{3m^{*2}} \sum_{\mathbf{k}} \tau k^2 (1 - \cos\vartheta)\delta(E(\mathbf{k}) - E_F) = \frac{ne^2\tau_m}{m^*}, \qquad (4.101)$$

so that we have now replaced the scattering time with the momentum relaxation time.

The result obtained here, with the approximations used, provides a connection with the semi-classical result for impurity scattering obtained from the Boltzmann equation. The complications are greater here, but this result also allows us to create a formalism that can be moved forward to treating other diagrammatic terms for new effects that are not contained in the Boltzmann equation. The angular dependence of the fraction that appears as the second term in the curly brackets of Eq. (4.100) provides the needed conversion from a simple scattering time to a momentum relaxation time. On the other hand, if we have a matrix element that really does not depend on **q**, such as a delta-function scattering potential or a heavily screened Coulomb potential where $V_0 \sim e^2/\varepsilon_s q_s^2$ (where q_s is the screening wavevector, either for Fermi–Thomas screening or for Debye screening), then the ladder correction terms do not contribute as the angular variation integrates to zero. Thus, scattering processes in which the matrix element is independent of the scattering wavevector do not contribute the angular correction. This simple fact is often overlooked in semi-classical treatments, as this latter angular variation is put in by hand, and its origin is not fully appreciated.

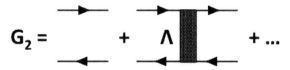

Figure 4.12 The leading correction term to the two-particle Green's function as represented in Eq. (4.96). The scattering vertex Λ will be approximated by the ladder diagrams—the upper left set in Fig. 4.10.

Finally, we return to the electron–electron interaction by once again pointing out that the conductivity bubble is similar to the polarization bubble that appeared in Fig. 4.6. The extension of the conductivity bubble via the Bethe–Salpeter equation can certainly be applied to the polarization bubble of Fig. 4.6. The full interaction is shown in Fig. 4.12. This leads to screening functions that go beyond the Lindhard dielectric function. Some of these are discussed in Mahan [11].

Problems

1. Using the computer code developed in Problem 1 of Chapter 3, modify the Hamiltonian to incorporate an on-site self-energy correction to account for impurity scattering. Vary the number of impurities and investigate the transmission through the structure. How does the transmission change?

2. Let us assume that a magnetic field is added to the Hamiltonian in an otherwise homogeneous steady-state system. How would this change the spectral density (using real-time Green's functions)? Assume that there is elastic scattering present in the system, such as impurity scattering.

3. Formulate the self-energy for impurity scattering and evaluate the less-than Green's function in the presence of a magnetic field, as discussed above.

4. We want to consider the conductivity of the channel in an Si MOSFET at low temperatures where the quantization becomes important. Show that the conductivity can be written as

$$\sigma_{2D} = \frac{2e^2}{h}\sum_n (E_F - E_n)\Lambda_n(E_F - E_n, E_F)\int_0^\infty \frac{dE_k}{2\pi}\frac{a_n(E_k, E_F)}{2\Gamma_m(E_k, E_F)}$$

where

$$\Lambda_n(E_k, \omega) = 1 + \sum_m \Lambda_n(\hbar\omega - E_m, \omega)$$

$$\times \iint \frac{d^2k_1}{(2\pi)^2}\frac{\mathbf{k}\cdot\mathbf{k}_1}{k^2}T_{nm}(\mathbf{k} - \mathbf{k}_1)\frac{a_m(E_k, \omega)}{2\Gamma_m(E_k, \omega)}$$

is the ladder bubble from the Bethe–Salpeter equation (discussed in class), $T_{nm}(\mathbf{q})$ equals the summation of the squared matrix elements of all scattering mechanisms in the model, $a_n(\mathbf{k},\omega) = -2\text{Im}[G^r_n(\mathbf{k},\omega)]$ is the spectral density for sub-band n, and Γ_{nm} is the matrix element for a scattering process between sub-bands n and m. At low temperature, the dominant scattering mechanisms are impurity and surface-roughness scattering.

References

1. A. Messiah, *Quantum Mechanics*, Vol. 1 (Interscience, New York, 1961), p. 442.

2. A. L. Fetter and J. D. Walecka, *Quantum Theory of Many-Particle Systems* (McGraw-Hill, New York, 1971).

3. G. C. Wick, *Phys. Rev.*, **80**, 268 (1950).

4. D. K. Ferry, *Semiconductors* (Macmillan, New York, 1991).

5. W.-Y. Chung and D. K. Ferry, *Solid-State Electron.*, **31**, 1369 (1988).

6. T. Ando, A. B. Fowler, and F. Stern, *Rev. Mod. Phys.*, **54**, 437 (1982).

7. D. K. Ferry, *Semiconductors* (Macmillan, New York, 1991).

8. R. Kubo, *J. Phys. Soc. Jpn.*, **12**, 570 (1957).

9. C. P. Enz, *A Course on Many-Body Theory Applied to Solid-State Physics* (World Scientific Press, Singapore, 1992).

10. A. A. Abrikosov, L. P. Gor'kov, and I. Ye. Dzyaloshinskii, *Quantum Field Theoretical Methods in Statistical Physics* (Pergamon Press, New York, 1965).

11. G. D. Mahan, *Many-Particle Physics* (Plenum Press, New York, 1981), Sec. 5.5C.

Chapter 5

Role of Temperature

When the temperature is raised above absolute zero, the amplitudes of many quantum interference effects are reduced below the nominal value e^2/h. This is usually attributed to the increase in the scattering by various processes, and this leads to a reduction in the coherence length, but this can arise as a consequence of either a reduction in the coherence time or a reduction in the diffusion coefficient. In fact, both these effects occur. As the mobility changes with temperature, it leads to an equivalent change in the diffusion constant, $D = v_F^2 \tau / d$, where d is the dimensionality of the system, through both a small temperature dependence of the Fermi velocity and a much larger temperature dependence of the elastic scattering rate. The temperature dependence of the phase coherence time is less well understood but generally is thought to be limited by electron–electron scattering, particularly at low temperatures. At higher temperatures, of course, phonon scattering can introduce phase breaking. We will deal with these two processes later in this chapter.

Another interaction, though, is treated by the introduction of another characteristic length, the thermal diffusion length. The source for this lies in the thermal spreading of the energy levels or, more precisely, in thermal excitation and motion on the part of the carriers. At high temperatures, of course, the lattice interaction

An Introduction to Quantum Transport in Semiconductors
David K. Ferry
Copyright © 2018 Pan Stanford Publishing Pte. Ltd.
ISBN 978-981-4745-86-4 (Hardcover), 978-1-315-20622-6 (eBook)
www.panstanford.com

becomes important, and energy exchange with the phonon field will damp the phase coherence. The basis of the thermal diffusion length is introduced through the assumed thermal broadening of a particular energy level (any level of interest, that is). This broadening of the energy level (discussed in Chapter 1) leads to a broadening of the available range of states into which a nearly elastic collision can occur, and this leads to wavefunction mixing and phase information loss. Imagine two interfering electron paths from which an interference effect such as the Aharonov–Bohm effect can be observed. We let the enclosed magnetic field (or the path lengths) be such that the net phase difference is δ. At $T = 0$, all of the motion is carried out by carriers precisely at the Fermi energy. When the temperature is nonzero, however, the motion is carried out by carriers lying in an energy width of about $3.5\,k_B T$ (this is the full width at half-maximum of the derivative of the Fermi–Dirac distribution), centered on the Fermi energy, where k_B is the Boltzmann constant. Thus, thermal fluctuations will excite carriers into energies near, but not equal to, the Fermi energy. The carriers with energy $E_F + k_B T$ will produce an additional phase shift $\delta\omega t$, where t is approximately the transit time along one of the interfering paths, and we also assume that $\delta\omega \sim k_B T/\hbar$. This leads to a total phase difference that is roughly $\delta + k_B T t/\hbar$. In fact, the actual energies are distributed over a range of energies corresponding to the width mentioned above, so there will be a range of phases in the interfering electrons. The phase difference is always determined modulo 2π, so that when the extra phase factor is near unity (in magnitude), there will be a de-correlation of the phase due to the distribution of the actual phases of individual electrons, which means some phase interference is destroyed. Thus, one can think about the time over which this thermally induced phase destruction occurs as $\tau \sim \hbar/k_B T$. This time is a sort of thermal phase-breaking time, and it can be used with the diffusion coefficient to define a thermal length $l_T = (D\hbar/k_B T)^{1/2}$. Thus, l_T is the length over which dephasing of the electrons occurs due to the thermal excitations in the system. We call this length the *thermal diffusion length.* The above process can be summarized by three steps: (1) the nonzero temperature means that a spread of energies, rather than a single well-defined energy, is involved in transport at the Fermi energy; (2) this spread in energies defines a dephasing time $\hbar/k_B T$, which describes interference among the ensemble of waves corresponding to the spread in energy; and (3) the normal phase coherence time τ_φ is replaced in the definition of the coherence length by this temperature-induced dephasing time. Thus, the thermal diffusion length describes dephasing introduced specifically by the spread in the energy of contributing states at the Fermi surface.

The introduction of the thermal diffusion length means that nanoscale devices are now characterized by a more complex set of lengths. The presence of an additional temperature-dependent quantity is just one of these new lengths. But with this thermal length, we can determine a nominal temperature at which we do not need to worry about thermal effects. If $l_T > l_\phi$, then the dominant dephasing properties are contained in the phase coherence length, and the thermal effects are minimal. On the other hand, if $l_T < l_\phi$, then the primary dephasing interaction is due to thermal excitations in the system, and this becomes the important length scale.

In this chapter, we will examine how the temperature variations arise. The simplest, of course, is just through the temperature dependence of the Fermi–Dirac distribution function, which has already appeared in previous chapters. We will examine this in connection with the Landauer formula. Then we will examine a new set of Green's functions—the Matsubara Green's functions, which are also called temperature Green's functions. We will examine how this leads to a slight difference in the dielectric function, from which the original version can be obtained via the process of analytic continuation. Then we turn to a discussion of the phase-breaking process and the corresponding coherence times. Finally, we discuss the electron–phonon interaction and its contributions to self-energy.

5.1 Temperature and the Landauer Formula

In the treatment of the transmission problems discussed in Chapters 2 and 3, we primarily considered the fate of individual modes in the source and drain contact regions. So far, the transmission probability from a particular mode in the source to another mode in the drain was an elastic process, although this does not have to be the case. And, of course, in talking about the transmission process, we also include the fact that tunneling from one side of the device to the other remains a study in the overall transmission probability from one side to the other.

In a real device, the current flow through the central region (the region where we describe the transport by the mode transmission) is related to the total transmission probability as well as to the total number of carriers which are available. Hence, we may write the current as

$$I_{LR} = 2eA \int \frac{d^3\mathbf{k}}{(2\pi)^3} v_z(k_z)T(k_z)f(E_L),$$

(5.1)

where A is the cross-sectional area of the device, $v_z(k_z)$ is the velocity in the z-direction (which is taken to be the current flow direction), T is the transmission, and $f(E)$ is the Fermi–Dirac distribution. The factor of 2 accounts for spin degeneracy in the density of states. Here, the transmission is defined from the equilibrium region in the left contact (source) to the equilibrium region in the right contact (drain). In a similar manner, we can write the current flowing from right to left as

$$I_{RL} = 2eA \int \frac{d^3\mathbf{k}}{(2\pi)^3} v_z(k_z')T(k_z')f(E_R),$$

(5.2)

where $v_z(k_z')$ is the velocity in the z-direction (which is taken to be the current flow direction), $T(k_z')$ is the transmission in the reverse direction. Now the momenta in these two equations are different due to the fact that there is usually a bias voltage applied to the drain contact, which makes $E_R < E_L$ and, therefore, $k_z' > k_z$. On the other hand, transmission is a reciprocal process, so that the two transmission terms must be the same, as the two terminal device is symmetric in its terminal properties. In addition, we note that the velocities can be put together as

$$dk_z v_z = dk_z \frac{1}{\hbar} \frac{dE}{dk_z} = \frac{1}{\hbar} dE,$$

(5.3)

which will simplify the equations. Hence, we can combine Eqs. (5.1) and (5.2) as

$$I = \frac{eA}{4\pi^3 \hbar} \int d^2\mathbf{k}_\perp \int dE T(E) \left[f_L(E) - f_R(E - eV_a) \right].$$

(5.4)

The formulation of Eq. (5.4) is really that for a large transverse area so that no quantization of these transverse dimensions is being considered. In fact, for the nanostructures we previously dealt with, the transverse directions are quantized and this leads to the modes that were considered in Chapters 1 and 2. This is accomplished by changing the integration over the transverse coordinates into a summation over the (two-dimensional) transverse modes, as

$$\frac{A}{4\pi^2} \int T(E) d^2\mathbf{k}_\perp \rightarrow \sum_{i,j} T_{ij}(E), \tag{5.5}$$

where T_{ij} is the transmission from mode i in the input contact to mode j in the output contact (or vice versa in various forms in the literature). That is, the summation insures that we sum over all the modes in the input contact, weighted by the Fermi–Dirac function, to all available modes in the output contact. If the input voltage is small, the second Fermi–Dirac function can be expanded in a Taylor series, leading to a variation of the standard form of the Landauer formula [1]

$$G = \frac{2e^2}{h} \int dE \sum_{i,j} T_{ij} \frac{\partial f(E)}{\partial E}. \tag{5.6}$$

Of course, various forms of this appear, and in many cases, the transmission must be weighted by the various velocities of each mode as these change the overall conductance that may be observed [2]. The implication of Eq. (5.6) is that the most important temperature variation, particularly in the situation in which there is no scattering in the conducting region, arises from the temperature dependence of the Fermi–Dirac function. Hence, the transmission can be computed over a range of energies, and then each value weighted by the derivative of the Fermi–Dirac function as shown in Eq. (5.6). The full width of the derivative at one-half of its peak value is about $3.5 k_B T$. So most of the conduction occurs in this relatively small energy window.

5.2 Temperature Green's Functions

At temperatures above absolute zero, there is an analogous Green's function to those that have been introduced in the previous chapters. However, this function is more detailed, and the calculation of the equilibrium properties and excitation spectrum also is more detailed. The first part, the determination of the equilibrium properties, is handled by the introduction of a *temperature Green's function,* often called the Matsubara Green's function, while the second step requires the computation of a time-dependent Green's function that describes the linear response of the system [3]. In this chapter, we will deal just with the first of these two complications. In principle, the definition of the temperature Green's function is made simply by taking account of the fact that there exists both a distribution function in the system and a number of both full and empty states. The first change is that the average over the basis states that we previously defined through the simple set of brackets $\langle .... \rangle$ becomes more complicated. This average will now be given by $Tr\{\hat{\rho}[....]\}$, where the trace remains the sum over the diagonal elements of the matrix inside the curly brackets. The second change is the introduction of the density matrix $\hat{\rho} = \exp[\beta(\Omega - K)]$, where $\beta = 1/k_B T$, Ω is the free energy (defined through the partition function), and $K = H - \mu N$ represents the grand canonical ensemble, which includes the chemical potential μ (usually equal to the Fermi energy) and total number of electrons N. One can understand this better if it is recognized that Ω is simply a normalization factor so that the $Tr\{\rho\} = 1$, and the addition term μN, which takes H into K, also provides the background energy of the electrons at zero temperature. Hence, K measures primarily the changes from this background energy that occurs with temperature and various excitations. With this new Hamiltonian, we introduce the modified Heisenberg picture

$$A(\mathbf{x}, \tau) = e^{K\tau/\hbar} A(\mathbf{x}) e^{-K\tau/\hbar}, \tag{5.7}$$

where a new complex variable τ has been introduced as a replacement for the time, but it has appeared in a very subtle change. The time does not occur in the density matrix defined above, but we have created it by the replacement of β by τ, by which we are claiming that, for the moment, the inverse temperature plays the role of an imaginary time. The real time can be recovered by making the analytic continuation of τ into it. In making this introduction, the goal is to be able to connect the Hamiltonian behavior of the Heisenberg picture with the density matrix, and with the thermal distribution $e^{-\beta H}$ (although we shall use the grand canonical ensemble mentioned above), and we will have to be quite careful about various trajectories and paths that are taken, particularly for factors such as time-ordering operators. This then leads to the single-particle Green's function

$$G(\mathbf{x},\mathbf{x}';\tau,\tau') = -Tr\left\{\hat{\rho}T_\tau\left[\hat{\psi}(\mathbf{x},\tau)\hat{\psi}^\dagger(\mathbf{x}',\tau')\right]\right\}, \tag{5.8}$$

where the analogy with Eqs. (3.16) and (4.28) is obvious. The spin indices have also been suppressed in this form for simplicity, and T_τ is the time-ordering operator in this new time coordinate. Similar to the behavior in the previous chapters, the time variation of the field operators can be defined with Eq. (5.7) as

$$\hat{\psi}(\mathbf{x},\tau) = e^{K\tau/\hbar}\hat{\psi}(\mathbf{x})e^{-K\tau/\hbar}$$

$$\hat{\psi}^\dagger(\mathbf{x},\tau) = e^{K\tau/\hbar}\hat{\psi}^\dagger(\mathbf{x})e^{-K\tau/\hbar}. \tag{5.9}$$

The temperature Green's function is useful because it will allow us to calculate the thermodynamic behavior of the system. This is a consequence of the introduction of the normal thermal equilibrium form of the density matrix, and this will be reflected below when the Fermi–Dirac distribution arises. If the Hamiltonian is time independent, then the Green's function depends only on the difference $\tau - \tau'$, as for the earlier case. Similarly, a homogeneous system will lead to the Green's function being a function of $x - x'$. One useful property is provided by

$$\sum_s G_{rs}(\mathbf{x},\mathbf{x}';\tau,\tau') = \sum_s Tr\left\{\hat{\rho}T_\tau\left[\hat{\psi}_s^\dagger(\mathbf{x}',\tau')\hat{\psi}_r(\mathbf{x},\tau)\right]\right\}$$

$$= e^{\beta\Omega}\sum_s Tr\left\{e^{-\beta K}e^{K\tau/\hbar}\hat{\psi}_s^\dagger(\mathbf{x}')\hat{\psi}_r(\mathbf{x})e^{-K\tau/\hbar}\right\} \tag{5.10}$$

$$= e^{\beta\Omega}\sum_s Tr\left\{e^{-\beta K}\hat{\psi}_s^\dagger(\mathbf{x}')\hat{\psi}_r(\mathbf{x})\right\} = \langle n(\mathbf{x})\rangle,$$

where we have used the anti-commutation relation for the fermion field operators in the first line, Eq. (5.7) in the second line, and the cyclic permutation properties of the trace in the last line. That is $Tr\{ABC\} = Tr\{CAB\} = Tr\{BCA\}$, and exponentials of the same operator naturally commute. This allowed us to move the last exponential to the front of the set of operators, and then it commutes with the first operator in the second line so that these two Heisenberg operators cancel under the trace. Finally, the total number of electrons arises by integrating the final term above over all space. Here we see the power of the Matsubara functions, in that it casts the temporal evaluation operators into the same form as the statistical density matrix itself, and allows directly for the commutation of these different functions of the Hamiltonian. This greatly simplifies the determination of functions that arise in the presence of a nonzero temperature.

To illustrate how this fits in with our earlier discussions, let us consider a simple non-interacting system, in which the creation and annihilation operators $a_{\mathbf{k}\lambda}^{+}$ and $a_{\mathbf{k}\lambda}$ create and destroy an electron in momentum state $\mathbf{k}$ with spin λ. Then, the field operators can be written as

$$\widehat{\psi}(\mathbf{x}) = \frac{1}{\sqrt{V}} \sum_{\mathbf{k},r} e^{i\mathbf{k}\cdot\mathbf{x}} \eta_r a_{\mathbf{k}r}$$

$$\widehat{\psi}^{\dagger}(\mathbf{x}) = \frac{1}{\sqrt{V}} \sum_{\mathbf{k},r} e^{-i\mathbf{k}\cdot\mathbf{x}} \eta_r^{\dagger} a_{\mathbf{k}r}^{\dagger},$$

(5.11)

where η_r is a spinor, or spin wavefunction, and V is the normalization volume. Using Eq. (4.2), suitably modified for the imaginary time (the factor of i has been absorbed in the τ and the resulting minus sign is absorbed in reversing the anti-commutator relationship for the fermion operators), we can write the equation of motion for the creation and annihilation operators as

$$\hbar \frac{\partial a_{\mathbf{k}r}}{\partial \tau} = e^{K\tau/\hbar} \left\{ K, a_{\mathbf{k}r} \right\} e^{-K\tau/\hbar}$$

$$= -\left[E(\mathbf{k}) - \mu \right] a_{\mathbf{k}r}, \quad E(\mathbf{k}) = \frac{\hbar^2 k^2}{2m^*},$$

(5.12)

or

$$a_{\mathbf{k}r}(\tau) = a_{\mathbf{k}r} e^{-[E(k)-\mu]\tau/\hbar}$$

$$a_{\mathbf{k}r}^{\dagger}(\tau) = a_{\mathbf{k}r}^{\dagger} e^{[E(k)-\mu]\tau/\hbar}.$$

(5.13)

With this we can write the temperature Green's function from Eq. (5.10) as

$$G_{rs}(\mathbf{x},\mathbf{x}';\tau,\tau')$$

$$= -\frac{1}{V} \sum_{\mathbf{k},\mathbf{k}',r,s} e^{i\mathbf{k}\cdot\mathbf{x}-i\mathbf{k}'\cdot\mathbf{x}'} \eta_r \eta_s^{\dagger} e^{-[E(k)-\mu]\tau/\hbar+[E(k')-\mu]\tau'/\hbar} \left\langle a_{\mathbf{k}r} a_{\mathbf{k}'s}^{\dagger} \right\rangle.$$

(5.14)

The spin wavefunctions are orthonormal, so that $r = s$. Similarly, the last average requires that the momenta states are the same, and $\left\langle a_{\mathbf{k}\lambda} a_{\mathbf{k}\lambda}^{+} \right\rangle = 1 - \left\langle a_{\mathbf{k}\lambda}^{+} a_{\mathbf{k}\lambda} \right\rangle = 1 - f_k$, where

$$f(k) = \frac{1}{1 + e^{\beta[E(k)-\mu]}} \tag{5.15}$$

is the Fermi–Dirac distribution with the Fermi energy being the electrochemical potential. Now we can write the resulting ground-state Green's function as

$$G^0(\mathbf{x},\mathbf{x}';\tau,\tau') = \begin{cases} -\dfrac{1}{V}\displaystyle\sum_{k} e^{i\mathbf{k}\cdot(\mathbf{x}-\mathbf{x}')-[E(k)-\mu](\tau-\tau')/\hbar}[1-f(k)], & \tau > \tau', \\[4mm] \displaystyle\sum_{k} e^{i\mathbf{k}\cdot(\mathbf{x}-\mathbf{x}')+[E(k)-\mu](\tau-\tau')/\hbar} f(k), & \tau < \tau'. \end{cases}$$

$$\tag{5.16}$$

An important property of these Green's functions is the periodicity in complex time of the functions themselves. One obvious reason for using the complex time version of the operators is that the Heisenberg temporal propagator is in a form that commutes with the density matrix itself. This practice creates some unusual, but very useful, behaviors. Consider the case in which, for convenience, we take $\tau = 0$, $\tau' > 0$, so that we can write Eq. (5.10) as (we take only the fermion case here), using the cyclic permutation of the trace and Eq. (5.7),

$$\begin{aligned} G_{rs}(\mathbf{x},\mathbf{x}';0,\tau') &= e^{\beta\Omega} Tr\left\{ e^{-\beta K}\,\widehat{\psi}_s^{\dagger}(\mathbf{x}',\tau')\widehat{\psi}_r(\mathbf{x},0) \right\} \\ &= e^{\beta\Omega} Tr\left\{ \widehat{\psi}_r(\mathbf{x},0)e^{-\beta K}\,\widehat{\psi}_s^{\dagger}(\mathbf{x}',\tau') \right\} \\ &= e^{\beta\Omega} Tr\left\{ \widehat{\psi}_r(\mathbf{x},0)e^{-\beta K}\,\widehat{\psi}_s^{\dagger}(\mathbf{x}',\tau')e^{-\beta K}e^{\beta K} \right\} \quad (5.17) \\ &= e^{\beta\Omega} Tr\left\{ e^{-\beta K}\,\widehat{\psi}_r(\mathbf{x},\beta\hbar)\widehat{\psi}_s^{\dagger}(\mathbf{x}',\tau') \right\} \\ &= -G_{rs}(\mathbf{x},\mathbf{x}';\beta\hbar,\tau'). \end{aligned}$$

Thus, it appears that the Green's function is periodic in the imaginary time with a period of $\beta\hbar$. Because of this periodicity, we have to assume that $0 < \tau' < \beta\hbar$, and similarly for τ. Hence, each time τ increases by $\beta\hbar$, there is a phase change of π in the Green's function, which means that the temperature Green's function is *anti-periodic* in $\beta\hbar$. (The boson version of the temperature Green's function is periodic in $\beta\hbar$.) This result is very important and builds the properties of the density matrix into the results for the Green's function. The importance of this is that the perturbation series is now integrated only over the range 0 to $\beta\hbar$ instead of 0 to ∞. In the usual situation in which the Hamiltonian is independent of time, the Green's function depends only on the difference in time coordinates, and the time is readily shifted so that

$$G_{rs}(\mathbf{x},\mathbf{x}';\tau - \tau' < 0) = -G_{rs}(\mathbf{x},\mathbf{x}';\tau - \tau' + \beta\hbar > 0). \qquad (5.18)$$

For the non-interacting Green's function (5.16), this leads to the important result

$$e^{\beta[E(k)-\mu]}f(k) = 1 - f(k), \qquad (5.19)$$

which is easily rearranged to yield the Fermi–Dirac distribution function. Hence, the anti-periodicity of the temperature Green's function for electrons insures that they are described (in equilibrium) by the Fermi–Dirac distribution.

With the above periodicity in mind, we can now introduce the Fourier transform representation (in imaginary time) for the temperature Green's function. We note that both the boson and fermion temperature Green's functions are fully periodic in $2\beta\hbar$. To proceed, we let $\tau'' = \tau - \tau'$, and define the inverse Fourier transform from

$$G(\mathbf{x},\mathbf{x}';\tau'') = \frac{1}{\beta\hbar}\sum_{n} e^{-i\omega_n\tau''}G(\mathbf{x},\mathbf{x}';\omega_n). \qquad (5.20)$$

Now there are a number of important points here. First, the periodicity, or anti-periodicity, is in the quantity $\beta\hbar$, and that is why it appears in the denominator of the prefactor instead of the more usual 2π. Secondly, we have given the frequency a subscript n. This is also because of this periodicity. The periodicity of the exponent is of course 2π, so that

$$\omega_n 2\beta\hbar = 2n\pi \rightarrow \omega_n = \frac{n\pi}{\beta\hbar}, \qquad (5.21)$$

which assures that if we only move $\beta\hbar$ in time, we pick up the minus sign from this exponential. These frequencies are called the Matsubara frequencies, and the values of n will have a special significance, which will be developed in the following paragraphs. In addition, we require that τ'' be restricted to lie within the principal periodic region (e.g., $-\hbar\beta < \tau'' < \hbar\beta$) since it is a complex quantity, and we must assure that the summation does not diverge because of this. Principally, this means that the real time is restricted to a finite range. This now leads to the transform itself

$$G(\mathbf{x},\mathbf{x}';\omega_n) = \frac{1}{2}\int_{-\beta\hbar}^{\beta\hbar} d\tau'' e^{i\omega_n\tau''} G(\mathbf{x},\mathbf{x}';\tau'') . \tag{5.22}$$

To proceed, we will split this integral into two parts for the negative values and the positive values as

$$G(\mathbf{x},\mathbf{x}';\omega_n) = \frac{1}{2}\int_{-\beta\hbar}^{0} d\tau'' e^{i\omega_n\tau''} G(\mathbf{x},\mathbf{x}';\tau'') + \frac{1}{2}\int_{0}^{\beta\hbar} d\tau'' e^{i\omega_n\tau''} G(\mathbf{x},\mathbf{x}';\tau'')$$

$$= -\frac{1}{2}\int_{-\beta\hbar}^{0} d\tau'' e^{i\omega_n\tau''} G(\mathbf{x},\mathbf{x}';\tau''+\beta\hbar) + \frac{1}{2}\int_{0}^{\beta\hbar} d\tau'' e^{i\omega_n\tau''} G(\mathbf{x},\mathbf{x}';\tau'')$$

$$= \frac{1}{2}\left(1-e^{-i\omega_n\beta\hbar}\right)\int_{0}^{\beta\hbar} d\tau'' e^{i\omega_n\tau''} G(\mathbf{x},\mathbf{x}';\tau'').$$

$$\tag{5.23}$$

From the first line to the second, we used Eq. (5.18) and then did a coordinate shift on the imaginary time, which leads to the last line. It is important to note that the prefactor vanishes for n even. Had we begun with the boson Bose–Einstein distribution, we would have found that the prefactor vanished for odd values of n. In fact, in Eq. (5.23), the prefactor is unity for odd values of n. What this means is that only the odd n Matsubara frequencies are used for electrons (fermions), while only the even n frequencies are retained for bosons. Hence, for fermions, Eq. (5.22) reduces to

$$G(\mathbf{x},\mathbf{x}';\omega_n) = \frac{1}{2}\int_{-\beta\hbar}^{\beta\hbar} d\tau'' e^{i\omega_n\tau''} G(\mathbf{x},\mathbf{x}';\tau''), \quad \omega_n = \frac{(2n+1)\pi}{\beta\hbar}, \tag{5.24}$$

and these are called the Matsubara Green's function.

To examine the nature of this function, let us expand the field operators in a set of basis functions $\{\varphi_m\}$, so that in Dirac notation, we can write Eq. (5.24) using Eq. (5.17) as

$$G(\mathbf{x};\omega_n) = -e^{\beta\Omega}\sum_{m,m'}\left|\langle m|\widehat{\psi}(\mathbf{x})|m'\rangle\right|^2 e^{-\beta E_m}\int_0^{\beta\hbar} d\tau'' e^{[i\omega_n + (E_m - E_{m'})/\hbar]\tau''}$$

$$= -e^{\beta\Omega}\sum_{m,m'}\left|\langle m|\widehat{\psi}(\mathbf{x})|m'\rangle\right|^2 \frac{e^{-\beta E_m} - e^{-\beta E_{m'}}}{i\omega_n + (E_m - E_{m'})/\hbar}.$$

$$(5.25)$$

In particular, if the basis set is the same as that for which the creation and annihilation operators are defined for the field operator, then the matrix elements are zero except for particular connections between the two coefficients; that is, $m' = m \pm 1$, with the sign determined by which of the two operators is used to define the matrix elements. However, there is no requirement that this basis set be the same as that used in the field operator definition. On the other hand, if we use momentum wavefunctions (the plane waves), then the difference in energies produces just the single momentum $E(k)$, where $\mathbf{k}$ is the Fourier transform for the difference in position $\mathbf{x} - \mathbf{x}'$. In comparing this result with that obtained in earlier chapters, it is clear that we move from the Matsubara Green's function to a normal equilibrium or time-dependent Green's function (in later chapters) through the analytic continuation of $i\omega_n \to \omega$. Finally, in direct analogy with the results of the previous two chapters, it is now possible to write out the retarded and advanced temperature Green's functions as

$$G^r(\mathbf{x},\mathbf{x}';\tau,\tau') = -\theta(\tau - \tau')Tr\left\{\widehat{\rho}T_\tau\left[\psi(\mathbf{x},\tau),\psi^\dagger(\mathbf{x}',\tau')\right]_+\right\}$$
$$G^a(\mathbf{x},\mathbf{x}';\tau,\tau') = \theta(\tau' - \tau)Tr\left\{\widehat{\rho}T_\tau\left[\psi(\mathbf{x},\tau),\psi^\dagger(\mathbf{x}',\tau')\right]_+\right\}.$$

$$(5.26)$$

It should be noted here that the curly brackets are now reserved for the argument of the trace operation and we can no longer use them to indicate the anti-commutator. Hence, we use the subscript "+" sign to indicate this anti-commutator. Of course, the anti-commutator is used for the fermions, while the regular commutator will be used for the bosons.

5.3 Spectral Density and Density of States

As we have discussed in the previous chapters, the spectral density function $A(\mathbf{k}, \omega)$ is another quantity of interest. In the temperature Green's function case, the frequency will be the Matsubara frequency, so that the spectral density is defined from the retarded or advanced temperature Green's functions (5.26) through the Fourier transform of the forms in this latter equation. Thus, we may write this as

$$A(\mathbf{k},\omega_n) = -2\text{Im}\left\{G^r(\mathbf{k},\omega_n)\right\} = 2\text{Im}\left\{G^a(\mathbf{k},\omega_n)\right\}. \tag{5.27}$$

The general Fourier transform form for the temperature Green's function, in a basis expansion, is given in Eq. (5.25). The key factor for us is the denominator of the last expression, and we analytically continue this via the change $i\omega_n \to \omega$, and then use the limiting expression for the retarded Green's function

$$\frac{1}{\omega + (E_m - E_{m'})/\hbar + i\eta/\hbar} = P\frac{1}{\omega + (E_m - E_{m'})/\hbar} \tag{5.28}$$
$$- i\pi\delta\big(\omega + (E_m - E_{m'})/\hbar\big).$$

We can now use this in Eq. (5.25) to find the spectral density as

$$A(\mathbf{k},\omega) = 2\pi e^{\beta\Omega} \sum_{m,m'} \left|\langle m|\hat{\psi}(x)|m'\rangle\right|^2 \left(e^{-\beta E_m} - e^{-\beta E_{m'}}\right)\delta\big(\omega - (E_m - E_{m'})/\hbar\big)$$
$$= 2\pi e^{\beta\Omega}\left(1 + e^{-\beta\omega\hbar}\right)\sum_{m,m'} \left|\langle m|\hat{\psi}(x)|m'\rangle\right|^2 e^{-\beta E_m}$$

$$\tag{5.29}$$

with the restriction that the two basis states, over which the matrix element is computed, must lie on the same energy shell. It is easy to show that this spectral density satisfies the normality imposed by [4]

$$\int \frac{d\omega}{2\pi} A(\mathbf{k},\omega) = 1, \tag{5.30}$$

as was done in the earlier chapters, so will not be repeated here.

In the non-interacting case, we expect the spectral density to be a delta function relating the energy to the momentum, as was discussed in the first chapter. The broadening of this function by the interaction eliminates the simple relationship between energy and momentum that exists in classical mechanics and gives us the off-shell contributions, which is the major introduction of quantum effects in this regard. Finally, we note that by comparing Eqs. (5.25) and (5.29), we can write the Green's functions in terms of the spectral density as

$$G^r(\mathbf{k},\omega_n) = \int \frac{d\omega}{2\pi} \frac{A(\mathbf{k},\omega)}{i\omega_n - \omega}$$

$$G^r(\mathbf{k},\omega) = \int \frac{d\omega'}{2\pi} \frac{A(\mathbf{k},\omega')}{\omega - \omega' + i\eta/\hbar} \tag{5.31}$$

for the two types of Green's functions we have been discussing. The connection between these two equations is an example of the process of analytic continuation. Here we take the complex frequency $i\omega_n$ and analytically continue it to the proper real frequency (with an imaginary convergence factor, as has been used in the previous chapters), $\omega + i\eta$.

In a classical system, the concept of the density of states (per unit energy per unit volume) arises from counting the number of states which lie on this energy shell. Hence, the density of states can be written as

$$\rho_d(\omega) = \frac{2}{\hbar} \int \frac{d^d\mathbf{k}}{(2\pi)^d} \delta(\omega - E(k)/\hbar), \tag{5.32}$$

where d is the dimensionality of the system, and the factor of two arises from a summation over the spins. For three dimensions, one recovers the familiar

$$\rho_3(\omega) = \frac{1}{2\pi^2} \left(\frac{2m^*}{\hbar^2} \right)^{3/2} \sqrt{\hbar\omega}, \tag{5.33}$$

where it is clear that the frequency plays the role of the energy (reduced by the reduced Planck constant).

In the quantum situation, a similar form arises except that we need to consider the off-shell contributions, and this is taken into account via the spectral density. Hence, we may write the density of states as

$$\rho_d(\omega) = \frac{2}{2\pi\hbar} \int \frac{d^d\mathbf{k}}{(2\pi)^d} A(\mathbf{k},\omega)$$

$$= -\frac{2}{\pi\hbar} \int \frac{d^d\mathbf{k}}{(2\pi)^d} \mathrm{Im}\{G^r(\mathbf{k},\omega)\}, \tag{5.34}$$

where again a factor of 2 represents the sum over spins. Now, by the one-particle density of states, we do not mean the quantity for a single electron, but rather the density of states that is appropriate for a single electron in the sea of other electrons and impurities. The temperature Green's function can be recognized through the reverse of the analytic continuation procedures, and this may be related to the non-interacting Green's function and the interaction-induced self-energy via Dyson's equation, given in the last chapter. By this approach, we find the density of states to be

$$\rho_d(\omega) = -\frac{2}{\pi\hbar}\int\frac{d^d\mathbf{k}}{(2\pi)^d}\,\text{Im}\left\{\frac{1}{\left[G^0(\mathbf{k},\omega)\right]^{-1} - \Sigma^r(\mathbf{k},\omega)}\right\}. \tag{5.35}$$

In essence, the result for the interacting Green's function is essentially the same as that of the last chapter for equilibrium Green's functions and arises from the properties of the linear perturbation expansion and resummation that leads to Dyson's equation itself, and not from any particular property of any one type of Green's function.

Thus, the spectral density in the interacting system represents the entire density of states, including both the single-particle properties and their modification that arises from the self-energy corrections that come from the presence of the interactions. In the next few sections, we will return to the electron–electron interaction and its modification in the presence of strong impurity scattering. This will entail an approach that is quite different from the one usually found in high-mobility materials, where the electron–electron interaction is thought to be the dominant interaction process.

5.4 Conductivity Again

Let us now examine the temperature dependence of the conductivity that arises merely from impurity scattering, without the complications of the electron–electron interaction. There is not much change in the self-energy given earlier, so the diagonal conductivity is still given as (with the temperature Green's functions inserted in place of the zero-temperature functions)

$$\sigma_{\alpha\beta} = \frac{e^2\hbar^2}{m^{*2}}\sum_{\mathbf{k}}\frac{k_\alpha k_\beta}{\beta\hbar^2}\sum_{\omega_n}$$

$$\times\frac{1}{i\omega_n - [E(k)-\mu]/\hbar + i/2\tau}\frac{1}{i\omega_n - [E(k)-\mu]/\hbar - i/2\tau}, \tag{5.36}$$

where we have assumed that the conductivity bubble is given merely by the two Green's functions as shown in Fig. 5.1. There are two aspects to this equation that are different than the earlier result. The first is that the frequency summation given here is actually the resolution of the delta function. The second is that the frequency sums are more difficult than those encountered in the previous section. This is because of the delta function that was incorporated into the earlier results. This delta function is broadened at finite temperature, and some effect must occur to lead to this behavior. The actual complication arises from the presence of the self-energy terms (the scattering terms) in the Green's functions. The presence of these energy-dependent self-energy terms means that the complex integration becomes more complicated in that there may be branch cuts that arise from the presence of the self-energy. Consequently, the best method of attacking the problem is to go back to the actual polarization bubble itself and use the limiting process

$$\sigma = - \lim_{\omega_n \to 0} \left[\frac{\text{Im}\{\Pi(\omega_n)\}}{\omega_n} \right].$$
(5.37)

Figure 5.1 Simplest bubble diagram for the polarization or the conductivity.

The product of the two Green's functions must first be rewritten as the convolution that it should be, and then the lowest-order bubble approximation (Fig. 5.1) to the polarization can be written as

$$\Pi(\omega_n) = \frac{e^2 \hbar^2}{m^{*2}} \sum_{\omega_m} \frac{\hbar}{i\hbar\omega_m - [E(k) - \mu] + i\hbar/2\tau}$$

$$\times \frac{\hbar}{i\hbar(\omega_m + \omega_n) - [E(k) - \mu] + i\hbar/2\tau}.$$
(5.38)

In this integration, however, the contributions from the simple pole of the Green's function that occurred earlier must be replaced by the contributions from the two branch cuts. To see this, we must use a relationship between the summation over frequency and a contour integral, which can be written as [3]

$$\lim_{\eta \to 0} \sum_{\omega_n} \frac{e^{i\omega_n \eta}}{i\omega_n - x} = \frac{\beta\hbar}{2\pi i} \oint_C \frac{dz}{e^{\hbar\beta z} - 1} \frac{e^{\eta z}}{z - x} = \frac{\beta\hbar}{2\pi i} \oint_C \frac{dz}{e^{\hbar\beta z} - 1} g(z). \quad (5.39)$$

With this conversion, we see that the poles will arise from the contour integral. Now we see that there are two branch cuts at $z = \xi$ and $z = \xi - i\omega_n$, where we have written $\xi = E(k) - \mu$, as shown in Fig. 5.2. The contour remains the basic circle of radius R in which the limit $R \to \infty$ is taken. This contour must be deformed, however, to create two line integrals, one above and one below each branch cut. Let us consider first the contribution from the branch cut at $z = \xi$, as shown in Fig. 5.3. The contribution of this quantity to Eq. (5.38) is given by

$$\frac{\beta\hbar}{2\pi i} \left\{ \int_{-\infty}^{\infty} d\xi G(\mathbf{k}, \xi + i\delta) G(\mathbf{k}, \xi + i\omega_n) - \int_{-\infty}^{\infty} d\xi G(\mathbf{k}, \xi - i\delta) G(\mathbf{k}, \xi + i\omega_n) \right\} f(\xi)$$

$$= \frac{\beta\hbar}{2\pi i} \int_{-\infty}^{\infty} d\xi \left[G(\mathbf{k}, \xi + i\delta) - G(\mathbf{k}, \xi - i\delta) \right] G(\mathbf{k}, \xi + i\omega_n) f(\xi).$$

$$(5.40)$$

where the distribution function arises from evaluating the first term in the contour integral in terms of the critical values of z, hence producing the Fermi–Dirac function. The first term in the square brackets is the retarded Green's function, and the second term is the advanced Green's function. The bracketed term is just the spectral density. Contribution of the second branch cut, at $z = \xi - i\omega'_m$ can similarly be calculated (in this case, the second Green's function is expanded on either side of the branch cut), which gives

$$\frac{\beta\hbar}{2\pi i} \int_{-\infty}^{\infty} d\xi A(\mathbf{k}, \xi) \left[G(\mathbf{k}, \xi + i\omega_n) + G(\mathbf{k}, \xi - i\omega_n) \right] f(\xi). \quad (5.41)$$

To proceed further, we take the analytic continuation of the last equation. This makes the change $i\omega'_m \to \omega + i\eta$, which again converts the two Green's functions in the square brackets into the retarded and advanced functions, respectively. Then the imaginary parts of these two Green's functions (everything else is real) contribute just another factor of the spectral density. We make a change of variables in the second term by shifting the axis of the ξ integration, and

$$\text{Im}\{I\} = -\frac{\beta\hbar}{2\pi i}\int_{-\infty}^{\infty} d\xi A(\mathbf{k},\xi)\text{Im}\big[G(\mathbf{k},\xi+\omega+i\eta)+G(\mathbf{k},\xi-\omega-i\eta)\big]f(\xi)$$

$$= \frac{\beta\hbar}{2\pi i}\int_{-\infty}^{\infty} d\xi A(\mathbf{k},\xi)\big[A(\mathbf{k},\xi+\omega)-A(\mathbf{k},\xi-\omega)\big]f(\xi)$$

$$= \frac{\beta\hbar}{2\pi i}\int_{-\infty}^{\infty} d\xi A(\mathbf{k},\xi)A(\mathbf{k},\xi+\omega)\big[f(\xi)-f(\xi+\omega)\big]$$

$$= -\frac{\beta\hbar}{2\pi i}\int_{-\infty}^{\infty} d\xi A(\mathbf{k},\xi)A(\mathbf{k},\xi+\omega)\frac{\partial f(\xi)}{\partial\omega}.$$

(5.42)

This now leads to the conductivity as

$$\sigma = -\frac{e^2\hbar^2}{m^{*2}}\sum_{\mathbf{k}}\frac{k^2}{4\pi\hbar}\int_{-\infty}^{\infty} d\xi A^2(\mathbf{k},\xi)\frac{\partial f(\xi)}{\partial\xi}.$$

(5.43)

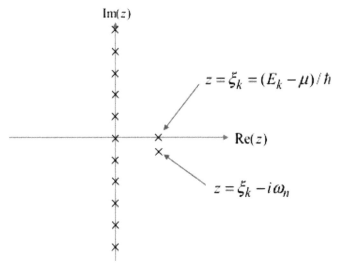

Figure 5.2 The poles from the Green's functions are indicated by the two arrows. The poles along the imaginary axis come from the exponential in Eq. (5.39).

The spectral density remains a fairly sharply peaked function if the scattering is not too strong, and this is the case in semiconductors.

Thus, the derivative of the distribution function can be brought out of the integral and evaluated at the shell (energy) value, and then the integral becomes

$$\frac{1}{\pi}\int_{-\infty}^{\infty} d\xi A^2(\mathbf{k},\xi) = \frac{1}{\pi}\int_{-\infty}^{\infty} d\xi \left(\frac{\Sigma_i}{\xi^2 + \Sigma_i^2}\right)^2 = \frac{1}{\Sigma_i} = \frac{2\tau}{\hbar}. \tag{5.44}$$

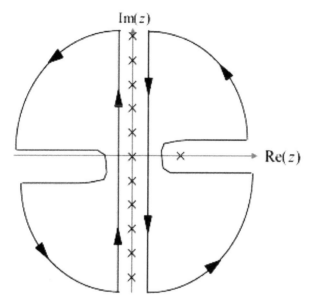

Figure 5.3 The integration contour around the branch cut at $z = \xi$, which lies on the real axis.

Hence, we can finally write the conductivity as

$$\sigma = -\frac{2e^2\hbar^2}{m^{*2}}\int \frac{k^2\tau d^d\mathbf{k}}{(2\pi)^d}\frac{\partial f(E)}{\partial E} = -\frac{2e^2}{m^*}\int dE\rho(E)E\tau(E)\frac{\partial f(E)}{\partial E}. \tag{5.45}$$

This last result is essentially the classical result obtained with the Boltzmann equation, which is reassuring, but it depends on a special interpretation of the result of using the frequency summations. We note that the angular integration factor normally found for impurity scattering, the factor $(1 - \cos\theta)$, which was shown in the last chapter to require the next higher-order terms, is missing here, since it must usually be put into the Boltzmann equation by hand. We can put the present result into the more usual form as

$$\sigma = \frac{ne^2 \langle \tau \rangle}{m^*} \quad , \quad \langle \tau \rangle = \frac{2}{n}\int dE \rho(E)E\tau \frac{\partial f}{\partial E} \ , \tag{5.46}$$

where

$$n = -2\int dE \rho(E)E \frac{\partial f}{\partial E} \tag{5.47}$$

is the form of the density used in Boltzmann transport theory.

It is important to note at this point that the main result of the integration contained above is to take the effective zero-temperature conductivity and average it over an energy width of the order of $k_B T$ at the Fermi energy. This latter is given by the width of the derivative of the Fermi function. In the low-temperature limit, this derivative approaches a delta function at the Fermi energy, and this leads to the low-temperature conductivity.

5.5 Electron–Electron Self-Energy

One of the more interesting quantities that arises in mesoscopic devices at low temperature is the phase-breaking time. We will discuss this further below, but the most common source of the phase-breaking process at low temperatures comes from the electrons (or holes) themselves. Scatterings from the screened potential of the other electrons and from the collective plasmon modes are both part of the total electron–electron interaction among the free carriers. Generally, it is not possible to consider the full Coulomb interaction beyond the lowest order of perturbation theory because of the long range of the potential associated with this interaction. As discussed above, the dielectric function has singularities at the plasmon frequency and at zero frequency, corresponding to the plasmon modes and the single-particle scattering, respectively. Formally, one can split the summation over $\mathbf{q}$ that appears in a Fourier transform of the potential into a short-range (in real space, which means large q) part, for which $q > q_c$, and a long-range (in real space, which means small q) part, for which $q < q_c$, where q_c is a cutoff wavevector defining this split. It was shown some time ago that the short-range part of the potential corresponds to the screened Coulomb interaction for single-particle scattering. The long-range part, on the other hand, is responsible for scattering by the collective oscillations of the electron gas, which describe the motion of the electrons in the field produced by their own Coulomb potential, the *plasmons*. In fact, these are just the modes we discussed in the plasmon-pole approximation in Section 4.3.2. These collective oscillations are bosons, so their scattering rate can be calculated in much the same way as phonons—the distribution function that we include is that of the plasmons rather than the free electrons. The only difference from the normal approach is that we will find the need to incorporate a maximum q (= q_c) that can be involved in the scattering, and this cutoff is essentially the Debye (or Fermi–Thomas) wavevector. We will find that, in the case of plasmons, which will be treated later, there are two regimes: a low-temperature regime where the phase-breaking time is independent of temperature and a high-temperature regime where the phase-breaking time decreases as a power of the temperature. The transition region is where the plasmon energy is comparable to the thermal energy. In the case of the single-particle scattering, we will not find this temperature independence at low temperatures, so that the experimental results dictate the importance of the plasmon contribution to the phase-breaking process. We begin with the latter single-particle case, which is evaluated from the self-energy contributions of the electron–electron scattering.

Before delving into the details of the self-energy, we want to talk a little further about dynamic screening and its effects with temperature. We will begin with the form (4.60) for the dielectric function from the last chapter. There are a number of factors that have to be considered. First, the product of the Fermi–Dirac functions for small $q \to 0$ gives essentially the derivative of the function, which is equivalent to a delta function at the Fermi energy (at sufficiently low temperature). But for this term, the angular average vanishes. If we wish to keep the frequency variables in the dielectric function, we have two approaches to consider. In one approximation, we expand the term in $f(k + q)$ and expand the energy differences, keeping only the lowest-order terms in q. The expansion of the energy around the momentum k yields a function of the angle between the two vectors, which is involved in the averaging process of the d-dimensional integration. In the second approximation, we use the expansion already found in Chapter 4. In the former case, we arrive at the derivative of the distribution function, which is useful for non-degenerate materials. However, in the case of degenerate materials, we must use the second approach, and we will follow this approach. Thus, we begin with the slightly modified Eq. (4.60), which leads to the dielectric function

$$\varepsilon(\mathbf{q},\omega) = \varepsilon_s + \frac{2e^2}{\hbar q^2} \int \frac{d^d \mathbf{k}}{(2\pi)^d} \frac{f(\mathbf{k}) - f(\mathbf{k}+\mathbf{q})}{\omega + [E(\mathbf{k}+\mathbf{q}) - E(k)]/\hbar - i/\tau}. \quad (5.48)$$

For a degenerate semiconductor, we will rewrite this expression as

$$\varepsilon(\mathbf{q},\omega) = \varepsilon_s + \frac{2e^2}{\hbar q^2} \int \frac{d^d \mathbf{k}}{(2\pi)^d} f(\mathbf{k})$$

$$\times \frac{E(k) - E(\mathbf{k}+\mathbf{q}) - E(\mathbf{k}-\mathbf{q})}{\{\omega - [E(k) - E(\mathbf{k}-\mathbf{q})]/\hbar - i/\tau\}\{\omega - [E(\mathbf{k}+\mathbf{q}) - E(k)]/\hbar - i/\tau\}}$$

$$(5.49)$$

We now expand the numerator as

$$2E(\mathbf{k}) - E(\mathbf{k}+\mathbf{q}) - E(\mathbf{k}-\mathbf{q}) \approx -\frac{\hbar^2 q^2}{m*}, \quad (5.50)$$

so that Eq. (5.49) becomes

$$\varepsilon(\mathbf{q},\omega_n) = \varepsilon_s - \frac{2e^2}{m*} \int \frac{d^d \mathbf{k}}{(2\pi)^d} f(\mathbf{k}) \frac{1}{(\omega + \mathbf{q} \cdot \mathbf{v} - i/\tau)^2}$$

$$= \varepsilon_s + \frac{2e^2 \tau^2}{m*} \int \frac{d^d \mathbf{k}}{(2\pi)^d} f(\mathbf{k}) \frac{1}{[1 + i(\omega + \mathbf{q} \cdot \mathbf{v})]^2}. \quad (5.51)$$

The fraction can be expanded as

$$F = \frac{1}{[1 + i(\omega + \mathbf{q} \cdot \mathbf{v})\tau]^2}$$

$$= \frac{1}{(1 + i\omega\tau)^2} \left[1 - 2i \frac{\mathbf{q} \cdot \mathbf{v}\tau}{1 + i\omega\tau} - 3 \frac{(\mathbf{q} \cdot \mathbf{v}\tau)^2}{(1 + i\omega\tau)^2} + \dots \right]. \quad (5.52)$$

The integration in Eq. (5.51) involves an integral over the angle, which can be rotated to be the angle involved in the dot product, and only the first and third terms survive this integration. After the integration, the terms can be resumed and the other integrals performed to give

$$\varepsilon(q,\omega) = \varepsilon_s + \frac{ne^2}{m*} \frac{\tau^2}{(1+i\omega\tau)^2 + Dq^2\tau}$$

$$= \varepsilon_s \left[1 + \frac{\omega_p^2\tau^2}{(1+i\omega\tau)^2 + Dq^2\tau} \right],$$

(5.53)

where $D = v^2\tau/d$ is the diffusion coefficient.

We would now like to calculate the self-energy for the single-particle electron–electron interaction to gain an insight into the scattering dynamics. For this, we will use the dynamic screening approximation above in the polarizability, and the effective interaction can then be written as

$$V_{eff} = \frac{e^2/2\varepsilon_s q}{1 + \frac{\omega_p^2\tau^2}{(1+i\omega\tau)^2 + Dq^2\tau}},$$

(5.54)

where we have written this for the case of a two-dimensional semiconductor, or electron gas, as is appropriate, e.g., at the GaAs/GaAlAs interface. We now want to reintroduce the Matsubara frequencies to go with the Green's functions we derived above. Then, this effective potential becomes

$$V_{eff} = \frac{e^2}{2\varepsilon_s q} \frac{(1-\omega_n\tau)^2 + Dq^2\tau}{(1-\omega_n\tau)^2 + Dq^2\tau + \omega_p^2\tau^2}.$$

(5.55)

Since both the Green's function and the interaction strength are functions of frequency, and since these would be a product in real space, it is necessary to convolve them in frequency space, so the self-energy can be written as

$$\Sigma_{ee}^{r,a}(\mathbf{k},\omega_n) = -\sum_{\omega_m} \frac{1}{\beta\hbar^2} \int \frac{d^2\mathbf{q}}{(2\pi)^2} G^{r,a}(\mathbf{k}-\mathbf{q},\omega_m+\omega_n)$$

$$\times \frac{e^2}{2\varepsilon_s q} \frac{(1-\omega_m\tau)^2 + Dq^2\tau}{(1-\omega_m\tau)^2 + Dq^2\tau + \chi_2 q\tau^2},$$

(5.56)

where the two-dimensional plasma frequency is given by

$$\omega_p^2 = \chi_2 q \quad , \quad \chi_2 = \frac{n_s e^2}{2m*\varepsilon_s}$$

(5.57)

and n_s is the two-dimensional sheet density. Fukuyama [5] has argued that we are really interested in cases in which the momentum change is quite small, so that the energy $E(\mathbf{k})$ in the Green's function differs little from the Fermi energy. Fukuyama makes the additional observation that since we are interested in a small frequency change, the Green's functions can be approximated as

$$G^{r,a}(\mathbf{k},\omega_n) = \frac{1}{i\omega_n - [E(k)-\mu]/\hbar \pm i/2\tau} \sim \mp 2i\tau \, , \qquad (5.58)$$

and this was certainly found to be the case in Section 4.5, where we computed the convolution integral of two Green's functions for the conductivity bubble. Here, τ is the broadening in the Green's function that arises from other scattering, such as from impurity scattering. With this approximation, the self-energy is momentum independent, since the last momentum variables will be integrated, as will the frequency dependence. However, this is not quite true if we take the zero of frequency at the Fermi energy for convenience; the range of the convolution integral is limited to the singular case for which one frequency is below the Fermi energy and the other is above the Fermi energy, or $\omega_n(\omega_n + \omega_m) < 0$. This limits the summation over frequency to those values for which $\omega_m < -\omega_n$. We can now rewrite the self-energy as

$$\Sigma_{ee}^{r,a}(\mathbf{k},\omega_n) \approx - \sum_{\omega_m < -\omega_n} \frac{2i\tau}{\beta\hbar^2} \int \frac{d^2q}{(2\pi)^2} \frac{e^2}{2\varepsilon_s q} \frac{Dq^2\tau}{(1-\omega_m\tau)^2 + Dq^2\tau + \chi_2 q\tau^2} \, .$$

$$(5.59)$$

It is clear that this formula has singularities. The self-energy is singular in two dimensions, in the limit of $q,\omega_n \to 0$, with a variation as $\ln(\omega_n) \sim \ln(T)$. Nevertheless, this integral can be rewritten as

$$\Sigma_{ee}^{r,a}(\mathbf{k},\omega_n) \approx - \sum_{\omega_m < -\omega_n} \frac{i\tau^2 e^2 D}{2\beta\hbar^2 \varepsilon_s \pi} \int_0^\infty \frac{q^2 dq}{(1-\omega_m\tau)^2 + Dq^2\tau + \chi_2 q\tau^2} \, .$$

$$(5.60)$$

There is a general problem with which integration should be pursued first: that over q or that from the summation over ω_m. Either leads to the need to introduce a cutoff into the actual integration, either a cutoff on the largest value for q or in the lower frequency limit. This is slightly easier in the case of the momentum, so we can rewrite the last equation in the leading terms as

$$\Sigma_{ee}^{r,a}(\mathbf{k},\omega_n) \approx -\sum_{\omega_m<-\omega_n} \frac{i\tau e^2}{2\beta\hbar^2\varepsilon_s\pi}\left\{ q_{max} - \frac{\chi_2\tau}{2D}\ln\left[1+\frac{q_{max}^2 D\tau}{(1-\omega_m\tau)^2}\right]\right.$$
$$\left. -\left[1+\frac{(\chi_2\tau)^2\tau/D}{(1-\omega_m\tau)}\right]\left[\frac{\pi}{2}-\tan^{-1}\left(\frac{D\tau}{1-\omega_m\tau}\right)\right]\right\}.$$

$$(5.61)$$

Generally, the second and third terms in the curly brackets are small and can be neglected. The value of q_{max} is often taken to be the effective diffusion length, and we can write the self-energy as

$$\Sigma_{ee}^{r}(\mathbf{k},\omega_n) \sim -i\frac{\tau e^2}{2\beta\hbar^2\varepsilon_s\pi\sqrt{D\tau}}. \qquad (5.62)$$

The scattering time in the function is the broadening of the spectral density and arises from the sum of all types of scattering processes— impurities, electron–electron, phonons, etc. On the other hand, we can define an electron–electron time from this self-energy as

$$\frac{1}{\tau_{ee}} = -\frac{1}{\hbar}\text{Im}\,\Sigma_{ee}^{r}(\mathbf{k},\omega_n). \qquad (5.63)$$

If we relate this to the phase-breaking process, then the factor of β assumes importance as it says that the phase-breaking rate increases linearly with temperature, or that the phase-breaking time decreases as $1/T$. Similar temperature dependences for the single-particle scattering have been found by Chaplik [6] and Giuliani and Quinn [7]. We will see later that the plasmon scattering terms also give this temperature behavior at high temperatures, but also produce a temperature-independent contribution at low temperatures.

5.6 Self-Energy in the Presence of Disorder

It is now clear that the nature of the temperature variations will be found to some extent in the self-energy, since it describes the important broadening of the single-particle density of states.

This, in turn, relates to the mixing of phase-coherent states due to broadening of the Fermi distribution function. Thus, the task *in disordered systems* is to define the self-energy in the system in which both impurities and electron–electron interactions (and possibly others as well) exist. In treating the self-energy, it is necessary to decide how one wants to resum the various diagrams for the interactions. For example, in disordered systems, it was assumed in the last chapter that the disorder-inducing impurity interactions are the dominant interactions, and the electron–electron interactions were ignored except as they screened the impurities. Above, the electron–electron interaction was assumed to be the dominant interaction. In this section, we will return to consider that the disorder-inducing impurity interactions are the dominant scattering process, and that the electron–electron interactions are a perturbation of this process. Now, clearly, the entire set of Feynman diagrams that relate to scattering by impurities and by the electron–electron Coulomb interaction can be separated into sets of terms that lead to a sum of two distinct self-energies. Thus, the impurity scattering processes are treated as previously, and we add a new self-energy term Σ_{ee}, which accounts for the interaction effects that are themselves mediated by the impurities. If these were the only diagrams in the perturbation expansion, life would indeed be quite easy. However, we must recall that, for the impurities alone, we wound up with a great many different types of diagrams. There was the simple summation of independent Coulomb scattering events, which were impurity averaged and which led to the simple impurity self-energy found in the last chapter. In addition, however, the two-particle Green's function contribution to the polarization in the Kubo formula led to a set of ladder diagrams, which led to a correction to the conductivity which actually accounted for the angular correction to the momentum relaxation time. We will also discover just below another set of impurity diagrams that will be important here and also lead to weak localization in the next section.

The problem with two perturbing species at hand is what to do with diagrams, where the two perturbations interfere with each other. The question is really to which of the self-energies will we ascribe such terms (or should they be ignored), and then how are they to be included in the selected self-energy. The normal approach (normal in the sense of semi-classical transport with the Boltzmann equation) is to assume that the electron–electron interaction is the dominant scattering process. This usually results in some sort of assumption of a drifted Maxwellian (or Fermi–Dirac) distribution, with the interaction being subsequently ignored except for its role in screening other scattering processes. This leads to a philosophy of the following strategic approach for quantum transport when the various perturbing interactions interfere with one another: (1) an assumption is made regarding the dominant perturbation interaction, and this self-energy is treated as if the other processes are absent; and (2) the other interactions are treated as if they are *screened* by the dominant perturbation through the introduction of *vertex corrections.* The vertex is the point (see Fig. 4.1, for example) where the interaction line meets the two Green's functions. Normal screening by the electrons is such a vertex correction, where, e.g., we use the screened impurity interaction rather than the bare Coulomb interaction of the impurity.

In disordered materials, where the disorder is induced by the heavy impurity scattering, the impurity scattering is the dominant perturbation. Thus, in this section, we will screen the electron–electron interaction by the impurity interaction, through the introduction of a vertex correction. That is, we calculate the interaction effects not between free electrons (which would be simple plane-wave states) but between the diffusive electrons. This is handled by the means of a vertex correction for the Coulomb interaction between the diffusing electrons. The impurity self-energy is still assumed to be given by its previous form, but with the temperature dependence included. This temperature dependence does not occur within the self-energy but arises from the energy (frequency) dependence of $\tau(\omega)$, which is coupled to an energy dependence when the integration over frequency is performed in the calculation of the conductivity (and, therefore, of an average relaxation time). Here, we are interested in investigating the self-energy term arising from the interaction between diffusing electrons in a sea of impurity scattering.

So the electron–electron interaction is going to be subject to a vertex correction. Let us examine just what this means. The structure that we imagine is shown in Fig. 5.4. The two momenta $\mathbf{k}$ and $\mathbf{k} + \mathbf{q}'$ on the left of (the right side of) the diagram (at the bottom) become the momenta $\mathbf{k} + \mathbf{q}$ and $\mathbf{k} + \mathbf{q} + \mathbf{q}'$, respectively, where $\mathbf{q}'$

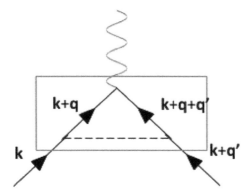

Figure 5.4 Diagram of one point where the photon line of the electron–electron interaction meets two Green's functions with the insertion (red box) of the impurity interaction.

is the momentum transferred via the electron–electron interaction. Thus, the two Green's function lines at the top of the two-particle structure are pulled together to meet at the vertex of the electron-electron interaction. Now, $\mathbf{q}$ is the total momentum transferred to the impurities in the ladder diagram summation. The vertex corrections then are the resummations of the ladder diagrams that represent the two-particle Green's function dressing of the vertex of the electron-electron interaction.

As a consequence, the vertex corrections lead to a Coulomb interaction not between two free electrons, but between two "dressed particles" in which the dressing is described by the impurity interaction. These interactions are appropriate only for diffusive transport when the energy exchange between the carriers is quite small; that is, the diffusive interactions with the impurities involve very small energy exchange (it was assumed to be zero in the last chapter). In essence, the Green's function and interaction in the red box of Fig. 5.4 constitute the second term in the two-particle Green's function of Fig. 4.12. We need to include two sets of diagrams for completeness in calculating the conductance. The first is the set of ladder diagrams at the upper left of Fig. 4.11 and treated in the last chapter to get the correction to the conductance. This correction term is referred to as the *diffusion* correction and leads to a contribution to the polarization as

$$\Pi(\mathbf{k},\omega_n) = \sum_{\omega_m} \int \frac{d^3q}{(2\pi)^3} \frac{1}{i\omega_m - [E(k)-\mu]/\hbar + i/\tau + i/\tau_\varphi}$$

$$\times \frac{1}{i(\omega_m + \omega_n) - [E(k)-\mu]/\hbar - i/\tau - i/\tau_\varphi}, \qquad (5.64)$$

and we have added a second time due to the phase-breaking process, which does not involve the impurities but will give additional broadening to the Green's functions. Computing the residues from the poles in the frequency space and integrating over the momentum, we find that the core of the two-particle Green's function is given as

$$\Lambda(\mathbf{q},\omega_n) = \frac{\tau_\varphi}{2\pi\hbar\rho(E_F)\tau^2 \left(1 + Dq^2\tau_\varphi + |\omega_n|\tau_\varphi\right)}. \qquad (5.65)$$

In the last chapter, we found the relationship

$$\Gamma_0 = \frac{1}{\hbar^2} n_i V_0^2 = \frac{1}{2\pi\hbar\rho(E_F)\tau}, \qquad (5.66)$$

so that the net screened electron–electron interaction term can be written as

$$V_{\text{eff}} = \frac{V_{ee}(\mathbf{q},\omega_n)}{\left[1/\tau_\varphi + Dq^2 + |\omega_n|\right]^2 \tau^2} \sim \frac{V_{ee}(\mathbf{q},\omega_n)}{\left[Dq^2 + |\omega_n|\right]^2 \tau^2}. \qquad (5.67)$$

The denominator here is squared since the vertex correction is applied at each end of the interaction line between the electrons. This is because the diffusions represent the modification of the matrix elements by the dressed diffusing particles.

To proceed, we are going to consider the two-dimensional case, since it is of interest in modern semiconductor devices. Since both the Green's function and the interaction strength are functions of frequency, and since these would be a product in real space, it is necessary to convolve them in frequency space, and the self-energy can be written (just as in the last section) as

$$\Sigma_{ee}^r(\mathbf{k},\omega_n) = -\sum_{\omega_m} \frac{1}{\beta\hbar^2\tau^2} \int \frac{d^2q}{(2\pi)^2} G^r(\mathbf{k}-\mathbf{q},\omega_m+\omega_n) \frac{V_{ee}(\mathbf{q},\omega_m)}{\left[Dq^2 + |\omega_m|\right]^2}. $$

$$(5.68)$$

Fukuyama [5] argues that we are really interested in cases in which the momentum change is quite small, so that the energy $E(k)$

differs little from the Fermi energy. Since we are interested in a small frequency change, Fukayama also observes that the Green's functions can be approximated as we have already done in Eq. (5.58). Then, we get

$$\Sigma_{ee}^r(\mathbf{k},\omega_n) = -\sum_{\omega_m} \frac{1}{\beta\hbar^2\tau} \int \frac{d^2q}{(2\pi)^2} \frac{V_{ee}(\mathbf{q},\omega_m)}{\left[Dq^2 + |\omega_m|\right]^2}. \qquad (5.69)$$

It is clear that this formula has singularities. To proceed, we will take the interaction potential as a constant, and then the momentum integration can be done to give

$$\Sigma_{ee}^r(\mathbf{k},\omega_n) = \sum_{\omega_m < -\omega_n} \frac{V_{ee}}{2\pi\beta\hbar^2 D\tau} \frac{1}{\left[Dq^2 + |\omega_m|\right]}$$

$$= \frac{1}{\tau} \frac{V_{ee}}{4\pi^2\hbar^2 D} G_2(\beta,\omega_n), \qquad (5.70)$$

where

$$G_2(\beta,\omega_n) = \frac{2\pi}{\beta\hbar} \sum_{\omega_m < -\omega_n} \frac{1}{|\omega_m|} = \ln\left(\frac{\beta\hbar}{2\pi\tau}\right) - \psi\left(\frac{\beta\hbar\omega_n + 1}{2}\right), \qquad (5.71)$$

and $\psi(x)$ is the digamma function of x. This self-energy can be added to that from impurity scattering to produce a correction to the conductivity by writing the net scattering rate as

$$\frac{1}{\tau} \rightarrow \frac{1}{\tau}\left[1 + \frac{V_{ee}}{4\pi^2\hbar^2 D} G_2(\beta,\omega_n)\right]. \qquad (5.72)$$

The second term has a logarithm divergence as the temperature is lowered and leads to a correction to the conductivity that has a similar behavior as that found in weak localization to be discussed next. At high temperatures, this term is usually ignored.

5.7 Weak Localization

In the presence of strong impurity scattering, an interesting effect can occur. An electron can scatter off a series of impurities in a manner in which its momentum is gradually reversed so that it winds up going in the opposite direction to that at the beginning. Time reversal symmetry tells us that it is possible to go around this scattering loop in two directions, one of which is the exact opposite of the other, as shown in Fig. 5.5. In fact, the two trajectories are time reversals of each other, and this means that there will be a continual interaction between these two electrons, which produces a reduction in the conductivity of the material, which is referred to as *weak localization*. An important issue here is the role of a magnetic field, which breaks time reversal symmetry. For example, if the magnetic field is in a direction for which the clockwise trajectory is gradually rotated to the left for each segment, the counter-clockwise trajectory will be rotated to the right in each segment. Then, the two trajectories are no longer the same and the interference will be broken up by the magnetic field. It is possible to actually compute the correction to the conductivity with our Green's functions, but the question is which set of diagrams are required to deal with this problem. In dealing with the corrections to the simple conductivity in Section 4.5, we used the upper left set of diagrams from Fig. 4.11. The next set of diagrams, which was ignored in this earlier treatment, is the set in the upper right of Fig. 4.11. These diagrams are referred to as the maximally crossed diagrams, and the series of such diagrams is shown in Fig. 5.6. It turns out that these diagrams, with a suitable approximation, give rise to the required conductivity correction for weak localization.

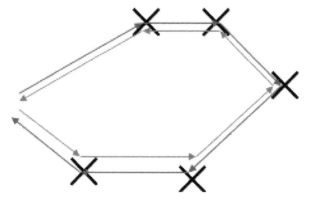

Figure 5.5 Two time-reversed paths by which an electron can scatter from multiple impurities in a manner that reverses its momentum.

Now, these diagrams look relatively intractable. However, a simple trick can produce an easier sum to evaluate in order to determine their contribution to the kernel Λ in the Bethe–Salpeter equation. It is the arrangement of the Green's function that must be addressed. What will be done is to reverse the direction of the Green's functions of one of the lines in the figure. This is usually done to the hole (bottom) line so that the diagram becomes an electron–electron interaction rather than an electron–hole interaction via the impurities. This changes each **k** into a –**k** in the Green's functions, which reverses the directions of one of the two particles, as shown in Fig. 5.7. Then the maximally crossed diagrams become just another ladder diagram, albeit with differences in the Green's functions. Such a reversal was first discussed in connection with superconductivity, so that the new set of diagrams is usually now called the *cooperon* contribution.

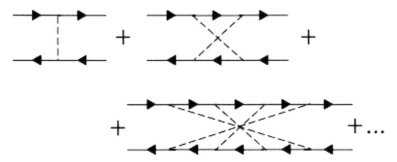

Figure 5.6 Some of the diagrams in the set of maximally crossed diagrams (note that the three-impurity diagram is skipped, but must be included of course).

Figure 5.7 The Green's function reversal, which unfolds the maximally crossed diagrams to produce a simpler ladder diagram. The new diagram is the *cooperon* contribution to the conductivity.

Rather than the entire two-particle Green's function, we are only interested in this correction term. In the earlier evaluation of the ladder diagrams, we let the frequency go to zero to achieve the static conductance. Here, however, we will find a divergence as $q, \omega \to 0$. This will be avoided by assuring that the phase-breaking process is incorporated into the polarization through the phase-breaking time as in Eq. (5.64) above. Now the contribution to the conductivity arises exactly as in the last chapter and the change in conductivity arises from Eq. (4.96) using only the last term of F in Eq. (4.97), although the contribution in Λ will come from the new cooperon correction. That is, we can now write the contribution as

$$\Delta\sigma = -\frac{e^2\hbar^2}{m^{*2}}\sum_{\mathbf{k}}\int\frac{d\omega}{2\pi}G^r(\mathbf{k},\omega)G^a(\mathbf{k},\omega)\sum_{\mathbf{k}'}\mathbf{k}\cdot\mathbf{k}'\Lambda_c(\mathbf{q},\omega)G^r(\mathbf{k}',\omega)G^a(\mathbf{k}',\omega).$$

(5.73)

Here we have analytically continued the frequencies to the real frequency for convenience, and the minus sign arises from the reversal in $\mathbf{k}'$. Here the scattering kernel is a function of the scattering wavevector, and the second summation involves the energy-conserving delta function, which sets $\mathbf{k}' = -\mathbf{k} + \mathbf{q}$. Then the frequency and momentum integrals can be written as

$$I = -\sum_{\mathbf{k}}\int\frac{d\omega}{2\pi}G^r(\mathbf{k},\omega)G^a(\mathbf{k},\omega)G^r(-\mathbf{k}+\mathbf{q},\omega)G^a(-\mathbf{k}+\mathbf{q},\omega)\delta(E-E_F)$$

$$= -\int\frac{d\omega}{2\pi}\int\rho(E)dE\frac{k^2}{d}\frac{\hbar}{\hbar\omega-E+i\hbar/2\tau}\frac{\hbar}{\hbar\omega-E-i\hbar/2\tau}$$

$$\times\frac{\hbar}{\hbar\omega-E+\hbar\mathbf{v}_F\cdot\mathbf{q}+i\hbar/2\tau}\frac{\hbar}{\hbar\omega-E+\hbar\mathbf{v}_F\cdot\mathbf{q}-i\hbar/2\tau}\delta(E-E_F).$$

(5.74)

The contour of integration is closed in the upper half plane, giving two poles as

$$\hbar\omega = E + i\hbar/2\tau$$
$$\hbar\omega = E + \hbar\mathbf{v}_F\cdot\mathbf{q} + i\hbar/2\tau,$$

(5.75)

which leads to

$$I = -\frac{k_F^2}{d}\rho(E_F)\frac{2\tau}{(\mathbf{v}_F\cdot\mathbf{q})^2-(1/\tau)^2}\sim\frac{2k_F^2\tau^3}{d}\rho(E_F),$$

(5.76)

where the last term arises for small q. Hence, the conductivity correction becomes

$$\Delta\sigma = -\frac{2e^2\hbar^2 k_F^2\tau^3}{dm^{*2}}\rho(E_F)\int\frac{d^d\mathbf{q}}{(2\pi)^d}\Lambda(\mathbf{q}).$$

(5.77)

Our task now becomes to evaluate the kernel.

Now it is time to turn our attention to the kernel. We cannot use our earlier evaluation of the polarization because we have reversed the directions of the Green's functions at the bottom of Fig. 5.7. The new version of the polarization can be written as

$$\Pi(\mathbf{q},\omega) = \int \frac{d^d\mathbf{k}}{(2\pi)^d} G^r(\mathbf{k},\omega)G^a(-\mathbf{k}+\mathbf{q},\omega)$$

$$= \int \frac{d^d\mathbf{k}}{(2\pi)^d} \left\{ \frac{\hbar}{\hbar\omega - E + i\hbar(1/\tau + 1/\tau_\varphi)} \frac{\hbar}{\hbar\omega - E - i\hbar(1/\tau + 1/\tau_\varphi)} \right\}.$$

(5.78)

As promised above, we have inserted the broadening due to the phase-breaking process along with the normal total scattering time. Once more, the energy integral is closed in the upper half plane, and evaluated at the pole $E = \hbar\omega + i\hbar(1/\tau + 1/\tau_\varphi)$. This gives us

$$\Pi(\mathbf{q}) = -2\pi i\rho(E_F) \int \frac{dS_k}{S_k} \frac{\hbar^2}{\hbar\mathbf{v}_F\cdot\mathbf{q} - 2i\hbar(1/\tau + 1/\tau_\varphi)}$$

$$= -2\pi i\rho(E_F)\hbar^2 \int \frac{dS_k}{S_k} \frac{i\tau}{2\hbar} \left[1 - i\frac{\tau v_F q}{2}\cos\vartheta - \frac{\tau}{\tau_\varphi} + \left(\frac{\tau v_F q}{2}\cos\vartheta\right)^2 + ... \right].$$

$$= \pi\rho(E_F)\hbar\tau \left[1 - \frac{\tau}{\tau_\varphi} + Dq^2\tau + ... \right]$$

(5.79)

The scattering kernel remains given by

$$\Gamma_0 = \frac{n_i V_0^2}{\hbar^2} = \frac{1}{2\pi\rho(E_F)\tau}$$

(5.80)

so that the kernel becomes

$$\Lambda = \frac{\Gamma_0}{1 - \Pi\Gamma_0} = \frac{\tau_\varphi'}{\pi\hbar\rho(E_F)\tau^2(1 + Dq^2\tau_\varphi')},$$

(5.81)

where

$$\tau_\varphi' = \tau_\varphi\left(\frac{\tau}{\tau + \tau_\varphi}\right).$$

(5.82)

We can now combine Eq. (5.81) with Eq. (5.77) and obtain the correction term to the conductance that is provided by the cooperon interaction to be

$$\Delta\sigma = -\frac{2e^2}{\pi\hbar}\frac{D\tau'_\varphi}{d}\int\frac{d^d\mathbf{q}}{(2\pi)^d}\frac{1}{1+Dq^2\tau'_\varphi}. \tag{5.83}$$

If the second term in the denominator within the integral is large, it is clear that the integral diverges. While the factor in the integral appears similar to a Green's function, it is not as it contains the diffusion constant and leads to what is called a *diffusion pole*. This correction term is itself known as the cooperon correction. In fact, we are primarily interested in small values of q, and consequently will limit the integration to $q < 1/\sqrt{D\tau}$, where the latter quantity is the diffusion length. Then the integration is straightforward and we find the conductance correction to be

$$\Delta\sigma = -\frac{e^2}{4\pi^2\hbar}\ln\left(1+\frac{\tau'_\varphi}{\tau}\right) \tag{5.84}$$

in two dimensions.

Earlier, we remarked that a magnetic field destroyed the time reversal symmetry and destroyed weak localization. This is because the vector potential of the magnetic field modifies the phase of the wavefunctions. In the Green's function, the phase differences of the wavefunctions lead to a phase change in the Green's function as

$$G(\mathbf{x},\mathbf{x}';B) = G(\mathbf{x},\mathbf{x}';0)\exp\left[\frac{ie}{\hbar}\int_\mathbf{x}^{\mathbf{x}'} A(\mathbf{x}'')\cdot d\mathbf{x}''\right]. \tag{5.85}$$

Because the electrons are propagating in opposite directions in the cooperon correction, the phases of the retarded and advanced function actually add together in the polarization to produce twice the phase shift shown in Eq. (5.85). Because of this phase doubling, we have an additional factor of 2, if we think of the behavior of a single particle in a magnetic field. Now, in the presence of the magnetic field in the phase-coherent regime for which the correction to the conductance occurs, the breakup of the weak localization arises because of the curvature in the trajectories induced by the field. In this phase-coherent region, we will take account of the magnetic field by assuming that there are a very large number of Landau levels occupied because the spacing of the levels goes to zero as the magnetic field goes to zero. In terms of the momentum change caused by the impurity scattering, we will replace the q^2 in Eq. (5.81) by q_n^2, where the latter is defined via the Landau energy

$$E_n = \frac{\hbar^2 q_n^2}{2m^*} = \hbar\omega_c\left(n + \frac{1}{2}\right) \quad , \quad \omega_c = 2\frac{eB}{m^*},$$ (5.86)

where the last factor of 2 is the doubling of the phase discussed above, so that the cyclotron frequency is twice its normal value. Then, Eq. (5.81) is replaced by

$$\Lambda(\mathbf{q}_n) = \frac{1}{\pi\hbar\rho(E_F)\tau^2}\frac{\tau'_\varphi}{\left(Dq_n^2\tau'_\varphi + 1\right)}.$$ (5.87)

The fact that there is a maximum q that is used to translate into a maximum value of n, which is also given by the inverse of the diffusion length as

$$n_{max} = \frac{\hbar}{4eBD\tau}.$$ (5.88)

Then, when the integration in Eq. (5.83) is replaced by a summation over n, there is an additional factor in the sum, which is the density of states per Landau level, which is $eB/\pi\hbar$. With these factors, Eq. (5.83) now becomes

$$\Delta\sigma = -\frac{e^2}{\pi\hbar}(D\tau)\frac{eB}{\pi\hbar}\sum_{n=0}^{n_{max}}\frac{1}{\dfrac{4eBD\tau}{\hbar}\left(n + \dfrac{1}{2}\right) + \dfrac{\tau}{\tau'_\varphi}}$$

$$= -\frac{e^2}{4\pi^2\hbar}\sum_{n=0}^{n_{max}}\frac{1}{n + \dfrac{1}{2} + \dfrac{\hbar}{4eBD\tau'_\varphi}}.$$ (5.89)

The summation is a difference in digamma functions, which can be written as

$$\Delta\sigma = -\frac{e^2}{4\pi^2\hbar}\left\{\psi\left[\frac{3}{2} + \frac{\hbar}{4eBD\tau}\left(1 + \frac{\tau}{\tau'_\varphi}\right)\right] - \psi\left(\frac{1}{2} + \frac{\hbar}{4eBD\tau'_\varphi}\right)\right\}.$$ (5.90)

By fitting to the shape of the weak localization contribution to the conductance in the above equation, one can estimate the values for the coherence length and the phase-breaking time from these measurements. For example, in Fig. 5.8, the conductance measured in a GaAlAs/GaAs heterostructure at low temperature is plotted [8]. There is basically a logarithmic rise of the conductance as the magnetic field is raised, but at low magnetic field, there tends to be a saturation. This increase in conductance with magnetic field is opposite to the normal magnetoresistance in such a structure and is the signal of the weak localization. The shape of these curves can be fit, and the main parameter is the phase-breaking time.

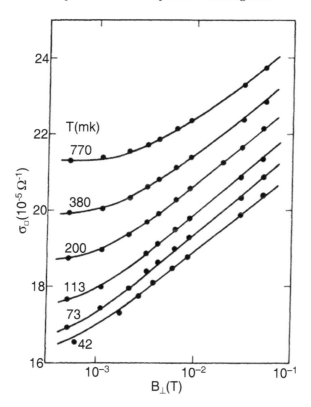

Figure 5.8 Conductance of the electrons in a GaAlAs/GaAs heterostructure as a function of the magnetic field normal to the heterostructure interface. Reprinted with permission from Ref. [8], Copyright 1984, American Physical Society.

5.8 Observations of Phase-Breaking Time

In small semiconductor structures, often called mesoscopic devices [9], the size of various effects often depends on the phase-breaking, or phase coherence, time τ_φ, or less directly on the phase coherence length $l_\varphi = \sqrt{D\tau_\varphi}$. For example, these quantities are crucial to the description of weak localization (which will be dealt with in the next chapter). In Fig. 5.9, the phase-braking time that is found from the curves is plotted as a function of temperature for three different two-dimensional densities in the heterostructure interface. It is the temperature dependence of this phase-breaking time that we want to study in the next section. There are a variety of contributors to the decay of the coherence length and the phase-breaking time with temperature. In this section, we review some of the measurements of the phase-breaking time. The solid lines in Fig. 5.9 are fits to the data and have the slope of $1/T$, which is the usual behavior for quasi-two-dimensional systems except at very low temperatures.

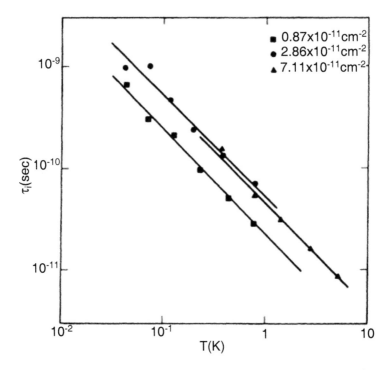

Figure 5.9 Phase-breaking time found for the electrons in a GaAlAs/GaAs heterostructure as a function of the temperature. Reprinted with permission from Ref. [8], Copyright 1984, American Physical Society.

In general, one sees a saturation of the phase coherence length and the phase-breaking time at low temperatures, both in metals

[10] and in semiconductors [11]. Usually, the transition from the temperature-independent behavior and the temperature-dependent behavior occurs in the vicinity of 0.1–1.0 K, depending on the material and the quality of the sample. There is just a hint that this might be occurring for the top curve in Fig. 5.9. An illustration of this is shown in Fig. 5.10 [12], where we show the phase-breaking time for an open quantum dot that has been fabricated on a AlGaAs/GaAs heterojunction. In these heterojunctions, a two-dimensional electron gas is formed at the interface, as mentioned above. The quantum dot is formed by depositing a set of Schottky barrier metallic gates, with two quantum point contacts to allow current to enter and leave the dot area where the electrons remain. These dots are called "open" when the quantum point contacts are not tunneling but conducting. Similar behavior is found in dot arrays [13] and silicon wires and two-dimensional regions [11]. The key point in this latter figure is that the phase-breaking time is independent of temperature for sufficiently small temperature. A transition temperature is indicated in the figure by the arrow. Above this temperature, the phase-breaking time decays with temperature as $1/T$ in two dimensions, and the data agree with this decay [9, 10]. We will describe the theory that leads to these behaviors in the next section.

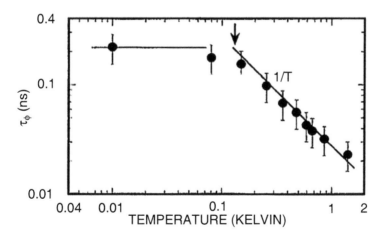

Figure 5.10   Phase-breaking time found in an open quantum dot formed in a GaAlAs/GaAs heterostructure as a function of the temperature. Reprinted with permission from Ref. [12], Copyright 1995, American Physical Society.

5.9 Phase-Breaking Time

As we pointed out above, the long-range part of the Coulomb interaction is responsible for scattering by the collective oscillations of the electron gas, which describe the motion of the electrons in the field produced by their own Coulomb potential, the *plasmons*. These are just the modes we discussed in the plasmon-pole approximation in the last chapter. Here, we want to consider the phase-breaking that occurs when coherent electrons interact with the plasmons in a low dimensional system. We will find that, in general, there are two regimes: a low-temperature regime where the phase-breaking time is independent of temperature and a high-temperature regime where the phase-breaking time decreases as a power of the temperature. The transition region is where the plasmon energy is comparable to the thermal energy.

5.9.1 Quasi-Two-Dimensional System

Electrons that are in excited states (or states with energy above the average energy) will on average lose energy to the overall electron gas, as discussed above. This is true regardless of the dimensionality of the semiconductor. Here, we want to discuss this energy exchange with the plasmon modes. In two dimensions, the plasma frequency is not constant, but is a function of the wavevector, hence approaches zero at $q = 0$. In the present section, we want to compute the scattering rate in this quasi-two-dimensional system for the electron–plasmon interaction. We will assume that the carriers are in the lowest sub-band in the two-dimensional carrier gas. An important factor is to limit the lower range of the integration over q (in a sense, this is using a long-range cutoff), just as was done in treating weak localization. Here we will limit this value to the inverse of the mean free path, or $1/v\tau = 1/\sqrt{D\tau}$ in two dimensions, predicated on the fact that the scattering will break up any process that would emit a plasmon with lower momentum value (or that the coherence cannot be maintained for more than a diffusion length).

We can now formulate the process of scattering by an electron in a quasi-two-dimensional electron gas, by treating a nearly free electron. Our treatment will be applicable to that of high-mobility carriers in, e.g., a heterostructure with little impurity scattering, even though we will find essentially the same result as that of the disordered system. We can write the scattering rate from the self-energy as

$$\frac{1}{\tau_\varphi} = \frac{2\pi}{\hbar} \sum_q \int \frac{d\omega}{2\pi} \coth\left(\frac{\hbar\omega}{2k_B T}\right) \left|\text{Im}\left\{\frac{V(q)}{\varepsilon(q,\omega)}\right\}\right| \delta\left(\omega - \frac{E(\mathbf{k}\pm\mathbf{q}) - E(k)}{\hbar}\right).$$

(5.91)

We have ignored the details of the Fermi factor that arises from the possibility that the final states are full, and we have combined the emission and absorption terms through

$$N_q(x) + \left(N_q(x) + 1\right) = \coth(x/2).$$

(5.92)

For the dielectric function, we take Eq. (5.53) in the form

$$\varepsilon(q,\omega) = \varepsilon_s \left[1 + \frac{\omega_p^2 \tau^2}{(1+i\omega\tau)^2 + Dq^2\tau}\right]$$

(5.93)

with

$$\omega_p^2 = \frac{n_s e^2 q}{2m^* \varepsilon_s} \equiv \chi_2 q$$

(5.94)

in two dimensions. We note here that the plasmon frequency goes to zero as the momentum exchange becomes small. In the low frequency limit, we can write the imaginary part of the inverse dielectric function as

$$-\frac{e^2}{2\varepsilon_s q} \frac{2\omega}{\chi_2 q\tau + Dq^2}.$$

(5.95)

The leading term in the denominator factor is the plasmon term and would have led to the plasmon-pole approximation. Here the diffusion provides a correction factor, and the scattering term τ provides the broadening of the plasmon pole, which is important here.

At the higher temperatures, we can use the small argument expansion for the hyperbolic cotangent as

$$\coth(x/2) \sim \frac{2}{x}.$$

(5.96)

With this approximation, we can now write the energy integral as

$$\int_{-\infty}^{\infty} \frac{d\omega}{2\pi} \frac{2k_B T}{\hbar\omega} \frac{e^2}{\varepsilon_s q} \frac{\omega}{\chi_2 q\tau + Dq^2} \delta\left(\omega - \frac{E(\mathbf{k}+\mathbf{q}) - E(\mathbf{k})}{\hbar}\right)$$

$$= \frac{e^2 k_B T}{\pi\hbar\varepsilon_s q(\chi_2 q\tau + Dq^2)}. \tag{5.97}$$

This can be used to write the scattering rate as

$$\frac{1}{\tau_\varphi} = \frac{2\pi}{\hbar} \sum_{\mathbf{q}} \frac{e^2 k_B T}{\pi\hbar\varepsilon_s q(\chi_2 q\tau + Dq^2)} = \frac{e^2 k_B T}{\pi\hbar^2\varepsilon_s} \int_{q_{min}}^{\infty} \frac{dq}{\chi_2 q\tau + Dq^2}$$

$$\sim \frac{e^2 k_B T}{\pi\hbar^2\varepsilon_s \chi_2\tau} \ln\left(1 + \chi_2\tau\sqrt{\frac{2\tau}{D}}\right), \tag{5.98}$$

where we have used the lower cutoff discussed above. The result shows that the phase-breaking time decays in two dimensions as $1/T$ at high temperatures, as found in the experiments discussed in the last section. We can estimate the value of τ_φ for a GaAs/AlGaAs heterostructure at low temperature. We assume that the density is 4×10^{11} cm^{-2}, and that the mobility is 10^6 cm^2/Vs, typical of a high mobility structure. Then, $\tau \sim 3.8 \times 10^{-11}$ s, and $\chi_2 \sim 8.5 \times 10^{18}$ cm/s^2. This gives the value of $\tau_\varphi \sim 4.4 \times 10^{-10}$ s at 1 K. This value is comparable to actual measured values in such semiconductor structures at low temperatures.

At the low temperature end of the range, we use the large argument expansion of the hyperbolic cotangent, where

$$Coth(x/2) \sim 1. \tag{5.99}$$

Then, the phase-breaking rate becomes

$$\frac{1}{\tau_\varphi} = \frac{2\pi}{\hbar} \sum_{\mathbf{q}} \int \frac{d\omega}{2\pi} \frac{e^2}{\varepsilon_s q} \frac{\omega}{\chi_2 q\tau + Dq^2} \delta\left(\omega - \frac{E(\mathbf{k}+\mathbf{q}) - E(\mathbf{k})}{\hbar}\right). \tag{5.100}$$

The obvious thing would be to integrate over the frequency, but this would give a cosine term, which would then integrate to zero in the $\mathbf{q}$ integration. So this has to be handled a little more carefully. So we integrate the angle within the $\mathbf{q}$ integration first, as

$$\int \frac{d\vartheta}{2\pi} \delta(\omega - qv\cos\vartheta) = \frac{1}{2\pi\sqrt{q^2 v^2 - \omega^2}}. \tag{5.101}$$

This can now be used in the frequency integral, where we will have a perfect differential. However, we need an upper cutoff for the frequency, and we take this as $1/\tau$. Then, the remaining integral is

$$
\frac{1}{\tau_\varphi} = \frac{e^2}{2\pi^2\varepsilon_s\hbar}\int\frac{dq}{q(\chi_2\tau+Dq)}\left[qv-\sqrt{q^2v^2-1/\tau^2}\right]
$$

$$
\sim \frac{e^2}{2\pi^2\varepsilon_s\hbar v\tau}\int\frac{dq}{q^2(\chi_2\tau+Dq)}.
$$

(5.102)

In this case, we are going to have to limit both the maximum value of q and the minimum value of q. For the former, we are guided by the equation itself, and take $q_{max} = \chi_2/D\tau$, but this will only be used in the logarithmic term that results from the integral. Hence, Eq. (5.102) now gives

$$
\frac{1}{\tau_\varphi} = \frac{e^2}{8\pi^2\varepsilon_s\hbar v\tau\chi_2}\left[\sqrt{\frac{D}{\tau}}+\frac{2D}{\chi_2\tau^2}\ln\left(\frac{\sqrt{D\tau}}{\chi_2}\right)\right].
$$

(5.103)

Here, it is clear that, in this limit, the phase-breaking time is independent of the temperature. This is an effect that is seen in most mesoscopic systems where the phase-breaking time has been measured, such as was discussed in the previous section.

5.9.2 Quasi-One-Dimensional System

In one dimension, the plasma frequency is not constant, but is a function of the wavevector, hence approaches zero at $q = 0$. In this present section, we want to compute the scattering rate in this quasi-one-dimensional system for the electron–plasmon interaction. We will assume that the carriers are in a single sub-band. As above, we will have to limit the range of some of the integrations, just as is done for inelastic phonon scattering. In the quasi-one-dimensional limit, we are really treating a quantum wire. Although we think of this as a one-dimensional system, in most cases they are really narrow two-dimensional systems, with the Fermi energy being determined by the two-dimensional reservoirs to which the wires are connected. In this case, the one-dimensional density is more properly given by $n_1 = n_2 W$, where W is the width of the wire. Nevertheless, we approach the electron–plasmon interaction as if it were a one-dimensional wire. We formulate the process of energy exchange between an electron and the plasmon by treating a nearly free electron, just as above. We write the scattering rate exactly as Eq. (5.91), as the dimensionality is not expressed in this equation. However, there are some changes for the quantum wire, as the Coulomb potential now appears with the Fourier transform as

$$V(q) = \frac{e^2}{4\pi\varepsilon_s \ln(1+q_0^2/q^2)} \sim \frac{e^2}{4\pi\varepsilon_s}, \tag{5.104}$$

and the plasmon frequency is now

$$\omega_p^2 = \frac{n_1 e^2 q^2}{2m^*\varepsilon_s} \equiv \chi_1 q^2. \tag{5.105}$$

With these new values, the low frequency limit of the imaginary part of the inverse dielectric function is now

$$-\frac{e^2}{4\pi\varepsilon_s q^2} \frac{\omega}{\chi_1\tau + D}. \tag{5.106}$$

In evaluating the phase-breaking time, it will prove more convenient to carry out the integration over q prior to that over ω. The former will entail the delta function, and we will need to invoke some cutoffs on the ω integration. In fact, we are interested in frequencies that lie between $1/\tau_\varphi$ and $1/\tau$. We will take these as the upper and lower cutoffs, respectively, when we need to utilize such in the evaluation of the integrals. We can now write the q integration as

$$\sum_\pm \int_0^\infty \frac{dq}{2\pi} \frac{e^2\omega}{2\pi\varepsilon_s q^2(\chi_1\tau+D)} \delta\left(\omega - \frac{\hbar q^2}{2m^*} \mp \frac{\hbar q k}{m^*}\right)$$
$$\sim \frac{e^2 m^*}{4\pi^2\varepsilon_s q^2(\chi_1\tau+D)}\left(\frac{\hbar}{2m^*}\right)^{3/2}\frac{1}{\sqrt{\omega}}. \tag{5.107}$$

Using this, our phase-breaking rate can now be written as

$$\frac{1}{\tau_\varphi} = \frac{e^2 m^*}{4\pi^2\hbar^2\varepsilon_s(\chi_1\tau+D)}\left(\frac{\hbar}{2m^*}\right)^{3/2}\int_{-\infty}^\infty \frac{d\omega}{\sqrt{\omega}}\coth\left(\frac{\hbar\omega}{2k_BT}\right). \tag{5.108}$$

First, we will consider the high-temperature limit, where the argument of the hyperbolic cotangent is not large, and use the approximation (5.96) so that Eq. (5.108) may be written as

$$\frac{1}{\tau_\varphi} = \frac{e^2 m^* k_B T}{2\pi^2 \hbar^3 \varepsilon_s (\chi_1 \tau + D)} \left(\frac{\hbar}{2m^*}\right)^{3/2} \int \frac{d\omega}{\omega^{3/2}}$$

$$\sim \frac{e^2 m^* k_B T}{\pi^2 \hbar^3 \varepsilon_s (\chi_1 \tau + D)} \left(\frac{\hbar}{2m^*}\right)^{3/2} \sqrt{\tau_\varphi},$$

(5.109)

where we have cut off the integration at a lower frequency defined by the inverse of the phase-breaking time. Clearly, this latter equation gives a temperature decay that varies as $1/T^{2/3}$.

At the low-temperature end of the range, we use the large argument limit for the hyperbolic cotangent and the approximation given by Eq. (5.99). Then, the phase-breaking time becomes

$$\frac{1}{\tau_\varphi} = \frac{e^2 m^*}{4\pi^2 \hbar^2 \varepsilon_s (\chi_1 \tau + D)} \left(\frac{\hbar}{2m^*}\right)^{3/2} \int_{-\infty}^{\infty} \frac{d\omega}{\sqrt{\omega}}$$

$$\sim \frac{e^2 m^*}{2\pi^2 \hbar^2 \varepsilon_s (\chi_1 \tau + D)\sqrt{\tau}} \left(\frac{\hbar}{2m^*}\right)^{3/2},$$

(5.110)

where we have cut off the integration at an upper frequency given by the inverse of the scattering time. Once again, we find that this low-temperature version is independent of the temperature, as observed in the experiments.

Problems

1. Use the measured data published in Ref. [8] to discuss the diffusion length, the mean time between collisions, the coherence length, the thermal diffusion length, etc.
2. What is the width of the derivative of the Fermi–Dirac function at half its maximum absolute value in terms of $k_B T$.
3. Using the properties of complex integration, confirm the relation found in Eq. (5.39). (Hint: There is a discussion of this relation in Ref. [3].)
4. Use the result (5.90) to evaluate the phase-breaking time that is determined from the curves of Fig. 5.8.
5. Use the results of Section 5.9.1 to compare with the curve in Fig. 5.10. Use the data for the samples that is reported in Ref. [13].

References

1. R. Landauer, *IBM J. Res. Dev.*, **1**, 223 (1957).

2. M. Büttiker, Y. Imry, R. Landauer, and S. Pinhas, *Phys. Rev. B*, **31**, 6207 (1985).

3. A. L. Fetter and J. D. Walecka, *Quantum Theory of Many-Particle Systems* (McGraw-Hill, New York, 1971).

4. D. K. Ferry, S. M. Goodnick, and J. P. Bird, *Transport in Nanostructures*, 2nd Ed. (Cambridge University Press, Cambridge, 2009), p. 525.

5. H. Fukuyama, in *Electron–Electron Interactions in Disordered Systems*, edited by A. L. Efros and M. Pollak (North-Holland, Amsterdam, 1985).

6. A. V. Chaplik, *Sov. Phys. JETP*, **33**, 997 (1971).

7. G. F. Giuliani and J. J. Quinn, *Phys. Rev. B*, **26**, 4421 (1982).

8. B. J. F. Lin, M. A. Paalanen, A. C. Gossard, and D. C. Tsui, *Phys. Rev. B*, **29**, 927 (1984).

9. D. K. Ferry, *Transport in Semiconductor Mesoscopic Devices* (Institute of Physics Publishing, Bristol, 2015).

10. R. A. Webb, S. Washburn, H. J. Haucke, A. D. Benoit, C. P. Umbach, and F. P. Milliken, in *Physics and Technology of Submicron Structures*, edited by H. Heinrich, G. Bauer, and F. Kuchar (Springer-Verlag, Berlin, 1988), pp. 98–107.

11. C. de Graaf, J. Caro and S. Radelaar, *Phys. Rev. B*, **46**, 12814 (1992).

12. J. P. Bird, K. Ishibashi, D. K. Ferry, Y. Ochiai, Y. Aoyagi, and T. Sugano, *Phys. Rev. B*, **51**, 18037 (1995).

13. C. Prasad, D. K. Ferry, A. Shailos, M. Elhassan, J. P. Bird, L.-H. Lin, N. Aoki, Y. Ochiai, K. Ishibashi, and Y. Aoyagi, *Phys. Rev. B*, **62**, 15356 (2000).

Chapter 6

Quantum Devices

It is now possible to approach the fundamental limiting scale of solid state electronics in which a single electron can, in principle, represent a single bit in an information flow through a device or circuit [1]. The burgeoning field of single-electron tunneling (SET), although currently operating at very low temperatures, has brought this consideration into the forefront. Indeed, the recent observations of SET effects in poly-Si structures *at room temperature* by a variety of authors have grabbed the attention of the semiconductor industry [2]. The resulting behavior has important implications to future semiconductor electronics, regardless of the final interpretation of the physics involved. Indeed, the semiconductor industry is rapidly carrying out its own advance, with transistor gate lengths in the 14 nm range in production in 2015 (the so-called 14 nm node).

We pointed out in Chapter 2 that the semiconductor industry is following a linear scaling law that is expected to be fairly rigorous. With dimensions approaching 10 nm within a year or so, there is a rapid search for possible new technologies that can supplement Si with the offer of improved performance. This means that we must seek the manner in which quantum effects in open device structures will carry over to these applications. In the earlier chapter mentioned, we also discussed the use of the wavefunction itself to simulate a small silicon MOSFET and to include the role of scattering via impurities and phonons. However, we have not fully developed the

An Introduction to Quantum Transport in Semiconductors
David K. Ferry
Copyright © 2018 Pan Stanford Publishing Pte. Ltd.
ISBN 978-981-4745-86-4 (Hardcover), 978-1-315-20622-6 (eBook)
www.panstanford.com

quantum nature and approach to the electron–phonon interaction. We begin this chapter by doing this in the next section. Then we discuss a variety of devices that have been the subject of theoretical investigations with quantum transport approaches.

The theoretical methods discussed so far in this book, other than the wavefunction approach in Chapter 2, have been described in terms of an equilibrium system, either at zero temperature or in thermal equilibrium with the surroundings. It is possible to describe the thermodynamics of an *open, nonequilibrium* system through the use of other methods, the most common of which is the nonequilibrium Green's functions, the so-called real-time Green's functions developed by Schwinger and colleagues [3, 4], by Kadanoff and Baym [5], and by Keldysh [6]. However, one needs to worry about how far we can go away from equilibrium [7], where, at least to lowest order, Liouville's equation remains valid as an equation of motion for the carrier transport under the applied fields. In *far-from-equilibrium* systems, however, there is usually strong dissipation, and the resulting statistical ensemble (even in steady state) is achieved as a balance between driving forces and dissipative forces; for example, it does not linearly evolve from the equilibrium state when the applied forces are "turned on." Enz [7] argues that there is no valid unitary operator (such as that which describes the impact of perturbation theory) to describe the evolution through this symmetry-breaking transition to the dissipative steady state. As a consequence, there is no *general* formalism at this time that can describe the all-important far-from-equilibrium devices. On the other hand, the application of these other approaches to what are surely very strongly far-from-equilibrium systems—the excitation of electron–hole pairs in intense femtosecond laser pulses for example—has yielded results that suggest that their use in these systems is quite reliable for studying both the transition to the semiclassical Boltzmann theory and to explain experimentally observed details [8–10]. As a consequence of these latter studies, as well as initial attempts to actually begin to model real devices with these approaches, the situation is believed to be much better than this pessimistic view would warrant. But to proceed with this, one must first understand the type of devices, and their physical principles, before judging the transport methods. These methods will be the subject of the next several chapters. However, as we remarked above, we now want to look at these devices.

6.1 Electron–Phonon Interaction

Probably the most important dissipative process in semiconductors is due to the electron–phonon interaction. The approach is quite similar to that for impurity scattering except for a few important differences [11, 12]. First, the Coulomb interaction is instantaneous, whereas the electron–phonon is not. This means that there will be an additional time integration and a propagator for the phonon itself. This propagator corresponds to the propagation of the phonon along the time path of this additional time integration. In Green's function parlance, the kernel of the impurity interaction, in the Fourier transform space, is replaced by

$$\frac{e^2}{\varepsilon_s q^2} \quad \rightarrow \quad \int dt_2 |M(q)|^2 D^0(\mathbf{q}, t_1 - t_2) \tag{6.1}$$

where D^0 is the phonon Green's function and $M(q)$ is the matrix element for the electron–phonon interaction. Finally, while the $q = 0$ mode is allowed for impurity scattering, it is not allowed for the phonons. Such a mode would just be a uniform displacement of the lattice, which does not change anything in the periodic lattice. Now, the acoustic mode has a linear dependence on the momentum q, the interaction vanishes at $q = 0$. In general, we must have a phonon momentum that changes the electron momentum, even though the matrix element may be independent of the momentum. This latter is the case for the non-polar optical phonon interaction. The lowest-order connected term in the electron–phonon interaction is shown in Fig. 6.1, where the dashed line represents the phonon propagator, or Green's function. If the interaction is screened, then the two vertexes will each have a corresponding vertex correction.

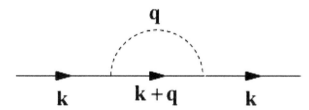

Figure 6.1 The lowest-order interaction Green's function diagram for the electron–phonon interaction. The dashed line represents the phonon Green's function.

The full development of the inclusion of the electron–phonon interaction will be treated in the next several chapters. Here, we want to develop the important matrix elements for the various phonon modes, so we get a unified set of expressions. In general, one can expand the electron–phonon interaction Hamiltonian as a power series in the wavevector q. There are terms that vary as $1/q$ such as the polar optical phonon and piezoelectric interactions. There are also terms that are independent of q such as the non-polar optical phonon interaction. Finally, there are terms linear in q, such as the acoustic phonon interaction. There are, of course, higher-order terms as well, but these low-order terms are the most important and give rise to the majority of scattering in semiconductors. We will begin with the last of these, treating these different terms in reverse order.

6.1.1 Acoustic Phonons

Scattering by the acoustic modes of the lattice vibrations is one of the most common interactions. The acoustic modes have frequencies that vanish as the wavevector $\mathbf{q}$ goes to zero and, as may be expected, the wave velocity corresponds to one of the sound velocities in the crystal. The sound velocity depends on the crystal direction and the polarization of the wave as it moves through the crystal. The wave creates a local strain in the crystal, and this strain perturbs the energy bands locally. This provides a scattering potential, which is referred to as the *deformation potential*. This is typically expressed as [13]

$$V_1 = \delta E = \Xi_1 \Delta = \Xi_1 \nabla \cdot \mathbf{u_q}, \tag{6.2}$$

where Ξ_1 is the deformation potential for a particular band in which the carrier is located, Δ is the dilation of the crystal lattice created by the acoustic wave of Fourier amplitude $\mathbf{u_q}$. Any static displacement of the lattice would be a movement of the crystal as a whole and does not contribute. The acoustic wave Fourier amplitude is a relatively uniform value over the entire crystal and the wave itself can be written as

$$\mathbf{u_q} = \left(\frac{\hbar}{2\rho_m \Omega \omega_\mathbf{q}} \right)^{1/2} \left[a_\mathbf{q} e^{i\mathbf{q}\cdot\mathbf{r}} + a_\mathbf{q}^\dagger e^{-i\mathbf{q}\cdot\mathbf{r}} \right] \mathbf{e_q} e^{-i\omega_\mathbf{q} t}, \tag{6.3}$$

where the plane wave factors have been incorporated along with the normalization factor for completeness. The factors $a_{\mathbf{q}}$ and $a_{\mathbf{q}}^{\dagger}$ are the annihilation and creation operators for the quantized harmonic oscillator representation of the phonon mode. The quantity $\mathbf{e}_{\mathbf{q}}$ is the polarization vector for the wave. The quantity ρ_m is the mass density of the crystal, and Ω is the volume of the crystal. Because the divergence operation in Eq. (6.2) produces a result only for propagation along the polarization direction, this gives rise normally to the longitudinal acoustic mode interaction. A different result can be obtained in ellipsoidal bands, but we ignore this complication. The fact that the interaction is now first-order in $\mathbf{q}$ leads to this term being called a first-order mode.

We can now calculate the matrix element by considering the proper sum over both the electron and the phonon wavefunctions. The second term of Eq. (6.3), the term for the emission of a phonon by the carrier, leads to the matrix element

$$M(\mathbf{k},\mathbf{q}) = i\mathbf{q}\cdot\mathbf{e}_{\mathbf{q}}\left(\frac{\hbar\Xi_1^2}{2\rho_m\Omega\omega_{\mathbf{q}}}\right)^{1/2}$$

$$\times \int d^3\mathbf{r}\int d^3\mathbf{r}'\psi_{\mathbf{k}'}^{\dagger}(\mathbf{r})\phi^{\dagger}(\mathbf{r}')a_{\mathbf{q}}^{\dagger}e^{-i\mathbf{qgr}-i\omega_{\mathbf{q}}t}\phi(\mathbf{r}')\psi_{\mathbf{k}}(\mathbf{r}), \quad (6.4)$$

where the ϕ are the harmonic oscillator functions for the lattice parts of the total wavefunction and the ψ are electronic Bloch functions. Normally, one might expect a single integration, since the space of the electrons and phonons is the same. However, the waves and the electrons are considered independent quantities. The creation operator excites the harmonic oscillator to a higher-energy state by one quantum of vibrational energy. This also produces the Bose–Einstein distribution to enter the expression as $N_{\mathbf{q}} + 1$, where $N_{\mathbf{q}}$ is this distribution. The next step is to split the integration over the lattice variables into a summation over the set of lattice cells and an integration over a unit cell itself. This summation produces the conservation of momentum, which causes the electron wavevector to be changed by the phonon wavevector, as shown in Fig. 6.1. When all of these factors are taken together, Eq. (6.4) reduces to

$$M(\mathbf{k},\mathbf{q}) = i\mathbf{q}\cdot\mathbf{e}_{\mathbf{q}}\left(\frac{\hbar\Xi_1^2}{2\rho_m\Omega\omega_{\mathbf{q}}}\right)^{1/2}\sqrt{N_q+1}\int u_{\mathbf{k}-\mathbf{q}}^{\dagger}(\mathbf{r})u_{\mathbf{k}}(\mathbf{r})d^3\mathbf{r} \quad (6.5)$$

with the last integration being carried out only over the unit cell, and the two functions are the cell periodic part of the Bloch function. This last integral is called the overlap integral and is usually unity in parabolic bands. In the presence of the **k•p** interaction, the cell periodic wavefunction has sp^3 hybrid admixtures, which are energy dependent, so that the overlap integral has an energy-dependent part. This latter interaction also leads to nonparabolic bands. With the nonparabolic assumption, the overlap integral (squared) becomes

$$|M(\mathbf{k},\mathbf{q})|^2 = \frac{\hbar \Xi_1^2 q^2}{2\rho_m \Omega \omega_{\mathbf{q}}}(N_q + 1).$$ (6.6)

The result for the absorption term is the same except that the factor of 1 in the parentheses is missing. Normally, in the acoustic mode interaction, the temperature is sufficiently high that $N_q \gg 1$, and the difference between the two situations can be ignored. The Bose–Einstein distribution itself can be approximated as

$$N_q = \frac{1}{\exp(\hbar\omega_q/k_B T)-1} \sim \frac{k_B T}{\hbar\omega_q} = \frac{k_B T}{\hbar q v_s} \gg 1$$ (6.7)

where we have introduced the linearity of the acoustic dispersion through the sound velocity v_s. Using these relations, we can now write Eq. (6.6) as

$$|M(\mathbf{k},\mathbf{q})|^2 = \frac{\Xi_1^2 q^2 k_B T}{\rho_m \Omega v_s^2}.$$ (6.8)

The approximation in Eq. (6.7) is known as the equipartition approximation and must be questioned at low temperatures. Typically, the acoustic phonon energy is at most a few millivolts, but when the temperature is low, the expansion of the exponential is not valid.

6.1.2 Piezoelectric Scattering

In materials such as GaAs, which lack a center of inversion symmetry, it is possible for the piezoelectric interaction to lead to scattering of carriers, particularly at low temperatures. This interaction arises from the distortion of the lattice by acoustic modes, which leads to the generation of an electric field via the piezoelectric tensor. This new electric field can act upon the carriers as a scattering process. Normal semiconductors such as Si have this inversion symmetry and thus show a piezoelectric interaction only under rare conditions. In the III-V semiconductors, it is primarily the d_{14} element of the piezoelectric tensor, which leads to the interaction [14, 15]. The interaction energy shift can be found from

$$\delta E = -\varepsilon_s \mathbf{F} \bullet \mathbf{P}, \tag{6.9}$$

where $\mathbf{F}$ is the electric field and $\mathbf{P}$ is the polarization of the lattice. The latter is determined by the stress–strain parameters and arises from the presence of the acoustic wave. The electric field is induced by this polarization via the piezoelectric interaction and is of Coulomb form as

$$\mathbf{F} = -i \frac{e}{\varepsilon_s} \frac{\mathbf{q}}{q^2 + q_{sc}^2}. \tag{6.10}$$

Putting this into Eq. (6.9), we obtain

$$\delta E \sim \frac{ed_{14}}{\varepsilon_s} \frac{q^2}{q^2 + q_{sc}^2} u_{\mathbf{q}}, \tag{6.11}$$

and the matrix element becomes

$$|M(\mathbf{k},\mathbf{q})|^2 \sim \frac{4e^2 d_{14}^2 k_B T}{\varepsilon_s^2 \rho_m \Omega v_s^2} \frac{q^2}{\left(q^2 + q_{sc}^2\right)^2}. \tag{6.12}$$

This result has also used the equipartition approximation to expand the Bose–Einstein distribution, and so is subject to the same discussion as followed in the above section.

6.1.3 Non-polar Optical and Intervalley Phonons

When there are more than a single atom per unit cell, the optical modes of the lattice vibration are allowed. In these vibrations, the two (or more) atoms per unit cell vibrate with a relative motion between them. As a result, these phonons are rather energetic, with energies on the order of 30–60 meV. The interactions of these phonons with the carriers are *inelastic* due to the large energy exchange from one particle to the other. Although one normally thinks of scattering occurring solely within the principle valley of the conduction (or valence) band, the zone edge modes can also contribute to scattering between inequivalent sets of valleys in the band. For example, these phonons also cause scattering among the six equivalent ellipsoids of the conduction band of Si, but also between the Γ and the L valleys of the conduction band of many III-V compounds.

The matrix element generally is found with the use of a deformable ion model, in which the two sublattices move relative to one another. This causes the potential field of each set of atoms to be displaced slightly, which causes a shift in the bond charges. This leaves a small excess of charge where the ions have moved apart and a small deficit of charge where they have moved together. These effects lead to a deformation field D, which is usually given in units of eV/cm. The interaction itself is a zero-order interaction in that there is no explicit term in the phonon wavevector in the interaction term itself, although there is of course still a momentum conservation condition, which can be seen in the energy-conserving delta function. Because the interaction is zero order, we can write the perturbing potential as

$$\delta E = D u_{\mathbf{q}}, \tag{6.13}$$

where, as before, $u_{\mathbf{q}}$ is the phonon amplitude. Then the square of the matrix element can be written as

$$\left|M(\mathbf{k},\mathbf{q})\right|^2 = \frac{\hbar D^2}{2\rho_m \Omega \omega_q} \left[N_q \delta(E(k) - E(\mathbf{k}+\mathbf{q}) + \hbar\omega_q) \right.$$
$$\left. + (N_q + 1)\delta(E(k) - E(\mathbf{k}-\mathbf{q}) - \hbar\omega_q) \right], \tag{6.14}$$

where the first term in the square brackets is for the absorption of a phonon by the electron (or hole) and the second term is for the emission of a phonon by the electron (or hole).

6.1.4 Polar Optical Phonons

When the two atoms per unit cell are different, particularly when they come from different columns of the periodic table, then the polar modes can take on a Coulombic nature as the effective charge on the two atoms per unit cell is different. This polarization is important in the dielectric function as well. Because the interaction is Coulombic, it is a strong interaction and is the dominant scattering processes in most III-V materials. The vibration of the two dissimilar atoms leads to a polarization, which can be expressed as

$$\mathbf{P_q} = \sqrt{\frac{\hbar}{2\gamma\Omega\omega_q}} \mathbf{e_q} \left(a_q^\dagger e^{-i\mathbf{q}\cdot\mathbf{r}} + a_q e^{i\mathbf{q}\cdot\mathbf{r}} \right) e^{-i\omega_q t} \tag{6.15}$$

where

$$\frac{1}{\gamma} = \omega_q^2 \left(\frac{1}{\varepsilon_\infty} - \frac{1}{\varepsilon(0)} \right) \tag{6.16}$$

is the effective coupling constant, which is written in terms of the high-frequency dielectric constant and the low-frequency dielectric constant. These two values differ by the polarization contribution that arises from this lattice vibration. Since these are known from experiment, they serve as a good value to determine the strength of the polar interaction with the carriers. The interaction energy is given by Eq. (6.9), but with this much stronger polarization term. In the presence of screening of the Coulombic potential, the square of the matrix element can be written as

$$|M(\mathbf{k},\mathbf{q})|^2 = \frac{\hbar e^2}{2\gamma\omega_q} \left(\frac{q}{q^2 + q_{sc}^2} \right)^2 \Big[N_q \delta(E(k) - E(\mathbf{k}+\mathbf{q}) - \hbar\omega_q) \tag{6.17}$$
$$+ (N_q + 1)\delta(E(k) - E(\mathbf{k}-\mathbf{q}) + \hbar\omega_q) \Big],$$

where again the first term is for the absorption of the phonon and the second term is for the emission of the phonon by the carrier.

6.1.5 Precautionary Comments

While the energy-conserving delta functions have been included within the matrix elements in the above portions of this section, this is the semiclassical approach. Instead, in keeping with Eq. (6.1), these functions need to be replaced by the integration over the Green's function for the phonons. These are naturally just the imaginary part that arises in the limit of the resonance within the Green's functions when we are on the energy shell. In actual usage in quantum transport, the replacement with the proper Green's function will lead to the inclusion of the important off-shell contributions.

6.2 Return to Landauer Formula

Beginning in the late 1950s, Rolf Landauer presented an approach to transport and the calculation of conductance that was dramatically different from the microscopic kinetic theory based on the Boltzmann equation that had been utilized previously (and is still heavily utilized in macroscopic conductors) [16, 17]. He suggested that one could compute the conductance of low dimensional systems simply by computing the transmission of a mode from an input reservoir to a similar mode in an output reservoir. The transmission probability from one mode to the other was then very similar to the computation of a tunneling probability, except that there was no requirement that the process be one of tunneling. The only real constraint was that of lateral confinement so that the two reservoirs could be discussed in terms of their transverse modes. We introduced this approach briefly in Chapter 2, where the conductance was written in Eq. (2.14) as

$$G = \frac{2e^2}{h} N , \tag{6.18}$$

where N is the number of channels that move through the one-dimensional conductor between the two contacts, or reservoirs. The key property of these two reservoirs was that they were in equilibrium with any applied potentials. That is, the electrons in the reservoirs were to be described by their intrinsic Fermi–Dirac distributions with any applied potentials appearing only as a shift of the relative energies (which would shift one Fermi level relative to the other). While he originally considered that the transport was ballistic, this is not required, as was demonstrated in Section 2.4 where a simple MOSFET structure was treated in the presence of scattering. Rather, the requirement is that we can assign a definitive mode to the electron when it is in either of the two reservoirs, which means that if scattering is present, it must be described specifically as a transfer of the electron from one internal mode to another, which of course may have a different transverse eigenenergy than the first mode.

In Fig. 6.2, we sketch a view of a small quantum device, in which the central constriction is connected to a pair of reservoirs that are maintained in equilibrium. As indicated, the bias is applied to the right reservoir so that the left Fermi energy can be used as the reference level for the applied potential (and indeed for the energies throughout the structure). The right contact now emits carriers into the constriction with energies up to the local Fermi level plus the applied bias, $E_F + eV$ (note that the energy eV will be negative for a positive voltage). The left contact emits electrons into the constriction with energies only up to E_F. By making the initial assumption that no scattering takes place within the constriction, we can make a definitive association between the energy of the carriers and their direction of propagation. That is, electrons injected into the constriction from the left reservoir have a positive momentum and travel from left to right in the figure. On the other hand, electrons injected into the constriction from the right reservoir have a negative momentum and thus travel from the right to the left. It is this fact that set up the discussion in Chapter 2 about MOSFET. But the steps in the conductance that are indicated in Eq. (6.18) lead to the observed conductance steps in the quantum point contact (QPC) discussed in this earlier chapter (and will be discussed further below).

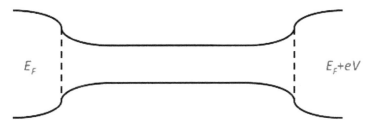

Figure 6.2 A sketch of a generic small quantum device. One can treat the region between the two reservoirs, which may be a quantum wire or a QPC as a ballistic constriction, although the ballistic requirement is not necessary. The bias is assumed to be applied to the right reservoir, so that the left reservoir provides the reference level for the energy.

At first thought, it sometimes seems that, when we have purely ballistic transport, the conductance should have been infinite rather than the finite value given by the Landauer relation (6.18). In fact, the transition between the reservoir and the channel has a nonzero resistance, which is a contact resistance between the one-dimensional channel and the reservoirs that give access to it. We have indicated the transition region as a dotted line in the figure, but it is normally much broader than this, as we shall see. Normally, the reservoirs are very wide regions, which are approximately quasi two dimensional, and the contact resistance arises from a mode mismatch between these latter regions and the channel. This occurs even when the transition is very smooth and adiabatic. This contact resistance goes directly to the question of just where the voltage drop in the device actually occurs. We have assumed that the voltage exists between the two reservoirs, but can we be more specific about this. In a normal resistive conductor, we would expect that this voltage should be dropped uniformly along the length of the constriction, just like a normal resistor, but this still requires unusual behavior in the contacts.

In the case in which there is no scattering within the constriction (the region between the dotted lines in the figure), the voltage drop across it must correspond to a constant electric field, which can be quite low as we will see. The fact that the voltage is not dropped smoothly across the device means that charge must accumulate at the contacts, which leads to the variations in the voltage. As a result, the field in the center of the constriction is quite small, and the voltage drops are forced to occur in the transition regions. To illustrate that this is the only proper solution, let us consider an alternative—we take the case of a linear voltage drop across the constriction so that there is a constant electric field throughout this region. Then a ballistic electron entering from the left contact would travel horizontally in an energy plot, but would gain kinetic energy from the constant electric field. That is, the ballistic electron travels at constant energy, and the potential energy at the cathode is converted to kinetic energy as it travels. By the time it reaches the anode at the right-hand side, it will have gained kinetic energy given by eV. A more important aspect is introduced by Kirchhoff's current law. As the electrons are accelerated by the electric field, their increase in velocity requires that fewer electrons are present as we move along the channel. That is, Kirchhoff's current law requires the current to be uniform throughout the structure. Since the cross-sectional area does not change, this requires the product of density and velocity to be a constant value. As the velocity rises, the density must decrease. But in a quantum semiconductor, the density is usually set by the doping and Fermi energy. As the number of carriers drops, this creates an additional space charge in the channel, which in turn requires that the potential be nonlinear; e.g., the potential drop cannot be a linear one. Thus, the linear potential drop is inconsistent with Kirchhoff's current law. The linear potential drop can occur only if there is sufficient scattering to assure that the carriers move with a near-equilibrium energy. Our conclusion then can only be that the electric field must essentially be quite near zero in the channel, if the carriers are to move via ballistic transport. As a result, the potential drop must divide between the cathode and the anode transition regions, as shown in Fig. 6.3.

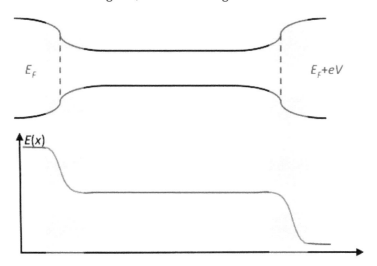

Figure 6.3 Potential drop through the device that is required for ballistic transport in a degenerate quantum device where the density is normally constant.

While we have drawn the two voltage drops as nearly the same, there is no real requirement that this be the case. In fact, it is usually assumed in the transport world that most of the discontinuity is related to the cathode. But now, ballistic transport can occur through the constriction without the carriers gaining excess energy from the applied bias. At the same time, we note that the potential drops in the transition regions now require that dipole charge densities exist at each transition (the shaded areas in Fig. 6.3). The potential drop can occur only through the existence of such dipole charge densities. Each transition region has its own dipole charge density, and this leads to the contact resistance. These results give rise to an important modification of the Landauer formula. The transmission from one mode to another is defined in the reservoirs, which means to the left of the cathode transition region and to the right of the anode transition region. What if we actually do measurements in the channel itself? If these are two terminal measurements, in which the same contacts are used to source the current and to measure the voltage, we will obtain an answer, but the voltage drop has to occur in these contacts, which have merely replaced the normal ones. On the other hand, if we use four-terminal measurements, then we will find that the Landauer formula becomes modified.

We can understand how the contact resistance becomes an important concept by rearranging Eq. (6.18) in the following manner. Let us first assume that only a few modes are occupied so that each mode has the same partial transmission $T_{pc} < 1$. Then we can invert the conductance to talk about the resistance of the device as

$$\frac{1}{G} = \frac{h}{2e^2} \frac{1}{NT_{pc}} = \frac{h}{2e^2} \frac{1}{N} + \frac{h}{2e^2} \frac{1-T_{pc}}{NT_{pc}} \equiv \frac{1}{G_C} + \frac{1}{G_D}, \qquad (6.19)$$

where the subscripts C and D refer to the contacts and the device (or channel), respectively. We can see now how the resistance can be divided into a contact resistance, which can never vanish, and a device resistance, which can go away in the absence of scattering. The contact resistance is connected with the manner in which the actual voltage drop is proportioned across the charge regions in the transitions between the reservoirs and the channel.

6.3 Landauer and MOSFET

In recent years, it has been popular to try to stretch the Landauer formula to more common everyday devices, such as the scaled CMOS transistor [18] (see also Chapter 2). It turns out that this is not a particularly useful or straightforward approach, but one can gain some insight into the behavior of the MOSFET with this approach. In the transmission approach for very small transistors, one often jumps to the conclusion that the potential will have a linear drop and that the transport is ballistic. For reasons discussed in the previous section, such an approach is incorrect. The linear potential drop can occur only with considerable scattering to keep the energy of the carriers from rising in the electric field. Rather than start with the transmission itself, let us look at the common equation for the MOSFET and see how we can connect with contact effects and the Landauer formula. This will shed light on just how the behavior of the common MOSFET is not that different from our channel discussed in the earlier parts of this chapter.

The common formula for MOSFET within the gradual channel approximation is given by an expression for the drain current in terms of the various potentials applied to the device, as

$$I_D = \frac{W\mu_e C_{ox}}{L}\left(V_G - V_T - \frac{V_D}{2}\right)V_D$$
$$= \frac{W\mu_e C_{ox}}{2L}\left[(V_G - V_T)^2 - (V_G - V_T - V_D)^2\right],$$

(6.20)

where V_G, V_D, and V_T are the gate and drain potentials and the threshold voltage, W is the width of the transistor, L is the gate (or channel) length, μ_e is the electron mobility, and C_{ox} is the gate capacitance per unit area. The threshold voltage is the gate voltage required to establish the channel between the source and drain contacts. In writing the equation in the manner shown in the second line of Eq. (6.20), we can immediately make contact with the Landauer formula. The first term in the square brackets represents the current entering the device from the source (left) electrode, while the second term represents the current entering the device from the drain (right) electrode. Thus, the source and drain are our reservoirs, and we connect with particle flow exactly as in the Landauer formula.

In thermal equilibrium, with no bias applied to the gate and drain electrodes, there is no current flow through the device, since the two terms cancel each other. This cancellation remains to be the case even with gate voltage applied, but with the drain voltage set to zero. The gate voltage changes the properties of the channel but does not upset the detailed balance of particle flow through the structure. However, when a small drain bias is applied, the second term in the square brackets is reduced relative to the first term, a small current begins to flow. As the drain bias is raised, the current increases. Once the second term is reduced to zero (it is not allowed to go negative, as this violates the conditions under which it was derived), the current saturates through the device at the value

$$I_{D,sat} = \frac{W\mu_e C_{ox}}{2L}(V_G - V_T)^2. \tag{6.21}$$

When a drain bias is applied, the drain region is lowered in energy relative to the source region, just as shown in Fig. 6.3. Saturation occurs when this energy difference is sufficiently large that electrons in the drain can no longer surmount the barrier to reach the source. No matter whether the transport is ballistic or diffusive (where there is lots of scattering), the energy regions in the source and drain will be the same. What will be different between these two cases is how the energy drops between the source and drain.

Let us now take the second line of Eq. (6.20) and write it in terms of an "equilibrium" current that depends on the gate voltage, as

$$I_{eq}^+ = I_{D,sat} = \frac{W\mu_e C_{ox}}{2L}(V_G - V_T)^2. \tag{6.22}$$

Then we can write the second line of Eq. (6.20) as

$$I_D = I_{eq}^+ \left[1 - \left(1 - \frac{V_D}{V_G - V_T}\right)^2\right]. \tag{6.23}$$

For small values of the drain voltage, we can expand the terms in the square brackets and define a resistance as

$$R_C = \frac{V_D}{I_D} \sim \frac{V_G - V_T}{2I_{eq}^+}. \tag{6.24}$$

Even for a very short device with perfect transmission, we cannot get around this minimum resistance. This resistance characterizes the transition region between the pure source and the channel and appears regardless of whether the transport is described by the mobility or is ballistic. In some sense, this resistance describes a barrier between the source and the channel [19]. The current flux of Eq. (6.22) is due to carriers from the source with sufficient kinetic energy to overcome this barrier between the source and the channel. It is important to note in Eq. (6.22) that the flux is actually metered by the gate potential, so that this is a gate-controlled barrier. Once the carriers supplant this barrier, they are going to flow downhill (down the potential drop) to the drain. They will flow to the drain regardless of the nature of the transport. The only effect that ballistic transport can play is to change the shape of the potential landscape between the source and the drain. In this sense, scattering actually affects this potential landscape and serves to isolate the drain from the source. In a ballistic device, this isolation does not occur and the source–channel barrier can be affected by the drain voltage, an effect known as *drain-induced barrier lowering*.

6.4   Quantum Point Contact

One of the simplest devices for the understanding of mesoscopic and nanoelectronic devices has been the observation of one-dimensional conductance quantization. In Chapter 2, we introduced the QPC in which the transport is not just along a single direction but is actually characterized as being transport in a narrow one-dimensional channel, not unlike a small pipe. The observed conductance is quantized when the transverse dimensions of this "pipe" are comparable to the Fermi wavelength, and the conductance is found to increase in steps of $2e^2/h$ as the size of the pipe is slowly increased. Generally, this phenomena is observed in short quantum wires at low temperatures when the electron transport is largely ballistic in nature. We will discuss the reason for this limitation later, but one can simply understand that scattering interferes with the general quantization process.

The first measurements through a QPC were made by van Wees et al. [20] and Wharam et al. [21]. In Fig. 6.4, we illustrate this behavior through two QPCs of different sizes, both defined by two electrostatic Schottky barrier gates [22], for a GaAs/AlGaAs heterostructure. We can see the physical structure in the insets, with the two split gates (gold regions) and source and drain contacts placed far away from the QPCs themselves. The overall behavior with gate bias is shown in the main panel and the discrete steps in conductance are easily observed. To within the accuracy that can be inferred from the figure, the steps correspond to plateaus with a conductance that is an integer multiplier of $2e^2/h$. In this structure, the negative gate voltage controls the electrostatic width W of the opening between the metal gates. As the gate voltage is made more negative, the width is reduced and the conductance is also reduced. The fact that the conductance goes down in steps is a property of the quasi-one-dimensional nature of the channel that passes through the QPC and the nearly ballistic nature of the transport, both of which are expressed easily with the Landauer formula. One may also see that varying the opening (differences in the two parts of the figure) changes the gate voltage range over which a single plateau exists, while changing the length of the QPC gives more distinct plateaus.

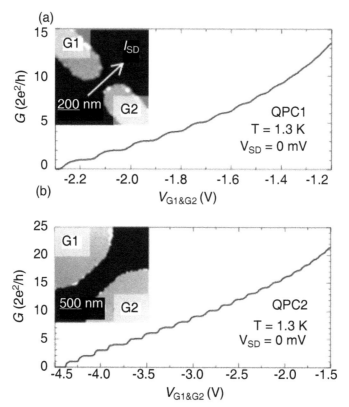

Figure 6.4 (a) Inset: atomic force micrograph of the sample surface. The two Schottky gates appear bright; the GaAs surface appears dark. The distance between the gates is w = 200 nm. When the gates are negatively biased, free electrons reside only in the electron gas underneath the dark areas. Main graph: differential conductance of QPC1, measured as a function of the voltage applied to gates G1 and G2. Quantized conductance in multiples of $G = 2e^2/h$ indicates the formation of discrete sub-bands between the tips of the gates. (b) Differential conductance of QPC2, which is w = 500 nm wide and $l \approx$ 1 μm long. Reprinted from Rössler et al. [22], with permission. Copyright IOP Publishing & Deutsche Physikalische Gesellschaft (CC BY-NC-SA).

The QPCs described above provide the prototypical system for the study of quasi-one-dimensional conductance that can be realized by defining narrow constrictions in a normal quasi-two-dimensional electron gas. For example, we recall that a quasi-two-dimensional electron gas exists at the interface between GaAs and GaAlAs. It has a quasi-two-dimensional behavior as the motion is restricted and quantized in the direction normal to the heterostructure layers. Hence, the electrons can move only in the plane of the heterostructure. The transverse gates provide an additional level of quantization, which further restricts the motion of the carriers, limiting this to be essentially in a single direction. Usually, these gates are used to define a constriction, as is observed in Fig. 6.4, whose size is less than a micron and hopefully sufficiently narrow to avoid the presence of any impurities, which can lead to scattering. At the center of the QPC, if we move toward the two gates, the potential increases due to the gate potential. However, if we move along the channel direction, the potential will decrease away from the center of the QPC. The highest potential in the channel (along the center line) provides a barrier to transport through the QPC. In most cases, it is reasonable to assume that the potential is parabolic in both directions, and this will make the Schrödinger equation exactly solvable. While one might expect that there are reasons for nonparabolic behavior of the potential, in fact the self-consistency that constrains the potential tends to drive it toward a quadratic, parabolic behavior [23]. As one moves away from the center line of the QPC, the energy potential rises as one is moving toward the metallic edges of the gates. Thus, the overall energy contour of the QPC looks like a saddle potential. The value of the energy at the saddle itself is given by $\alpha|V_G|$ (we use the magnitude as the voltage is negative, but the potential is positive relative to the bottom of the conduction band), where α is a lever arm whose value is determined by the movement of the saddle potential for a given gate voltage change. At a gate voltage of about −2.6 V on QPC1 and about −4.38 V on QPC2, the conductance drops to zero, and this corresponds to the saddle minimum being at the Fermi energy, so no carriers can get over the barrier. Then as the negative gate voltage is reduced, carriers start to propagate through the QPC. More modes are excited as the saddle potential is gradually reduced through raising the gate voltage.

When a source–drain bias is applied, this changes the Fermi potential at one contact, as discussed with the Landauer formula above, and this shifts the gate voltages at which the transitions occur. That is, if the saddle sits at an energy of $E_s < E_F$, then conduction can occur through the QPC. If now we apply a drain potential, the Fermi energy of the drain becomes $E_F - eV_D$, where the reference E_F is taken at the source contact. In the center of the QPC, we expect that the conduction band edge is pulled down by about one-half of the applied voltage. This makes the bias between the semiconductor and the gate itself larger in magnitude and thus works to pinch off the QPC a little more. On the other hand, if we apply a negative bias to the drain, the opposite occurs and the QPC becomes more open. Tracking the plateaus while one varies the source–drain bias allows a method of spectroscopy in which we may analyze the sub-band spacing in the device. This is illustrated in Fig. 6.5, where the transconductance is plotted as the two voltages are varied [22]. The plot uses the source–drain voltage for the vertical axis and the gate voltage in the horizontal axis, with the transconductance color coded. Bright yellow is the largest value and black is the smallest value. Hence, the plateaus appear as dark regions between the brighter "lines," which indicate the transitions between the plateaus. It can be easily seen how the integer plateaus gradually change with bias into half-integer ones, and then change again into integer ones with increasing the source–drain bias. Observation of these returning integer plateaus can be difficult, and the very high mobility in these samples makes this possible. This high mobility manifests itself as a reduction in the back-scattering of various modes at the transitions and leads to enhanced ability to do spectroscopy in this nonlinear transport regime.

Also indicated in Fig. 6.5 is how one can get analytical results. Three white circles are indicated at the transition between the fourth and fifth plateaus at different values of source–drain bias. On the dashed line between the circles, the observed change while the source–drain bias is increased obviously measures the distance between the sub-bands. That is, as one moves across the plateau from low bias to the bias corresponding to the next transition, one is clearly coupling exactly two new sub-bands between the Fermi energies. The transition occurs as the sub-band crosses one Fermi level, so clearly one-half of this applied bias corresponds to the sub-band spacing, as indicated in the figure. This suggests that, at this bias, the sub-band spacing is about 2.25 meV. But one can go further, as each of the transitions allows us to study the sub-band spacing as a function of the two bias voltages. Thus, one can see just how the eigenenergies of the sub-bands vary with gate voltage and with source–drain bias voltage. Thus, for the devices discussed here, the authors have found that the sub-band spacing varies from almost 5 meV down to about 1 meV when the QPC is sufficiently open that 10 modes are propagating through it [22]. It is clear that, with good quality material, one can fully characterize both the device under study and the models used to explain the results.

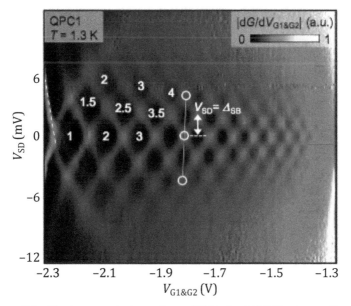

Figure 6.5 The transconductance $G_{TC} = dG/dV_G$ for QPC1 (the two split gates are biased equally). The source–drain voltage is plotted vertically and the gate voltage horizontally, with the transconductance color coded according to the scale at the upper right corner. Plateaus in conductance (small transconductance) appear as dark areas, and the various ones are labels accordingly. Reprinted from Rössler et al. [22], with permission. Copyright IOP Publishing & Deutsche Physikalische Gesellschaft (CC BY-NC-SA).

6.5   Resonant-Tunneling Diode

Perhaps the best studied transport phenomenon associated with quantum transport is that of tunneling. In general, the term "tunneling" refers to particle transport through a classically forbidden region, where we mean a region in which the total energy of a classical point particle is less than its potential energy. Quantum mechanically, the underlying equation of motion is the Schrödinger equation, in which the role of the potential is analogous in electromagnetics to that of a spatially varying permittivity. In electromagnetics, the solution to the wave equation must satisfy certain boundary conditions at the abrupt interface between two dielectrics of different permittivity, which leads to a certain portion of an incident wave being transmitted and a certain portion reflected. Likewise in quantum mechanics, the wavefunction and its normal derivative must be continuous across a boundary of two regions of different potential energy, which similarly leads to reflection and transmission probability waves at the boundary. The wavefunction associated with a particle incident from the left on a potential barrier has nonzero solutions inside the barrier. Because the square of the wavefunction represents the probability density for finding a particle in a given region of space, it follows that quantum mechanically a particle incident on a potential barrier has a finite probability of *tunneling* through the barrier and appearing on the other side.

Historically, the phenomenon of tunneling was recognized soon after the founding of quantum theory in connection with field ionization of atoms and nuclear decay of alpha particles. Shortly thereafter, tunneling in solids was studied by Fowler and Nordheim [24] in the field emission of electrons from metals, that is, the electric field-aided thermionic emission of electrons from metal into vacuum. Later, interest developed in tunneling through thin insulating layers (such as thermally grown oxides) between metals (MIM), and between semiconductors and metals (MIS). After the development of the band theory of solids, Zener [25] proposed the concept of *interband* tunneling, in which electrons tunnel from one band to another through the forbidden energy gap of the solid. The time period of the late 1940s and early 1950s saw tremendous breakthroughs in the development of semiconductor device technology, and conditions favorable to the experimental observation of Zener tunneling in *p-n* junction diodes were realized. In the late 1950s, Esaki [26] proposed the so-called Esaki diode, in which negative differential resistance (NDR) is observed in the I–V characteristics of heavily doped *p-n* diodes due to interband Zener tunneling between the valence and conduction bands. The Esaki diode continues to be important technologically and finds many applications in microwave technology. Recent theories associated with *single-electron charging* in quantum dots are descendants of the transfer Hamiltonian method and will be revisited in the next section in connection with single-electron phenomenon. The transfer Hamiltonian model [27] was also utilized quite extensively in the study of independent particle tunneling [28–30], which became the basis for interpreting a host of experimental tunneling studies in normal metals and semiconductors. A thorough review of the status of experimental and theoretical tunneling related research prior to the 1970s is given by Duke [31].

During the 1970s, advances in epitaxial growth techniques such as MBE increasingly allowed the growth of well-controlled heterostructure layers with atomic precision and low background impurity densities. In their pioneering work in this field, Tsu and Esaki [32, 33] at IBM predicted that when bias is applied across the structure, the current–voltage (I–V) characteristics of GaAs/$Al_xGa_{1-x}As$ double and multiple barrier structures should show NDR similar to that in Esaki diodes. However, NDR in this case occurs due to resonant tunneling through the barriers within the same band. Resonant tunneling refers to tunneling in which the electron transmission coefficient through a structure is sharply peaked about certain energies, analogous to the sharp transmission peaks as a function of wavelength evident through optical filters, such as a Fabry–Perot étalon consisting of two parallel dielectric interfaces. This is shown in Fig. 6.6, where simulations from the Resonant Tunneling Diode Tool at Nanohub.org [34] are given. Here, two different tunnel barriers, each 5 nm thick, are separated by a 5 nm potential well. The structure is shown in panel (a), while the transmission is shown in panel (b). The peak transmission of the lower energy peak is 0.758, while the transmission of the higher energy quasi-bound state is 0.709.

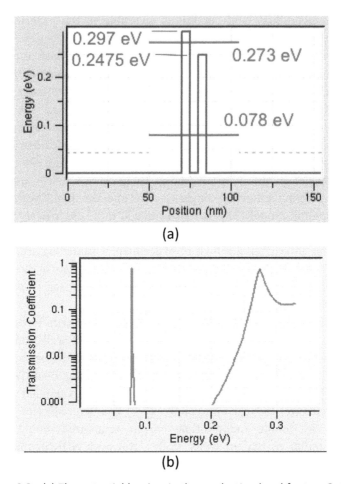

Figure 6.6 (a) The potential barriers in the conduction band for two GaAlAs barriers of 5 nm thickness each, separated by the 5 nm well. The peak energies (blue) of the two barriers and the bound states (red) are shown. (b) The transmission through the structure as a function of the incident energy. The simulations were done with the Resonant Tunneling Diode tool available at NanoHub.org [34].

The first experimental evidence for resonant tunneling in double barrier structures was reported by the IBM group in MBE-grown structures [33] where weak NDR was observed in the I–V characteristics. More effective devices were subsequently developed by Sollner et al. [35]. This particular structure consists of two $Al_xGa_{1-x}As$ barriers ($x \sim 0.25$–0.30) separated by a thin

GaAs quantum well. For the Al mole fraction of this structure, the estimated barrier height due to the conduction band offset, ΔE_c, is approximately 0.23 eV. The thickness of the barriers (here 5 nm) is sufficiently thin that tunneling through the barriers is significant. The energy E_1 corresponds to the lowest resonant energy, which as discussed above is the energy where the transmission coefficient is very peaked, and in fact may approach unity in some cases. This energy may qualitatively be thought of as the bound state associated with the quantum well formed between the two confining barriers. However, since the electron may tunnel out of this bound state in either direction, there is a finite lifetime τ associated with this state, and the width of the resonance in energy (i.e., the energy range in which the transmission coefficient is sizable) is inversely proportional to this lifetime, approximately as $\hbar/\tau$. Depending on the well width and barrier heights, there may exist several such *quasi-bound states* in the system. The double barrier structure is surrounded by heavily doped GaAs layers, which provide low-resistance emitter and collector contacts to the tunneling region, forming the RTD structure. With a positive bias applied to the right contact relative to the left, the Fermi energy on the left is pulled through the resonant-level E_1. As the Fermi energy passes through the resonant state, a large current flows due to the increased transmission from left to right. At the same time, the back flow of carriers from right to left is suppressed as electrons at the Fermi energy on the right see only a large potential barrier, as shown in the figure. Further bias pulls the bottom of the conduction band on the left side through the resonant energy, which cuts off the supply of electrons available at the resonant energy for tunneling. The result is a marked decrease in the current with increasing voltage, giving rise to a region of NDR as shown schematically by the I–V characteristics. A simulation of this behavior is shown in Fig. 6.7 for the structure of Fig. 6.6. The sharp drop in current at 0.16 V occurs when the resonant level 1 is pulled below the conduction band edge of the left-hand GaAs layer.

A considerable volume of research into RTDs has evolved since the MIT group's successful demonstration of NDR effects. Interest stems not only from the fundamental physics aspects of this deceptively simple structure, but also from the potential practical applications in high-speed microwave systems and novel digital logic circuits. One advantage of the RTD for electronic applications is that the fundamental time associated with the intrinsic tunneling process itself may be quite short, often taken as the lifetime of the quasi-bound state (i.e., the inverse of the resonance width). In their original work, Sollner et al. [35] demonstrated NDR up to frequencies of 2.5 THz, which qualitatively implies charge transport on the order of $\tau = 6 \times 10^{-14}$ s. In reality, there are several time constants that come into play in the frequency response of an RTD, including the transit time across the nontunneling regions of the device and the RC time constant associated with the capacitance of the structure. With proper design, the various time constants can be minimized, and high-frequency performance may be obtained in analog applications such as oscillators operating up to 420 GHz in GaAs/AlAs double barrier structures [36].

Current flow through the resonant-tunneling diode has been calculated many times by semiclassical approaches. These appear in many books. Of interest here, and in the later chapters, is the simulation by full quantum mechanical means. We will discuss these approaches as we come to the techniques in the following chapters.

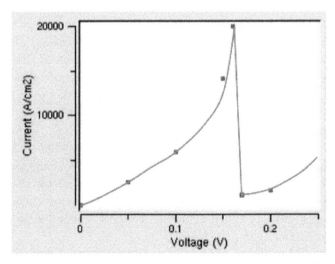

Figure 6.7 The simulated current–voltage curve for the structure of Fig. 6.6. The line is a guide to the eye. The simulation was done with the Resonant Tunneling Diode tool available at NanoHub.org [34].

6.6   Single-Electron Tunneling

When the dielectric material in a capacitor becomes very thin, it is possible for the charge on the capacitor plate to tunnel through the insulator. Under some mesoscopic device concepts, this tunneling is a desirable effect and can be used to create some interesting devices. In other cases, such as the gate oxide in a MOSFET, this is an undesirable effect. When the capacitor is made with a small area, then another effect begins to occur. The change in electrostatic potential due to a change in the charge on an ideal conductor is associated with the linear relationship between the charge and the voltage, $Q = CV$, where C is the capacitance, Q is the charge on either of the two plates, and V is the electrostatic potential that exists across the capacitor. For highly conducting electrodes, any charge added to either of the electrodes will rearrange itself such that the internal electric field vanishes, and the surface of the electrode becomes an equipotential surface. Therefore, the electrostatic potential associated with the electrode relative to its reference is uniquely defined. In our capacitor, we consider two conductors connected by a DC voltage source. This leads to a charge $+Q$ on one conductor and a charge $-Q$ on the other. The capacitance of the two-conductor system is then defined as $C = Q/V_{12}$. The electrostatic energy stored in the two-conductor system is the work done in building up the charge Q on the two conductors and is given by

$$E = \frac{Q^2}{2C}. \qquad (6.25)$$

In the case of very small capacitors, the charging energy given by Eq. (6.25) due to a single electron, $e^2/2C$, becomes comparable to the thermal energy, $k_B T$. The transfer of a single electron between conductors, therefore, results in a voltage change that is significant compared to the thermal voltage fluctuations and creates an energy barrier to the transfer of electrons. This barrier remains until the charging energy is overcome by sufficient bias. How small must the capacitor be for such effects to become important? If the energy stored in the capacitor is about the same as the thermal energy, then the capacitor has a value of 3×10^{-18} F, or 3 attofarad, at room temperature. Of course, at very low temperatures, the capacitance can be significantly larger. This blockage of the charge tunneling is known as the *Coulomb blockade.*

Historically, Coulomb blockade effects were first predicted and observed in small metallic tunnel junction systems. As mentioned already, the conditions in metallic systems of high electron density, large effective mass, and short phase coherence length (compared to semiconductor systems) usually allow us to neglect size quantization effects. The dominant single-electron effect for small metal tunnel junctions is, therefore, the charging energy due to the transfer of individual electrons, $e^2/2C$. The effects of single-electron charging in the conductance properties of very thin metallic films were recognized in the early 1950s by Gorter [37] and Darmois [38]. It was found that metal films formed arrays of small islands, and conduction occurs due to tunneling between these islands. Since the island size is small, the tunneling electron has to overcome an additional barrier due to the charging energy, which leads to an increase in resistance at low temperature. Such discontinuous metal films show an activated conductance, similar to an intrinsic semiconductor. Neugebauer and Webb [39] developed a theory of activated tunneling in which the activation energy resembles an energy gap and is, therefore, referred to as a *Coulomb gap.*

There have been many experimental studies of the transport properties of metal clusters or islands imbedded in an insulator that are then contacted by conducting electrodes. More interest in the area developed in studies of superconducting tunnel junctions, where the Coulomb blockade interacted with the normal Josephson tunneling. Very soon, however, the Coulomb blockade and the tunneling of single electrons were observed for normal metal systems. What is normally seen in these experiments is that no, or a very low, current flows until the applied voltage reaches $e/2C$, then the current begins to rise. So this gives a plateau around zero bias, and the width of this plateau is typically e/C, corresponding to the energy in Eq. (6.25). But our interest is in semiconducting systems, where the tunneling action is normally in the plane of the sample, usually fabricated in a heterostructure system such as GaAlAs/Gas.

A single capacitor is seldom used to make a structural device. Certainly, a single capacitor is used in a DRAM, but there it is coupled to a transistor, which controls the charging and discharging of the capacitor. In small semiconductor devices, the idea of a *single-electron device*, or single-electron transistor, has appeared, and this uses (at least) two capacitors connected in series. Between the two capacitors is a region that can accumulate charge. Typically, this small region is termed a quantum dot. In metals, it may contain more than a thousand electrons, but in semiconductors the charge states may have their own quantization due to the small size of the dot and the number of electrons can be few, even down to zero.

A single-electron circuit can be created by placing two capacitors in series, so that an "island" is created between them. Electrons that tunnel through one capacitor or the other are assumed to immediately relax due to carrier–carrier scattering, so that resonant tunneling through both barriers is simultaneously neglected. This assumption is made as we are interested in the charge that can accumulate in/on the quantum dot. That is, charge may tunnel through only one of the two capacitors at a time, and this will change the amount of charge on the dot. Tunneling represents the injections of single particles, which involve several characteristic time scales. The tunneling time (the time to tunnel from one side of the barrier to the other) is the shortest time (on the order of 10^{-14} s), whereas the actual time between tunneling events themselves is on the order of the current divided by *e,* which for typical currents in the nA range, implies a mean time of several hundred picoseconds between events. The time for charge to rearrange itself on the electrodes due to the tunneling of a single electron will be something on the order of the dielectric relaxation time, which also can be very short. Therefore, for purposes of analysis, we can consider that the junctions in the regime of interest behave as ideal capacitors through which charge is slowly leaked.

While this simple structure is interesting, it was quickly realized that having an additional bias voltage to control the actual charge on the quantum dot would be beneficial to creating a single-electron transistor. Consider the single-electron transistor circuit shown in Fig. 6.8. In this circuit, a separate voltage source, V_g, is coupled to the island through an ideal (infinite tunnel resistance) capacitor, C_g. This additional voltage modifies the charge balance on the island so that an additional polarization charge arises from this new bias source and its coupling capacitance. (As in the previous section, we pursue the simple classical equations here, leaving the full quantum transport for discussion in the following chapters.) The net charge on the electrodes of the individual capacitors is given by

$$Q_1 = C_1 V_1$$
$$Q_2 = C_2 V_2 \tag{6.26}$$
$$Q_g = C_g (V_g - V_2).$$

The net charge on the quantum dot island may be defined via

$$Q_{dot} = -ne = Q_2 - Q_1 - Q_g. \tag{6.27}$$

The voltages across the two tunneling capacitors may be written as

$$V_1 = \frac{1}{C_T} \left[(C_g + C_2)V_a - C_g V_g + ne \right]$$

$$V_2 = \frac{1}{C_T} \left[C_1 V_a + C_g V_g - ne \right] \tag{6.28}$$

$$C_T = C_1 + C_2 + C_g.$$

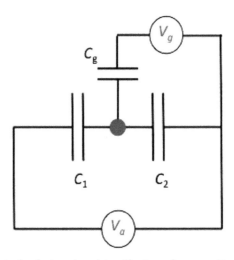

Figure 6.8 A single-electron transistor. The tunneling capacitors are indicated as C_1 and C_2 and the applied bias is that of V_a. The storage node is the quantum dot and is indicated by the solid dot, and the charge on this island is controlled by the gate voltage V_g through the gate capacitor C_g.

With these equations, we can write the energy due to the charges that exist on each of the three capacitors as

$$E = \frac{1}{2C_T} \left[C_g C_1 (V_a - V_g)^2 + C_1 C_2 V_a^2 + C_g C_2 V_g^2 + Q_{dot}^2 \right]. \tag{6.29}$$

The work performed by the voltage sources during the tunneling through junctions 1 and 2 now includes both the work done by the gate voltage and the charge flowing onto the gate capacitor electrodes. We write this in terms of the number of electrons n_1 and n_2 that exist on capacitors C_1 and C_2, respectively, as

$$W_a(n_1) = -\frac{n_1 e}{C_T}\left[C_2 V_a + C_g(V_a - V_g)\right]$$
$$W_a(n_2) = -\frac{n_2 e}{C_T}\left[C_1 V_a + C_g V_g\right].$$

(6.30)

These three contributions to the total energy can now be combined. However, it is more important to determine the *change* in the energy that occurs for a change in the charge on the dot by adding or subtracting a single-electron charge from one of the two capacitors. These changes in energy are defined by

$$\Delta E_1^{\pm} = E(n_1, n_2) - E(n_1 \pm 1, n_2)$$
$$= \frac{e}{C_T}\left\{-\frac{e}{2} \mp \left[ne + (C_g + C_2)V_a - C_g V_g\right]\right\}$$

(6.31)

for capacitor 1 and

$$\Delta E_2^{\pm} = E(n_1, n_2) - E(n_1, n_2 \pm 1)$$
$$= \frac{e}{C_T}\left\{-\frac{e}{2} \pm \left[ne - C_1 V_a - C_g V_g\right]\right\}$$

(6.32)

for capacitor 2. The gate bias now allows us to change the charge on the island and, therefore, to shift the region of Coulomb blockade. Thus, a stable region of Coulomb blockade may be realized for $n \neq 0$. As before, the condition for tunneling at low temperature is that $\Delta E_{1,2} > 0$ so that the system goes to a state of lower energy after tunneling. The conditions for forward and backward tunneling then become

$$-\frac{e}{2} \mp \left[ne + (C_g + C_2)V_a - C_g V_g\right] > 0$$
$$-\frac{e}{2} \pm \left[ne - C_1 V_a - C_g V_g\right] > 0.$$

(6.33)

The four equations (6.33) for each value of n may be used to generate a stability plot in the $V_a - V_g$ plane, which shows stable regions corresponding to each n for which no tunneling may occur.

That is, the set of four lines defines a set of diamonds, which lie along the gate voltage axis. At the values where lines of opposite slope cross the axis simultaneously current is allowed to flow. The lines represent the boundaries for the onset of tunneling given by these equations for different values of n. The trapezoidal shaded areas correspond to regions where no solution satisfies the equations, and hence where Coulomb blockade exists. Each of the regions, termed a *Coulomb island*, corresponds to a different integer number of electrons on the island, which is "stable" in the sense that this charge state cannot change, at least at low temperature when thermal fluctuations are negligible. The gate voltage then allows us to tune between stable regimes, essentially adding or subtracting one electron at a time to the island.

In Section 1.4, we discussed the single-atom transistor. We can now understand this device better in terms of this single-electron transistor. If we refer to Fig. 1.3, the quantum dot is the single phosphorous atom sitting at the center of the transistor. Capacitors 1 and 2 are the gaps that exist between this atom and the heavily doped source and drain regions. The gate capacitor is the gap between the atom on the two gates G1 and G2. The behavior of the transistor itself is easily understood from the above set of equations. But, prior to this, there were many simulations of such a structure with complex equations of quantum transport, as we will see later.

There are extensions of these simple approaches, such as using a double quantum dot, which gives rise to a two-dimensional array of Coulomb islands. In Fig. 6.9, we illustrate such a double quantum dot that is used in studies of spin–spin interactions and readout of the spin [40]. This paper represents the first experimental readout of the spin in a quantum dot structure. The interactions are between the spins in the two separate quantum dots. The experiments are complicated, so we only discuss them briefly here, in order to understand the structure that is fabricated. The light areas in the figure are metal lines on the semiconductor surface, while the semiconductor itself is the black region. The grey region over the right-hand side is a deposited micro-magnet, in this case made of Ni. A cross section is shown in the lower panel, and one can see the dots (the two white circles in each image) formed at the interface between the AlGaAs and the GaAs. The narrow-wide metal at the left is used as a microwave strip line in order to introduce the microwave modulation signal to the dots. On the right-hand side, between gates G3 and G4, we see a QPC. The presence of charge in QD2 will affect the opening of this QPC, and this can be monitored by the conductance of the QPC. The microwave signal leads to photon-assisted tunneling of an electron from one dot to the other. But this can occur only if the two electrons have opposite spin (assuming they go to the same energy level of the dot). The micro-magnet induces a particular spin state in QD2, and by adjusting the various biases, one can probe this spin state according to whether or not the tunneling is permitted. In addition, of course, one can study a great deal about the physics of double quantum dots. This is beyond the level we wish to treat in this introduction to a double quantum dot device.

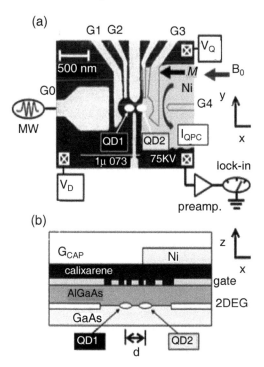

Figure 6.9 (a) A scanning electron micrograph of a lateral double quantum dot device, which includes an Ni micromagnet (grey area). M and the arrow denote the direction of magnetization of the magnet (and correspondingly for the magnetic field B_0). Two white circles, labeled QD1 and QD2, denote the quantum dots. (c) A cross section of the structure to show how the dots are located at the interface between the AlGaAs and the GaAs. Figure adapted with permission from Ref. [40], Copyright 2010, American Physical Society.

Problems

1. A rectangular quantum wire with cross section 15×15 nm^2 is realized in GaAs. (a) Write an expression for the sub-band threshold energies of the quantum wire. (b) Write a general expression for the electron dispersion in the quantum wire. (c) In a table, list the quantized energies and their degeneracies (neglecting spin) for the first 15 energetically distinct sub-bands. (d) Plot the density of states of the wire over an energy range that includes the first 15 sub-band energies.

2. In Problem 1, we solved for the density of states in a rectangular GaAs quantum wire with cross section 15×15 nm^2. For this same wire: (a) Write an expression for the electron density (per unit length) as a function of the Fermi energy of the wire. (b) Plot the variation of the electron density as a function of energy for a range that corresponds to filling the first five distinct energy levels, taking proper account of level degeneracies.

3. When a voltage (V_{sd}) is applied across a QPC, it is typical to assume that the quasi-Fermi level on one side of the barrier is raised by $\alpha e V_{sd}$, while that on the other drops by $(1 - \alpha)$ eV_{sd}, where α is a phenomenological parameter that, in a truly symmetrical structure, should be equal to $\frac{1}{2}$. If we consider a device in which only the lowest sub-band contributes to transport, then the current flow through the QPC may be written as

$$I_{sd} = \frac{2e}{h} \left[\int_L T(E)dE - \int_R T(E)dE \right]$$

where $T(E)$ is the energy-dependent transmission coefficient of the lowest sub-band and L and R denote the left and right reservoirs, respectively. If we assume low temperatures, we can treat the transmission as a step function, $T(E) = \theta(E - E_1)$, where E_1 is the threshold energy for the lowest sub-band. (a) Write this integral with limits appropriate to determine the current. (b) Use this information to obtain an expression for the current flowing through the QPC when the source–drain voltage is such that both reservoirs populate the lowest sub-band, and when it populates the sub-band from just the higher-energy reservoir.

4. At nonzero temperature, the conductance through a nanodevice is determined by weighting the transmission at each energy by the value of the (negative of the) derivative of the Fermi–Dirac distribution. This leads to the fact that the conductance takes place near the Fermi level. Compute the derivative of the Fermi–Dirac function and evaluate the full-width at half-maximum for this function.

5. In treating the QPC with the harmonic oscillator approximation, the different energy levels are all equally spaced at a given value of the curvature, which is determined by the gate potential. Using the data shown in Fig. 6.5, estimate the energy level spacing as a function of the gate voltage (horizontal axis) in this device. Include at least the first nine transitions above the first plateau.

6. Consider a SET whose individual tunnel barriers have capacitances $C_1 = C$ and $C_2 = 2C$ and whose gate capacitance $C_g = C/3$. You may assume that any background charge can be taken to be zero in this problem. (a) Write the set of four equations that can be used to define the charge–stability diagram for this device. (b) Plot the charge–stability diagram for this device and indicate the regions where current flow is Coulomb blockaded. (c) An important parameter for any transistor is its voltage gain, i.e., the change in the source–drain voltage arising from a change in the gain voltage. For the SET considered here, what can you say about its voltage gain?

7. Consider a parabolic quantum dot implemented in a GaAs 2DEG with $\hbar\omega_0 = 2$ meV, where ω_0 is characterized by the harmonic oscillator itself. Now consider the situation in which 12 electrons are present in this dot and assume also that spin splitting of any electron states can be neglected. Plot the energy levels of the dot as a function of magnetic field (for $0 < B < 4$ T) and indicate the variation of the highest filled electron state as a function of magnetic field in the situation where the electrons occupy the ground state of the dot. (Hint: This problem involves the Darwin–Fock spectrum of a two-dimensional quantum dot.)

References

1. M. Fuechsle, J. A. Miwa, M. Mahapatra, H. Ryu, S. Lee, O. Warschkow, L. C. L. Hollenberg, G. Klimeck, and M. Y. Simmons, *Nat. Nanotechnol.*, **7**, 242 (2012).

2. S. Oda and D. K. Ferry, Eds., *Nanoscale Silicon Devices* (CRC Press, New York, 2015).

3. P. C. Martin and J. Schwinger, *Phys. Rev.*, **115**, 1342 (1959).

4. J. Schwinger, *J. Math. Phys.*, **2**, 407 (1961).

5. L. P. Kadanoff and G. Baym, *Quantum Statistical Mechanics* (Benjamin, New York, 1962).

6. L. V. Keldysh, *Sov. Phys. JETP*, **20**, 1018 (1965).

7. C. P. Enz, *A Course on Many-Body Theory Applied to Solid-State Physics* (World Scientific Press, Singapore, 1992), p. 76.

8. H. Haug, in *Quantum Transport in Ultrasmall Devices*, edited by D. K. Ferry, H. L. Grubin, C. Jacoboni, and A.-P. Jauho (Plenum Press, New York, 1995).

9. T. Kuhn and F. Rossi, *Phys. Rev. B*, **46**, 7496 (1992).

10. H. Haug and A.-P. Jauho, *Quantum Kinetics in Transport and Optics of Semiconductors* (Springer, Berlin, 2004).

11. G. D. Mahan, *Many-Particle Physics* (Plenum, New York, 1981).

12. A. L. Fetter and J. D. Walecka, *Quantum Theory of Many-Particle Systems* (McGraw-Hill, New York, 1971).

13. W. Shockley and J. Bardeen, *Phys. Rev.*, **77**, 407 (1950); **80**, 72 (1950).

14. C. Kittel, *Introduction to Solid State Physics*, 6th Ed. (Wiley, New York, 1986).

15. K. Seeger, *Semiconductor Physics*, 4th Ed. (Springer-Verlag, Berlin, 1989).

16. R. Landauer, *IBM J. Res. Dev.*, **1**, 223 (1957).

17. R. Landauer, *Philos. Mag.*, **21**, 863 (1970).

18. S. Datta, F. Assad, and M. S. Lundstrom, *Superlattices Microstruct.*, **23**, 771 (1998).

19. H. Kroemer, *IEEE Trans. Electron Dev.*, **15**, 819 (1968).

20. B. J. van Wees, H. van Houten, W. J. Beenakker, J. G. Williamson, L. P. Kouwenhoven, D. van der Marel, and C. T. Foxon, *Phys. Rev. Lett.*, **60**, 848 (1988).

21. D. A. Wharam, T. J. Thornton, R. Newbury, M. Pepper, H. Ahmed, J. E. F. Frost, D. G. Hasko, D. C. Peacock, D. A. Ritchie, and G. A. C. Jones, *J. Phys. C*, **21**, L209 (1988).

22. C. Rössler, S. Baer, E. de Wiljes, P.-L. Ardelt, T. Ihn, K. Ensslin, C. Reichl, and W. Wegscheider, *New J. Phys.*, **13**, 113006 (2011). http://dx.doi.org/10.1088/1367-2630/13/11/113006.

23. S. E. Laux, D. J. Frank, and F. Stern, *Surf. Sci.*, **196**, 101 (1988).

24. R. H. Fowler and L. Nordheim, *Proc. Roy. Soc. (London)*, **119**, 173 (1928).

25. C. Zener, *Proc. Roy. Soc. (London)*, **A145**, 523 (1943).

26. L. Esaki, *Phys. Rev.*, **109**, 603 (1957).

27. J. Bardeen, *Phys. Rev. Lett.*, **6**, 57 (1961).

28. M. H. Cohen, L. M. Falicov, and J. C. Phillips, *Phys. Rev. Lett.*, **8**, 316 (1962).

29. W. A. Harrison, *Phys. Rev.*, **123**, 85 (1961).

30. E. O. Kane, *J. Appl. Phys.*, **32**, 83 (1961).

31. C. B. Duke, *Tunneling in Solids*, *Solid State Physics*, **10** (Suppl.), Academic Press, New York (1969).

32. R. Tsu and L. Esaki, *Appl. Phys. Lett.*, **22**, 562 (1973).

33. L. L. Chang, L. Esaki, and R. Tsu, *Appl. Phys. Lett.*, **24**, 593 (1974).

34. Resonant Tunneling Diode Tool, NanoHub.org. doi:10.4231/D35D8NG1G.

35. T. C. L. G. Sollner, W. D. Goodhue, P. E. Tannenwald, C. D. Parker, and D. D. Peck, *Appl. Phys. Lett.*, **43**, 588 (1983).

36. E. R. Brown, T. C. L. G. Sollner, C. D. Parker, W. D. Goodhue, and C. L. Chen, *Appl. Phys. Lett.*, **55**, 1777 (1989).

37. C. Gorter, *Physica*, **17**, 777 (1951).

38. E. Darmois, *J. Phys. Radium*, **17**, 210 (1956).

39. C. A. Neugebauer and M. B. Webb, *J. Appl. Phys.*, **33**, 74 (1962).

40. Y.-S. Shin, T. Obata, Y. Tokura, M. Pioro-Ladrière, R. Brunner, T. Kubo, K. Yoshida, and S. Tarucha, *Phys. Rev. Lett.*, **104**, 046802 (2010).

Chapter 7

Density Matrix

In Chapter 1, we developed some important concepts that distinguish quantum transport from what we call semi-classical transport. The latter is usually distinguished by the use of the Boltzmann transport equation. The general scope of Boltzmann transport theory is a non-ideal theory that fails on the short-time and short-distance scales [1]. Yet, it is conceptual and possesses relative mathematical simplicity. In addition, it tends to work much better than one might expect from its classical origins. On the other hand, quantum transport enjoys no such status. It is quite often neither conceptually nor mathematically simple. The equations can be quite complicated, and often after a significant amount of work has been done, the result reduces to that expected from the Boltzmann equation. This was certainly the case with impurity scattering in Chapter 4. Still, the quantum transport that we have discussed in the preceding chapters is necessary to explain how the Boltzmann picture arises and can be justified.

In Chapter 2, we have already introduced briefly the various forms of distributions that are used in quantum transport. With this chapter, we begin to delve deeper into the understanding and use of these distributions. Here we begin with the density matrix. The density matrix was apparently first proposed by von Neumann [2]. To see how it can be developed, consider beginning with a wavefunction $\psi(x)$. The density matrix in the *position representation* can be written as

An Introduction to Quantum Transport in Semiconductors
David K. Ferry
Copyright © 2018 Pan Stanford Publishing Pte. Ltd.
ISBN 978-981-4745-86-4 (Hardcover), 978-1-315-20622-6 (eBook)
www.panstanford.com

$$\rho(\mathbf{r}, \mathbf{r}') = \psi(\mathbf{r})\psi^\dagger(\mathbf{r}').$$ (7.1)

The diagonal elements, where $\mathbf{r} = \mathbf{r}'$, provide us with the probability density as a function of position, $\rho(\mathbf{r}, \mathbf{r}') = |\psi(\mathbf{r})|^2$. That is, if we integrate over the spatial coordinate, we have

$$\int_{-\infty}^{\infty} \rho(\mathbf{r}, \mathbf{r})d^3\mathbf{r} = \int_{-\infty}^{\infty} |\psi(\mathbf{r})|^2 \, d^3\mathbf{r} = 1.$$ (7.2)

This has certain constraints if we expand the wavefunction in a basis set as

$$\psi(\mathbf{r}) = \sum_n a_n \varphi_n(\mathbf{r}),$$ (7.3)

then we must have

$$\int_{-\infty}^{\infty} \rho(x, x)dx = \sum_{n,m} a_n a_m^* \int_{-\infty}^{\infty} \varphi_n(x)\varphi_m^\dagger(x)dx$$

$$= \sum_{n,m} a_n a_m^* \delta_{nm} = \sum_n |a_n|^2 = 1.$$ (7.4)

The first line of this latter equation requires that the density matrix be Hermitian. This implies that we can use the density matrix quite generally to evaluate the expectation values of various operators. For example, we can write this expectation value as

$$\langle A \rangle = \int_{-\infty}^{\infty} \psi^\dagger(\mathbf{r})A\psi(\mathbf{r})d^3\mathbf{r} = \sum_{n,m} a_m^* a_n \int_{-\infty}^{\infty} \varphi_m^\dagger(\mathbf{r})A\varphi_n(\mathbf{r})d^3\mathbf{r}$$

$$= \sum_{n,m} a_m^* a_n A_{mn} = \sum_{n,m} A_{mn}\rho_{nm},$$ (7.5)

where ρ_{mn} are the matrix elements for the density matrix in the basis function expansion.

The time evolution of the density matrix was indicated in the brief discussion of Chapter 2, but let us expand this just a bit. This time evolution is determined by the time evolution of the two wavefunctions with which the density matrix is formed. Then we can directly form the proper products with

$$\frac{\partial \rho(\mathbf{r}, \mathbf{r}')}{\partial t} = \frac{\partial \psi(\mathbf{r})}{\partial t}\psi^\dagger(\mathbf{r}') + \psi(\mathbf{r})\frac{\partial \psi^\dagger(\mathbf{r}')}{\partial t}$$

$$= -\frac{i}{\hbar}\left\{[H\psi(\mathbf{r})]\psi^\dagger(\mathbf{r}') + \psi(\mathbf{r})[H\psi(\mathbf{r}')]^\dagger\right\} \qquad (7.6)$$

$$= -\frac{i}{\hbar}[H, \rho(\mathbf{r}, \mathbf{r}')] = -\frac{i}{\hbar}\hat{H}\rho(\mathbf{r}, \mathbf{r}').$$

The last line introduces the Liouville equation, which is the equation of motion for the density matrix. The last form, however, is somewhat different in that a *superoperator* has been introduced. In this case, the superoperator is a "commutator generating superoperator," but in general these functions are higher-order operators that reside in a Hilbert space of operators. This space is often called the *Liouvillian space*. Now it is important to note from the second line of Eq. (7.6) that the Hamiltonian operates on different wavefunctions in the two terms of the commutator that follows. This means that the derivatives operate on specific wavefunctions, and the explicit form of Eq. (7.6) is given by Eq. (2.69), which is

$$i\hbar\frac{\partial \rho}{\partial t} = \left[-\frac{\hbar^2}{2m*}\left(\frac{\partial^2}{\partial \mathbf{r}^2} - \frac{\partial^2}{\partial \mathbf{r}'^2}\right) + V(\mathbf{r}) - V(\mathbf{r}')\right]\rho(\mathbf{r}, \mathbf{r}', t) \qquad (7.7)$$

in three dimensions.

As discussed in Chapter 2, a different form of the equation arises when we introduce the imaginary time $t \to -i\hbar\beta$, $\beta = 1/k_B T$. Then, Eq. (7.7) becomes

$$-\frac{\partial \rho}{\partial \beta} = \left[-\frac{\hbar^2}{2m*}\left(\frac{\partial^2}{\partial \mathbf{r}^2} - \frac{\partial^2}{\partial \mathbf{r}'^2}\right) + V(\mathbf{r}) - V(\mathbf{r}')\right]\rho(\mathbf{r}, \mathbf{r}'). \qquad (7.8)$$

This expression is known as the Bloch equation for the density matrix. In a sense, the Bloch equation is a quasi-steady-state, or quasi-equilibrium, form in which the time variation is either nonexistent or sufficiently slow as to not be important in the form of the statistical density matrix. It is important to note that the Bloch equation possesses an adjoint equation, which arises from the anti-commutator form of Eq. (7.6), and is given by

$$-\frac{\partial \rho}{\partial \beta} = \left[-\frac{\hbar^2}{2m*}\left(\frac{\partial^2}{\partial \mathbf{r}^2} + \frac{\partial^2}{\partial \mathbf{r}'^2}\right) + V(\mathbf{r}) + V(\mathbf{r}')\right]\rho(\mathbf{r}, \mathbf{r}'). \qquad (7.9)$$

The importance of the adjoint equation is that it allows one to find the equilibrium form of the density matrix, and this can be used as the initial condition for the time evolution of Eq. (7.7).

Before proceeding too far, it is useful to discuss the properties of the superoperators. Let us consider the matrix elements of the last form that appears in Eq. (7.6), which become

$$\left(\hat{H}\rho\right)_{kn} = \sum_{r,s}(H_{kr}\rho_{rn} - \rho_{ks}H_{sn}) = \sum_{r,s}(H_{kr}\delta_{sn} - H_{sn}\delta_{kr})\rho_{rs}$$

$$= \sum_{r,s}\hat{H}_{kn,rs}\rho_{rs}. \tag{7.10}$$

Thus, the matrix representation of the superoperator is a tetradic, or fourth-rank tensor. If the operator itself is diagonal, then

$$\hat{H}_{kn,rs} = (H_{kk} - H_{nn})\delta_{kr}\delta_{ns}. \tag{7.11}$$

In fact, for any function of a diagonal operator, we can write the properties

$$\left[f(\hat{A})B\right]_{kn} = f(A_{kk} - A_{nn})B_{kn}$$

$$\left(\frac{1}{\hat{A}-z}B\right)_{kn} = \frac{1}{A_{kk} - A_{nn} - z}B_{kn} \tag{7.12}$$

$$\left[e^{i\hat{A}}B\right]_{kn} = \left[e^{iA}Be^{-iA}\right]_{kn}.$$

These expressions will be particularly useful for the kinetic equation development that arises in the next section.

7.1 Quantum Kinetic Equation

The above equations give the solution to the density matrix for an arbitrary potential that is imposed on the system. But the equations do not contain any dissipation. On the other hand, the Boltzmann equation is a kinetic equation, which describes the dynamic evolution of a distribution function under the influence of external fields and collisional forces. In this light, the resulting transport is a balance between the driving forces of the fields and the relaxation forces of the collisions. We can develop an analogous kinetic equation, which is derived from the density matrix and describes the transport in equivalent fields and relaxation forces. We will also introduce a more extensive approach, which gives a generic prescription for the general treatment. We describe a semiconductor system composed of electrons and phonons, and the interaction between these two subsystems, although we neglect the interactions among the electrons (which, of course, could be added). The system is described by the Hamiltonian

$$H = H_0 + H_L + H_F + H_{eL},$$ (7.13)

where the terms on the right-hand side describe the electrons, the lattice, the external fields, and the electron–lattice interaction, respectively. The external field will be taken in the scalar potential gauge, and the electron interaction could contain all of the appropriate many-body interactions if we chose. The driving field is expressed, for a homogeneous field, as

$$H_F = -\mathbf{E} \cdot \mathbf{r}.$$ (7.13)

The reader will note that this bold-faced **E** is the field and not the energy.

The total density matrix ρ is defined over the entire system. If there were no interactions between the electrons and the lattice, then the density matrix could be written as a tensor product of the density matrix for the electrons and the density matrix for the lattice. In this situation, a trace Tr_L over the lattice coordinates would yield just the density matrix for the electrons. Similarly, a trace Tr_e over the electron coordinates would yield just the density matrix for the lattice. As we will see, when the interactions are present, then the simple division into two distinct density matrices will not be possible. Nevertheless, we can still define the relevant density matrix for the electrons by the partial trace over the lattice variables as

$$\rho_e = Tr_L\{\rho\}.$$ (7.14)

But in the interacting case, this will not be the simple density matrix for the electrons alone described above. Instead, it will incorporate the effect that the lattice has on the electrons due to the interactions between the two systems. There are, of course, some limitations upon this approach and the resulting Eq. (7.14). These will not concern us here, but they have been addressed in a more general approach dealing in particular with the effect the field may have on the lattice as well [3]. Note also in the above equations, we have invoked the effective mass approximation, so that transport within a single band is being studied.

The approach with the Hamiltonian (7.12) is followed by introducing this into the Liouville equation (7.6), and then tracing over the lattice variables to produce the partial density matrix for the electrons, as

$$i\hbar\frac{\partial\rho_e}{\partial t}=\left[H_0+H_F,\rho_e\right]+Tr_L\left\{[H_{eL},\rho]\right\}. \tag{7.15}$$

Clearly, the first term on the right-hand side is the electronic motion within the effective mass approximation under the influence of the applied field. The second term represents the electron–phonon interaction and the effect that this has upon the resulting electronic density matrix. In the subsequent developments, this trace over the lattice variables corresponds to the summation over the phonon momentum vector **q** that was used in previous chapters.

To proceed beyond Eq. (7.15), we will introduce the use of projection superoperators, which will project the one-electron density matrix out of the many-body multi-electron density matrix. This approach is fully equivalent to the BBGKY hierarchy of equations, but is somewhat simpler to work through [4–7]. To proceed, we Laplace transform Eq. (7.15) to give

$$\left(s+\frac{i}{\hbar}\hat{H}_e\right)\tilde{\rho}_e=-\frac{i}{\hbar}Tr_L\left\{\hat{H}_{eL}\tilde{\rho}\right\}+\rho_e(0), \tag{7.16}$$

where $\hat{H}_e=\hat{H}_0+\hat{H}_F$, in superoperator notation, and the tilde over the density matrix refers to the Laplace-transformed form. We now introduce the projection superoperator [8–10] as

$$\tilde{\rho}_1=\hat{P}\rho_e, \hat{P}^2=\hat{P}, \hat{Q}=1-\hat{P}. \tag{7.17}$$

The last two expressions tell us that the projection superoperator is idempotent and that there exists a projection operator onto the complement space. Now what do we mean with this projection operator and the one-electron density matrix? In general, there is a very large number of electrons in the semiconductor, say N. Then, the total phase space, or configuration space, has $6N$ coordinates (plus time). What we want is a typical single-electron density matrix, which has six coordinates plus time. This is achieved via the BBGKY hierarchy with one complication in which the one-electron density matrix results with one term that involves a two-electron function, just as in Section 3.5. To proceed, one must make some approximations to this two-electron function. With the projection operator defined above, we do not miss this term, but it is buried within the scattering function. This particular projection superoperator commutes with the trace operation in Eq. (7.16), so that we can introduce the scattering superoperator through the definition

$$\hat{\Sigma}\tilde{\rho}_1 \equiv Tr_L\left\{\hat{P}\hat{H}_{eL}\hat{\rho}\right\}. \tag{7.18}$$

We will explore this latter quantity more deeply below. With these definitions, Eq. (7.16) can now be rewritten in the form

$$\tilde{\rho}_1 = \hat{P}\frac{1}{i\hbar s - \hat{H}_e - \hat{\Sigma}}i\hbar\rho_e(0) = \hat{P}\frac{1}{i\hbar s - \hat{H}_e - \hat{\Sigma}}i\hbar\left[\hat{P}\rho_e(0) + \hat{Q}\rho_e(0)\right]. \tag{7.19}$$

It is clear at this point that one needs only products of various projections of the resolvent operator (the term following the first $\hat{P}$ on the right-hand side of the equation).

At this point, we need to develop an expansion identity for the resolvent operator, which we write as [11]

$$\hat{R}(s) = \frac{1}{i\hbar s - \hat{H}_e - \hat{\Sigma}} = \frac{1}{i\hbar s - \hat{H}'}. \tag{7.20}$$

To begin, we note that the denominator can be rewritten as

$$i\hbar s - \hat{H}' = i\hbar s - \left(\hat{P} + \hat{Q}\right)\hat{H}'\left(\hat{P} + \hat{Q}\right)$$
$$= i\hbar s - \hat{P}\hat{H}'\hat{P} - \hat{P}\hat{H}'\hat{Q} - \hat{Q}\hat{H}'\hat{P} - \hat{Q}\hat{H}'\hat{Q}. \tag{7.21}$$

This result can now be used to expand Eq. (7.20) in two different ways:

$$\hat{R}(s) = \frac{1}{i\hbar s - \hat{P}\hat{H}'\hat{P}}\left[1 - \left(\hat{P}\hat{H}'\hat{Q} + \hat{Q}\hat{H}'\hat{P} + \hat{Q}\hat{H}'\hat{Q}\right)\hat{R}(s)\right]$$
$$= \frac{1}{i\hbar s - \hat{Q}\hat{H}'\hat{Q}}\left[1 - \left(\hat{P}\hat{H}'\hat{Q} + \hat{Q}\hat{H}'\hat{P} + \hat{P}\hat{H}'\hat{P}\right)\hat{R}(s)\right]. \tag{7.22}$$

We now operate on the first of these equations with $\hat{P}$ and on the second of these equations with $\hat{Q}$. This gives us the two equations

$$\hat{P}\hat{R}(s) = \frac{1}{i\hbar s - \hat{P}\hat{H}'\hat{P}}\left[1 - \left(\hat{P}\hat{H}'\hat{Q}\right)\hat{Q}\hat{R}(s)\right]$$
$$\hat{Q}\hat{R}(s) = \frac{1}{i\hbar s - \hat{Q}\hat{H}'\hat{Q}}\left[1 - \left(\hat{Q}\hat{H}'\hat{P}\right)\hat{P}\hat{R}(s)\right]. \tag{7.23}$$

Solving these two equations for the terms on the left, we can then recombine them to give us a new formulation for the resolvent, which is

$$\hat{R}(s) = \left(\hat{P} + \hat{Q}\frac{1}{i\hbar s - \hat{Q}\hat{H}\hat{Q}}\hat{Q}\hat{H}\hat{P}\right)\frac{1}{i\hbar s - \hat{P}\hat{H}\hat{P} - \hat{C}}\left(\hat{P} + \hat{P}\hat{H}\hat{Q}\frac{1}{i\hbar s - \hat{Q}\hat{H}\hat{Q}}\hat{Q}\right)$$

$$+ \hat{Q}\frac{1}{i\hbar s - \hat{Q}\hat{H}\hat{Q}}\hat{Q}, \tag{7.24}$$

where the collision term is

$$\hat{C} = \hat{P}\hat{H}\hat{Q}\frac{1}{i\hbar s - \hat{Q}\hat{H}\hat{Q}}\hat{Q}\hat{H}\hat{P}, \tag{7.25}$$

which only involves the terms that connect a "diagonal" element to an off-diagonal element and reconnects them by the conjugate operation. This is just the type of term that leads to the Fermi golden rule, and the leading term in the electron–phonon interaction must be of this type. Thus, our crude approximation (7.18) is found to be of this more correct latter version. Let us probe this a little further. We can rewrite the expression (7.18), using the result (7.25), as

$$\hat{\Sigma}\hat{\rho}_1(s) = Tr_L\left\{\hat{C}\left(\hat{H}_{eL} + \hat{H}_e\right)\tilde{\rho}\right\}$$

$$= Tr_L\left\{\hat{P}\left(\hat{H}_{eL} + \hat{H}_e\right)\hat{Q}\frac{1}{i\hbar s - \hat{Q}\left(\hat{H}_{eL} + \hat{H}_e\right)\hat{Q}}\hat{Q}\tilde{\rho}\right\}. \tag{7.26}$$

This complicated structure tells us many things. First, we should include the electron–electron interactions in the Hamiltonian, although we are projecting out the single-electron density matrix. Second, the electron–phonon interactions can lead to real energy shifts, which are the self-energy corrections to the single-particle energies. This is in addition to the normal dissipative scattering processes.

We can now use these various operators, together with the form of the collision operator, to rewrite Eq. (7.15) as

$$i\hbar s\tilde{\rho}_1(s) = i\hbar\rho_1(0) + \hat{P}\hat{H}_e\hat{P}\tilde{\rho}_1 + \hat{\Sigma}\tilde{\rho}_1$$

$$+ \hat{P}\left(\hat{H}_{eL} + \hat{H}_e\right)\hat{Q}\frac{1}{i\hbar s - \hat{Q}\left(\hat{H}_{eL} + \hat{H}_e\right)\hat{Q}}\hat{Q}\tilde{\rho}_e(0). \tag{7.27}$$

The last term has been thought to produce a number of effects, including the random force used for the Langevin equation at the lowest order of the electron–phonon interaction [12] and a screening of the driving field in higher order [11]. The temporal equation can now be obtained by retransforming this last equation, which results in

$$\frac{\partial \rho_1}{\partial t} = -\frac{i}{\hbar} \hat{P} \hat{H}_e \hat{P} - \frac{1}{\hbar^2} \int_0^t \hat{\Sigma}(t-t') \rho_1(t') dt' , \qquad (7.28)$$

and the last term of Eq. (7.27) has been ignored. If the system is homogeneous, the only contribution from the first term on the right-hand side is from the accelerative electric field, which produces

$$\frac{i}{\hbar}[H_F, \rho_1] = e\mathbf{E} \cdot \frac{\partial \rho_1}{\partial \mathbf{p}}, \qquad (7.29)$$

so that the final quantum kinetic equation for the homogeneous system is just

$$\frac{\partial \rho_1}{\partial t} + \frac{e\mathbf{E}}{\hbar} \cdot \frac{\partial \rho_1}{\partial \mathbf{k}} = -\frac{1}{\hbar^2} \int_0^t \hat{\Sigma}(t-t') \rho_1(t') dt' . \qquad (7.30)$$

This form of the quantum kinetic equation was first derived by Barker [13], but its form is quite analogous to the Prigogine–Resibois equation [14]. It is also in the same form as derived by Levinson [15], and this will be explored in the next chapter. Except for the convolution form of the collision term, this equation is essentially the same as the Boltzmann transport equation, which means that some quantum effects, such as quantization of the states, have been left out or are buried in the form of the one-electron density matrix.

The resolvent in the scattering operator (7.26) produces an exponential form in the scattering term in the convolution integral. This can be expanded as

$$\hat{Q}_e - \frac{it}{\hbar} \hat{Q} \left(\hat{H}_e + \hat{H}_{eL} \right) \hat{Q}$$

$$= \hat{Q} \left[1 - \frac{it}{\hbar} \hat{Q} \left(\hat{H}_e + \hat{H}_{eL} \right) \hat{Q} - \frac{t^2}{\hbar^2} \hat{Q} \left(\hat{H}_e + \hat{H}_{eL} \right) \hat{Q} \left(\hat{H}_e + \hat{H}_{eL} \right) \hat{Q} + \ldots \right] \hat{Q}$$

$$= \hat{Q} e^{-\frac{it}{\hbar} \left(\hat{H}_e + \hat{H}_{eL} \right) \hat{Q}} \qquad (7.31)$$

where the effect of the trailing projection operator (and that leading in the exponential) is superfluous. The importance of this form is that the entire Hamiltonian arises in the exponential operator or propagator, which in turn generates the full perturbation series if needed. Further, the leading and trailing terms are of the general form

$$\left(\hat{P}\hat{H}\hat{X}\right)_{nnmm} = \sum_{r,s} P_{nnrs}\left(\sum_{p,q} H_{rspq}X_{pqmm}\right) = \sum_{p,q} H_{nnpq}X_{pqmm} \quad .(7.32)$$

Since the non-scattering terms of the Hamiltonian involve diagonal terms, these vanish due to the properties of the teradic formed from diagonal matrices. A similar result arises from examining the elements of $\hat{Q}\hat{H}$. Hence, the scattering function involves only the off-diagonal parts of the electron Hamiltonian, which are the carrier–carrier scattering parts and the electron–phonon scattering parts. Hence, this term, Eq. (7.31), is truly a scattering operator. Hence, the scattering kernel in Eq. (7.30) may be written as

$$\hat{\Sigma}(t) = \hat{P}\hat{H}_{eL}\hat{Q}_e^{-\frac{it}{\hbar}\left(\hat{H}_0 + \hat{H}_F + \hat{H}_{eL}\right)}\hat{Q}\hat{H}_{eL}\hat{P} . \tag{7.33}$$

Clearly, the lowest-order term arises from neglecting the interaction term in the exponential, and this will lead to the Fermi golden rule. However, the presence of the field in the exponential leads to what is known as the intra-collisional field effect [16, 17]. In addition, the presence of the interaction term in the exponential will lead to higher-order scattering events consistent with the interaction representation series.

7.2 Quantum Kinetic Equation 2

The previous section introduced us to the quantum kinetic equation. Some view the approach as being dense and not very comprehensible, but this can be said about any quantum transport approach. Nevertheless, we present a different approach to the quantum kinetic equation, which may be a little more comprehensible [18]. It is based on a time-convolutionless method, which can be arrived at without the use of the partial traces inherent in the previous section [19]. In this approach, a method is given, which avoids the operator inversion typical for time-convolutionless approaches, yet produces an effective memory interaction, which enables us to highlight the so-called memory dressing. This memory dressing separates the effective from the real physical interaction and has its own nonlinear equation of motion.

Consider an open system *S*, interacting with its environment *E*, so that the system environment (*S+E*) is closed. This closed system still could be influenced by external driving fields that we take to be known and are unaffected by any feedback from *S+E*. The Hilbert spaces of the environment and the system, Ω_E and Ω_S, respectively, each has a finite dimension, which we denote as d_E and d_S, respectively. Thus, the total Hilbert space is a tensor product of the two spaces for the system and environment as $\Omega_{S+E} = \Omega_S \otimes \Omega_E$. Correspondingly, the total Hamiltonian $H(t)$ consists of the system part and the environment part as well as an interaction part. These must be put into the tensor product forms, so this is done through the use of unit matrices that have the dimensions of either the system or the environment, as [20]

$$H_S' = \mathbf{1}_E \otimes H_S$$
$$H_E' = H_E \otimes \mathbf{1}_S, \tag{7.34}$$

whereas the interaction H_{int} spans both spaces. With these preliminaries, we can write the Liouville equation as in Eq. (7.6), in which the Liouville superoperator is denoted by the caret over the Hamiltonian term. We will also find useful the time-ordering operator T and the anti-time-ordering operator $T^\dagger$. As in earlier chapters, we will also have use of the Heaviside step function θ. For convenience, the initial time is taken to be $t = 0$, although it could be any arbitrary time. The initial density matrix is assumed to be known at this initial time, so that we can make use of the evolutionary operator U introduced earlier via

$$\rho(t) = U(t, 0)\rho(0). \tag{7.35}$$

Since the Hamiltonian is assumed to be time varying, the evolution operator (1.11) must be rewritten as

$$U(t,t') = \theta(t-t')e^{-\frac{i}{\hbar}\int_{t'}^{t} L(t'')dt''} + \theta(t'-t)e^{-\frac{i}{\hbar}\int_{t}^{t'} L(t'')dt''}. \tag{7.36}$$

As previously, the evolution of the open system S is described by the reduced density matrix $\rho_s = Tr_E\{\rho\}$. To gain an understanding of how this evolution progressed, we introduce the projection operators (7.17) with a little more explicit form as

$$\hat{P} = \rho_E \otimes Tr_E, \quad \hat{Q} = 1 - \hat{P}, \tag{7.37}$$

and these operate on any quantity that exists in the Liouville space. It is important to note that this latter space corresponds to the tetradic space discussed above, so that it can be described as Ω_{S+E}^2. Note that this projection operator is now defined so as to separate out the environment partial density matrix, with the remaining part being an entangled portion of the density matrix. The advantage here is that this partial trace is really unnecessary [19]. The projection operator (7.36) is different from that introduced in the previous section. There, we used the projection operator to reduce the many-body problem to an effective one-electron problem. Here, we want a projection operator to work on the various parts of the Liouville space. To achieve this, we will use another particular property of the operator, and that is the idempotent property that $\hat{P}^2 = \hat{P}$. To see the importance of this, we note that any operator can be diagonalized by the choice of a particular basis set. When this is done, the matrix representation of the operator is diagonal, which makes it easy to see the importance of the idempotent, as the eigenvalues all lay on the diagonal. For the idempotent property to be true, these diagonal eigenvalues must be either 0 or 1. These are the only values for which $\hat{P}^2 = \hat{P}$. Thus, the projector (7.37) has eigenvalues of either 1 or 0, with corresponding eigenspaces of the Liouville space, which have dimensions of d_S^2 and $d_S^2(d_S^2 - 1)$, respectively. The total Hilbert space can then be represented as a direct sum, or tensor product, of these two eigenspaces

$$\Omega_{S+E}^2 = \left(\Omega_{S+E}^2\right)_{\hat{P}=1} \otimes \left(\Omega_{S+E}^2\right)_{\hat{P}=0} . \tag{7.38}$$

While these statements hold true for any projection operator, the choice (7.37) has special features, which will be useful for this discussion. With this definition, if we take any basis

$$|\alpha\beta\rangle \text{ for } \alpha, \beta = 1,..., d_S, \tag{7.39}$$

in the Liouville space of the system Ω_S^2, then there is a simply constructed preferred basis set

$$|\overline{\alpha}\overline{\beta}\rangle \text{ for } \overline{\alpha}, \overline{\beta} = 1,..., d_S, \tag{7.40}$$

for which the projection of any arbitrary operator in Ω_S^2 may be expressed

$$(Tr_E\{X\})_{\alpha\beta} = \sqrt{d_E} \, (\hat{P}X)_{\overline{\alpha}\overline{\beta}} . \tag{7.41}$$

This means that the basis for the set of states with an eigenvalue of 1 is going to be orthonormal to an equivalent basis for the set of states with an eigenvalue of 0. This follows from the decomposition (7.38) and means that any vector X that exists in the total space Ω^2_{S+E} can be represented on these basis sets in a column vector form as

$$X = \begin{bmatrix} x_1 \\ x_0 \end{bmatrix}.$$ (7.42)

The projectors themselves can also be represented in this composite space as

$$\hat{P} = \begin{bmatrix} 1 & 0 \\ 0 & 0 \end{bmatrix}, \quad \hat{Q} = \begin{bmatrix} 0 & 0 \\ 0 & 1 \end{bmatrix}.$$ (7.43)

Now, if one were to take the partial trace over the environment variables, as done in the previous section, of any operator or variable that exists in the entire $S+E$ space, we would obtain just what part of this operator is seen by the system. We achieve the same result here with the projection operators, so we need make no distinction between $x_s = Tr_E\{x\}$, and its representation column in the basis $|\alpha\beta\rangle$ of Ω^2_{S+E}. When we do this in the present approach, Eq. (7.41) leads to

$$x_S = x_1 \sqrt{d_E}.$$ (7.44)

In general, then, any superoperator that works in the Liouville space Ω^2_{S+E} can be represented by a block-matrix form

$$\hat{A} = \begin{bmatrix} A_{11} & A_{10} \\ A_{01} & A_{00} \end{bmatrix}.$$ (7.45)

If the operator is a system operator, it can be given as $\hat{A}_S = I_E \otimes A_S$, where A_S acts only in the subspace Ω^2_S. Then, it commutes with the projection operator and is, therefore, represented by a block diagonal form of Eq. (7.45) and can be written as

$$\hat{A}_S = I_E \otimes A_S = \begin{bmatrix} A_S & 0 \\ 0 & A_0 \end{bmatrix}.$$ (7.46)

We can now return to the evolution of the system and the environment, described by Eq. (7.6) and its solution Eq. (7.36). We can write these in terms of the eigenbasis of the projection operator described above as

$$\rho = \begin{bmatrix} \rho_1 \\ \rho_0 \end{bmatrix}, \quad \rho_S = \rho_1 \sqrt{d_E}.$$ (7.47)

The Liouville operator and the evolution operator are also given in these block forms as

$$\hat{L}(t) = \begin{bmatrix} L_{11}(t) & L_{10}(t) \\ L_{01}(t) & L_{00}(t) \end{bmatrix} \tag{7.48}$$

and

$$\hat{U}(t,t') = \begin{bmatrix} U_{11}(t,t') & U_{10}(t,t') \\ U_{01}(t,t') & U_{00}(t,t') \end{bmatrix}. \tag{7.49}$$

As $\hat{L}$ is Hermitian, we have $L_{12} = (L_{21})^\dagger$ and $\hat{U}$ is unitary. When Eqs. (7.6) and (7.36) are written out in their matrix representations, we obtain

$$\frac{d\rho_1}{dt} = -\frac{i}{\hbar}L_{11}(t)\rho_1(t) - \frac{i}{\hbar}L_{10}(t)\rho_0(t)$$

$$\frac{d\rho_0}{dt} = -\frac{i}{\hbar}L_{01}(t)\rho_1(t) - \frac{i}{\hbar}L_{00}(t)\rho_0(t) \tag{7.50}$$

and

$$\rho_1(t) = U_{11}(t,t')\rho_1(t') + U_{10}(t,t')\rho_0(t')$$

$$\rho_0(t) = U_{01}(t,t')\rho_1(t') + U_{00}(t,t')\rho_0(t'). \tag{7.51}$$

From the above discussion, it is clear that we have two classes of states in the complete Liouville space. Due to the isomorphism given by Eq. (7.36), which tells us that the states from the space with eigenvalues of 1 faithfully reproduce what goes on with the system, we will call this set "purely system states." With the definitions that are present in Eqs. (7.45) and (7.46), it is clear that any superoperator of this form will have the upper left block representing only these purely system states and is thus the "purely system part" of the operator. Consequently, the upper left block of the Liouvillian (7.48) is of this form and thus corresponds to the effective system Hamiltonian

$$H_{S,\text{eff}} = H_S + Tr_E\{H_{\text{int}}\}/\sqrt{d_E}. \tag{7.52}$$

This accounts for the well-known first-order correction to the system energy spectrum that arises from coupling to the environment [21, 22]. States that arise in the ortho-complement space, the space with the basis corresponding to the eigenvalues of 0, are referred to as the "entangled states." In quantum information theory, entanglement has a precise definition, and a composite system is said to be in a non-entangled (separable) state, if its density matrix can be written as a linear combination of tensor products of sub-system density matrices (here those of the system and environment). Otherwise, the composite system is said to be in an entangled state. Here, this means that the purely system states are states that are depleted of any information about the environment. On the other hand, the entangled states are those states that also include some system states, which are rich in information about the environment. Zurek [23] would call the purely system states the pointer states, whereas the entangled states may well become decoherent over time due to the interaction with the environment. The terms in the Liouvillian that effectively account for the system–environment coupling are the off-diagonal terms L_{10} and L_{01}. It is worth noting that if properties of the system are to be measured, then this coupling must exist because physical measurements are usually made in the environment and not in the system. If we want the system to evolve in a manner that is decoupled from the environment, then we must have

$$|| L_{10}\rho_0 ||<<|| L_{11}\rho_1 ||, \tag{7.53}$$

where the double vertical lines refer to a norm of the elements, essentially a form of average over time and space. This is obviously the case if the interaction term is set equal to zero. However, it can also be the case when a balance between driving and dissipation forces is achieved, so that the system is in a stationary steady state [24].

Solving the above equations becomes quite tedious and results in very long sets of expressions. To ease the problem, we introduce some auxiliary operators, which are defined by

$$H_{00}(t,t')=T^c e^{-\frac{i}{\hbar}\int_{t'}^{t} L_{00}(t'')dt''} \tag{7.54}$$

and by the differential equations

$$\frac{dK_{01}(t,t')}{dt}=-\frac{i}{\hbar}L_{00}(t)K_{01}(t,t')+\frac{i}{\hbar}K_{01}(t,t')L_{11}(t)+\frac{i}{\hbar}K_{00}(t,t')L_{01}(t)$$

$$\frac{dK_{00}(t,t')}{dt}=-\frac{i}{\hbar}L_{00}(t)K_{00}(t,t')+\frac{i}{\hbar}K_{00}(t,t')L_{00}(t)+\frac{i}{\hbar}K_{01}(t,t')L_{10}(t)$$

$$\tag{7.55}$$

subject to the initial conditions

$$K_{01}(t', t') = 0, K_{00}(t', t') = 1. \tag{7.56}$$

Strictly speaking, the second time variable in these new functions is a parameter and not a real variable as it merely labels the initial time, which is fixed. With these auxiliary functions, we can now solve for the differential equation for the system density matrix as

$$\frac{d\rho_1(t)}{dt} = -\frac{i}{\hbar}\left[L_{11}(t) - L_{10}(t)K_{00}^{-1}(t,t')K_{01}(t,t')\right]\rho_1(t)$$

$$-\frac{i}{\hbar}L_{10}(t)K_{00}^{-1}(t,t')H_{00}(t,t')\rho_0(t) \qquad (7.57)$$

$$\rho_0(t) = -K_{00}^{-1}(t,t')\left[K_{01}(t,t')\rho_1(t) - H_{00}(t,t')\rho_0(t)\right].$$

Before combining these two equations, we want to introduce two short-hand definitions to simplify the equations. These are

$$R(t,t') = K_{00}^{-1}(t,t')K_{01}(t,t'), \quad R(t',t') = 0$$

$$S(t,t') = K_{00}^{-1}(t,t')H_{00}(t,t'), \quad S(t',t') = 1. \qquad (7.58)$$

Equations (7.57) can be combined now to yield a single equation for the system density matrix, as

$$\frac{d\rho_1(t)}{dt} = -\frac{i}{\hbar}\left[L_{11}(t) - L_{10}(t)R(t,t')\right]\rho_1(t)$$

$$+\frac{i}{\hbar}L_{10}(t)S(t,t')\left[1_0 - S(t,t')\right]^{-1}R(t,t')\rho_1(t). \qquad (7.58)$$

This should be compared with Eq. (7.27) or Eq. (7.28), and the similarities in the last term on the right-hand side are obvious.

There are now two approaches to finding the effective system density matrix in an entangled system plus environment. The end results are quite similar. Nevertheless, neither Eq. (7.58) nor Eq. (7.28) are usable until a very specific system and environment Hamiltonian is defined. In this sense, Eq. (7.28) is a little cleaner in that the driving field and the scattering operator are clearly seen and can be connected to the classical Boltzmann equation. We will try to illustrate Eq. (7.58) a little later in this chapter.

7.3 Barker–Ferry Equation [25]

The most successful solution technique for the Boltzmann transport equation over the past few decades has been the ensemble Monte Carlo technique. This approach is based on converting the equation into a path integral, which is now susceptible to the use of random number techniques with weightings based on the nature of the physical processes involved [1, 26, 27]. The difference between the Boltzmann transport equation and the quantum equivalent (7.30) lies in the convolution integral for the scattering function on the right-hand side of the latter equation. However, this should not be a hindrance to the development of a path integral formulation that would allow for an ensemble Monte Carlo treatment of quantum transport. Here, we proceed to do this for Eq. (7.30). To begin, we can expand the retarded collision term in Eq. (7.30), so that it is more amenable to understanding. This leads to

$$\frac{\partial \rho_1(\mathbf{p},t)}{\partial t} + e\mathbf{E}(t) \cdot \frac{\partial \rho_1(\mathbf{p},t)}{\partial \mathbf{p}} = \int_0^t dt' \sum_{\mathbf{p}'} \{ S(\mathbf{p},\mathbf{p}';t,t')\rho_1(p',t')$$
$$-S(\mathbf{p}',\mathbf{p};t,t')\rho_1(\mathbf{p},t')\} \tag{7.59}$$

where $\mathbf{p}$ is the single-electron momentum. The two momenta in Eq. (7.59) are explicit functions of the retarded time t' on the right-hand side through the accelerative expressions

$$\mathbf{p}(t') = \mathbf{p} - \int_{t'}^t e\mathbf{E}(t'')dt''$$

$$\mathbf{p}'(t') = \mathbf{p}' - \int_{t'}^t e\mathbf{E}(t'')dt'' \tag{7.60}$$

and the transition scattering rates S take the form, for inelastic phonon scattering,

$$S(\mathbf{p},\mathbf{p}';t,t')$$

$$= \mathrm{Re} \left\{ \frac{2\pi}{\hbar} \sum_{\mathbf{q}} \frac{1}{\pi\hbar} e^{-\frac{t-t'}{t_\Gamma}} \left(N_\mathbf{q} + \frac{1}{2} \pm \frac{1}{2} \right) \delta_{\mathbf{p},\mathbf{p}'\pm\mathbf{q}} |V(\mathbf{q})|^2 e^{-\frac{1}{\hbar} \int_t^{t'} dt'' \zeta(\mathbf{p},\mathbf{p}';t',t'')} \right\},$$

$$\tag{7.61}$$

where

$$\zeta(\mathbf{p},\mathbf{p}';t',t'') = E(\mathbf{p}(t'')) - E(\mathbf{p}'(t'')) \pm \hbar\omega_\mathbf{q}. \tag{7.62}$$

In the classical limit, the last exponential in Eq. (7.61), which represents the joint spectral densities of the two states, becomes the energy-conserving delta function. In the above equations, the upper sign refers to the emission of a phonon, while the lower sign refers to the absorption of a phonon. Note that in Eq. (7.59), the momentum arguments in S are reversed. The first term in the curly brackets refers to scattering *in* to the state with momentum $\mathbf{p}$ from all the possible states with momentum $\mathbf{p}'$. This process is weighted by the occupancy of the initial momentum through the density matrix at this momentum. On the other hand, the second term corresponds to scattering *out* of the state **p** to all the possible states with momentum **p'**, and is weighted by the density matrix of this initial state.

Equation (7.59) is a very nonlocal equation and cannot directly be put into the path integral form because of the inherent retardation of the out-scattering term. To address this, we can generalize the concept of self-scattering [27, 28]. To achieve this, we add and subtract identical terms to the right-hand side of Eq. (7.59), so that we can redefine the scattering function as

$$S^*(\mathbf{p},\mathbf{p}';t,t')=S(\mathbf{p},\mathbf{p}';t,t')+\left[\Gamma(t,t')-\Gamma_{out}(t,t')\delta(\mathbf{p}\cdot\mathbf{p}')\right] \quad (7.63)$$

where

$$\Gamma_{out}(t,t')=\sum_{\mathbf{p}''}S(\mathbf{p},\mathbf{p}'';t,t'),\quad \Gamma(t,t')=\Gamma_0\delta(t-t'). \quad (7.64)$$

The terms in the square bracket of Eq. (7.63) make no contribution to the scattering integral, but the term in Γ_{out} plays a very important role. This term projects out of Γ the contribution due to uncompleted out-scattering processes, and the retardation in the out-scattering term is accounted for here rather than in the equivalent term in Eq. (7.59). With this introduction, Eq. (7.59) can be rewriten as

$$\left(\frac{\partial}{\partial t}+e\mathbf{E}(t)\cdot\frac{\partial}{\partial \mathbf{p}}+\Gamma_0\right)\rho_1(\mathbf{p},t)=\int_0^t dt'\sum_{\mathbf{p}''}S*(\mathbf{p},\mathbf{p}';t,t')\rho_1(\mathbf{p}',t').$$
$$(7.64)$$

This may now be solved by the method of characteristics [29] to obtain the path–variable structure

$$\rho_1(\mathbf{p},t)=\int_0^t dt' e^{-\Gamma_0(t-t')}G(\mathbf{p};t,t'), \quad (7.65)$$

where the kernel is

$$G(\mathbf{p};t,t')=\int_0^{t'} dt''\sum_{\mathbf{p}'}S^*(\mathbf{p}(t''),\mathbf{p}'(t'');t',t'')\rho_1(\mathbf{p}'(t''),t''). \quad (7.66)$$

It is reassuring that, in the limit of long times and instantaneous collisions and no quasi-particle effects from the field, this result reduces to that found for the Boltzmann transport equation.

The evolutionary properties included in Eq. (7.60) lead to an intra-collisional field effect [16, 17]. This effect is very important when the field is large or when the collision duration is a significant fraction of the mean-time between collisions and has previously been analyzed for steady-state transport in uniform, time-independent fields [30, 31]. In the steady-state case, the intra-collisional field effect induces a broadening and skewing of the usual energy-conserving delta functions. This leads, for example, to a lowering of the threshold for phonon emission due to acceleration during the collision and to a lowering of the scattering strength at high fields.

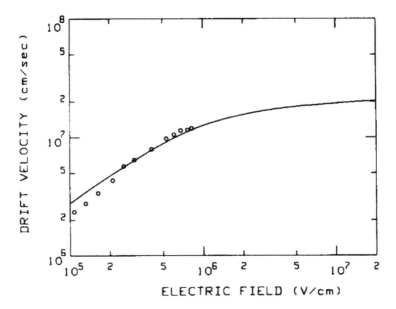

Figure 7.1 Velocity–field curve computed by an ensemble Monte Carlo technique based on Eq. (7.64). Experimental data from Hughes [36, 37] are shown as the open circles. Reprinted with permission from Ref. [33], Copyright 1985, American Physical Society.

As remarked above, this path integral form is the basis for simulations of the transport with ensemble Monte Carlo techniques. Such codes were first developed primarily in the past for studying high-field transport in SiO_2, in studies of the mechanisms by which dielectric breakdown would occur in this material [32, 33]. These two approaches both used the polar optical phonons of SiO_2 as the main scattering process but also included the acoustic phonons of low energy. However, it was found that the polar optical phonons could not stabilize the distribution function above fields of about 2–3 MV/cm [34]. If no other process occurred, this would lead to a much lower breakdown field for the oxide than is observed experimentally. In both of the simulations mentioned at first, such an additional process was introduced. Fischetti [32] introduced a high-energy acoustic process, which involved Umklapp scattering across the zone boundary. On the other hand, Porod and Ferry [33] introduced non-equivalent intervalley scattering with the second set of valleys lying some 3 eV above the principal minimum of the conduction band. Both of these processes are successful in controlling the distribution function and pushing the breakdown field somewhat above 7–10 MV/cm. The experiments and knowledge of the properties of the oxide at high energies are so limited that it is not possible to select which of the two processes actually is the most important [35]. In Fig. 7.1, the simulated velocity–field curve is shown for an oxide conduction effective mass of $0.5m_0$. Experimental data from Hughes are also shown for comparison [36, 37].

7.4 An Alternative Approach [38, 39]

While the previous sections have been devoted to developing a kinetic equation for the one-electron density matrix, it is possible to work directly with the Liouville-von Neumann equation (7.6) (sometimes, this equation is also referred to as the Pauli master equation [40] when limited to the diagonal terms). An alternative approach is to try to solve this equation directly. However, it is not obvious that one can solve this easily in an application for very small semiconductor devices. One problem that faces such an attempt is that we cannot just do the diagonal terms in the density matrix, but must also properly account for the off-diagonal terms. Nevertheless, there is a long history of this approach. Van Hove has shown that it may be used to find the irreversible approach to the steady state for a system that is subjected to a perturbation [41]. Along this line, Frensley has shown that the inclusion of the off-diagonal terms is a necessary ingredient of any solution for a transport process [42]. While some have suggested that the off-diagonal terms are unnecessary [41], the presence the reservoirs, or contacts, generally leads to an interference that appears within the off-diagonal terms [43–46]. However, if what localizes the electrons in the reservoirs is a dephasing process, which is characterized by a coherence length l_φ, then devices whose length is smaller than this will see the incoming wave packets as totally delocalized, and so essentially as plane waves. In this case, off-diagonal terms of the density matrix are not injected from the reservoirs. Then one can develop the device basis functions as the natural basis states that are arrived at by the self-consistent solutions to the Schrödinger and Poisson's equations, which include the interaction terms at least through the Hartree approximation. In a sense, the solutions to Eq. (7.6) are treated as perturbations away from the ballistic transport model, where the perturbations are a result of the interactions between the electrons and the phonons. Instead of initial conditions, one now must look at the conditions that the device/reservoir interactions must meet in order to preserve a large ratio between the on-diagonal and the off-diagonal terms of the density matrix. If we can keep the reservoir–device interactions (approximately) within the diagonal terms, then we can solve the transport problem with these alone. This is the approach that is followed in this alternative methodology.

We define a small region of semiconductor material as the device, which is connected to large reservoirs in which the particles are in thermal equilibrium with the lattice, although each reservoir may have a different chemical and electrostatic potential due to an applied bias voltage. Within the device itself, we write a Hamiltonian of the form

$$H = H_e - eV + H_L + H_{eL} + H_{res},\qquad(7.67)$$

where e is the electron charge, V is the applied bias voltage, $H_e, H_L,$ and H_{eL} are the electronic, lattice, and electron–phonon coupling terms described previously. The last term, H_{res}, describes the interaction between the device and the reservoirs to which it is connected. *An explicit form for the commutator between this term and the density matrix will not be explicitly given, as it involves the full complexity of the previous sections.* Instead, a phenomenological approach will be followed [42, 44], attempting to assess the conditions for which this interaction makes a negligible contribution to the off-diagonal terms of the density matrix. In this sense, we will retain only terms of lowest order in this interaction energy. In addition, only stationary (steady-state) solutions will be considered here.

To begin, let us take a complete set of basis functions of the form (7.3) for the single-particle Hilbert space associated with the device. If we have N particles, each is assumed to exist in its own state characterized by a superscript n, so that the particular state for the nth electron is given by

$$\psi^{(n)}(\mathbf{r}) = \sum_j a_j^{(n)} \varphi_j(\mathbf{r}) \tag{7.68}$$

at time $t = 0$. The time evolution of the density matrix is given by a modification of Eq. (7.6) as

$$\frac{\partial \rho}{\partial t} = \frac{i}{\hbar}\left[\rho, H - H_{\text{res}}\right] + \left(\frac{\partial \rho}{\partial t}\right)_{\text{res}}, \tag{7.69}$$

where the last term separates out the time variation of the density matrix as a result of its explicit interaction with the reservoirs. In terms of the basis set, the density matrix may be described by the expansion on this basis set as

$$\rho_{kl} = \sum_{h=1}^{N} a_k^{(n)}(t) a_l^{\dagger(n)}(t). \tag{7.70}$$

Here the density matrix is that for the one-electron picture. The lattice (phonons) is supposed to remain in equilibrium, and this density matrix follows the procedure by which the lattice variables have been traced out, such as is done in Section 7.1. Further, we proceed assuming that the device carrier density is in the non-degenerate limit. The full form of the last term of Eq. (7.69) is equivalent to Eq. (7.15), but with the electron–lattice interaction replaced with H_{res}.

We will simplify the problem a little further and assume that the device is homogeneous and of infinite extent in the x- and y-directions. We also assume that the device is in contact with reservoirs at $z = 0$ and $z = W$. Then, within the effective mass approximation, we can write the Hamiltonian (7.67) as

$$H = -\frac{\hbar^2}{2m^*}\frac{\partial^2}{\partial z^2} - eV(z) + H_{eL} + H_{\text{res}} = H_0 + H_{eL} + H_{\text{res}}. \tag{7.71}$$

In semiconductors, it is generally sufficient to acknowledge that the electron–phonon interaction term is small and can be treated by perturbation theory yielding, e.g., the Fermi golden rule, and this has been done by, e.g., Kohn and Luttinger [47]. Here, however, instead of using free electron plane waves, we take the eigenstates of H_0. These states are the eigenstates

$$H_0 |\mu\mathbf{k}\rangle = \left[E_\mu + \frac{\hbar^2 k^2}{2m^*}\right]|\mu\mathbf{k}\rangle = E_{\mu k} |\mu\mathbf{k}\rangle, \tag{7.72}$$

for which the eigenstates can be written as

$$|\mu\mathbf{k}\rangle = \frac{1}{2\pi}\zeta_\mu(z)e^{i\mathbf{k}\cdot\mathbf{r}}, \tag{7.73}$$

where $\mathbf{k}$ and $\mathbf{r}$ are two-dimensional vectors corresponding to the (x,y) plane. The z-directed function is a solution to the one-dimensional Schrödinger equation

$$-\frac{\hbar^2}{2m^*}\frac{d^2\zeta_\mu(z)}{dz^2} - eV(z)\zeta_\mu(z) = E_{\mu\mathbf{k}}\zeta_\mu(z). \tag{7.74}$$

The Hartree potential is found from the Poisson's equation subject to the boundary conditions $V(0) = V_L$ and $V(W) = V_R$. In this scenario, we typically consider that electrons enter the device on the left from the left reservoir and exit the device on the right via the right reservoir, so that $V_R > V_L$. It is convenient to write waves in the reservoirs that are injected from the left and then partially transmitted and reflected at the boundaries. This gives

$$\zeta_\mu^+(z) = \begin{cases} A_\mu^+ e^{ik_\mu^- z} + A_\mu^+ r_\mu^+ e^{-ik_\mu^- z} & z < 0, \\ A_\mu^+ t_\mu^+ e^{ik_\mu^- z} & z > W, \end{cases} \tag{7.75}$$

where

$$\hbar k_\mu^- = \sqrt{2m^*(E_{\mu\mathbf{k}} + eV_L)}, \quad \hbar k_\mu^+ = \sqrt{2m^*(E_{\mu\mathbf{k}} + eV_R)}. \tag{7.76}$$

Correspondingly, there is a wave traveling in the opposite direction that appears in the reservoirs as

$$\zeta_\mu^-(z) = \begin{cases} A_\mu^- e^{ik_\mu^+ z} + A_\mu^- r_\mu^- e^{-ik_\mu^+ z} & z > W, \\ A_\mu^- t_\mu^- e^{ik_\mu^+ z} & z < 0. \end{cases} \tag{7.77}$$

The various constants A are normalization constants that arise by normalizing the eigenstates (7.73) over the length W of the device. The reflection and transmission coefficients satisfy

$$1 - \left|r_\mu^+\right|^2 = \left(\frac{k_\mu^+}{k_\mu^-}\right)^2 \left|t_\mu^+\right|^2 = \left(\frac{k_\mu^-}{k_\mu^+}\right)^2 \left|t_\mu^-\right|^2 = 1 - \left|r_\mu^-\right|^2, \tag{7.78}$$

which follows from the general properties of the Wronskian of the Schrödinger equation [48]. As with the Landauer formula, only states whose energies lie above the lower of the reservoir chemical potentials are consider in the transport. To denote the left- and right-directed waves, we will use the pseuodo-spin index $\sigma = \pm$, such that $\zeta_\mu^\sigma(z) = \langle z | \mu\sigma \rangle$.

With these preliminaries, it is now possible to write the equation for the diagonal terms of the density matrix in this wave formulation as

$$\frac{\partial \rho_{\mu k \sigma}}{\partial t} = \sum_{\mu' k' \sigma'} \left[S(\mu k \sigma, \mu' k' \sigma') \rho(\mu' k' \sigma') \right.$$

$$\left. - S(\mu' k' \sigma', \mu k \sigma) \rho(\mu k \sigma) \right] + \left(\frac{\partial \rho_{\mu k \sigma}}{\partial t} \right)_{res}, \tag{7.79}$$

where the scattering kernels S are given by Eq. (7.61) in the limit of $t = t'$, e.g., the Fermi golden rule approximation with energy-conserving delta functions.

Now we need to turn to the effect of the reservoirs. Recall that the requirement is that the reservoirs do not inject appreciably into the off-diagonal elements of the density matrix. This will be the case for very delocalized wave packets entering the device from the reservoirs, and as discussed previously, this means mainly waves that are nearly plane waves. Hence, we may consider that the left reservoir attempts to inject a flux of particles of any $\mathbf{k}$ into the right-bound component of the states associated to waves incident from the left, $\zeta_\mu^+(z)$, and this flux must be in equilibrium with the Fermi level in this left reservoir, as

$$\left(\frac{\partial \rho_{\mu k +}}{\partial t} \right)_{res, in} = v_k^- \left[\left| A_\mu^+ \right|^2 f_{eq, L}(E_{\mu k}) - \left| A_\mu^- t_\mu^- \right|^2 \rho_{\mu k -} \right], \tag{7.80}$$

where $v_k^+ = \hbar k_\mu^+ / m^*$ and each of the two terms corresponds to the out of equilibrium occupation of the two reservoirs in each of the sub-bands μ. Since the dependence on wave momentum $\mathbf{k}$ is trivial, we can drop this subscript by using the total occupation

$$\rho_{\mu +} = \sum_k \rho_{\mu k +}. \tag{7.81}$$

The occupation of the states in equilibrium is determined by the effective density of states

$$f_{eq, L}(E_\mu) = \frac{m^* k_B T}{\pi \hbar^2} \ln \left[1 + e^{-\eta_{\mu, L}} \right], \quad \eta_{\mu, L} = \frac{E_{\mu k} - E_{F, L}}{k_B T}. \tag{7.82}$$

However, this is not a current-carrying distribution. Instead, the reservoirs must be allowed to inject carriers from a distribution that differs from the thermal equilibrium one, and a drifted distribution must be used for this in order to impose current continuity throughout the device [49]. This can be accomplished by a modification of the sub-band energy to account for this shift. Accordingly, we write the energy as

$$E_{\mu\mathbf{k}} = \frac{\hbar^2 \left(\mathbf{k}_\mu^\pm - \mathbf{k}_d^\pm\right)^2}{2m*} - eV_{L,R} \, , \tag{7.83}$$

where the second term in parentheses is the drift wave number, corresponding to the average velocity of the carriers through the structure. This drift term must be determined as part of the self-consistent loop through the Poisson–Schrödinger iterations, along with the need for current continuity through the entire device–reservoir structure. Corresponding to the entering flux (7.80), the flux out of the device through the two reservoirs is

$$\left(\frac{\partial \rho_{\mu+}}{\partial t}\right)_{\text{res, out}} = \left[v_\mathbf{k}^- \rho_{\mu+} \left|r_\mu^+\right|^2 - v_\mathbf{k}^+ \rho_{\mu+} \left|t_\mu^+\right|^2\right] \left|A_\mu^+\right|^2. \tag{7.84}$$

Using the relationships expressed in Eq. (7.78), we can rewrite the various expressions in the form

$$\left(\frac{\partial \rho_{\mu+}}{\partial t}\right)_{\text{res}} = v_\mathbf{k}^- \left[\left|A_\mu^+\right|^2 [f_L(E_{\mu\mathbf{k}}) - \rho_{\mu+}] - \left|A_\mu^- A_\mu^-\right|^2 \rho_{\mu-}\right]. \tag{7.85}$$

It seems that the left reservoir tries to maintain a semblance of equilibrium by injecting charge at a rate that is determined by the deficiency in the device near the cathode, which is an attempt to restore charge neutrality. We have a similar expression for states incident from the right as

$$\left(\frac{\partial \rho_{\mu-}}{\partial t}\right)_{\text{res}} = v_\mathbf{k}^- \left[\left|A_\mu^-\right|^2 [f_R(E_{\mu\mathbf{k}}) - \rho_{\mu-}] - \left|A_\mu^+ t_\mu^+\right|^2 \rho_{\mu+}\right]. \tag{7.86}$$

Finally, the net current through the device may be written as [38]

$$j = -e \sum_\mu \left\{\rho_{\mu+} v_\mathbf{k}^+ \left|t_\mu^+ A_\mu^+\right|^2 - \rho_{\mu-} v_\mathbf{k}^- \left|t_\mu^- A_\mu^-\right|^2\right\}, \tag{7.87}$$

which has an obvious connection to the Landauer formula.

Fischetti has applied this approach to first simulate an *n-i-n* GaAs semiconductor device [38]. Here the entire structure was $L = 300$ nm long, and reservoirs were $D = 100$ nm at each end. This left the device region at $W = 100$ nm. The doping throughout the structure was distributed according to

$$N_D(z) = N_{D0}\left\{1 + \frac{1}{2}\left[\tanh\left(\frac{D-z}{l}\right) + \tanh\left(\frac{z-L+D}{l}\right)\right]\right\} + n_i, \quad (7.88)$$

where $l = 2.5$ nm and the range of z is over the entire 300 nm device. Of course, the electrons will diffuse into the device region, so that there is a large charge density at each end with leads to a smoother variation in the potential and actual electron density. In Fig. 7.2, we show the various parameters that result from the application of a bias of 0.25 V across this structure. Most of this potential drop is across the central device region, as can be seen. Thus, this leads to electric fields of the order of 25 kV/cm in the device region, and peak velocities of 2×10^7 cm/s are reached, while the average energy of the electrons peaks at about 0.2 eV.

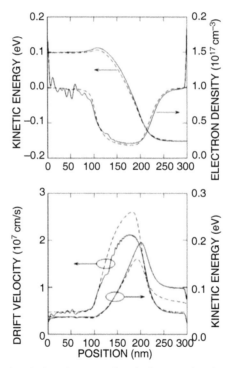

Figure 7.2 (Top) Calculated potential and electron distribution in the *n-i-n* structure. (Bottom) The average electron velocity and average electron energy as a function of position. In each case, the solid curve is the solution to the quantum density matrix equations, while the dashed curve is a conventional (classical) ensemble Monte Carlo solution to the Boltzmann transport equation. Reprinted with permission from Ref. [38], Copyright 1998, AIP Publishing LLC.

7.5   Hydrodynamic Equations

In the presence of transport in the semiconductor, whether this is electrical transport or thermal transport, one of the major concerns is to find the proper values for the distribution function, or, in the present case, the appropriate density matrix. When the transport is electrical transport, the density matrix needs to be found from Eq. (7.59). In the semi-classical world, it is nice to develop a set of so-called balance equations for the density, the momentum, and the energy, for example [1]. The quantum case can be more exacting, however. Nevertheless, a similar approach can be followed. For general transport, a full extended density matrix in terms of the constants of the motion (density, momentum, and energy, for example) has been developed by Zubarev [50]. The key factor is usually the introduction of the retardation of the collisions that appear in Eq. (7.59), and we will address a simpler approach here to the development of such a set of balance equations [51, 52]. The retardation becomes particularly important at high electric fields when the mean-time between collisions becomes small, and the finite duration of the scattering becomes important. This is also the case in fast laser excitations of the semiconductor [53]. The problem is complicated by the fact that, on the time scale involved, the transport coefficients should actually involve time-convolution functions [54]. This is because, on the short-time scale, a truly causal theory introduces memory effects, which lead to convolution integrals in the transport coefficients.

In the approach to be followed here, the moment equations will be developed in a straight-forward manner from Eq. (7.59). The important modifications arise from the intra-collisional field effect [16, 17], which both broadens and shifts the collision resonances, effectively lengthening the collision duration and weakening the effect of the collision itself. That a memory effect should be included in general is evident from simple arguments based on Langevin-type equations, of which the moment equations are typical examples (this has clearly been shown for the semi-classical case [55]). If the damping terms in the moment equations evolve on the same time scale as, e.g., the velocity or the energy, then the product of these quantities with the damping terms must appear as a time-convolution function.

The beginning point is Eq. (7.59) along with the description of the scattering function (7.61). The two momenta evolve in the presence of the electric field according to Eq. (7.60). As mentioned in this previous section, the two exponentials that appear in Eq. (7.61) arise from the joint spectral density functions and reduce to an energy-conserving delta function in the case of an instantaneous collision, as assumed in classical transport. In small semiconductor devices, the carrier density generally will be rather high. This is significant, as Fröhlich [56] first pointed out that the isotropic part of the distribution will be nearly Maxwellian, as observed in Chapter 5 for the Matsubara Green's functions. This is the case when the inter-carrier scattering drives the system toward spherical symmetry. In these cases, we can introduce a parameterized density matrix and then evaluate the parameters from the appropriate moments of the transport equation [49, 54]. Thus, we could approximate the density matrix as (for the non-degenerate case)

$$\rho(E) = \exp[-\beta_e(E - \mathbf{v}_d \cdot \mathbf{p})], \tag{7.89}$$

where $\mathbf{v}_d$ is the drift velocity and related to the average value of the momentum via $m^*\mathbf{v}_d = \langle \mathbf{p} \rangle$, and β_e is the inverse of the effective electron temperature T_e. However, this form, suggested by the classical approach, is instantaneous in time and we need a proper retarded function. Consequently, we modify Eq. (7.89) with the ansatz

$$\rho(E) = \int_0^t dt' \left[\delta(t - t') - \mathbf{v}_d(t - t') \cdot \frac{d}{dt'} \left(\mathbf{p} \frac{\partial}{\partial E} \right) \right] \rho_0(E, t'), \tag{7.90}$$

where the two derivatives are required, respectively, to preserve the dimensionality and to bring β_e into the second expression in the square bracket. Here, $\rho_0(E, t)$ is an evolving function of the electron temperature and time, but is spherically symmetric.

To begin the approach, we once again Laplace transform Eq. (7.59), assuming that we are dealing with a single isotropic conduction band, to obtain

$$s\tilde{\rho}(\mathbf{p}, s) - \rho(\mathbf{p}, 0) - e\mathbf{E} \cdot \frac{\partial \tilde{\rho}(\mathbf{p}, s)}{\partial \mathbf{p}}$$
$$= \frac{2}{\hbar^2} \sum_{\mathbf{p}'} \left\{ S(\mathbf{p}, \mathbf{p}') \tilde{\rho}(\mathbf{p}', s) - S(\mathbf{p}', \mathbf{p}) \tilde{\rho}(\mathbf{p}, s) \right\} G(s), \tag{7.91}$$

where

$$G(s) = \text{Re}\left\{ \int_0^\infty e^{-st} e^{i(E-E'\pm\hbar\omega_0)t/\hbar + i\frac{e\mathbf{E}\cdot(\mathbf{p}-\mathbf{p}')t^2}{2m^*\hbar} - t/t_\Gamma} \, dt \right\}, \tag{7.92}$$

and the time dependences in Eq. (7.60) have been expanded. The second exponential in the integral is a very rapidly oscillating function, and we can generate a first-order estimate by an asymptotic approximation to $G(s)$. We do this within the spirit of the method of stationary phase [57]. The first exponential can be brought outside the integral and evaluated at τ_c, an effective collision duration for which the phase in the second exponential is zero. Then, $s\tau_c \ll 1$, especially as we need to take s itself small to insure the long-time limit for which the collisions are complete. Thus, we may expand this first exponential as

$$e^{-s\tau_c} \sim 1 - s\tau_c \sim \frac{1}{1 + s\tau_c}. \tag{7.93}$$

The remaining integral is just the field shifted and broadened joint spectral density function, which would approach the delta-function limit at small fields and long times. While the exact form of the integral is not easily analyzed in physical terms (we will do so in a later chapter), a simple perturbative expansion allows us to model all but the irrelevant fast oscillatory behavior by

$$\delta_E = \frac{1}{\pi} \frac{\gamma}{\gamma^2 + \eta^2}, \tag{7.94}$$

where

$$\gamma = \frac{\hbar}{\tau_\Gamma} + \frac{\hbar}{\tau_c}$$

$$\eta = E(\mathbf{p} + e\mathbf{E}\tau_c') - E(\mathbf{p}' + e\mathbf{E}\tau_c') \pm \hbar\omega_0 \tag{7.95}$$

$$\tau_c' = \left| \frac{2m^*\pi\hbar}{e\mathbf{E}\cdot(\mathbf{p} - \mathbf{p}')} \right|^{1/2}.$$

The effective collision duration τ_c is found from the point of constant phase as $\tau_c = \eta\tau_c^2/2\pi\hbar$. Thus, the intra-collisional field effect produces a broadening of the resonance and introduces a shift that can actually reduce the threshold for phonon emission or absorption, and this affects the velocity–field relationship in the semiconductor.

Using the definitions and approximations above, we can rewrite Eq. (7.91) in the more compact form as

$$s\tilde{\rho}(\mathbf{p}, s) - \rho(\mathbf{p}, 0) - e\mathbf{E} \cdot \frac{\partial\tilde{\rho}(\mathbf{p}, s)}{\partial\mathbf{p}}$$

$$= \frac{2}{\hbar} \sum_{\mathbf{p}'} \left\{ S(\mathbf{p}, \mathbf{p}')\tilde{\rho}(\mathbf{p}', s) - S(\mathbf{p}', \mathbf{p})\tilde{\rho}(\mathbf{p}, s) \right\} \frac{\delta_E(\mathbf{p}, \mathbf{p}')}{1 + s\tau_c'}. \quad (7.97)$$

We can now multiply by an arbitrary function $\phi(\mathbf{p})$ and we will eventually then sum over all states $\mathbf{p}$. This will allow us to make a change in the variables of the first scattering term in the curly brackets, as both arguments are being summed over. With this change, the right-hand side becomes

$$\frac{2\pi}{\hbar} \sum_{\mathbf{p}'} S(\mathbf{p}', \mathbf{p})[\phi(\mathbf{p}') - \phi(\mathbf{p})] \frac{\delta_E(\mathbf{p}', \mathbf{p})}{1 + s\tau_c'}. \quad (7.98)$$

In the semi-classical limit, this expression reduces to the same as found for the Boltzmann transport equation [1].

For the energy balance equation, the major relaxation process is via the optical phonons for which $\Delta E = \pm\hbar\omega_0$, where the upper sign is for the absorption of an optical phonon and the lower sign is for the emission of an optical phonon. The energy average involves only the isotropic part of the density matrix (7.90), and we may write the summation over the density matrix and momentum $\mathbf{p}$ as

$$\frac{\langle \Gamma_E(s) \rangle}{1 + s\bar{\tau}_c} = \frac{1}{n} \sum_{\mathbf{p}} \Gamma_E(\mathbf{p}) \frac{\rho_0(\mathbf{p}, s)}{1 + s\tau_c'}, \quad (7.99)$$

where $\bar{\tau}_c$ is an appropriate average of the collision duration, and

$$\Gamma_E(\mathbf{p}) = \frac{2\pi}{\hbar} \sum_{\mathbf{p}', \pm} S(\mathbf{p}', \mathbf{p})(\pm\hbar\omega_0)\delta_E(\gamma, \eta_\pm). \quad (7.100)$$

Inverting the Laplace transform gives us the final form, recognizing that the average energy is related to the electron temperature in three dimensions,

$$\frac{3k_B}{2} \frac{\partial T_e}{\partial t} = e\mathbf{E} \cdot \mathbf{v}_d - \frac{1}{\bar{\tau}_c} \int_0^t e^{-t'/\bar{\tau}_c} \Gamma_E(t - t')dt'. \quad (7.101)$$

If we now take $\phi = \mathbf{p}$, then the average value of this quantity is just the drift momentum. However, the relevant term in the density matrix is the anisotropic term in Eq. (7.90), and the contribution of the drift term is of the form

$$\rho_1(\mathbf{p}, s) = -s\mathbf{v}_d \cdot \mathbf{p} \frac{\partial \rho_0(\mathbf{p}, s)}{\partial E} \tag{7.102}$$

to the lowest order. Then, the evaluation of Eq. (7.98) for the momentum exchange is simply the analogy to Eq. (7.99), as

$$\frac{\langle \Gamma_m(s) \rangle}{1 + s\bar{\tau}_c} = \frac{1}{n} \sum_{\mathbf{p}} \frac{m^* v_d^2}{1 + s\tau_c'} \cdot \Gamma_m(\mathbf{p}) \frac{\partial \rho_0(\mathbf{p}, s)}{\partial E}, \tag{7.103}$$

with

$$\Gamma_m(\mathbf{p}) = \frac{2\pi}{\hbar} \sum_{\mathbf{p}', \pm} S(\mathbf{p}', \mathbf{p})(\pm m^* \mathbf{v}_d) \delta_E(\gamma, \eta_\pm). \tag{7.104}$$

Inverting the Laplace transform then leads to the equivalent momentum balance equation

$$m^* \frac{\partial \mathbf{v}_d}{\partial t} = e\mathbf{E} - \frac{m^*}{\bar{\tau}} \int_0^t \mathbf{v}_d(t-t') \int_0^t \langle \Gamma_m(t'-t'') \rangle e^{-t''/\bar{\tau}_c} dt'' dt', \tag{7.105}$$

reduces to the Boltzmann transport equation in the limit of vanishingly small collision duration and very slow variation of the drift velocity so that the memory effect can be ignored.

The collision duration approximated by Eq. (7.95) is a very crude approximation and likely overestimates the value for this quantity. We will see later in our nonequilibrium Green's function discussion how this can be determined relatively exactly and is a few femtoseconds for the polar optical phonons in GaAs [58]. In comparison, the value determined from Eq. (7.95) for the intervalley optical phonons in Si is some tens of femtoseconds [52], which seems especially large. In addition, we have been rather *laissez faire* in the above treatment. The density matrix we defined in the earlier sections of this chapter is expressed in terms of two spatial quantities and is not a phase–space function. In this representation, the momentum is a differential operator and care must be taken to ensure the proper ordering of various terms. In the next chapter, we will discuss the methods of transforming the density matrix into a true phase–space function, but at the cost of this function not being at least positive semi-definite. We will see there that the negative excursions of the phase–space function entail the limitations of the exclusion principle.

Other forms for the various balance equations have also been developed, particularly by Grubin [59, 60]. Often, these approaches bring the explicit quantization in a confined structure into the problem via an effective potential. Such approaches can be quite useful [61], and we will develop the structure for various effective potentials in the next section. The role of the collision duration on the transport has been studied in ensemble Monte Carlo approaches as well. The influence of the collision duration has been studied for both the transient response of bulk semiconductors [62] and the response of the semiconductor to femtosecond laser excitation [53].

7.6 Effective Potentials

In Section 2.7, we separated the wavefunction into its amplitude and its phase, which led to two separate equations, and the appearance of a new form of potential. That is, if we wrote the wavefunction as

$$\psi(\mathbf{r}, t) = A(\mathbf{r}, t) e^{iS(\mathbf{r}, t)/\hbar}, \tag{7.106}$$

we could then write the Bohm, or quantum [63], potential as in Eq. (2.95)

$$U_B = -\frac{\hbar^2}{2m^* A(\mathbf{r}, t)} \nabla^2 A(\mathbf{r}, t). \tag{7.107}$$

This is one form of what we call an effective potential. Since we usually relate the squared magnitude of the wavefunction, A^2, to the local electron density, this potential is often called the density gradient potential [64], expressed as

$$U_B = -\frac{\hbar^2}{2m^* \sqrt{n(\mathbf{r}, t)}} \nabla^2 \sqrt{n(\mathbf{r}, t)}. \tag{7.108}$$

There have been many versions of such an effective potential that have been developed over the years. In this section, we review a number of these approaches.

7.6.1 Wigner Form

The interesting aspect of the two arguments for the density matrix is the form of the potential(s) to which the density matrix must respond. For example, the potentials in the two coordinates may be taken from Eq. (7.7) and written as

$$V(\mathbf{r}, \mathbf{r}', t) = V(\mathbf{r}, t) - V(\mathbf{r}', t). \tag{7.109}$$

This can be expanded by first going over to the so-called center-of-mass, or Wigner coordinates [65], as

$$\mathbf{r} = \mathbf{R} + \frac{\mathbf{x}}{2}, \quad \mathbf{r}' = \mathbf{R} - \frac{\mathbf{x}}{2}, \tag{7.110}$$

and then expanding the potential in a Taylor series, as

$$V(\mathbf{r}, t) - V(\mathbf{r}', t) = V\left(\mathbf{R} + \frac{\mathbf{x}}{2}\right) - V\left(\mathbf{R} + \frac{\mathbf{x}}{2}\right)$$

$$= \mathbf{x} \cdot \nabla V(\mathbf{R}) + \frac{\mathbf{x}^3}{6} \cdot \nabla^3 V(\mathbf{R}) + \dots \tag{7.111}$$

The first term is just the classical electric field times the position coordinate, while the next term provides some correction terms. When the density matrix equation (7.7) is subjected to the same coordinate transformation, and moment equations taken, one can find an effective potential correction which appears as [65, 66]

$$U_W = \frac{\hbar^2}{8m^* k_\mathrm{B} T} \nabla^2 V(\mathbf{r}) \rightarrow -\frac{\hbar^2}{8m^*} \nabla^2 \ln(n). \tag{7.112}$$

The leading term arises from integrating out the primed coordinate and the last form arises from assuming a thermal density matrix. This form has been used in semiconductor simulations to indicate the leading order quantum corrections that can arise from quantum effects [67]. We will return to the Wigner form in the next chapter where we examine the detailed properties of the Wigner distribution function and clearly develop the form in Eq. (7.112).

7.6.2 Spatial Effective Potential

In quantum semiconductor devices, which are typically nanoscale devices, one must begin to question the relative size of the electron (or hole) compared to the device itself. Classically, the electron is miniscule and, for all practical purposes, is zero in comparison to the real world. However, in quantum mechanics, the size is set by the wavefunction and the de Broglie wavelength of this wave. Hence, it is unlikely that the size of this particle can be ignored in nanoscale devices. Generally, when dealing with quantum transport in a hypothetical device, it is assumed that an initial wave packet arises in the source and exits in the drain. Yet this over-constrains the wavefunction, which must satisfy the Schrödinger equation—a single initial condition in time is the only allowed temporal constraint. In trying to arrive at a proper wavefunction description (the argument is easily extended to density matrices, Green's functions, and/or Wigner functions), a different approach is required in which the wavefunction is required to be variationally minimized subject to approaching *two positions*, one in the source and one in the drain, in a time-independent manner [68]. This introduces the *quantum localization* problem to device physics—just how small can the two regions be in this formulation (and, more importantly, how do these affect the transport?). One approach is to identify a minimum area in which a single electron can be localized. Previously [69], I have discussed the arguments for various sizes for electrons in semiconductor devices. In particular, in the case of a thermal distribution of electrons, it can be argued that this minimum packet is approximately 60% of the thermal de Broglie wavelength. In this section, we will review this approach and show how it leads to smoothing of the self-consistent potential within the device in a manner that gives a different view of the effective potential.

In computing an estimate of the size of the electron's wave packet, we follow an approach similar to that of Wannier functions in determining impurity ionization energies. A Wannier function is a localized wavefunction that is obtained by summing over all Bloch functions; the Bloch functions are taken from a single band in the effective mass approximation [70]. This approach finds its most notable use in determining the wavefunction for an electron trapped on an impurity. In this latter case, the wavefunction radius is found to be essentially the Bohr radius in the semiconductor (about 3.3 nm in Si). This wavefunction, in real space, is the "amplitude" function describing the expansion in Wannier functions and is termed *the envelope function*. We follow this procedure by describing the occupancy of the Bloch functions by the appropriate distribution function and then computing the proper envelope function.

For a degenerate two-dimensional electron gas in semiconductors, the probability for occupancy of a particular state is described by the Fermi–Dirac distribution function. As the temperature is increased, higher momentum states become occupied as the distribution spreads under the influence of the temperature. A wider momentum space distribution means a tighter distribution in real space for the electron wave packet. At low temperature in our two-dimensional gas, all states up to the Fermi energy are occupied, and the Fermi wavevector is defined by the sheet carrier density n_s as

$$k_F = \sqrt{2\pi n_s} \; . \tag{7.113}$$

This means that, in the momentum representation, all states up to this wavevector are occupied, and we may write a wavefunction in this representation as

$$\varphi_m(k) = \frac{\sqrt{2\pi}}{k_F} u_0(k_F - k), \tag{7.114}$$

where u_0 is the Heaviside function. From this momentum space representation, we can now define a real-space wave packet (centered at the origin as a reference point) by taking the Fourier transform of this latter equation, which leads to

$$\psi(r) = \frac{1}{r}\sqrt{\frac{2}{\pi}} J_1(k_F r), \tag{7.115}$$

where J_1 is the first-order Bessel function of the first kind. If we remember that **k** and **r** are really two-dimensional vectors, then we can use these wavefunctions to determine the uncertainties

$$\Delta p = \frac{\hbar k_F}{\sqrt{2}}, \quad \Delta r = \frac{1}{k_F} \; . \tag{7.116}$$

Hence, the spatial extent of the real-space wave packet for the electron can be estimated to be the full-width at half-maximum value, or roughly twice the uncertainty in position, which leads to

$$\delta r \sim 2\Delta r = \frac{2}{k_F} = \frac{\lambda_F}{\pi} \; . \tag{7.117}$$

If we have an electron gas in the AlGaAs/GaAs heterostructure system, with a density of 10^{12} cm^{-2}, then the Fermi wavelength is 2.5×10^6 cm^{-1}, and the spatial extent of the wave packet is approximately 8 nm. This is certainly an important percentage of a nanoscale device. Other approximations, such as non-degenerate statistics, give different values by a numerical factor of order 2–3 [71], and certainly not an order of magnitude.

While we found a Bessel function shape for the wave packet above, the more normal approach is to use a Gaussian wave packet for the minimum uncertainty. Let us consider such a wave packet as the magnitude squared of the wavefunction to represent the size of the electron. We can write this wave packet as

$$\psi(r) = \frac{1}{(2\pi)^{1/4}\sqrt{\lambda}} \exp\left(-\frac{r^2}{4\lambda^2}\right) = C \exp\left(-\frac{r^2}{4\lambda^2}\right), \tag{7.118}$$

where λ is the effective size of the electron, somewhere in the range of 4–8 nm. This can be used to generate an effective potential, since the total energy includes the integral of the potential over the local charge density [72]. Hence, the potential term can be written as

$$V = \int d\mathbf{r} V(\mathbf{r}) \sum_i n_i(\mathbf{r}) \tag{7.119}$$

and the sum runs over the set of particles. We can introduce the Gaussian into the density as

$$n_i(\mathbf{r}) = C^2 \exp\left(-\frac{|\mathbf{r}-\mathbf{r}_i|^2}{2\lambda^2}\right) = C^2 \int d\mathbf{r}' \exp\left(-\frac{|\mathbf{r}-\mathbf{r}'|^2}{2\lambda^2}\right) \delta(\mathbf{r}'-\mathbf{r}_i), \tag{7.120}$$

and this can be introduced into Eq. (7.119) to give

$$V = \int d\mathbf{r} V(\mathbf{r}) \sum_i \int d\mathbf{r}' C^2 \exp\left(-\frac{|\mathbf{r}-\mathbf{r}'|^2}{2\lambda^2}\right) \delta(\mathbf{r}'-\mathbf{r}_i)$$

$$= \int d\mathbf{r}' V(\mathbf{r}') \sum_i \int d\mathbf{r} C^2 \exp\left(-\frac{|\mathbf{r}-\mathbf{r}'|^2}{2\lambda^2}\right) \delta(\mathbf{r}-\mathbf{r}_i) \tag{7.121}$$

$$= \sum_i \int d\mathbf{r} \delta(\mathbf{r}-\mathbf{r}_i) \int d\mathbf{r}' V(\mathbf{r}') C^2 \exp\left(-\frac{|\mathbf{r}-\mathbf{r}'|^2}{2\lambda^2}\right).$$

In the second line, we have changed the two variables. Since they are both being integrated out of the problem, this creates no problems. The end result is the fact that the third line says we treat each electron as a *classical* electron of zero size, but it sees a smoothed potential. This Gaussian smoothing of the actual (self-consistent) potential leads to the effective potential in the problem. For device simulations, this is an alternative to including the Bohm or Wigner corrections to the potential, but has been shown to give very equivalent results.

The Gaussian smoothing of the self-consistent potential is especially important in MOSFETs or HEMTs. There are two important modifications of the potential that arise from this smoothing. First, the sharp barrier between the semiconductor channel and the oxide (or undoped barrier layer in the HEMT) is smoothed, which allows the wavefunction to penetrate into the barrier slightly, but this is not the most important effect. Rather, the smoothing moves the minimum of the potential away from the interface itself. This causes the centroid of the inversion density to be moved away from the interface, and produces a small additional capacitance, which modifies the gate capacitance, as shown in Fig. 7.3. This movement would naturally occur if one used the wavefunctions themselves and reflects the quantum effect in the channel. In addition, the minimum of the potential lies above the actual minimum of the triangular potential and thus approximates the confinement energy of the lowest sub-band in the channel [73]. These effects all change the performance of the device [74].

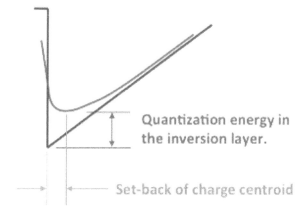

Quantization energy in the inversion layer.

Set-back of charge centroid

Figure 7.3 A typical triangular potential well as found in a MOSFET or a HEMT is shown in red. The smoothed effective potential is shown in blue. The two most important quantum effects are shown: the quantization energy and the charge set-back from the interface.

Before proceeding too far, let us try to connect this effective potential with some of the ones discussed previously. We can change the variables in Eq. (7.121) and then expand the potential as

$$V_{\text{eff}}(\mathbf{r}) = \frac{1}{\sqrt{2\pi}\lambda} \int d\mathbf{r}' V(\mathbf{r}+\mathbf{r}') \exp\left(-\frac{r'^2}{2\lambda^2}\right)$$

$$= \frac{1}{\sqrt{2\pi}\lambda} \int d\mathbf{r}' \left[V(\mathbf{r})+\mathbf{r}'\cdot\nabla V(\mathbf{r})+\frac{r'^2}{2}\nabla^2 V(\mathbf{r})...\right] \exp\left(-\frac{r'^2}{2\lambda^2}\right)$$

$$= V(\mathbf{r})+\alpha^2\nabla^2 V(\mathbf{r})+..., \tag{7.122}$$

which can be compared with the Wigner form (7.111). If we assert that the density varies approximately as a thermal density matrix, then we can say that

$$V_{eff}(\mathbf{r}) \sim V(\mathbf{r}) + \frac{\alpha^2}{\beta} \ln[n(\mathbf{r})] \,, \tag{7.123}$$

which, within the leading constant, gives the Wigner correction to the potential (7.112). The result is similar to the Bohm potential correction (7.108), but the details have slightly different sets of derivatives involved. For example, if we change the argument of the natural logarithm to the square root of the density, then we can convert Eq. (7.123) into

$$V_{eff}(\mathbf{r}) \sim V(\mathbf{r}) + \frac{\alpha^2}{2\beta} \left[\frac{1}{\sqrt{n(\mathbf{r})}} \nabla^2 \sqrt{n(\mathbf{r})} - \frac{1}{2n^{3/2}(\mathbf{r})} \left(\nabla \sqrt{n(\mathbf{r})} \right)^2 + \cdots \right], \tag{7.124}$$

from which we can recognize the Bohm potential as the leading correction. So these effective potentials seem to all be related closely to one another. Let us continue this with another form.

7.6.3 Thermodynamic Effective Potential

The thermodynamic approach has been developed by Feyman and Hibbs [75]. Their earliest approach gives an effective potential like Eq. (7.112), but with the factor of 8 replaced by 24. Their later approach follows the introduction of an approximate partition function Z, which is connected to a Helmholtz free energy F through $Z = \exp(-\beta F)$. They then develop a path integral representation of the partition function, which is evaluated to give

$$e^{-\beta F} = \sqrt{\frac{m^* k_B T}{2\pi \hbar^2}} \int e^{-\beta V_{eff}(r)} dr \,, \tag{7.125}$$

where the effective potential is expressed as

$$V_{eff}(r) = \sqrt{\frac{6m^* k_B T}{\pi \hbar^2}} \int_{-\infty}^{\infty} V(r + r') e^{-6r'2m^* k_B T/\hbar^2} \,. \tag{7.126}$$

The form of this potential is clearly quite close to that of Eq. (7.122). As these authors point out, the meaning is that we can calculate an approximate free energy in a classical manner and then arrive at a good quantum mechanical estimate by smoothing this classical potential with a Gaussian function.

Many people have extended the Feynman approach to the case of bound particles, which arrive with quantum confinement [76–80] and to the case of particles at interfaces [81]. These approaches use the fact that the most-likely trajectory in the path integral no longer follows the classical path when the electron is bound inside a potential well. The introduction of the effective potential and its effective Hamiltonian is closely connected to the return to a phase–space description. This can be done at present only for Hamiltonians containing a *kinetic* energy quadratic in the momenta and a coordinate-only dependence in the potential energy. That is, it is clear that some modifications will have to be made when non-parabolic energy bands, or a magnetic field, are present. However, the Gaussian approximation is well established as the method for incorporating the purely quantum fluctuations around the resulting path.

Let us first consider the bound states. The key new ingredient for bound states (such as in the potential well at the interface of a MOSFET) is the need to determine variationally the dominant path and hence the "correct" value for the parameter λ. For the case in which the bound states are well defined in the potential, both Feynman and Kleinert [77] and Cuccoli et al. [80] find

$$\lambda^2 = \frac{\hbar^2}{4m^*k_{\rm B}T}\left[\frac{\coth(f)}{f} - \frac{1}{f^2}\right], \quad f = \frac{\hbar\omega}{2k_{\rm B}T}, \qquad (7.127)$$

where $\hbar\omega$ is the spacing of the sub-bands. If we take the high-temperature limit, as appropriate for semiconductor devices at room temperature, we can expand for small f and arrive at

$$\lambda^2 = \frac{\hbar^2}{12m^*k_{\rm B}T} \qquad (7.128)$$

to leading order. In Si, this gives a value for the smoothing distance of 0.52 nm normal to the oxide interface, which is much smaller than the values discussed above. For transport along the channel, a different mass appears, and this gives a value of 1.14 nm.

7.6.4   A More Formal Approach

The natural place for starting a discussion of a statistical ensemble is just the density matrix as shown in the previous sections. We would like to now develop from this basis a more formal method of obtaining the effective potential that modifies the classical density matrix with the onset of quantum phenomena. Our beginning point is the adjoint of the Bloch equation, which was given in Eq. (2.71). To get to our point of interest [82], we introduce the Wigner coordinate transformation of Eq. (7.110), so that the adjoint equation can be expressed as

$$\frac{\partial \rho}{\partial \beta} = \left[\frac{\hbar^2}{8m^*} \nabla^2 + \frac{\hbar^2}{2m^*} \frac{\partial^2}{\partial \mathbf{x}^2} \right] \rho - \left[\cosh\left(\frac{1}{2}\mathbf{x}\cdot\nabla \right) V(\mathbf{R}) \right] \rho \quad (7.129)$$

where the gradient operation is with respect to the center-of-mass coordinate $\mathbf{R}$ and the last term is a short-hand notation

$$\cosh\left(\frac{1}{2}\mathbf{x}\cdot\nabla \right) V(\mathbf{R}) = \frac{1}{2}\left[V\left(\mathbf{R}+\frac{\mathbf{x}}{2} \right) + V\left(\mathbf{R}-\frac{\mathbf{x}}{2} \right) \right] \quad (7.130)$$

which expresses the fact that the expansion of the potential terms in a Taylor series can be resummed as hyperbolic cosine of the differential operator. We note that the latter term tells us that we are looking for an appropriate average of the two nonlocal potential terms, and this average is the spirit that led to the forms of the effective potentials discussed above. In solving Eq. (7.129), we use Eq. (7.130) as an integrating factor, with which we can rewrite Eq. (7.129) as

$$\frac{\partial}{\partial \beta}(e^{\beta W}\rho) = \frac{\hbar^2}{2m^*} e^{\beta W} \left[\frac{1}{4}\nabla^2 + \frac{\partial^2}{\partial \mathbf{x}^2} \right] \rho \quad (7.131)$$

with $W(\mathbf{R}, \mathbf{x})$ being Eq. (7.130). The term in brackets on the right-hand side, along with the prefactor constants, can be written as a new differential operator expression

$$F(\mathbf{R}, \mathbf{x})\rho = \frac{\hbar^2}{2m^*} \left[\frac{1}{4}\nabla^2 + \frac{\partial^2}{\partial \mathbf{x}^2} \right] \rho. \quad (7.132)$$

Using this expression, we can then solve Eq. (7.131) for the density matrix as

$$\rho(\mathbf{R}, \mathbf{x}) = A\exp\left[-\beta W(\mathbf{R}, \mathbf{x}) + \int_\beta F(\mathbf{R}, \mathbf{x})d\beta' \right]. \quad (7.133)$$

The problem now has been reduced to finding the function F.

To proceed, it is convenient to divide the function F into its "potential" parts and its "dynamic" parts, through the defined separation

$$F(\mathbf{R}, \mathbf{x}) = -Q(\mathbf{R}, \mathbf{x}) + S(\mathbf{R}, \mathbf{x}). \tag{7.134}$$

Let us deal with the second term first, as we assert that this term is the dynamic portion of the function. In general, the dynamic terms are related to the integral invariants of the motion, and may be written in terms of a sum over these quantities as [83]

$$\int_\beta F(\mathbf{R}, \mathbf{x}) d\beta' = -\sum_{k=0}^n \zeta_k P_k, \tag{7.135}$$

where ζ_k in the integral invariant and P_k is the conjugate quantum mechanical operator. The most common application is to let ζ_0 be β so that P_0 is H, the total Hamiltonian used to solve the Schrödinger equation. Here, however, it is the off-diagonal difference variable $\mathbf{x}$ that transforms, through a Wigner–Weyl transform (to be discussed in the next chapter), into the equivalent momentum, so that the form should be somewhat different in the current application. Nevertheless, it is easy to formulate the form that will be useful here, since it must satisfy certain limitations that follow the derivatives with respect to this difference variable. Hence, we are guided by the knowledge that follows from the moment equations in Section 7.5. Thus, we will let the integration over the function S take the form

$$\int_\beta S(\mathbf{R}, \mathbf{x}) d\beta' = J(\mathbf{R}, \mathbf{x}) = \frac{i}{\hbar} \mathbf{p}_d \cdot \mathbf{x} - \frac{m^*}{2\beta} \left(\frac{\mathbf{x}}{\hbar} \right)^2 - \frac{1}{2} \ln(\beta). \tag{7.136}$$

This follows from considerations of the drifted Maxwellian introduced in Section 7.5, and from the form of the Schrödinger equation when separated into its magnitude and phase as done in Section 2.7, so that the second term on the right is effectively the kinetic energy of the system. The last term is a normalization, which leads to adding the term $-d/2\beta$ to S, where d is the dimensionality of the system. After differentiating with respect to β, we note that the second term is equivalent to adding a harmonic oscillator potential, which is normalized by the temperature, in the transverse coordinate. This is reassuring in that its form is reminiscent of that used by Feynman and Kleinert [77]. While the drift momentum term is relatively normal in its form, it does not involve β and will, therefore, vanish in S. While this satisfies the general requirements of Eq. (7.135), it does not allow for the replication that we expect to find in solving the differential equation for the quantum potential, but we only need to incorporate the approximations that will be required in this task. So the present form that is now achieved for Eq. (7.134) may be written as

$$-Q(\mathbf{R}, \mathbf{x}) + S(\mathbf{R}, \mathbf{x}) = -Q(\mathbf{R}, \mathbf{x}) + \frac{m^* \mathbf{x}^2}{2\hbar^2 \beta^2} - \frac{3}{2\beta} = \frac{\hbar^2}{2m^*} \left[\frac{1}{4} \nabla^2 \rho + \frac{\partial^2 \rho}{\partial \mathbf{x}^2} \right].$$

(7.137)

The last expression is just a restatement of Eqs. (7.132) and (7.134). If we now insert Eq. (7.133), using Eq. (7.134) and the first two terms in Eq. (7.136), we can write the above equation as

$$-Q(\mathbf{R}, \mathbf{x}) + S(\mathbf{R}, \mathbf{x}) = \frac{\hbar^2}{8m^*} \nabla^2 \rho - \frac{\hbar^2 \beta}{2m^*} \frac{\partial^2 (W + Q)}{\partial \mathbf{x}^2} + \frac{\hbar^2}{2m^*} \frac{\partial^2 J}{\partial \mathbf{x}^2}$$

$$+ \frac{\hbar^2}{2m^*} \left[-\beta \frac{\partial (W + Q)}{\partial \mathbf{x}} + \frac{\partial J}{\partial \mathbf{x}} \right]^2.$$

(7.138)

Using the values of S in Eq. (7.137), this simplifies to

$$-Q(\mathbf{R}, \mathbf{x}) = \frac{\hbar^2}{8m^*} \nabla^2 \rho - \frac{\hbar^2 \beta}{2m^*} \frac{\partial^2 (W + Q)}{\partial \mathbf{x}^2} + \frac{\hbar^2}{2m^*} \left[\beta \frac{\partial (W + Q)}{\partial \mathbf{x}} \right]^2$$

$$- \frac{\hbar^2 \beta}{2m^*} \frac{\partial (W + Q)}{\partial \mathbf{x}} \cdot \frac{\partial J}{\partial \mathbf{x}}.$$

(7.139)

In general, the term involving J can be taken to be small and thus ignored. Also consistency means that the derivatives of both W and Q with respect to $\mathbf{x}$ must be small and vanish in the limit of $\mathbf{x}$ going to zero. This is the limit in which the density matrix becomes diagonal and basically classical. This means that we can ignore, for practical purposes, the nonlinear quadratic term. Now we also have to restore a few terms that have been omitted in this last form, and when this is done, we find the resulting differential equation for the sum of the applied and quantum potentials to be

$$-\frac{\hbar^2 \beta}{2m^*} \frac{\partial^2 (W + Q)}{\partial \mathbf{x}^2} + \mathbf{x} \cdot \frac{\partial (W + Q)}{\partial \mathbf{x}} + (W + Q) = -\frac{\hbar^2}{8m^* \rho} \nabla^2 \rho + W.$$

(7.139)

Hence, it is now clear that the total effective potential includes the quantum potential defined earlier as a driving force as well as a driving force from the applied potential. To simplify the form, let us scale this displacement by the thermal wavelength as

$$\mathbf{s} = \sqrt{\frac{2m^*}{\hbar^2 \beta}} \mathbf{x}, \tag{7.140}$$

so that our differential equation is now

$$\frac{\partial^2 (W+Q)}{\partial \mathbf{s}^2} + \mathbf{s} \cdot \frac{\partial (W+Q)}{\partial \mathbf{s}} + (W+Q) = -\frac{\hbar^2}{8m^* \rho} \nabla^2 \rho + W. \tag{7.141}$$

This latter equation can be solved through the use of a Green's function, which itself is found from the adjoint equation [84]

$$\frac{\partial^2 G}{\partial \mathbf{s}^2} + \frac{\partial (sG)}{\partial \mathbf{s}} - G = -\delta(\mathbf{s} - \mathbf{s}') \tag{7.142}$$

under the restrictions that both the Green's function and its derivatives vanish at infinity. Then by the normal methods, the Green's function may be found to be [82]

$$G(\mathbf{s}, \mathbf{s}') = \frac{1}{4\pi |\mathbf{s} - \mathbf{s}'|} e^{-|\mathbf{s} - \mathbf{s}'|^2 / 2}, \tag{7.143}$$

and we can now find the quantum potential from

$$Q(\mathbf{R}, \mathbf{s}) = \int \frac{d^3 \mathbf{s}'}{4\pi |\mathbf{s} - \mathbf{s}'|} \left[-\frac{\hbar^2}{8m^* \rho} \frac{\partial^2 \rho}{\partial \mathbf{s}'^2} + W(\mathbf{R}, \mathbf{s}') \right] e^{-|\mathbf{s} - \mathbf{s}'|^2 / 2} - W(\mathbf{R}, \mathbf{s}). \tag{7.144}$$

Now we remember that Q is the correction to the potential, so that the sum of Q and W gives the new total effective potential. Thus, there are two parts to this effective potential. The first is the smoothed correction that arises from the quantum potential, while the second is the smoothing of the potential itself.

7.7 Applications

Perhaps the most studied quantum device is the resonant-tunneling diode, in which a quantum well is bounded by two tunnel barriers, with the entire structure placed between two normal semiconductor layers. The conduction profile for such a structure is shown in Fig. 7.4b [85]. The barriers are formed in this case by $Al_{0.33}Ga_{0.67}As$ layers with a thickness of 4 nm. The quantum well is undoped GaAs with a thickness of 5 nm. The exterior GaAs layers are heavily doped, except for the 6 nm closest to the barriers, which is also undoped.

These latter regions are termed spacer layers and are used to focus the applied bias potential on the active double barrier structure. The resonant-tunneling diode was first conceived by Esaki and Tsu [86]. Subsequent studies have shown that the device is very good as an oscillator in the THz regime [87]. Since this time, the device has been the subject of a relatively large number of quantum simulations, such as this one using the density matrix [85, 88]. In Fig. 7.4a, the real part of the density matrix is plotted for the device in the absence of any applied bias. One can clearly see the small density that accumulates in the quantum well due to tunneling from the cladding layers. In the figure, the potential plot of (b) has been aligned with the density plot to clearly indicate the locations of the barriers and the well. The simulations assume that the device is at room temperature,

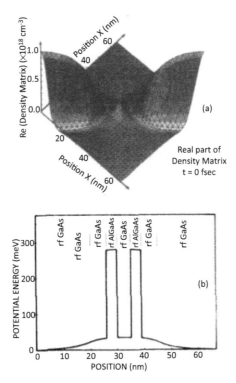

Figure 7.4 (a) The equilibrium density matrix and (b) conduction band profile for a resonant-tunneling diode at 300 K. The structure is discussed in the text. One can see the small density that exists in the quantum well around the bound state. Reprinted with permission from Ref. [85], Copyright 1991, IOP Publishing.

and an energy mesh of 0.5 meV was used to compute and store the wavefunctions that exist in the device up to 200 meV. The electrons in the quantum well are localized around the bound state that exists in the quantum well at about 60 meV above the conduction band edge in the well. The upward shift in the potential around the double barrier is a result of the band movement in the undoped regions that is necessary to allow the Fermi energy to be flat throughout the device, as required by the zero current through the device in equilibrium.

The simulation was carried out by discretizing Eq. (7.7) and introducing a relaxation time approximation for the scattering through a τ_s determined by the mobility of the GaAs used in the structure, which was about 2600 cm^2/Vs [85]. The authors calculated the transient response of the device, using an explicit integration technique, which usually requires extremely small time steps (of the order of femtoseconds for such a device). The discretization of Eq. (7.7) appears as

$$
\frac{\rho_{i,j}(t')-\rho_{i,j}(t')}{t'-t}
$$

$$
= C \left\{ \frac{i\hbar}{2} \left[\frac{1}{\Delta x_i} \left(\frac{1}{m^*_{i+1/2}} \frac{\rho_{i+1,j}(t')-\rho_{i,j}(t')}{\Delta x_i} - \frac{1}{m^*_{i-1/2}} \frac{\rho_{i-1,j}(t')-\rho_{i,j}(t')}{\Delta x_{i-1}} \right) \right. \right.
$$

$$
\left. - \frac{1}{\Delta y_j} \left(\frac{1}{m^*_{j+1/2}} \frac{\rho_{i,j+1}(t')-\rho_{i,j}(t')}{\Delta y_j} - \frac{1}{m^*_{j-1/2}} \frac{\rho_{i,j-1}(t')-\rho_{i,j}(t')}{\Delta y_{j-1}} \right) \right]
$$

$$
\left. + \frac{1}{i\hbar}[V_i(t)-V_j(t)]\rho_{i,j}(t') + \frac{\rho_{i,j}(t')-\rho_{0,i,j}}{\tau_s} \right\}
$$

$$
+ (1-C) \left\{ \frac{i\hbar}{2} \left[\frac{1}{\Delta x_i} \left(\frac{1}{m^*_{i+1/2}} \frac{\rho_{i+1,j}(t)-\rho_{i,j}(t)}{\Delta x_i} - \frac{1}{m^*_{i-1/2}} \frac{\rho_{i-1,j}(t)-\rho_{i,j}(t)}{\Delta x_{i-1}} \right) \right. \right.
$$

$$
\left. - \frac{1}{\Delta y_j} \left(\frac{1}{m^*_{j+1/2}} \frac{\rho_{i,j+1}(t)-\rho_{i,j}(t)}{\Delta y_j} - \frac{1}{m^*_{j-1/2}} \frac{\rho_{i,j-1}(t)-\rho_{i,j}(t)}{\Delta y_{j-1}} \right) \right]
$$

$$
\left. + \frac{1}{i\hbar}[V_i(t)-V_j(t)]\rho_{i,j}(t) + \frac{\rho_{i,j}(t)-\rho_{0,i,j}}{\tau_s} \right\}
$$

$$
(7.145)
$$

Here the equation is written for both an inhomogeneous effective mass, due to the various materials being used, and an inhomogeneous grid. The parameter C is a numerical constant, which depends on the implicit scheme being used. When C is 0.5, this becomes the normal Crank–Nicholson approach, whereas when C is 1.0, we have a backward Euler scheme. The two times differ by the time step

$$\Delta t = t' - t, \tag{7.146}$$

and a value between 1 and 2 fs was used.

In Fig. 7.5, the real part of the density matrix is shown at a time of 1 ps, which is essentially the steady-state condition. There is an evident depletion of carriers in the drain layer (right-hand side of the figure) and an accumulation in the source layer (left-hand side of the figure). Also evident are oscillations in the off-diagonal part of he density matrix, which extend from the source to the drain. These oscillations are thought to represent correlations between the electrons in the source and drain regions of the device. In Fig. 7.6, the imaginary part of the density matrix, for the same conditions as Fig. 7.5, is shown. Again, the correlations between the electrons in the source and drain appear and are labeled as C in the figure. In addition, oscillations are seen in the quantum well region, labeled B, and are thought to arise from interference between the resonant-tunneling

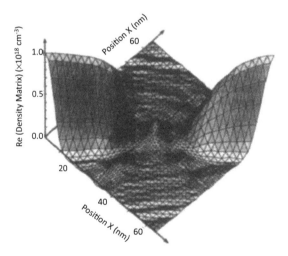

Figure 7.5 The real part of the density matrix under bias and current flow. This is at a time of 1 ps after the voltage is applied and represents a steady state. The details are discussed in the text. Reprinted with permission from Ref. [85], Copyright 1991, IOP Publishing.

electrons and the free electrons in the source and drain. It is thought that the phase coherence of the resonant electrons is maintained within the device even though there is decoherence through the scattering events. The current–voltage curves of the device show the traditional negative differential conductivity when the bias is such that the bound state energy passes below the conduction band edge of the source layer.

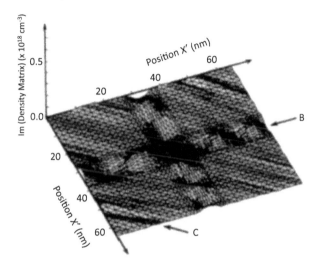

Figure 7.6 The imaginary part of the density matrix under an applied voltage and current flowing. This corresponds to a time of 1 ps and is the same density matrix as Fig. 7.5. Point B indicates phase-coherent resonant electrons that are correlated with the electrons in the source and drain. Point C is thought to show correlation between the electrons in source and drain. Reprinted with permission from Ref. [85], Copyright 1991, IOP Publishing.

7.8 Monte Carlo Procedure

Evaluating the complex integrals that arrive in either the equation of motion for the density matrix or in evaluation of the moments of the density matrix is difficult, especially if one goes beyond the simplest first-order perturbation theory. On the other hand, using a Monte Carlo procedure to evaluate charge transport based on the Boltzmann equation has been around for quite some time [89]. In the semi-classical case, an ensemble of particles is allowed to drift and then scatter. The length of the drift time is taken as a random variable, weighted according to the total scattering rate. In addition, the scattering process is also selected randomly from the possible scattering processes at the given energy of a particle. This approach is allowed to evolve in time with ensemble averages of the velocity, position, and energy of the carriers taken at convenient intervals. Such a process is not possible for the quantum mechanical density matrix, since we do not have a convenient set of paths, or trajectories, which can be adopted. Nevertheless, it is possible to develop a Monte Carlo procedure for the evaluation of the transport equation for the density matrix [90].

Generally, when one works with a Monte Carlo procedure for the Boltzmann equation, this equation is transformed into a path integral equation, which forms the basis for the subsequent Monte Carlo evaluation [91]. In this situation, there is a single time and a single position in the distribution function. The density matrix is more complicated as there are two positions that must be considered. Although there is a single time, the time along either of these positions must be considered, with only the final evolution time being the same for both positional paths. But a path integral equation can be constructed from the equation of motion for the density matrix [90]. This path integral serves as the basis for developing a Monte Carlo procedure. More importantly, this procedure is not limited to just the lowest-order scattering processes. This numerical procedure starts, of course, with an initial equilibrium density matrix. Then the numerical procedure starts with the selections, via random numbers, according to the probabilities for each process, with

i. The order of the perturbative correction that is to be evaluated,
ii. A given sequence of distinct processes that together correspond to one of the possible contributions to the path integral,
iii. Initial and final times that correspond to the process selected, and
iv. The wavevectors of the phonons involved in each quantum event in the process selected.

The times mentioned in point (iii) correspond to times at which the integrand functions are sampled. One unique feature of the density matrix in this form is that it is a backward propagation procedure, whereas normal Monte Carlo transport approaches are forward in time. By this, it is meant that one starts with a value of the state momentum $\mathbf{k}(t)$ at which the density matrix is to be evaluated, and both indices of the density matrix are then translated backward in time, in the presence of any electric fields, to the time of the latest vertex. At this point, the interaction process is evaluated and the value of the momentum changed accordingly. The procedure is repeated until the initial condition is reached.

There are many types of interactions that can occur. The simplest is just evolution via the presence of the electric field. Then there can be emission and absorption of phonons, and these can occur on either of the two positional indices of the density matrix. There can also be emission and reabsorption of a phonon. As the system starts from a diagonal, equilibrium density matrix, it may be expected that the physically real final density matrix should also be diagonal, or at least the meaningful parts should be the diagonal parts. Thus, as we saw in an earlier chapter with the Feynman diagrams, the final parts of the density matrix must have paired vertices, which correspond to completed scattering events. However, there can be virtual processes, in which a phonon tries to be emitted (or absorbed), but reverts with an absorption vertex (or emission vertex) to the original state with the scattering a virtual process. Since the vertices are separated by a time variation, which leads to the intra-collisional field effect, these virtual processes need to be considered. One might consider these virtual processes to not be important in the steady state, but they will certainly have an effect on the transient evolution of the density matrix.

One of the most important effects that the authors found in their simulations was that the violation of energy conservation by off-shell contributions to the spectral density (see Chapter 1) was very important in that observable effects arose from this [90]. In fact, at the very short times in a transient, the off-shell contributions were very important and were more effective than the normal energy-conserving interactions. This might be explained by the fact that the energy-conserving delta function normally found in the Fermi golden rule is a long-term limit, which is certain to be violated on a short-time basis. This means that the selection of a single, well-defined phonon energy is less likely on the short-time basis.

Problems

1. (a) Assume that the carrier distribution may be given by a Maxwellian distribution with a temperature of 300 K. Plot the density matrix in the two-dimensional space determined by the two arguments. Assume that this is being applied to a one-dimensional structure. (b) Now assume that the carrier distribution is a Fermi–Dirac with a Fermi energy given by 50 meV. Plot the density matrix in the two-dimensional space determined by the two arguments. Assume that this is being applied to a one-dimensional structure.

2. Show that, for three square matrices A, B, and C, the trace satisfies the cyclic permutation

   $Tr\{ABC\} = Tr\{CAB\} = Tr\{BCA\}$.

3. Consider GaAs at 300 K with a field of 3 kV/cm. Assume that one has two carriers, both at an energy of 35 meV. Compute the collision duration τ_c' from Eq. (7.95) and show how it depends on the angle between the electric field and the two momenta (there are two angles in this problem).

4. Suppose we have a density that varies as $e^{-x^2/2\lambda^2}$ for x > 0. Take $\hbar = m^* = 1$ and plot the Bohm potential (7.108) and the Wigner potential (7.112). Explain your results.

5. Assume that the system of Problem 4 also has a step in potential of 0.5 eV at x = 0. Compute the effective potential of Eq. (7.121) and compare with the results of this previous problem.

References

1. D. K. Ferry, *Semiconductors* (Macmillan, New York, 1991).

2. J. von Neumann, *Mathematical Foundations of Quantum Mechanics*, English translation (Princeton University Press, Princeton, NJ, 1955).

3. J. B. Krieger and G. J. Iafrate, *Phys. Rev.*, **B33**, 5494 (1986); **B35**, 9644 (1987).

4. N. N. Bogoliubov, *J. Phys. (U.S.S.R.),* **10**, 256 (1946).

5. M. Born and H. S. Green, *Proc. Roy. Soc. London*, **A188**, 10 (1946).

6. J. G. Kirkwood, *J. Chem. Phys.*, **14**, 180 (1946).

7. J. Yvon, *Act. Sci. Ind.*, **542–543** (Herman, Paris, 1937).

8. S. Nakajima, *Prog. Theor. Phys.*, **20**, 948 (1958).

9. R. Zwanzig, *J. Chem. Phys.*, **33**, 1338 (1960).

10. H. Mori, *Prog. Theor. Phys.*, **33**, 423 (1965).

11. J. R. Barker, in *Physics of Nonlinear Transport in Semiconductors*, edited by D. K. Ferry, J. R. Barker, and C. Jacoboni (Plenum, New York, 1980), pp. 127–151.

12. N. Pottier, *Physica*, **A117**, 243 (1983).

13. J. R. Barker, *Solid-State Electron.*, **21**, 197 (1978).

14. R. Balescu, *Equilibrium and Nonequilibrium Statistical Mechanics* (John Wiley, New York, 1975).

15. I. B. Levinson, *Sov. Phys. JETP*, **30**, 362 (1970).

16. K. K. Thornber, *Phys. Rev. B*, **3**, 1929 (1971).

17. J. R. Barker, *J. Phys. C*, **6**, 2663 (1973).

18. I. Knezevic and D. K. Ferry, *Phys. Rev. A*, **69**, 012104 (2004).

19. I. Knezevic and D. K. Ferry, *Phys. Rev. E*, **66**, 016131 (2002).

20. U. Fano, *Rev. Mod. Phys.*, **29**, 74 (1957).

21. J. R. Barker and D. K. Ferry, *Solid-State Electron.*, **23**, 531 (1980).

22. M. Saeki, *Prog. Theor. Phys.*, **67**, 1313 (1982); **79**, 396 (1988); **89**, 607 (1993).

23. W. H. Zurek, *Rev. Mod. Phys.*, **75**, 715 (2003).

24. I. Knezevic and D. K. Ferry, *Phys. Rev. E*, **67**, 066122 (2003).

25. J. R. Barker and D. K. Ferry, *Phys. Rev. Lett.*, **42**, 1779 (1979).

26. T. Kurosawa, *J. Phys. Soc. Jpn.*, Suppl. **21**, 424 (1966).

27. H. D. Rees, *J. Phys. C*, **5**, 641 (1972).

28. P. Price, *Solid-State Electron.*, **21**, 9 (1976).

29. H. F. Budd, *Phys. Rev.*, **158**, 798 (1967).

30. J. R. Barker, *Solid-State Electron.*, **21**, 267 (1978).

31. D. K. Ferry, in *Physics of Nonlinear Transport in Semiconductors*, edited by D. K. Ferry, J. R. Barker, and C. Jacoboni (Plenum, New York, 1980), pp. 577–588.

32. M. Fischetti, *Phys. Rev. Lett.*, **53**, 1755 (1984).

33. W. Porod and D. K. Ferry, *Phys. Rev. Lett.*, 54, 1189 (1985).

34. H.-J. Fitting and J.-U. Friemann, *Phys. Status Solidi A*, **63**, 349 (1982).

35. D. K. Ferry, in *The Physics and Technology of Amorphous SiO₂*, edited by R. A. B. Devine (Plenum, New York, 1988), pp. 365–373.

36. R. C. Hughes, *Phys. Rev. Lett.*, **30**, 1333 (1973); **35**, 449 (1975).

37. R. C. Hughes, *Solid-State Electron.*, **21**, 251 (1978).

38. M. V. Fischetti, *J. Appl. Phys.*, **83**, 270 (1998); **83**, 6202 (1998).

39. M. V. Fischetti, *Phys. Rev. B*, **59**, 4901 (1999).

40. W. Pauli, in *Festschrift zum 60. Geburtstage A. Sommerfeld*, edited by A. Sommerfeld and P. J. W. Debye (Verlag Hirzel, Leipzig, 1928), p. 30.

41. L. van Hove, *Physica*, **XXI**, 517 (1955).

42. W. Frensley, *Rev. Mod. Phys.*, **62**, 745 (1990).

43. M. Saeki, *J. Phys. Soc. Jpn.*, **55**, 1846 (1986).

44. W. Pötz, *J. Appl. Phys.*, **66**, 2458 (1989).

45. D. Ahn, *Phys. Rev. B*, **50**, 8310 (1994).

46. H. L. Grubin, D. K. Ferry, and R. Akis, *Superlattices Microstruct.*, **20**, 531 (1996).

47. W. Kohn and J. M. Luttinger, *Phys. Rev.*, **108**, 590 (1957).

48. A. Messiah, *Quantum Mechanics*, Vol. 1 (North Holland, Amsterdam, 1970).

49. R. Peierls, in *Lecture Notes in Physics 31: Transport Phenomena*, edited by J. Ehlers, K. Hepp, and H. A. Weidenmüller (Springer-Verlag, Berlin, 1974).

50. D. N. Zubarev, *Nonequilibrium Statistical Mechanics* (Consultants Bureau, New York, 1974), particularly Secs. 19ff.

51. D. K. Ferry and J. R. Barker, *Solid State Commun.*, **30**, 301 (1979).

52. D. K. Ferry and J. R. Barker, *J. Phys. Chem. Sol.*, **41**, 1083 (1980).

53. D. K. Ferry, A. M. Kriman, H. Hida, and S. Yamaguchi, *Phy. Rev. Lett.*, **67**, 633 (1991).

54. R. Zwanzig, *Phys. Rev.*, **124**, 983 (1961).

55. J. Zimmermann, P. Lugli, and D. K. Ferry, *Solid-State Electron.*, **26**, 233 (1983).

56. H. Fröhlich, *Proc. Roy. Soc.* (*London*), **A188**, 521 (1947).

57. G. G. Stokes, *Mathematical and Physical Papers*, Vol. 2 (Cambridge University Press, Cambridge, 1883), p. 329.

58. P. Bordone, D. Vasileska, and D. K. Ferry, *Phys. Rev. B*, **53**, 3846 (1996).

59. H. L. Grubin, T. R. Govindan, B. J. Morrison, and M. A. Stroscio, *Semicond. Sci. Technol.*, **7**, B434 (1992).

60. H. L. Grubin, T. R. Govindan, J. P. Kreskovsky, and M. A. Stroscio, *Solid-State Electron.*, **36**, 1697 (1993).

61. D. K. Ferry, R. Akis, and D. Vasileska, in *Proc. Intern. Electron Dev. Mtg.* (IEEE Press, New York, 2000), p. 287.

62. D. K. Ferry and J. R. Barker, *Phys. Status Solidi B,* **100**, 683 (1980).

63. C. Phillipidis, D. Bohm, and R. D. Kaye, *Il Nuovo Cim.*, **71B**, 75 (1982).

64. M. G. Ancona and G. J. Iafrate, *Phys. Rev. B*, **39**, 9536 (1989).

65. E. Wigner, *Phys. Rev.*, **40**, 749 (1932).

66. G. J. Iafrate, H. L. Grubin, and D. K. Ferry, *J. Physique*, **41**, C7–307 (1981).

67. J. R. Zhou and D. K. Ferry, *IEEE Trans. Electron Dev.*, **39**, 473 (1992); **39**, 1793 (1992).

68. L. S. Schulman, *Time's Arrow and Quantum Measurement* (Cambridge University Press, Cambridge, 1997), p. 18.

69. D. K. Ferry and H. L. Grubin, *Proc. IWCE-6, Osaka* (IEEE Press, New York, 1998), p. 84.

70. R. A. Smith, *Wave Mechanics of Crystalline Solids* (Chapman and Hall, London, 1961).

71. D. K. Ferry, *Superlattices Microstruct.* **27**, 61 (2000).

72. A. L. Fetter and J. D. Walecka, *Quantum Theory of Many-Particle Systems* (McGraw-Hill, New York, 1971).

73. D. K. Ferry, R. Akis, and D. Vasileska, in *Proc. Intern. Electron Dev. Mtg.* (IEEE Press, New York, 2000), p. 287.

74. C. Heitzinger, C. Ringhofer, S. Ahmed, and D. Vasileska, *J. Comp. Electron.*, **6**, 15 (2006).

75. R. P. Feyman and A. R. Hibbs, *Quantum Mechanics and Path Integrals* (McGraw-Hill, New York, 1965).

76. R. Giachetti and V. Tognetti, *Phys. Rev. Lett.*, **55**, 912 (1985).

77. R. P. Feynman and H. Kleinert, *Phys. Rev. A*, **34**, 5080 (1986).

78. J. Cao and B. J. Berne, *J. Chem. Phys.*, **92**, 7531 (1990).

79. G. A. Voth, *J. Chem. Phys.*, **94**, 4095 (1991).

80. A. Cuccoli, A. Macchi, M. Neumann, V. Tognetti, and R. Vaia, *Phys. Rev. B*, **45**, 2088 (1992).

81. A. Kriman and D. K. Ferry, *Phys. Lett. A*, **285**, 217 (1989).

82. D. K. Ferry and J.-R. Zhou, *Phys. Rev. B*, **48**, 7944 (1993).

83. D. N. Zubarev, *Nonequilibrium Statistical Thermodynamics* (Consultants Bureau, New York, 1970).

84. P. M. Morse and H. Feshbach, *Methods of Theoretical Physics* (McGraw-Hill, New York, 1953), Chapter 7.

85. H. Mizuta and C. J. Goodings, *J. Phys.: Condens. Matter*, **3**, 3739 (1991).

86. R. Tsu and L. Esaki, *Appl. Phys. Lett.*, **22**, 562 (1973).

87. T. L. C. G. Sollner, W. D. Goodhue, P. E. Tannenwald, C. D. Parker, and D. D. Peck, *Appl. Phys. Lett.*, **43**, 588 (1983).

88. H. L. Grubin, T. R. Govindan, B. J. Morrison, and M. A. Stroscio, *Semicond. Sci. Technol.*, **7**, B434 (1992).

89. T. Kurosawa, *J. Phys. Soc. Jpn.*, Suppl. **21**, 424 (1966).

90. R. Brunetti, C. Jacoboni, and F. Rossi, *Phys. Rev. B*, **39**, 10781 (1989).

91. H. Budd, *J. Phys. Soc. Jpn.*, Suppl. **21**, 420 (1966).

Chapter 8

Wigner Function

In Chapter 2, we introduced a number of quantum functions, including the density matrix discussed in the last chapter, and the Wigner function. Alone among the quantum functions we have discussed so far, the Wigner function has the convenient phase–space formulation. Wigner gave us his formulation already in 1932 [1] and demonstrated its usefulness particularly when we are interested in spanning the transition from the classical to the quantum world. The Wigner function has been studied for more than eight decades. Historically, it was developed from the Schrödinger wave equation, although it could have been developed on its own right. The Wigner function approach is generally felt to offer a number of advantages for use in modeling the behavior of physical processes in many fields of science [2–4]. First, it is a phase–space formulation, similar to classical formulations based on the Boltzmann equation. By this, we mean that the Wigner function involves both real space and momentum space variables, distinctly different from the Schrödinger equation. In this regard, modern approaches with the Wigner function provide a distinct formulation that is recognized as equivalent to, but is a different alternative to normal operator-based quantum mechanics [5]. Because of the phase–space nature of the distribution, it is now conceptually possible to identify where quantum corrections enter a problem by comparing with the classical version, an approach that

An Introduction to Quantum Transport in Semiconductors
David K. Ferry
Copyright © 2018 Pan Stanford Publishing Pte. Ltd.
ISBN 978-981-4745-86-4 (Hardcover), 978-1-315-20622-6 (eBook)
www.panstanford.com

has been used to provide an effective quantum potential that can be used as a correction term in classical simulations, as discussed in the last chapter.

At the boundaries, the phase–space distribution allows one to separate the incoming and outgoing components of the distribution, which now allows one to model both the contact itself and the entire open quantum system. Moreover, the Wigner function is entirely real, which simplifies both the calculation and the interpretation of results. For this reason, it was a natural choice for simulation of quantum transport in devices such as the resonant-tunneling diode [6–8]. But, the Wigner function is not the only possible phase–space distribution, although it is certainly relatable to other such approaches. We will examine the similarities and differences between the Wigner function and other approaches in phase–space later in this chapter.

Let us begin the discussion with the Wigner center-of-mass coordinates, introduced in the last chapter. These involve introducing the average and the difference coordinates as follows

$$\mathbf{x} = \frac{\mathbf{r} + \mathbf{r}'}{2}, \quad \mathbf{s} = \mathbf{r} - \mathbf{r}'. \tag{8.1}$$

We can now introduce the phase–space Wigner function as the Fourier transform on the difference coordinate as [1]

$$f_w = \frac{1}{h^3} \int d\mathbf{s} \rho(\mathbf{x}, \mathbf{s}, t) e^{i\mathbf{p} \cdot \mathbf{s}/\hbar} = \frac{1}{h^3} \int d\mathbf{s} \rho\left(\mathbf{x} + \frac{\mathbf{s}}{2}, \mathbf{x} - \frac{\mathbf{s}}{2}, t\right) e^{i\mathbf{p} \cdot \mathbf{s}/\hbar}. \tag{8.2}$$

Coupling to this new function is an equation of motion, which has great similarities with the Boltzmann transport equation. If we incorporate the coordinate transformations of Eq. (8.1) into Eq. (2.69), we arrive at a new equation of motion for the density matrix as

$$i\hbar \frac{\partial \rho}{\partial t} = \left[-\frac{\hbar^2}{m^*} \frac{\partial^2}{\partial \mathbf{x} \partial \mathbf{s}} + V\left(\mathbf{x} + \frac{\mathbf{s}}{2}\right) - V\left(\mathbf{x} - \frac{\mathbf{s}}{2}\right) \right] \rho(\mathbf{x}, \mathbf{s}, t). \tag{8.3}$$

The transform in Eq. (8.2), although introduced by Wigner, is often called the Weyl transform [1, 9–11]. When this transformation is applied to the above equation, we get a new equation of motion for the Wigner function as

$$\frac{\partial f_w}{\partial t} + \frac{\mathbf{p}}{m^*} \cdot \frac{\partial f_w}{\partial \mathbf{x}} - \frac{1}{i\hbar}\left[V\left(\mathbf{x}+\frac{i\hbar}{2}\frac{\partial}{\partial \mathbf{p}}\right) - V\left(\mathbf{x}+\frac{i\hbar}{2}\frac{\partial}{\partial \mathbf{p}}\right)\right]f_w = 0 \,. \quad (8.4)$$

In the absence of any dissipative processes, this can be rewritten in a more useful form as

$$\frac{\partial f_w}{\partial t} + \frac{\mathbf{p}}{m^*} \cdot \frac{\partial f_w}{\partial \mathbf{x}} - \frac{1}{h^3}\int d\mathbf{P} W(\mathbf{x}, \mathbf{P}) f_w(\mathbf{x}, \mathbf{p}+\mathbf{P}) = 0 \,, \quad (8.5)$$

where

$$W(\mathbf{x}, \mathbf{P}) = \int d\mathbf{s}\sin\left(\frac{\mathbf{P} \cdot \mathbf{s}}{\hbar}\right)\left[V\left(\mathbf{x}+\frac{\mathbf{s}}{2}\right) - V\left(\mathbf{x}-\frac{\mathbf{s}}{2}\right)\right]. \quad (8.6)$$

The use of the Wigner function is particularly important in scattering problems [12] and clearly shows the transition to the classical world.

Let us examine a simple wavefunction that has a propagation velocity associated with it. The form of this is a modified Gaussian so that there is an imaginary part that contains the velocity. Such a wave packet is often called a coherent wave packet, and we can write it as

$$\psi(x) = \frac{1}{(2\pi\sigma^2)^{1/4}}e^{-x^2/4\sigma^2 - ik_0 x}, \quad (8.7)$$

in which the velocity is given as $v_g = \hbar k_0/m^*$. To proceed, we will actually use the Fourier transform version of Eq. (8.2), where the wavefunctions are in momentum space, so that we use the transform of Eq. (8.7). The Fourier transform version of (8.2) is obtained by inserting the transforms of the two wavefunctions in (8.2), and evaluating the integrals. We will see in a moment why this approach is taken. Using this approach, we find that the Wigner function is now given by

$$f_w(x, k) = \frac{2}{h}e^{-2\sigma^2(k-k_0)^2 - x^2/2\sigma^2}. \quad (8.8)$$

We note that this is a positive definite function, which is appropriate for what is basically a ground-state function.

In general, the Wigner function is not a positive definite function. This is a consequence of the uncertainty relationship. On the other hand, if we integrate the Wigner function over either position or momentum, we will get a positive definite function. In fact the negative excursions of the function exist over phase–space regions whose volume is $\hbar^3$, so that smoothing the function over a volume corresponding to the uncertainty principle will produce a positive definite function. We can illustrate the presence of these negative excursions by considering an example. We consider an infinite barrier for $x < 0$ and examine the Wigner function for a finite density in the region $x > 0$ [13]. The result is shown in Fig. 8.1. Near the barrier, the density approaches zero quadratically. This is not a depletion layer in the normal sense but is a result of the requirement that the wavefunctions vanish at the barrier, which may be called a quantum repulsion. Also near the barrier, the distribution exhibits a number of non-classical features, one of which is the regions where the distribution is negative, a result of increasing uncertainty in position at the higher momenta values where the negative excursions develop.

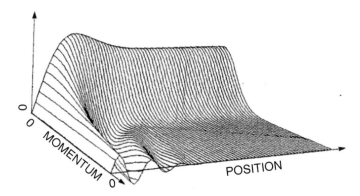

Figure 8.1 The Wigner distribution for an infinite barrier, in arbitrary units. The barrier exists for negative position. The perspective is chosen to display the quantum repulsion near the barrier and the negative values that exist at higher momentum values near the barrier. Reprinted with permission from Ref. [13], Copyright 1987, American Physical Society.

8.1 Generalizing the Wigner Definition

One of the great advantages of the Wigner form of a phase–space function is that it conserves normalization of the wavefunctions. That is, if we integrate the Wigner function (8.2) over all momenta, we obtain

$$\int d^3\mathbf{p} f_w(\mathbf{x},\mathbf{p}) = \frac{1}{h^3}\int d^3\mathbf{p}\int d^3\mathbf{s}\,\psi^*(\mathbf{x}+\mathbf{s}/2)\psi(\mathbf{x}-\mathbf{s}/2)e^{i\mathbf{p}\cdot\mathbf{s}/\hbar}$$

$$= \frac{1}{\hbar^3}\int d^3\mathbf{s}\,\psi^*(\mathbf{x}+\mathbf{s}/2)\psi(\mathbf{x}-\mathbf{s}/2)\delta(\mathbf{s}/\hbar)$$

$$= \psi^*(\mathbf{x})\psi(\mathbf{x}). \qquad (8.9)$$

Similarly, if we introduce the Fourier transform of the wavefunction

$$\varphi(\mathbf{p}) = \frac{1}{h^{3/2}}\int d^3\mathbf{x}\,\psi(\mathbf{x})e^{i\mathbf{p}\cdot\mathbf{x}/\hbar}, \qquad (8.10)$$

and use this in Eq. (8.2), integration over all position also conserves the normalization in the momentum space, as

$$\int d^3\mathbf{x} f_w(\mathbf{x},\mathbf{p}) = \varphi^*(\mathbf{p})\varphi(\mathbf{p}). \qquad (8.11)$$

This is an important result, as not all suggested phase–space formulations of quantum transport preserve this normalization. This has an important generalization. If we have a function $F(\mathbf{x},\mathbf{p})$, which is either a function of the position operator alone or of the momentum operator alone, or of any *additive* combination of these two operators, the expectation value of this function is given by

$$\langle F \rangle = \int d^3\mathbf{x}\int d^3\mathbf{x} F(\mathbf{x},\mathbf{p})f_w(\mathbf{x},\mathbf{p}). \qquad (8.12)$$

This is completely analogous to the equivalent classical expression for such an average. One of the interesting aspects of the Wigner function is the ability to transfer many of the results of classical transport theory into quantum approaches merely by replacing the Boltzmann distribution with the Wigner function. However, there is a caveat that must be understood in this regard. And this is the fact that the Wigner function may not be a positive semi-definite function. That is, there are cases in which the Wigner function possesses negative values, and these values can extend over a phase–space region of the extent of Planck's constant. If one smoothes over a region of this size, the negative excursions will go away. Clearly, these negative excursions represent the appearance of uncertainly in the quantum realm.

A further resemblance of the Wigner function to the classical distribution can be seen if we expand the potential term in Eq. (8.4). This last term can be expressed as

$$\tilde{\Theta}f_w = -\frac{1}{i\hbar}\left[V\left(\mathbf{x}+\frac{i\hbar}{2}\frac{\partial}{\partial\mathbf{p}}\right)-V\left(\mathbf{x}-\frac{i\hbar}{2}\frac{\partial}{\partial\mathbf{p}}\right)\right]f_w(\mathbf{x},\mathbf{p})$$

$$= -\frac{2}{\hbar}\sum_{n=0}^{\infty}(-1)^n\frac{(\hbar/2)^{2n+1}}{(2n+1)!}\frac{\partial^{2n+1}V(\mathbf{x})}{\partial\mathbf{x}^{2n+1}}\frac{\partial^{2n+1}f_w(\mathbf{x},\mathbf{p})}{\partial\mathbf{p}^{2n+1}}.$$

(8.13)

If we let $\hbar \to 0$, then only a single term survives, and this is the classical term

$$\tilde{\Theta}f_w = -\frac{\partial V}{\partial\mathbf{x}}\cdot\frac{\partial f_w}{\partial\mathbf{p}}.$$

(8.14)

Then, Eq. (8.4) reduces to just the streaming terms of the Boltzmann equation. The leading correction is of order $\hbar^2$, which is a term involving the third derivatives of the potential in position and the third derivatives of the Wigner function in momentum.

Further insight may be obtained by returning to the generalization with the Weyl operator [14]. To ease the reading, we pursue this in one dimension, but the generalization to the full three dimensions is straightforward. We begin by defining a characteristic function derived from the expectation value of an exponential operator with respect to the wavefunction. This is given by

$$C(\tau,\vartheta)=\int_{-\infty}^{\infty}dx\psi^*(x)e^{i(\tau p+\vartheta x)}\psi(x),$$

(8.15)

where p and x in the exponential are normal quantum mechanical operators. That is, p is a differential operator that works on the last wavefunction. Then the Wigner distribution can be obtained from this characteristic function as

$$f_w(x,p)=\frac{1}{4\pi^2}\int_{-\infty}^{\infty}d\tau\int_{-\infty}^{\infty}d\vartheta C(\tau,\vartheta)e^{-i(\tau p+\vartheta x)},$$

(8.16)

where p and x are now the c-numbers associated with the Wigner distribution. To see how the normal Wigner function arises from this approach, we can rewrite the exponential in Eq. (8.15) using the Baker–Hausdorf formula [15] as

$$e^{-i(\tau p+\vartheta x)}=e^{i\tau p/2}e^{i\vartheta x}e^{i\tau p/2}.$$

(8.17)

Using this result in Eq. (8.15) leads to

$$C(\tau,\vartheta) = \int_{-\infty}^{\infty} dx \left[e^{-i\tau p/2} \psi(x) \right]^* e^{i\vartheta x} e^{i\tau p/2} \psi(x)$$

$$= \int_{-\infty}^{\infty} dx \psi^*(x - \tau\hbar/2) e^{i\vartheta x} \psi(x + \tau\hbar/2). \tag{8.18}$$

Now we can put this back into Eq. (8.16) to yield

$$f_w(x,p) = \frac{1}{4\pi^2} \int_{-\infty}^{\infty} d\tau \int_{-\infty}^{\infty} d\vartheta e^{-i\tau p} \int_{-\infty}^{\infty} dx' \psi^*(x' - \hbar\tau/2) e^{iv(x'-x)} \psi(x' + \hbar\tau/2)$$

$$= \frac{1}{2\pi} \int_{-\infty}^{\infty} d\tau \int_{-\infty}^{\infty} dx' \psi^*(x' - \hbar\tau/2) e^{i\tau p} \psi^*(x' - \hbar\tau/2) \delta(x - x')$$

$$= \frac{1}{2\pi\hbar} \int_{-\infty}^{\infty} ds \psi^*(x' - s/2) e^{isp/\hbar} \psi(x' + s/2). \tag{8.19}$$

Clearly, this last form is the one-dimensional version of Eq. (8.2).

The use of the characteristic function to arrive at the Wigner function allows a more general approach, which we will return to later. Now, however, we want to show that the characteristic function can be generalized to a wider usage [16]. It can be used in connection with an observable, but does so in connection with a specific state $|\psi\rangle$. That is, the characteristic function of an observable A with respect to this state may be expressed as

$$C_A(\xi) = \langle \psi | e^{i\xi A} | \psi \rangle, \tag{8.20}$$

where ξ is a real parameter. If the operator A has an eigenvalue spectrum, which may be defined via the equation

$$A|k\rangle = A_k|k\rangle, \tag{8.21}$$

then the characteristic function can be evaluated in this representation as

$$C_A(\xi) = \sum_k \sum_{k'} \langle \psi | k \rangle \langle k | e^{i\xi A} | k' \rangle \langle k' | \psi \rangle$$

$$= \sum_k \sum_{k'} \langle \psi | k \rangle e^{i\xi k'} \delta_{kk'} \langle k' | \psi \rangle = \sum_k e^{i\xi k} |\psi_k|^2. \tag{8.22}$$

An equivalent version for continuous eigenvalues may be obtained. Here we see that the characteristic function for an operator A is the

Fourier transform of the probability distribution for the eigenvalues of that operator.

While the above approach suggests other types of distributions based on replacing the operator A with those in the Weyl operator appearing in Eq. (8.15), such an approach is hindered by the non-commutative nature of most conjugate operators. To illustrate this point, let us assume a characteristic function for two non-commuting operators A and B to be [17]

$$C_{AB}(\xi_1, \xi_2) = \langle \psi | e^{i(\xi_1 A + \xi_2 B)} | \psi \rangle. \tag{8.23}$$

Let us assume that each of the two operators has a valid basis set, which are formed from the natural eigenvalues of these operators, so that these basis sets may be defined via

$$A|k\rangle = A_k |k\rangle$$
$$B|k'\rangle = B_{k'} |k'\rangle. \tag{8.24}$$

Now we also want to impose a special relationship between the two operators. That is, we require that the two commutators satisfy the relations $[A, [A, B]] = [B, [A, B]] = 0$ so that we can utilize the identity [15]

$$e^{i(\xi_1 A + \xi_2 B)} = e^{i\xi_1 A} e^{i\xi_2 B} e^{-\xi_1 \xi_2 [A, B]/2}. \tag{8.25}$$

Inserting this relation in Eq. (8.21) and using the various basis sets, we have

$$C_{AB}(\xi_1, \xi_2) = e^{-\xi_1 \xi_2 [A, B]/2} \sum_{k, k'} e^{i(\xi_1 A + \xi_2 B)} \langle \psi | k \rangle \langle k | k' \rangle \langle k' | \psi \rangle. \tag{8.26}$$

Note that since the two basis sets do not have a mutual orthogonality relation, the central expectation is not a delta function. In general, we can assume that the commutator in the leading exponential is a c-number independent of the various eigenvalues. Now we can define the set of expectations as a generalized phase–space distribution

$$f(A_k, B_k) = \langle \psi | k \rangle \langle k | k' \rangle \langle k' | \psi \rangle$$
$$= \frac{1}{(2\pi)^2} \int d\xi_1 \int d\xi_2 e^{\xi_1 \xi_2 [A, B]/2} C_{AB}(\xi_1, \xi_2) e^{-i(\xi_1 A_k + \xi_2 B_{k'})} \tag{8.27}$$

The actual form that the result takes is dependent on the manner in which the last exponential in Eq. (8.23) is expanded. The present result depends on the approximations and assumptions detailed above. The first line of Eq. (8.27) differs from the previous definition of the Wigner function due to the presence of the first exponential. If we take A to be x and B to be p, then this last expression can be shown to reduce to [17]

$$f(x,p)\frac{1}{h}\int\limits_{-\infty}^{\infty} ds\psi^*(x)\psi(x-s)e^{ips/\hbar} = \frac{1}{\sqrt{h}}\psi^*(x)e^{ipx/\hbar}\varphi(p). \quad (8.28)$$

Here the Fourier transform of one wavefunction has given us the momentum wavefunction in Eq. (8.28). This remains a bilinear wavefunction in phase–space, but it is not the Wigner function proper. We will return to discuss other phase–space functions below.

8.2 Other Phase–Space Approaches

While the Wigner function has a lot of advantages, it is not the only phase–space function that has been used in quantum transport. There have been several alternative formulations, most of which depend on the Weyl exponential form [14, 18], which we have already used several times above. This form is given by

$$e^{i(\xi_1 q+\xi_2 p)} = \frac{1}{2}(e^{i\xi_1 q}e^{i\xi_2 p} + e^{i\xi_2 p}e^{i\xi_1 q}). \quad (8.29)$$

As discussed above, it is well known that the ordering of operators is quite important in quantum mechanics, as different ordering produces different expectation values [19]. It is well known that non-commuting operators must be carefully ordered to assure the desired result, as is seen from just px and xp. But how should one deal with a term such as $(\xi q)^2(\eta p)^3$ that arises in the expansion of Eq. (8.29). There is no single preferred ordering for these operators, such as *normal ordering* (where each operator is written in terms of its creation and annihilation operators, with all of the latter moved to the right-hand side of the expression) used in Green's functions [20]. Lacking any recognized protocol for choosing an ordering, many different ones have appeared in the literature, each of which has led to its own version of a phase–space distribution function [18].

Mehta [21] has given a standard review of the Wigner function but then explored what he called "standard ordering," in which he retains only the first term on the right-hand side of Eq. (8.29). By standard ordering, we mean that if one expands the exponential term into a power series, all powers of q precede all powers of p, where q and p are the position and momentum. This ordering occurs whether we are referring to the c-numbers that appear in the Wigner function or the operators that appear in the quantum functions. With this exponential term, he finds the characteristic function to be

$$C_s(\xi_1,\xi_2) = \int \psi^*(q)e^{i\xi_1 q}e^{i\xi_2 p}\psi(q)dp = \int \psi^*(q)e^{i\xi_1 q}\psi(q+\hbar\xi_2)dq .$$

(8.30)

If we now use Eq. (8.16) to compute the phase–space distribution function from the characteristic function, we find

$$
\begin{aligned}
f_s(x,p) &= \frac{1}{4\pi^2}\int\int d\xi_1 d\xi_2 C_s(\xi_1,\xi_2)e^{-i(\xi_2 p+\xi_1 x)} \\
&= \frac{1}{4\pi^2}\int\int d\xi_1 d\xi_2 \int \psi^*(q)e^{i\xi_1 q}\psi(q+\hbar\xi_2)dq e^{-i(\xi_2 p+\xi_1 x)} \\
&= \frac{1}{2\pi}\int d\xi_2 \int \psi^*(q)e^{-i\xi_2 p}\psi(q+\hbar\xi_2)dq\delta(q-x) \\
&= \frac{1}{2\pi}\int d\xi_2 \psi^*(x)e^{-i\xi_2 p}\psi(x+\hbar\xi_2).
\end{aligned}
$$

(8.31)

This form differs from Eq. (8.28) by a factor of $\hbar$, which can be recovered by a change in the integration variable. As is the case of Eq. (8.28), it is not clear whether this distribution function preserves normalization of the wavefunction (we have not included the factor of 2 that appears in Eq. (8.28), and this would indicate that the normalization may not be carried through properly. In fact, Lee [18] suggests that normalization is not maintained with this distribution function, as will be the case with most others that do not use the full form on the left of Eq. (8.28).

Kirkwood used the anti-standard ordering [22], in which just the last term of Eq. (8.29) is kept, and considered the many-body statistics of the distribution. By anti-standard ordering, we mean that if one expands the exponential term into a power series, all powers of p precede all powers of q, where q and p are position and momentum. The single-particle form differed little from the Wigner function, or from the above form. We can see this, as the characteristic function becomes

$$C_{AS}(\xi_1, \xi_2) = \int \psi^*(q)e^{i\xi_2 p}e^{i\xi_1 q}\psi(q)dq = \int \psi^*(q - \hbar\xi_2)e^{i\xi_1 q}\psi(q)dq. \tag{8.32}$$

If we now use Eq. (8.16) to compute the phase–space distribution function from the characteristic function, we find

$$
\begin{aligned}
f_{AS}(x, p) &= \frac{1}{4\pi^2}\int\int d\xi_1 d\xi_2 C_{AS}(\xi_1, \xi_2)e^{-i(\xi_2 p + \xi_1 x)} \\
&= \frac{1}{4\pi^2}\int\int d\xi_1 d\xi_2 \int \psi^*(q - \hbar\xi_2)e^{i\xi_1 q}\psi(q)e^{-i(\xi_2 p + \xi_1 x)}dq \\
&= \frac{1}{2\pi}\int d\xi_2 \int \psi^*(q - \hbar\xi_2)e^{-i\xi_2 p}\psi(q)dq\delta(q - x) \\
&= \frac{1}{2\pi}\int d\xi_2 \psi^*(x - \hbar\xi_2)e^{-i\xi_2 p}\psi(x). \tag{8.33}
\end{aligned}
$$

Again the result bears significant similarity to Eq. (8.28) except that it is the complex conjugate wavefunction, which is shifted by the parameter ξ_2.

Mehta has also examined the use of normal-ordered products of the quantum operators through the introduction of the creation and annihilation operators [21]. A somewhat similar discussion of the use of these operators and the connection with classical statistical mechanics for quantum optics has been carried out as well [23, 24]. For this purpose, these creation and annihilation operators may be defined as [21]

$$a = \frac{1}{\sqrt{2\hbar}}(q + ip), \quad a^\dagger = \frac{1}{\sqrt{2\hbar}}(q - ip), \tag{8.34}$$

where, as before, q and p are the quantum mechanical operators. Within some constants, these operators are the same as those used for the harmonic oscillator or for angular momentum raising and lowering operators. Here the adjoint (second) term is the creation operator, while the first form is the annihilation operator. If we use the full form of the left-hand side of Eq. (8.29) and expand the exponential, then the various terms in the exponential can be rearranged and resummed into two exponentials, each of which contains one of the two operators of Eq. (8.34). Specifically, Mehta [21] has shown that Eq. (8.29) can be rewritten as

$$e^{i(\xi_2 p + \xi_1 q)} = e^{ia^\dagger(\xi_1 + i\xi_2)\sqrt{\hbar/2}}e^{ia(\xi_1 - i\xi_2)\sqrt{\hbar/2}}. \tag{8.35}$$

The phrase "normal ordering" implies that all the creation operators $(a^\dagger)$ are kept to the left of all the annihilation operators. When this expansion is used to generate the characteristic function, the resulting distribution function is given by

$$f_N(x, p) = \exp\left[-\frac{\hbar}{4}\left(\frac{\partial^2}{\partial p^2} + \frac{\partial^2}{\partial x^2}\right)\right] f_W(x, p). \tag{8.36}$$

Thus, this normal-ordered distribution is an infinite series of corrections to the Wigner function itself. It is not clear under which circumstances one may prefer this latter distribution.

Finally, there is a further phase–space function, which has been quite useful not only in optics but also in quantum chaos [25]. This is the so-called Husimi function [26]. Husimi refers to our coherent wave packet (8.7) as a Heisenberg wave packet and points out that these are not mutually orthogonal. Under certain conditions, they will form a complete, normalized, almost orthogonal basis set (he attributes this to work of von Neumann [27]). Generally, the required conditions are a discretization of phase–space according to which the coherent wave packets are spaced according to

$$\Delta k = \frac{\sqrt{\pi}}{\sigma} \tag{8.37}$$

$$\Delta x = \sqrt{\pi}\sigma,$$

where σ is the standard deviation of the wave packet, as indicated in Eq. (8.7). The prescription given by Husimi is to first smooth the density matrix, the kernel of the integral in Eq. (8.2) via

$$\bar{\rho}(x') = \frac{1}{\sqrt{2\pi}\sigma}\int d\xi \psi^*\left(\xi - \frac{x'}{2}\right)\psi\left(\xi + \frac{x'}{2}\right)e^{-(\xi-x')^2/2\sigma^2}, \tag{8.38}$$

which provides a Gaussian smoothing of the density matrix. Then, this smoothed density matrix is then transformed using a coherent wave packet as

$$f_H(x, k) = \int \bar{\rho}(x')e^{-\frac{x'^2}{8\sigma^2} + ikx'} dx'. \tag{8.39}$$

When this procedure is done with the sample coherent wave packet introduced in Eq. (8.7), we find the Husimi function as

$$f_H(x, k) = e^{-\frac{x^2}{4\sigma^2} - \sigma^2(k-k_0)^2}, \tag{8.40}$$

which has some differences from the previous result of Eq. (8.8). To begin, the Husimi function is broader in position space. But these small differences mask some fundamental differences between the Wigner and Husimi functions. It is well known that the Wigner function can be negative, but these regions disappear when it is averaged over a region the size of an uncertainty box in phase–space. On the other hand, the Husimi function is already averaged via the prescription Eq. (8.38) and is, therefore, a positive semi-definite and semi-classical distribution. In fact, Takahashi has pointed out that the Husimi distribution is a coarse-graining of the Wigner function with a Gaussian smoothing [28]. Moreover, the Husimi function does not produce the correct probability functions [29]. That is, when the momentum is integrated out of the Husimi function, the resulting probability distribution in real space is broader than that from the Wigner function. Correspondingly, when the Husimi function is integrated over real space, the resulting momentum probability function is broader than that resulting from the Wigner function. Both of these results arise from the coarse-graining that is inherent in the Husimi function and is apparent when Eq. (8.40) is compared with Eq. (8.8). It should be remarked that a different prescription for obtaining the Husimi function has appeared in the literature. In this form, we find the Husimi function from [30–32]

$$ f_H = \left| \int dx' \psi(x', k_0) \varphi_T(x' - x, k) \right|^2, \qquad (8.41) $$

where $\psi(x', k_0)$ is our specific wavefunction with which we want to obtain the Husimi function. The second wavefunction φ_T is the transforming function by which the wavefunction is modified. This is a generalized coherent state written in terms of the two coordinates and a general wavevector k, which can be the exponential term in Eq. (8.39). This form gives the same result Eq. (8.40) for our initial coherent state. This particular formulation is actually just a reworking of the integrals in Eqs. (8.38) and (8.39) and has found wide application in quantum optics. In addition to the optical applications, as we remarked above, the Husimi function has found a great deal of usage in the arena of quantum chaos [25]. The reasons for this are obvious. In classical chaos, the natural simulation space is a Poincarè section of classical phase–space. Since the quantum simulation can be done for the same (equivalent) system to yield the wavefunction, the projection of this onto the classical Poincarè section gives direct information on the classical-to-quantum crossover. The use of Eq. (8.41) gives a quick transformation from the wavefunction to the Husimi distribution for this purpose and can be done quite easily numerically [33].

8.3 Moments of the Wigner Equation of Motion

A great deal of resemblance between the equation of motion (8.4) for the Wigner distribution function and the classical equivalent Boltzmann equation can be found by mere inspection. But it is also useful to extract the moment equations, or hydrodynamic equations, as was done for the density matrix in Section 7.5. For this, we will deal primarily with the streaming terms of Eq. (8.4), leaving the collision term for later. Here we will eventually deal with the relaxation-time approximation for the Wigner equation. While we begin with Eq. (8.4), we will use Eq. (8.13) for the potential term. The moment equations are obtained, as in Section 7.5, by multiplying the equation of motion by an arbitrary function of momentum $\varphi(\mathbf{p})$ and then integrating over the momentum. Using Eqs. (8.4) and (8.13), we then find

$$\frac{\partial \langle \varphi(\mathbf{p}) \rangle}{\partial t} + \frac{1}{m^*} \nabla \cdot \langle \varphi(\mathbf{p})\mathbf{p} \rangle = \left\langle \varphi(\mathbf{p}) \frac{\partial f_W}{\partial t} \Big|_{\text{coll}} \right\rangle$$

$$+ \frac{2}{\hbar} \sum_{s=0}^{\infty} \frac{(-1)^s (\hbar/2)^{2s+1}}{(2s+1)!} \frac{\partial^{2s+1} V(\mathbf{x})}{\partial \mathbf{x}^{2s+1}}$$

$$\times \int \varphi(\mathbf{p}) \frac{\partial^{2s+1} f_W}{\partial \mathbf{p}^{2s+1}} d^3\mathbf{p}, \qquad (8.42)$$

where, as before,

$$\langle \varphi(\mathbf{p}) \rangle = \frac{1}{n} \int f_W(\mathbf{x}, \mathbf{p}) d^3\mathbf{p}. \qquad (8.43)$$

Let us first begin by taking the function $\varphi(\mathbf{p})$ as just the c-number 1. This produces the simple

$$\frac{\partial n(\mathbf{x})}{\partial t} + \frac{1}{m^*} \nabla \cdot (n \langle \mathbf{p} \rangle) = \frac{\partial n(\mathbf{x})}{\partial t} + \nabla \cdot (n(\mathbf{x})\mathbf{v}_d) = 0, \qquad (8.44)$$

which is just a form of the continuity equation. We note here that the single value of momentum that exists in the second term of Eq. (8.44) corresponds exactly to the quantum mechanical current, and this term just takes the average of that quantity. In a single valley situation, which we consider here, the scattering conserves the density, although non-local scattering could change this with a broadening effect. This does not occur with the relaxation-time approximation. In addition, the first derivative with respect to the momentum of the Wigner function, in the last term, yields an asymmetric argument of the integral and this also vanishes. In fact, only the odd powers of the derivatives appear in the integral of the last term, so all of these terms vanish, and this last term makes no contribution to this moment equation.

Now let us turn to the next possible term, which occurs when we let $\varphi(\mathbf{p}) = \mathbf{p}$. This leads us to the equation

$$\frac{\partial(n\langle\mathbf{p}\rangle)}{\partial t} + \frac{1}{m^*}\nabla\cdot(n\langle\mathbf{pp}\rangle) - \frac{\partial V(\mathbf{x})}{\partial\mathbf{x}}\cdot\int\mathbf{p}\frac{\partial f_W}{\partial\mathbf{p}}d^3\mathbf{p} = \left\langle\mathbf{p}\frac{\partial f_W}{\partial t}\Big|_{\text{coll}}\right\rangle. \tag{8.45}$$

While there are higher-order terms in the potential term, these vanish so that we are left with only the $s = 0$ term. One can understand this by integrating by parts in the integral, and only the first derivative of the momentum does not vanish. Thus, the potential term can be taken to the voltage and then to the electric field, so that this term becomes $e\mathbf{E}n$. If we introduce the relaxation-time approximation as

$$-\frac{f_W(\mathbf{x},\mathbf{p}) - f_{W0}(\mathbf{x},\mathbf{p})}{\tau_m}, \tag{8.46}$$

where the second distribution is the equilibrium value, then the relaxation term just becomes

$$-\frac{n\langle\mathbf{p}\rangle}{\tau_m} = -\frac{nm^*\mathbf{v}_d}{\tau_m}. \tag{8.47}$$

The second term on the left-hand side of Eq. (8.45) is more problematic. The average produces a second-rank tensor, with the diagonal elements being proportional to the energy of the distribution. This is a major point of coupling between this equation and the next higher order. Even if we keep just the diagonal elements, they go beyond the classical energy, and we can show this by expanding as [34]

$$\langle p^2 \rangle = \left\langle -\frac{\hbar^2}{4} \left(\psi * \frac{d^2\psi}{dx^2} - 2\frac{d\psi*}{dx}\frac{d\psi}{dx} + \frac{d^2\psi*}{dx^2}\psi \right) \right\rangle. \tag{8.48}$$

If we now assumed that we can write the wavefunction as

$$\psi(x) = A(x)e^{iS(x)/\hbar}, \tag{8.49}$$

then Eq. (8.48) becomes

$$\langle p^2 \rangle = (m*v)^2 n(x) - \frac{\hbar^2}{4}n(x)\frac{d^2(\ln n)}{dx^2}, \tag{8.50}$$

where the last term is recognized as containing the Wigner potential correction to the semi-classical potential found from, e.g., solving the Poisson's equation. Finally, if we fold in the derivatives of the density from Eq. (8.44) with the approximations found above, we find the new form of the time derivative of the drift velocity to be

$$\frac{\partial v_d}{\partial t} = \frac{eE}{m*} - \frac{v_d}{\tau_m} + \frac{v_d}{n}\nabla\cdot(n v_d) - \frac{1}{nm*}\nabla\cdot(nE) + \frac{\hbar^2}{4m*}\frac{d^2(\ln n)}{dx^2}. \tag{8.51}$$

Only the last term differs from what would be found if we had used the classical Boltzmann equation. So it is somewhat natural that the Wigner equation approach brings in the Wigner potential as a quantum correction.

We want to do one more moment to account for the continuity of energy in the system, since the relaxation time τ_m is an energy-dependent quantity. For this purpose, we let $\varphi(\mathbf{p}) = p^2/2m*$. Again, we multiply the basic transport equation with this quantity and integrate over the momentum. This leads us to the equation

$$\frac{1}{2m*}\frac{\partial(n\langle p^2\rangle)}{\partial t} + \frac{1}{2m*}\nabla\cdot(n\langle p^2\mathbf{p}\rangle) + \frac{n}{m*}\nabla V\cdot\langle\mathbf{p}\rangle = \left\langle \frac{p^2}{2m*}\frac{\partial f_w}{\partial t}\Big|_{coll} \right\rangle. \tag{8.52}$$

Again, the last term on the left-hand side results from the fact that only the $s = 0$ term arises from the summation in Eq. (8.42). In fact, it would be at the next higher-order equation where the $s = 1$ term first appears. As before, the integral in this summation is integrated by parts (twice), which leaves just this simple term. The second term on the left-hand side can be evaluated just as was done in the momentum equation. The term is written out in terms of the wavefunction, and using Eq. (8.49), we arrive at

$$\langle p^2 \mathbf{p} \rangle = n(m*v)^2 m * \mathbf{v}_d - \frac{n\hbar^2}{4} \left[3m * \mathbf{v}_d \nabla^2 (\ln n) + \nabla^2 (m * \mathbf{v}_d) \right].$$

(8.53)

As before, the terms involving an explicit dependence on Planck's constant are quantum corrections to the potential that appears in the equation. One further note is that, in the generalized expansions of a parameterized distribution that we used in the last chapter, the leading term involves the time derivative

$$\frac{1}{2m*} \frac{\partial \left(n \langle p^2 \rangle \right)}{\partial t} = \frac{\partial}{\partial t} \left[n \left(\frac{3k_B T_e}{2} + \frac{1}{2} m * v_d^2 \right) \right].$$

(8.54)

Not only does this term fold Eq. (8.44) into the present equation, the derivative of the drift term also folds the complete Eq. (8.51) into the present equation. This will lead to a very long and complicated final equation for the time variation of the electron temperature. For brevity, we will omit this final expansion of Eq. (8.52), leaving it for the student to explore.

8.4 Scattering Integrals

To develop the form of the scattering integrals for the Wigner equation of motion, we begin with the form developed for the density matrix in Eq. (7.30). If we unfold the various superoperators, then the collision term may be written as

$$\frac{\partial f}{\partial t}\Big|_{\text{coll}} = -\frac{1}{\hbar^2} \int_0^t dt' Tr_L \left\{ [H_{eL}(t), U(t, t')[H_{eL}(t - t'), \rho]U^{\dagger}(t, t')] \right\}.$$

(8.55)

In this expression, the operators U are unitary time translation operators given by

$$U(t, t') = e^{-iH_e(t - t')/\hbar},$$

(8.56)

in which the evolution is governed by the electron part of the Hamiltonian. The trace is over the lattice states, or the phonon modes, and introduces the creation and annihilation operators for the phonons. Levinson [35] has shown that the right-hand side can be written as the sum of two terms as

$$\frac{\partial f}{\partial t}\Big|_{coll} = \Xi^{(+)}(t/\rho_1) + \Xi^{(-)}(t/\rho_1),$$ (8.57)

where the first term corresponds to the emission of a phonon while the last term corresponds to the absorption of a phonon. We will treat the first term, and the results are easily extended to the second term. The emission term can be written as [35]

$$\Xi^{(+)}(t/\rho_1) = \sum_q \frac{|M_q|^2}{\hbar^2}(N_q+1)\int_0^t dt' e^{-i\omega_q(t-t')}\left[U(t,t')a_q^\dagger \rho_1(t')U^\dagger\right.$$

$$\left.\times(t,t'), a_q\right] + h.c.$$ (8.58)

The quantity M_q is the appropriate matrix element and N_q is the Bose–Einstein distribution for the phonons. The two terms of the commutator represent the role of "in" scattering and "out" scattering where the pronouns refer to the state of interest. That is, the "in" terms represent electrons coming into the state by the emission of a phonon while the "out" terms represent electrons leaving the state by the emission of a phonon. With this recognition, the two terms can be rewritten as

$$\Xi^{(+)}(t/\rho_1) = -A^{(+)}(t/\rho_1) + B^{(+)}(t/\rho_1)$$ (8.59)

where the out-scattering term is

$$A^{(+)}(t/\rho_1) = \sum_q \frac{|M_q|^2}{\hbar^2}(N_q+1)\int_0^t dt' e^{-i\omega_q(t-t')}a_q U(t,t')a_q^\dagger \rho_1(t')U^\dagger(t,t') + h.c.$$ (8.60)

and the in-scattering term is

$$B^{(+)}(t/\rho_1) = \sum_q \frac{|M_q|^2}{\hbar^2}(N_q+1)\int_0^t dt' e^{-i\omega_q(t-t')}U(t,t')a_q^\dagger \rho_1(t')U^\dagger(t,t')a_q + h.c.$$ (8.61)

As we are interested in developing this collision term for the Wigner equation of motion, we must now Wigner transform each of these integrals into the appropriate phase–space formulation. For simplicity, we will treat only the term (8.60), since the extension to the other terms is relatively easy once we have this one result. For this, we follow the straightforward approach of Eq. (8.2) and write the integral in Eq. (8.59) as

$$I_{W,A}^{(+)} = \int d^3s\, e^{is\cdot p/\hbar} \int_0^t dt' \left\langle \mathbf{r} - \frac{\mathbf{s}}{2} \middle| a_q U(t,t') a_q^\dagger \middle| r - \frac{\mathbf{s}}{2} \right\rangle + h.c. \quad (8.62)$$

At this point, we need to introduce the resolution of unity via an integral over projection operators, but this is done with the Wigner equivalent

$$R(\mathbf{r},\mathbf{p}) = \int d^3s\, e^{is\cdot r/\hbar} \left| \mathbf{p} - \frac{\mathbf{s}}{2} \right\rangle \left\langle \mathbf{p} + \frac{\mathbf{s}}{2} \right| \quad (8.63)$$

$$R_W(\mathbf{r},\mathbf{p}) = 2\pi\delta(\mathbf{r}-\mathbf{r}')\delta(\mathbf{p}-\mathbf{p}').$$

This resolution thus maps into a delta function in the phase–space representation. We must remember that this is just a heuristic device to ease the computation of various integrals that we are going to encounter.

With the spectral operators from above, arbitrary introductions of delta functions can be made, each accompanied by an integral over one of the arguments. The advantage of this is that we can develop the appropriate phase–space function for each of the individual terms in the integral (8.61). This allows us to write this latter equation as

$$I_{W,A}^{(+)} \int_0^{t'} dt' \int dz\, e^{ipz/\hbar} e^{i\omega_q(t-t')} \int dr_n \int dp_n \int dz_n f_W(p_4,x_4,t')$$

$$\times \left\langle x - \frac{z}{2} \middle| a_q \middle| x_1 + \frac{z_1}{2} \right\rangle e^{ip_1 z_1/\hbar} \left\langle x_1 - \frac{z_1}{2} \middle| U(t,t') \middle| x_2 + \frac{z_2}{2} \right\rangle e^{ip_2 z_2/\hbar}$$

$$\times \left\langle x_2 - \frac{z_2}{2} \middle| a_q^\dagger \middle| x_3 + \frac{z_3}{2} \right\rangle e^{ip_3 z_3/\hbar} \left\langle x_3 - \frac{z_3}{2} \middle| U^\dagger(t,t') \middle| x + \frac{z}{2} \right\rangle. \quad (8.64)$$

Here, we have condensed several integrals, over the dimensions, and over the set into three with a short-hand notation. Using the orthogonality of the wavefunctions, the fact that integrations over isolated exponentials yield delta functions, and that the exponentials can represent displacement operators, and with some change of variables, Eq. (8.63) can be reduced to

$$I_{W,A}^{(+)} \int_0^{t'} dt' \int dz e^{ipz/\hbar} e^{i\omega_q(t-t')} \int dx' \int dx'' f_W \left(p_4, \frac{x'+x''}{2}, t' \right)$$

$$\times e^{i[(p-p_4-q)x''+(p-q)z]/\hbar} \left\langle -\frac{z_1}{2} \middle| U(t,t') \middle| \frac{x'}{2} \right\rangle e^{i(p_4+q)x'/\hbar} \left\langle -\frac{x'}{2} \middle| U^\dagger(t,t') \middle| \frac{z}{2} \right\rangle.$$

$$(8.65)$$

To proceed further, we are going to need to invoke translation invariance in the system, although the presence of the electric field during the collision means that this invariance is broken on the microscale, even for a homogeneous semiconductor. This leads to the so-called intra-collisional field effect. The unitary translation operators imply that the two parts of the interaction (which normally gives the magnitude squared of the matrix element) can occur at different positions quantum mechanically. Hence, the field can interact with the particle during the collision process for the time required to complete the collision. We will meet this again in the next two chapters. So if we can invoke at least coarse-grained homogeneity within the semiconductor, the last integral can be reduced further, so that we finally get

$$A^{(+)}(t/f_W) = \sum_q \frac{2\pi |M_q|^2}{\hbar} (N_q+1) \int_0^t dt' \int d^3k' f_W(\mathbf{r}, \mathbf{k}, t')$$

$$\times \Delta \left(\mathbf{k} - \frac{\mathbf{q}}{2}, \mathbf{k}' - \frac{\mathbf{q}}{2}, t, t' \right),$$

$$(8.66)$$

where

$$\Delta(\mathbf{k}, \mathbf{k}', t, t') = \frac{1}{\hbar} \int d^3x \int d^3x' \delta \left[\mathbf{p} - \mathbf{p}' - \frac{e}{\hbar} \mathbf{E}(t-t') \right] e^{-i\omega_q(t-t')}$$

$$\times e^{-i\mathbf{q}\cdot(\mathbf{x}-\mathbf{x}')/2\hbar - i(\mathbf{p}+\mathbf{p}')\cdot(\mathbf{x}-\mathbf{x}')/2\hbar} \left\langle \frac{\mathbf{x}}{2} \middle| U \middle| -\frac{\mathbf{x}}{2} \right\rangle$$

$$\times \left\langle \frac{\mathbf{x}'}{2} \middle| U^\dagger \middle| -\frac{\mathbf{x}'}{2} \right\rangle + h.c.$$

$$(8.67)$$

It is obvious that this last function replaces the traditional energy-conserving delta function and represents a convolution of the spectral densities of the various states. This term accounts for spatially and temporally broadening the scattering process. That is, it no longer occurs at a single point in space nor at a single point in time. We will return to this in our Green's function treatment in the next two chapters.

8.5 Applications of the Wigner Function

In this section, we want to consider a few applications of the Wigner function to typical devices. As in the previous chapter, the "fruit fly" is the resonant-tunneling diode (RTD), with its inherent negative differential conductivity. Many people have modeled it, so there will be no attempt here to give a definitive and complete review of these approaches.

As we pointed out in the last chapter, the RTD was conceived by Esaki and his colleagues, but became much more interesting after it was demonstrated that oscillations in the THz regime could be expected [36]. The device is built around two tunneling barriers (typically GaAlAs) separated by a quantum well (typically GaAs). This creates a bound state within the quantum well, which leads to unity tunneling coefficient for an incident electron whose energy lies at the energy of the bound state. The basic structure is then clad with outer layers of GaAs. For a given potential structure, such as that found in the RTD, the tunneling coefficient can be calculated as a function of the incident energy using an exact formulation or the WKB approximation. This can then be used for a simple analytical model. But this is not a good approach to use, as the tunneling probability will change with bias applied across the structure. For example, the unity tunneling coefficient is obtained only when the two barriers are exactly equal in shape, and this is no longer the case as soon as a bias is applied. For this reason, better simulations have been developed, which solve Poisson's equation for a self-consistent potential, and coupling this with a transport equation using either the density matrix (last chapter) or the Wigner function [37–39].

Now it is perfectly correct to assume that a correct quantum mechanical steady-state initial condition for the Wigner equation of motion (8.5) may be found merely by setting the time derivative to zero and seeking a solution. The fallacy in this approach lies in the fact that the correct boundary conditions presuppose some knowledge of the state of the system at regions near the boundaries. This knowledge must be based on the full knowledge of the internal potentials. Hence, one must know the solution prior to trying to find the solution. The correct manner in which to approach this problem is to find the initial condition from the adjoint equation [40]. One approach to this is to use scattering wavefunctions, as discussed in Chapter 2, and this is the method we will use [13, 39].

In the scattering wavefunction approach, we begin with forming the density matrix with these waves. It is important to note that if the potential is constant far from the active double barrier region, then we can use plane waves to build up the density matrix. These plane waves are chosen with the Fermi–Dirac probability function; that is, the number of plane waves at any incident energy is proportional to the value of this function at that energy. For states that are incident from the left of the barrier (these are waves traveling in the positive x-direction, with $k > 0$ in one dimension), we may write

$$\psi_k(x) = \begin{cases} \dfrac{1}{\sqrt{2\pi}}\left[e^{ikx} + r(k)e^{-ikx}\right] & x < x_{\min} \\ \dfrac{1}{\sqrt{2\pi}}t(k)e^{ikx} & x > x_{\max}, \end{cases} \tag{8.68}$$

where $x_{\min}$ and $x_{\max}$ are the left and right sides of the barrier–well region. The quantity $r(k)$ is the reflection coefficient of the wave from the barrier structure, and $t(k)$ is the transmission coefficient of the wave through the barrier region. These two quantities are calculated from solving the Schrödinger equation for the barrier–well structure. In a similar manner, waves incident from the right of the barriers can be defined by

$$\psi_k(x) = \begin{cases} \dfrac{1}{\sqrt{2\pi}}\left[e^{-ikx} + r(k)e^{ikx}\right] & x > x_{\min} \\ \dfrac{1}{\sqrt{2\pi}}t(k)e^{-ikx} & x < x_{\max}. \end{cases} \tag{8.69}$$

Note that here, $k > 0$ as well. Since the system is in equilibrium, it possesses a left–right symmetry, so that the reflection and transmission coefficients are the same regardless of the origin of the wave. This is generally true of tunneling structures in any situation, but the left–right symmetry that is upset by the bias translates into different probabilities for the number of waves from each side. With these definitions, the density matrix may then be defined by

$$\rho(x, x') = \frac{1}{Z}\left\{\sum_n \psi_n(x)\psi_n^*(x')f(E_n) + \int dk\psi_k(x)\psi_k^*(x')f(E_k)\right\},$$

(8.70)

where Z is the partition function. The sum over n is over the bound states that exist in the system and have an energy level given by E_n. The integral is over the propagating waves that enter from each side of the device.

One could also use unnormalized basis functions and the translation matrices. These states can be normalized by the use of the scattering theory in which the wavefunction in the presence of the scatterer (our barrier-tunneling region) is compared to those of a reference space. These are related through the Lippmann–Schwinger equation, which provides a very useful consequence—the scattering states we use above precisely satisfy the orthonormality relations as unperturbed states [13]. The partition function may then be found from

$$\frac{1}{Z} = 2\sqrt{\pi}\beta e^{\beta V} \lim_{x\to-\infty} \rho(x, x).$$

(8.71)

The Wigner function is then obtained by the Wigner–Weyl transformation of the density matrix given as Eq. (8.2). The resulting Wigner distribution is characterized by a thermal distribution far from the barriers, although there are oscillations near the barriers similar to those that appear in Fig. 8.1. As mentioned, these oscillations are a result of quantum repulsion from the barrier. The quantum reflection causes a depletion of the carriers from the regions near the barriers, although this depletion is not a depletion layer in the classical sense.

This depletion arises from the repulsion and not from band bending. The charge then accumulates a short distance from the barrier, and overall charge neutrality in the device is maintained. In Fig. 8.2, we show the Wigner distribution that is computed for a typical RTD in equilibrium. The charge accumulation and depletion may be seen quite easily in the figure.

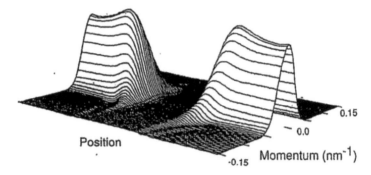

Figure 8.2 The Wigner distribution computed from a scattering state basis for a typical RTD. Charge accumulation and depletion may be observed in the region adjacent to the barriers in the center of the device. This is for a case of 5 nm barriers (AlGaAs) and well (GaAs). Reprinted with permission from Ref. [39], Copyright 1989, American Physical Society.

The question of discretization of the spatial and momentum variables is an important one, as these discretizations are coupled in the same manner in which the periodicity of a lattice defines the discretization of momentum within the reciprocal lattice. When the Wigner function is discretized with a spatial increment Δx, the momentum space becomes periodic with period $\pi\hbar/\Delta x$. Since one wants to treat both positive and negative values of momentum, it is then convenient o use

$$-\frac{\pi}{\Delta x} < k \leq \frac{\pi}{\Delta x}. \tag{8.72}$$

Once the spatial discretization is set, the discretization in momentum naturally follows. In treating the RTD, one is faced with the fact that the barriers and well are only a few lattice constants in extent. So in order to carefully treat the regions around the barriers with some care, it is often convenient to use a discretization of Δx ~ 0.25 nm [39]. The Wigner equation of motion (8.4) is the time derivative, which is usually evaluated from the discretized position and momentum via the second-order Lax–Wendorf method [41]. A second-order finite difference scheme for a point x_i involves the points x_{i+1} and x_{i-1}. This is no problem within the interior of the device, or even in the incoming boundary edge where the contact distribution function is specified. But it does lead to a problem with the outgoing boundary where the contact distribution function is not known. For this reason, it is convenient to split the overall grid into two regions, one for positive (right-going) momentum and one for negative (left-going) momentum. Then, a first-order upwind differencing scheme [41] can be used to propagate the distribution to the outgoing boundary along the characteristic direction as

$$\frac{\partial f}{\partial x} = \begin{cases} \dfrac{f(x_i) - f(x_{i-1})}{\Delta x} & \text{if } k(x_i) > 0 \\ \dfrac{f(x_{i+1}) - f(x_i)}{\Delta x} & \text{if } k(x_i) < 0. \end{cases} \tag{8.73}$$

The time step itself is limited by the need to satisfy the Courant–Friedrichs–Lewy (CFL) condition, which requires that

$$\frac{\Delta x}{\Delta t} \le v, \tag{8.74}$$

where v is the fastest characteristic velocity in the system [42]. In addition, it is convenient as well to assure that the time step suggested by the CFL condition is also smaller than the scattering time.

With a device like the RTD, we also have to deal with the self-consistent potentials in the system. This is handled by the additional solving of Poisson's equation. The self-consistent Wigner distribution is first calculated for the equilibrium state with no bias applied but with the presence of the ionized donors. Knowledge of the Fermi energy from the temperature and doping level, as well as the equilibrium temperature itself, is sufficient to determine this equilibrium state. Nevertheless, since the electron density is a function of the local potential, and the ionized-donor charge and the potential are interdependent, an iterative procedure is required to find the correct initial distribution, and this couples the Poisson's equation to the technique for developing the Wigner function and density matrix with Eqs. (8.66) and (8.67). The presence of bias can be incorporated by slowly varying the applied bias and allowing the time-dependent solution to evolve from the zero bias equilibrium state. If the general solution technique is stable, ramping the bias back down to zero should return the system to the equilibrium state, and this is one method of measuring stability in the simulation. In the simulation discussed here [39], scattering is incorporated via the relaxation-time approximation (8.46). The scattering time is taken to be the momentum relaxation time given by the mobility in GaAs at the appropriate doping level. Using 5 nm barriers and wells, with the barriers 0.3 eV high, and a doping level of 10^{18} cm^{-3}, which is heavily degenerate, good simulation results can be obtained. The system shows transients with a characteristic relaxation time of less than 700 fs [43]. Once steady state is reached at each bias level, the current through the device can be determined. This produces a typical RTD curve with a peak current density of ~0.18 mA/cm^2 at an applied voltage of ~0.28 V, and a peak-to-valley ratio of 2:1, all at a temperature of 300 K.

Interestingly enough, the I–V curve shows hysteresis in the negative differential conductance regime, which is believed to arise from the self-consistent potential being different in the up-sweep and the down-sweep. In the up-sweep, there is essentially no charge in the quantum well as the voltage is increasing. However, in the down-sweep, the well is fully charged and this difference in charge leads to a difference in the self-consistent potential, and of course in the current. It was also found that performance was improved if a lightly doped (10^{16} cm^{-3}) region of some 5 nm extent was placed adjacent to the barriers outside the nominal RTD barrier–well structure. This concentrated the potential to the active region, with less drop in the heavily doped parts of the structure.

Others do not see the hysteresis under d.c. conditions, but do see it under transient switching of the device [44, 45]. These authors observe that the size of the hysteresis tends to increase with the frequency of the device. They conclude that for low enough frequency, the charging and discharging of the quantum well can occur fast enough to avoid seeing any hysteresis. This would suggest that the sweeps discussed in the device above were in fact too fast to allow complete charge relaxation, and so hysteresis was observed. These authors have also coupled the RTD to a resonant circuit so as to allow self-excited oscillations [45]. The RTD appears to be essentially the same physical structure discussed above; e.g., 5 nm barriers and well. They have observed that the oscillatory behavior and device current waveforms remain quite similar up to about 120 GHz.

In Section 2.7, we had discussed the use of Bohmian trajectories to study quantum transport. There, we also saw the results for simulation of an RTD structure. In this case, the barriers were 3 nm thick and the quantum well was 5.1 nm thick, and the device was studied at 77 K [46]. The GaAs cladding regions were doped to 5×10^{18} cm^{-3}, which leads to a carrier concentration of 1.5×10^{17} cm^{-3} because of the degeneracy. In Fig. 8.3 (which is also Fig. 2.8b), we show the phase–space distribution of the carriers for an applied bias of 0.39 V, which is near the peak of the current. At this low temperature, they find in the absence of scattering that there is a very large peak-to-valley ratio. The electron concentration profile shows oscillations, as discussed previously, and an extensive charging in the quantum well.

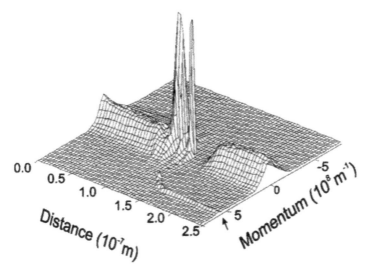

Figure 8.3 Phase–space distribution for an RTD with full density under a bias of 0.39 V. Reprinted with permission from Ref. [46], Copyright 1998, AIP Publishing LLC.

8.6 Monte Carlo Approach

There are many approaches to transport and quantum simulation, which describe themselves as Monte Carlo approaches. Here we are interested really in kinetic Monte Carlo, either as simulation with particles or as Monte Carlo approaches to the kinetic transport equation for the Wigner function. In Section 7.8, we discussed the expansion of the kinetic equation for the density matrix, after converting this kinetic equation into a path variable equation. Here the various orders of scattering, which contributed to the overall integral equation, were sampled by a Monte Carlo process. This same approach has been developed for the Wigner equation of motion [47–49]. That is, the Wigner equation of motion is converted into an integral equation. Then the Neumann series representing the integrand that is developed over many scattering processes (again for a backward propagation form of solution) is solved by a Monte Carlo sampling process, in the same manner as discussed in this previous section. Another approach is to consider the RTD as being split into three regions, the quantum barriers and well and the two cladding regions outside [50]. The central region, containing the barriers and wells, is also extended into the cladding regions. Only the central region is treated as a quantum region, and this is solved via the Wigner equation of motion. The two cladding regions are treated by classical kinetic Monte Carlo procedures, and the two types of distribution functions are matched at the interfaces between the various regions. Neither of these approaches are true particle simulations of quantum transport phenomena.

8.6.1 Wigner Paths

A number of early steps toward a full particle simulation of the Wigner equation in a complex device such as an RTD have appeared. In one approach, the simulation uses particles to handle the transport, but the quantum effects of tunneling are introduced by solving for the energy-dependent tunneling probabilities and then using these to decide whether or not a particle passes the barrier [51]. In this regard, the tunneling is treated as a scattering process, as has been suggested by Barker [52]. Another approach used the particles to simulate the Wigner equation of motion and study a process by which particles could be created (multi-particle production) [53]. In this approach, they weighted the individual particles, by a process that was introduced earlier in classical Monte Carlo to study rare events [54]. While these approaches bring particles into the simulation, they do not fully solve the quantum coherence problem in quantum transport.

How do we know that it is valid to impose classical trajectories onto the Wigner equation of motion? It turns out that this has been studied under the guise of Wigner *paths* [55–58]. The concept of a Wigner path is useful in that it provides a pictorial representation of the quantum evolution and is quite useful for numerical simulation of the Wigner equation of motion. Such a path is the path followed by a small sample of the Wigner function as it evolves through phase–space. Moreover, this path evolves quite like a classical group of particles, except for scattering and for rapidly varying potentials. That is, like classical particles, this small sample evolves with a conservation of its phase–space volume, except in these occasions. But the existence of these paths and their conservation properties means that we can really use a kinetic Monte Carlo approach to simulate the evolution of the Wigner function as described by the kinetic equation [58]. As before, the usual integro-differential equation that includes the full scattering integral is converted to a path integral formulation, which gives an integral equation in place of the standard integro-differential equation. In this integral equation, we no longer have "in" and "out" scattering, but only the incoming contributions from a variety of sources. Because this is a quantum system evolution, the integrand assumes complex values as a result of the quantum interference and the fact that the system is subjected to the uncertainty principle. As we recall from our earlier discussions, the negative parts of the Wigner function are direct indications of the regions where uncertainty is important. An extended version of the weighted Monte Carlo [54] is now used to generate a set of Wigner paths. The sum over these paths is a numerical estimate of the Wigner function itself. In order to generate a Wigner path, there is an entire sequence of random selections that have to be made, each subject to an appropriate probability distribution. Some of these are as follow [58]:

- The number of interactions to be considered for the particular path. For electron–phonon interactions, or potential interactions (such as from impurities), the number of vertices must be even, as described in Chapter 4. The scattering rate depends on the magnitude squared matrix element, and each vertex involves the matrix element.

- The times at which each of the interaction vertices occurs. These must lie between the initial time and the observation time at which the Wigner function is being considered.
- The type of the interaction at each vertex during the simulated path.
- The momentum change at each vertex. For the electron–phonon interaction, one-half of the phonon momentum modifies the wavevector at each vertex. This is described further below.
- The initial value of the path, along with the "free flights" between the vertices. This can be done for either forward or backward (in time) evaluations of the path. In addition, it is critical to consider carefully the phase of the wavefunction that is the basis of the path. Again it is the phase interference that gives rise to the negative parts of the Wigner function.

As an example, let us consider the second-order vertices that may contribute. In Fig. 8.4, the paths that can arise are shown. There are only five of these diagrams because it is the diagonal terms of the density matrix that properly correlate with the density. These five paths are (a) two interactions with a local potential, such as an impurity; (b) a path with phonon emission at both vertices; (c) a path with phonon emission at the first vertex and phonon absorption at the second vertex; (d) a path with phonon absorption at both vertices; and (e) a path with phonon absorption at the first vertex and phonon emission at the second vertex. Paths (c) and (e) are virtual processes, in which no phonon is really emitted or absorbed. In the classical world, where the collision occurs instantaneously, the two vertices would coalesce and the process would be non-existent. But in the quantum world, it is possible for the scattering to begin at one vertex, and then decide not to occur with the second vertex [59]. These virtual events still contribute important phase shifts and lead to the result that can give a negative value for the Wigner function. Of course, for a given path and a longer time of propagation, a great many scattering events can occur during the path. Importantly, it is possible to also consider multiple scattering processes, such as some of the diagrams in Fig. 4.3. That is, there may be a path for the emission of phonon A and then one for the absorption of phonon B followed by these same events in the same sequence, which would be the emission of two phonons of different energy and momentum in a high-order process.

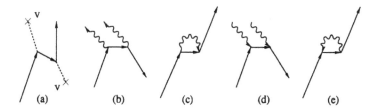

Figure 8.4 Diagrams representing the various types of second-order terms to the path integral expansion for the Wigner function evolution. These terms represent (a) potential scattering, (b) phonon emission (which accounts for one completed phonon scattering), (c) a virtual phonon emission, (d) phonon absorption, and (e) a virtual phonon absorption. Reprinted from Ref. [58], Copyright 2001, with permission from Elsevier.

Earlier, it was remarked that one-half of the phonon momentum was passed to the path at each vertex. For the virtual process, this one-half of the momentum is then withdrawn at the second vertex, so no net momentum is passed to the path. But if the scattering process is completed, the net momentum change is $\hbar q$, just as in the classical case. The passing of half the momentum at each vertex can be seen in the Levinson formulas of Section 8.3. So this is a correct interpretation of the manner in which the scattering events are inserted into the paths.

8.6.2 Modified Particle Approach

Now let us turn to another version of the weighted Monte Carlo approach to simulations of the Wigner function transport. In principle, this approach is close to the weighted Monte Carlo approach discussed above [54]. It is also similar to the use of the Wigner paths as well [58]. But there are some differences here, especially as we will consider paths with negative weights [60]. Because some of our Monte Carlo particles carry these negative weights, we need to consider carefully how to keep the density at its nominal value, as there may well be more simulation particles than real particles. This is achieved through the concept of a *maximal envelope*. Simply speaking, the maximal envelope at a point in space is the momentum distribution of the magnitude of the Wigner function at that point in space. In detail, this envelope is the maximum value of the total simulation particles before any weights are assigned. Usually, this exists in the contact regions, where the number of electrons is larger than that in the active region of the device. While this may not really be needed, it is a method to assure that there are adequate particles in each of the subsequent critical regions of the device. For example, when simulating a Gaussian Wigner wave packet impinging upon a barrier, we must assure ourselves that we have enough particles present to account for the large negative distributions that can occur at the barrier. Thus, a significant fraction of the associated particles carries negative weight as a means of giving these negative values of the Wigner distribution. These negative weights are a different approach to representing the phase interference that leads to these negative excursions of the Wigner function. Now the Wigner function at a point in phase–space can be described by a sum over the particles as

$$f_W(x,k) = \sum_{i=1}^{N} A(i)\delta(x - x_i)\delta(k - k_i),\tag{8.75}$$

where N is the number of simulation particles and $A(i)$ is the weight, or *affinity*, attached to each of these particles. While we have discussed negative values of the affinity, we note that this quantity can also be 0 for some particles. This allows us to use absorbing boundary conditions on the exit boundaries from the device in a very easy manner. An important aspect of the affinity is its change during the simulation. We calculate the non-local Wigner potential, and this determines the change in the affinity that each particle possesses. That is, the wave nature of the particles is maintained through the variation of each particle's affinity.

A well-known problem is the propagation of a single Gaussian wave packet through a single tunneling barrier. This was simulated with Bohmian trajectories as a solution to the Schrödinger equation by Dewdney and Hiley [61], and since has become a test case for other simulation approaches. In the present case, we will compare solutions found with the Monte Carlo procedure to that obtained by direct integration of the Wigner function itself. Here we will assume the GaAs/AlGaAs system to produce a barrier, which is 2 nm thick and 0.3 eV in amplitude. The initial Gaussian wave packet has a mean spread of 4 nm and a mean momentum wavevector k of 4×10^8/m. In Fig. 8.5, we plot the carrier density distributions at various instants of time for the direct solution to the Wigner distribution. Each panel has a width of approximately 45 nm, and the barrier is located in the center of the panel (just to the right of the Gaussian in the upper left panel). In Fig. 8.6, we plot the solution obtained by the weighted Monte Carlo technique for the same instants of time as for the previous figure. Comparison of these two figures reveals that this approach is a good representation of the direct integration of the Wigner distribution function. One knows that it is the leading parts of the Gaussian that tunnel most readily, as these have a higher velocity (and energy) and, therefore, a higher tunneling coefficient. This may be ascertained in both the figures. Of importance here is that the weighted Monte Carlo simulation has not changed the quantum nature of the system.

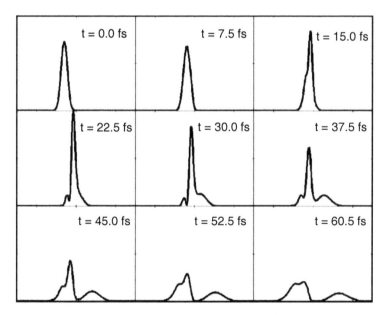

Figure 8.5 Direct integration of the Wigner equation of motion for a Gaussian, in both position and momentum, tunneling through a barrier. The density at different times is plotted in the individual panels. Reprinted from Ref. [60], Copyright 2001, with permission from Elsevier.

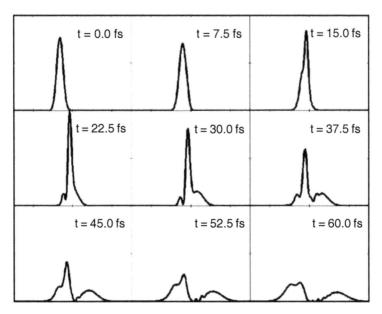

Figure 8.6 Simulation with a weighted Monte Carlo of a Gaussian Wigner distribution tunneling through a single barrier. The densities at different instances of time are plotted in the different panels. Reprinted from Ref. [60], Copyright 2001, with permission from Elsevier.

A natural question to ask, when looking at Figs. 8.5 and 8.6, is where is the negative part of the Wigner distribution that has been so prominent is the above discussions. It fact, it is there, but these figures plot the magnitude squared value of the Wigner function, as they plot the actual density rather than the Wigner function itself. In Fig. 8.7, we plot the Wigner function itself as it interacts with the tunneling barrier [62]. In panel (a), the Gaussian Wigner function has just begun to interact with the tunnel barrier and corresponds approximately to the t = 22.5 fs result in Fig. 8.6. One can make out the reflecting wave packet just beginning to build at the lower right of the Wigner function. While not quite evident in this figure, the reflecting parts seem to hit the tunnel barrier and slide along it, in momentum space, to the proper momentum for the backward wave and then build into a Gaussian packet (although distorted significantly, as can be seen in the previous figures). In panel (b) of Fig. 8.7, the total wave packet is plotted. This total includes the transmitted packet as well as the reflected packet, and an additional term centered on the barrier. This latter piece of the packet is rapidly oscillating in space with a period smaller than that corresponding to uncertainly. This piece of the total Wigner function represents the correlation between the transmitted and reflected wave packets. That is, these two packets originated from a single wave packet incident on the tunnel barrier. If we were to erase this correlation and then reverse the direction of propagation (time reversal), each of the transmitted and reflected packets would create its own transmitted and reflected packets. On the other hand, if just take the result in Fig. 8.7b and reverse the direction of propagation, these features all coalesce back into the original Gaussian wave packet. In that sense, the correlation is the memory of the system that it was created from a single Gaussian packet.

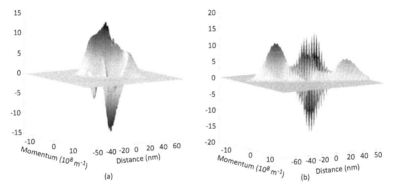

(a) (b)

Figure 8.7 (a) The Wigner function for a Gaussian packet incident on a tunnel barrier, just as it begins to build the reflected packet. (b) The Wigner function after the transmitted and reflected packets have separated from the barrier. The rapidly oscillating part that remains centered at the barrier is the correlation between the two outward going packets. Reprinted from Ref. [62], with permission from author.

Next we show simulation results for an RTD in the GaAs/AlGaAs system [63]. The structure for this has a 5 nm thick undoped GaAs quantum well, with AlGaAs barriers of 3 nm thickness and 0.3 eV height. A 5 nm undoped spacer layer on each outer side of the barriers was used to focus the applied bias effectively onto the quantum region. The doping in the remainder of the system was set to 10^{18} cm^{-3}. The total device length was 75 nm, and the system is simulated at a temperature of 300 K. No scattering was included so as to illustrate the basic quantum properties of the device. Particles are initially distributed uniformly throughout the system and the simulation is run with no bias for several picosecond until it reaches equilibrium. Then the bias is incrementally applied in steps of 10 mV until the maximum bias of 0.5 V is reached. During each increment, the device is run until a steady-state current is reached. The resulting I–V curve is shown in Fig. 8.8a, and the expected negative differential conductance is clearly evident. The peak-to-valley ratio is just over 2 for this device. In Fig. 8.8b, the Wigner distribution is plotted for a bias near the peak of the I–V curve, just before the resonant tunneling is cut off. The large buildup of negative correlation on the input side (left side) of the barrier–well structure is clearly evident in the picture, while the charge in the well may also be seen. At the highest applied bias (0.5 V), there is no charge remaining in the well, nor is there significant negative correlation in the resulting Wigner distribution function.

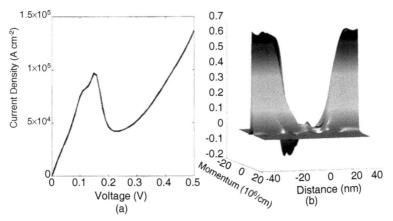

Figure 8.8 (a) The I–V characteristic for a typical RTD obtained by the simulation via a weighted Monte Carlo technique. (b) The Wigner distribution function for a bias near the peak of the I–V curve. Reprinted from Ref. [63], Copyright 2002, with permission from Elsevier.

Following the above simulation, the role of scattering was investigated by treating the detailed emission and absorption of polar optical phonons in the GaAs and GaAlAs regions [64]. This scattering was easily introduced since the transit of the various particles was handled by a standard ensemble Monte Carlo procedure. Boundary conditions remain critical, and here it was ensured that the current and electron number are conserved in the simulation. This means that the current must be continuous across both boundaries, and the net space–charge neutrality must be maintained. Self-consistency is maintained by solving the Poisson's equation with a direct matrix inversion approach. Much of this is accomplished by randomly, but carefully, distributing the particles according to the maximal envelope and assigning the particle affinities based on a thermal distribution in the contacts so that charge neutrality at the contacts is easily met in equilibrium. Once again, the applied bias was swept from 0 to 0.5 V in 25 mV steps. Each bias point is allowed to reach self-consistency before the next bias step. The bias is then reduced stepwise back to zero to check for any hysteresis, especially in the negative differential conductance regime. No significant hysteresis was found, which is consistent with the results found earlier [44, 45]. The resulting I–V characteristics are shown in Fig. 8.9a. Now, the peak-to-valley ratio has been significantly reduced by the presence of the scattering processes. This is approximately 1.5, which is closer to the values found experimentally in similar device structures. Moreover, the positions of the peak current and the valley current (in voltage) are where they are expected to be by a simple study of the structure. The cutoff of tunneling as the voltage is increased beyond that of the peak current leads to an accumulation of charge before the barrier. This is evident in Fig. 8.9b, where we plot the Wigner function at the voltage corresponding to the peak current. There is also a large negative correlation due to an increase in tunneling through the quantum structure at the peak current. Another interesting picture is shown in Fig. 8.10. Here we plot the energy distribution of the carriers as a function of position in the device. Red corresponds to a high density and blue to a low density. Clearly most of the carrier remains near the conduction band edge, although the distribution in the GaAs regions is degenerate. Notice the set of carriers tunneling through the barrier region at nonzero energy.

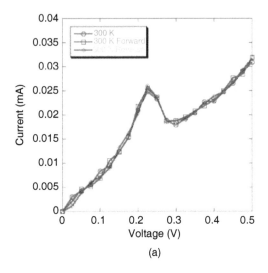

(a)

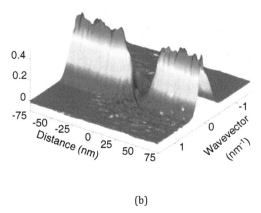

(b)

Figure 8.9 (a) The I–V curves for the RTD with the presence of scattering due to the polar optical phonons. Three curves are overlapped, including both an up-sweep and a down-sweep of the bias potential. (b) The Wigner distribution function at the peak current of the device. Reprinted from Ref. [62], with permission from author.

In this approach, there is no apparent need for the virtual collisions, as the interference is arrived at through the particle affinities, rather than dependence on the phase of a Wigner path. However, the intra-collisional field effect must be put in by hand in order to affect the simulation. The intra-collisional field effect is the result of the fact that the time required for an electron to emit, or absorb, a phonon is nonzero. In the Wigner paths discussed above, this was seen as the fact that two vertices were required to complete the emission/absorption process, and these two vertices were separated by a propagation path. It is the time between the two vertices that is important, as the "particle" can be accelerated by the field for this time, thus the field effect between the vertices, or intra-collisional field effect. This is an important quantum effect. For example, it has clearly been seen that this causes displacements of the particle both in real space and momentum space [65, 66]. We can begin to see this effect by considering a Gaussian Wigner function centered at a relatively high energy [65], as shown in Fig. 8.11. In the semi-classical simulation, and that at zero applied bias, there are a series of phonon peaks below the initial Wigner distribution. These peaks represent the emission of 1, 2, 3, and 4 optical phonons in GaAs. A similar behavior is observed in the femtosecond laser spectroscopy of GaAs and semi-classical Monte Carlo simulations of this process [67]. In this latter case, the phonon replicas are eventually washed out due to electron–electron and electron–hole interactions. These processes were not present in the Wigner function study. Instead, as can be seen in the 12 kV/cm curves, the general well-formed peaks are blurred together by the momentum space shifts caused by the intra-collisional field effect. Of great significance is the fact that the first phonon peak is shifted in position relative to the field-free case, and this is a direct result of the increase in the emission energy as the field is in the direction of the phonon emission process. As pointed out, there is also a spatial displacement of the electrons as well. This occurs since the carrier is accelerated, or decelerated, by the electric field, and its spatial position will be different from that expected for the classical particle. We can illustrate this for the RTD simulation discussed above. We will turn to the implementation below, but first we look at the results for this displacement due to the intra-collisional field effect. This is shown for the RTD in Fig. 8.12. Here the displacement is relative to that expected for a classical particle, and the distribution for the displacement is plotted along the axis of the RTD (zero corresponds to the barrier region). The enhancement at the barrier is a result of the buildup of charge at this point as well as an increase in the displacement.

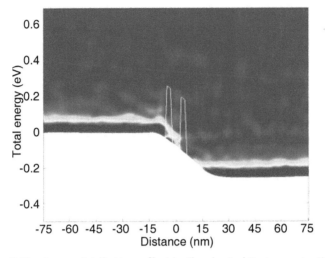

Figure 8.10 Energy distribution, offset by the classical Hartree potential, for the RTD with a bias of 0.225 V, which corresponds to the peak of the I–V curve in Fig. 8.9a. Reprinted from Ref. [62], with permission from author.

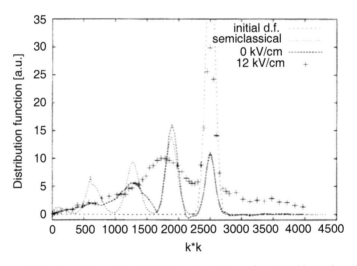

Figure 8.11 Simulation of an initial Gaussian Wigner function (dashed curve) after 300 fs. The solid curve is the simulation with a semi-classical distribution, while the dotted curve is the Wigner function with zero applied electric field. The crosses represent the Wigner distribution when a field of 12 kV/cm is applied. The broadening and shift is a result of the intra-collisional field effect. Reprinted from Ref. [65], Copyright 2002, with permission from Elsevier.

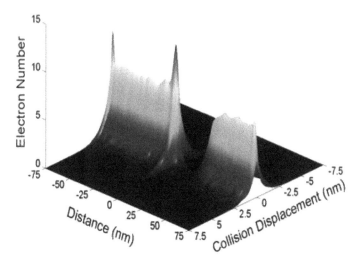

Figure 8.12 A histogram of the collision-induced displacement (from the classically expected position) is plotted as a function of the position in an RTD. Reprinted from Ref. [62], with permission from author.

Many methods have been suggested to include the collisional displacement in normal semi-classical simulations with the Monte Carlo method [55, 56, 68]. Some of these have been discussed above. In the case of Ref. [68], it was suggested to replace the energy-conserving delta function with the joint spectral density, represented by a Lorentzian line centered on the classical energy conservation point. This was not very successful due to the very long tails of this function, which led to arbitrarily high energies. There are also approaches using Green's functions, which will be discussed in the next two chapters. To include the role of the finite collision duration, we need to account for two effects: (1) the collision duration itself and the resulting energy shift, and (2) the position change during the collision. These are the momentum space shift and the real space shift, respectively.

The collision duration itself is typically only a few femtoseconds [69]. In this time, the change in position that occurs as a result of the collision duration may be written as

$$\delta x = \left(v(t) + \frac{eE}{m^*} \frac{\tau_c}{2} \right) \tau_c,$$

(8.76)

where E is the electric field that is experienced during the collision, and τ_c is the collision duration. During the interaction, there is an overlap between the wave packet at x and that at $x + \delta x$. We can assume that these wavefunctions are Wigner packets that are Gaussian in shape, which leads to an overlap that varies as

$$I(\delta x) = \frac{1}{\sqrt{4\pi\sigma^2}} e^{-\delta x^2/4\sigma^2} \, , \tag{8.77}$$

where

$$\sigma = \sqrt{\frac{\hbar^2}{12m^* k_B T}} \sim 1.9\,\text{nm} \tag{8.78}$$

is the spread expected for a particle with the thermal de Broglie wavelength. The numerical factors come from the same discussion as that of the effective potential in the last chapter. The probability for a given shift in position is then

$$P(\delta x) = \left| I(\delta x) \right|^2 = \frac{1}{4\pi\sigma^2} e^{-\delta x^2/2\sigma^2} \, . \tag{8.79}$$

Thus, as the expected displacement becomes large, the likelihood of finding the appropriate state becomes small.

The nonzero collision duration and the resulting position displacement can lead to some interesting effects beyond those normally expected. Consider, for example, the structure in Fig. 8.13. In panel (a), we have a particle approaching a barrier of modest height, such as may be found in the RTD. If there is no collision, then it is likely that the particle will be reflected backward, although there is a small probability that it will penetrate the barrier. In the absence of any positional displacement, this particle could also absorb or emit a phonon without affecting the fact that it will likely be reflected. In panel (b), however, we consider the effect of a particle undergoing the intra-collisional field effect near the barrier. Now if the field is sufficiently high, the particle may gain enough energy to surpass the barrier and wind up in a state that goes directly over the barrier. This may be accompanied by a positional displacement, which is indicated in the figure. During this collision, the particle may emit or absorb a phonon. If it emits a phonon, then it is obvious that it will wind up with a lower kinetic energy in its final state. But it may also absorb a phonon and still wind up with a lower kinetic energy in its final state. The difference in these two processes is merely a difference in the energy that must be absorbed from the electric field during the collision. Thus, our energy conservation requirement becomes

$$E(\mathbf{k} \pm \mathbf{q}) - E(\mathbf{k}) \mp \hbar\omega_0 = V(x) - V(x + \delta x) \qquad (8.80)$$

where $E(k)$ is the particle energy, $\hbar\omega_0$ is the phonon energy, and $V(x)$ is the local potential, and in this case is related to the barriers of the RTD.

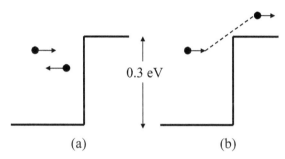

0.3 eV

(a)   (b)

Figure 8.13 (a) Electrons classically impinging upon a barrier and being reflected. (b) Electrons hopping over the barrier due to energy gained during the intra-collisional field effect and the emission or absorption of an optical phonon. Reprinted from Ref. [62], with permission from author.

The affinity approach has been extended in more recent studies. One perceived problem is the computation of average quantities over the various cells of the real and momentum space grids. If there are no particles in the cells, then convergence and noise can become significant problems. Hence, it has been suggested that one inject particles with zero affinity into cells that are normally empty to improve the simulation approach [70–72]. Whereas the earlier work primarily injected particles from the contacts, this new approach smoothens the simulation with injection of these zero-weight particles throughout the computational domain. These authors have then used this approach to simulate more meaningful devices, such as the double-gate MOSFET [70] and nanowire MOSFETs [72], as well as RTDs. Several reviews of this work have appeared [73, 74], and they have also studied carefully the phonon-induced decoherence of the memory effects in the Wigner function [75]. The affinity method has also been adopted in Japan [76].

8.6.3   More Recent Approaches

In the above, we first talked about the Wigner path approach, in which generally a back propagation of the paths in time was followed. In this approach, the interaction with the non-local potential could be treated as a scattering process instead of incorporating the full non-local potential. The quantum information in this approach was carried by the particle weight. We then turned to the ensemble Monte Carlo approach, in which particles were used, although these tended to follow something like the Wigner path, in a forward propagation scheme. In this latter case, the quantum information was carried by a particle affinity. Both the weight and the affinity are really artificial numerical quantities whose purpose is to simulate the quantum phase interference that occurs in the Wigner function during real propagation with a semiconductor device. A melding of these two concepts was pursued in an alternative approach [77]. In this approach, the weights were accompanied with signs associated with them. This led to high variance in the results, so that a weight decomposition approach, which limits the value of a weight by storing part of it on a phase–space grid, was necessary. This significantly improved the variance during simulations with this method.

The variance problem mentioned in the last paragraph finally led to the adoption of a pure sign convention [78, 79]. This creates a model in which the Wigner function is considered to include a Boltzmann-like scattering term, but which now includes a generation term. The quantum information is carried by the sign of the quasi-particles. This approach is most useful when treating the interaction with the non-local potential as a scattering event, as in the Wigner path method. When a scattering occurs even from the potential, two new particles are created, one with the momentum increased by q and one where it is reduced by q, with q determined randomly from the probability distribution of the potential's spatial Fourier transform. The sign on one of these new particles is taken to be the same as the incident particle, while the sign on the other is the opposite. These signs are taken into account in each averaging process that is used to find average values. Equivalent particles with opposite signs annihilate one another when they meet in phase–space.

The signed particle Monte Carlo approach has been used to study a number of small semiconductor devices, including the role of interface roughness [80]. In later work, they also adopted a potential decomposition in order to improve the method [81]. Here, the actual non-local potential was decomposed into its classical part and a remaining quantum mechanical non-local part. The classical part gives just the classical electric field that enters the Boltzmann equation, and this can lead to a normal ensemble Monte Carlo approach. The quantum part of the potential is then the difference between the actual non-local potential and the classical potential and is treated by the scattering method discussed above. Scattering by the phonons and the quantum potential both involve the generation of particles and the use of the particle sign to impart quantum interference. This approach has been extended to the many-body problem in which one deals with interacting electrons [82]. For example, consider two identical electrons trapped in a one-dimensional box. These two electrons interact with one another through the Coulomb potential. The system starts at $t = 0$ with the Wigner function being generated by a proper anti-symmetric wavefunction. If we take the positions of the two particles as x_1 and x_2, then we can write the Wigner function as

$$f_W(x_1, x_2; p_1, p_2) = \frac{1}{(\pi\hbar)^2} \int\int dx_1' dx_2' e^{i(x_1' p_1 + x' p_2)}$$
$$\times \psi_0^*\left(x_1 + \frac{x_1'}{2}, x_2 + \frac{x_2'}{2}\right) \psi_0\left(x_1 - \frac{x_1'}{2}, x_2 - \frac{x_2'}{2}\right),$$

$$(8.81)$$

where

$$\psi_0(x_1, x_2) = \begin{vmatrix} \varphi_1(x_1) & \varphi_2(x_1) \\ \varphi_1(x_2) & \varphi_2(x_2) \end{vmatrix} \tag{8.82}$$

is a proper Slater determinant for the two electron system. The individual wavefunctions are given by

$$\varphi_1(x) = N_1 e^{-(x-x_{10})^2/2\sigma^2} e^{ip_{10}x/\hbar}, \quad \varphi_2(x) = N_2 e^{-(x-x_{20})^2/2\sigma^2} e^{ip_{20}x/\hbar}. \tag{8.83}$$

Here, N_1 and N_2 are normalization factors, x_{10} and x_{20} are the initial positions of the two particles, and p_{10} and p_{20} are their initial momenta. In the particular simulation, the two electrons are initially 0.1 nm apart, and they are taken to have equal and opposite momenta of the value $\pi\hbar/\lambda$ with $\lambda \sim 0.36$ nm, corresponding to a velocity of 1 nm/fs.

As the particles are allowed to evolve, the effective probability density of the two particles at a particular value of time, t, is formed by integrating the evolving Eq. (8.79) over x_2 and p_2 to get the probability density for particle 1, and integrating over x_1 and p_1 to get the probability density for particle 2. Of course, this will not give simple Gaussians, as the correlation between the two particles will give dramatic effects. For example, in Fig. 8.14, the sum of these two probability densities, in phase–space, is shown at 2.5 fs and at 3.5 fs [82]. At the earlier time, a clear Fermi exchange hole has formed between the two particles. By the later time, the hole has broken up and the two particles have begun to separate into their own independent regions. This is a clear indication of how the exclusion principle enters into the simulations.

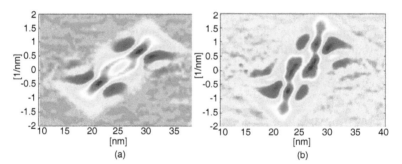

(a) (b)

Figure 8.14 (a) Plot of the Wigner distribution, and the two-particle distributions at 2.5 fs. (b) Plot of the two-particle distribution at 3.5 fs. A Fermi exchange hole is clearly seen in the plot of (a). Reprinted from Ref. [82], Copyright 2015, with permission from Elsevier.

There has also been a simulation of an Si nanowire transistor using the full Wigner function [83]. In this work, the transport along the channel is solved with a one-dimensional Wigner function approach, while the transverse quantization is handled by a two-dimensional Schrödinger solver. A relaxation-time approximation form for the scattering within the channel was assumed.

8.7   Entanglement

Entanglement, both as a concept and as an idea, seems to have occurred with the famous Einstein–Podolsky–Rosen paper [84]. In this seminal paper, the authors consider that there are two systems that are allowed to interact. This interaction continues for a period of time, after which they no longer interact. They then consider the use of Schrödinger's equation to compute the state of the *combined* system at any later time. This is a crucial point. In Schrödinger's words [85] "...between these two systems an entanglement can arise." Now, Schrödinger points out that once the interaction occurs (and the entanglement occurs), there are no longer two separated systems; rather, there is only a single combined system, which is described by a single wavefunction. It is no longer possible, even after the interaction is turned off, to separate this wavefunction into parts describing each of the two initial systems.

Now the reader may wonder why we are discussing this topic in this book and, more to the point, why in this chapter. In fact, even seeing the entanglement in a wavefunction is a difficult task and is usually only found with tensor product Hilbert spaces. This point suggests that it is difficult to actually see the existence of entanglement because one cannot define an operator whose expectation value is a measure of the entanglement. But we have seen that the Wigner function can clearly display properties, such as the negative excursions and the rapid oscillations, which are not measurable. In fact, we are quite often able to see identifiable features of the Wigner function, which can be connected to the entanglement. And that is why we are discussing this topic in this chapter.

8.7.1   Simple Particles

Let us now explore how the two particles described above would appear in the Wigner formulation. To describe these two particles, we must use a single wavefunction, as it is assumed that they have either interacted in the past, or were created together via another process. In any case, we assume that they are entangled. Let us assign each of the two particles a coherent wave packet, such as Eq. (8.7). We write the total wave packet as

$$\psi(x) = \frac{1}{\sqrt{2}}[\psi_1(x-x_0,k_0)+\psi_2(x+x_0,-k_0)]. \tag{8.84}$$

Thus, one particle is assumed to be at x_0 and moving in the positive x-direction, while the other particle is at $-x_0$ and moving in the negative x-direction. Of course, we really do not know which of the two particles is which, and this is preserved by the inversion symmetry of the combined wavefunction. The two particles do not have separate realities; only the single combined wavefunction description has a reality. We do know that the two particles exist and are moving apart with the same velocity, or momentum. It is well known that when one generates the Wigner function from such a composite wavefunction, extra terms arise [29]. If we apply the transformation (8.2) to (8.82), we arrive at the composite Wigner function

$$f_W(x,k) = \frac{1}{h}\left\{e^{-(x-x_0)^2/2\sigma^2 - 2\sigma^2(k-k_0)^2} + e^{-(x+x_0)^2/2\sigma^2 - 2\sigma^2(k+k_0)^2}\right.$$
$$\left. + 2e^{-x^2/2\sigma^2 - 2\sigma^2 k^2}\cos(x_0 k)\right\}. \tag{8.85}$$

The first two exponentials are clearly identified with the Wigner functions of the individual wave packets in Eq. (8.82). But in addition, the third term describes a Gaussian peak, centered on $x = 0$ and $k = 0$, which has an oscillatory behavior in momentum and the initial offset of the Gaussians (see Fig. 8.15). The fact that this extra term has both positive and negative values is clear evidence that it is not related to any probability function. Moreover, as the value x_0 increases, the oscillation becomes faster, but its amplitude is not affected. Hence, in this case, it becomes hard to argue that the Wigner function approaches a classical phase–space distribution. The motion of the particles can be recovered, or introduced, by letting $x_0 = vt$, so that they interacted (or were created) at the origin at $t = 0$. Just what does the extra term in Eq. (8.83) represent? Well, if we ignored this term, then we would have just two independent Gaussian coherent packets going on their merry way. But we are dealing with the entangled particles, and this extra term must correspond to the memory that these two packets are entangled. If we average over the momentum, this term vanishes and makes no contribution to the average momentum (or to the average position). Hence, it is not a measurable contribution to the probability density. We have to conclude that this extra term represents an embodiment of the entanglement between the two particle.

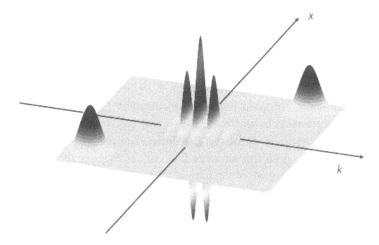

Figure 8.15 The Wigner transform of the composite wavefunction (8.82). The rapidly oscillating part centered on the origin is thought to be the representation of the entanglement between the two "particles."

We can examine the meaning of the extra term in another way. Consider a single Gaussian wave packet encountering a tunneling barrier, such as shown in Fig. 8.7. The single Gaussian decays into a pair of terms that represent the transmitted and reflected wave packets. This has a close similarity to the ideas of EPR and particularly Bohm's version of this experiment [86]. In spite of the breakup into two packets, the resultant Wigner function also displays an extra term, which remains centered at the barrier. If this term is erased, and the time is then reversed, the two packets do not reform into a single packet. Rather, each packet generates a new pair of transmitted and reflected packets. This yields four wave packets. But the proper resummation does occur if the extra term is retained. Hence, this extra term contains the memory that the two wave packets came from a single packet and must be correlated to one another. This fits the basic definition of entanglement—the information that the two particles were created together and must be correlated. Perhaps the philosophical interpretation of our ideas about tunneling should be reconsidered. In our normal course in quantum mechanics, we talk about the incoming wave packet producing two wave packets, one going forward and one being reflected. We never discuss the fact that these two packets are not isolated from one another. Rather, the decay of the original wave packet generates a correlated set of packets, which must be described as a single wavefunction. This is not normally done, and it is possible to get away with this treatment provided one does not ask about the reversibility that is inherent in the Schrödinger equation. There is no dissipation in the system, as it is closed to the outer world. But if we are to ensure that time reversibility is enforced, we have to account for this possibility. If we merely reverse the time, these two wave packets each will generate a new pair. That is, if the coherence of the wave packets is not ensured, then this process fails. In the Wigner function, the memory of the creation of the pair of packets is retained in the extra term at x and $k = 0$. The onset of decoherence in the system succeeds by destroying this term and thus breaking time reversal symmetry. This leads us to reiterate an important result for the use of Wigner functions. These phase–space representations contain information about the quantum nature that does not have a counterpart in the classical phase–space function. The extra term in Eq. (8.83) clearly corresponds to entanglement for those systems that should exhibit such an effect. Even lacking that designation, it corresponds to a quantum correlation, which does not exist in the equivalent classical dynamics.

An interesting numerical experiment has been carried out with Wigner wave packets [87]. This experiment is depicted in Fig. 8.16a. A waveguide form of an Aharonov–Bohm ring is created using potentials (in red) to define the structure, as shown in the figure. The overall simulation area is 60 nm wide and 65 nm tall. The waveguides (in blue) are 16 nm wide. A Gaussian wave packet in the Wigner function form (such as the first term in Eq. (8.83)) is sent into the structure from the bottom. This packet has a full-width spread of 4 nm ($\sigma = 2$ nm) and an initial momentum of $k_0 \sim 9.4 \times 10^7$ m^{-1}. A coherence length of 30 nm was assumed for the simulation, which means that the entanglement will be relatively heavily damped by the time the wave packet reaches the output of the A-B ring. We note that this dephasing is more the result of discretization in the system rather than an actual scattering process, as the effective coherence length is determined by the grid size in the momentum coordinate. The simulation is performed using the signed Monte Carlo particle simulation discussed in the last section [78, 79, 88]. In Fig. 8.16b, the two wave packets that have formed from the initial single packet are depicted, at a time of about 45 fs after the start. As these two packets were created from a single input wave packet and remain described by a single Wigner function, they are entangled, and this may be seen by the rapid oscillatory features that exist outside the ring (see the arrow in this panel). Some of the penetration into this area could arise from the wavefunction penetration into the barrier, but this would not be expected to show the transverse oscillation that appears to be here. Figure 8.16c depicts the two packets as they are making the turn at the ring corners, at a time of 65 fs. Here there are additional interesting features that arise as these points are turning points of the classical trajectories. One notes that fringes have formed, much in the manner discussed by Berry [89]. In a sense, this arises from the need for the transverse momentum to reverse at these turning points, much like the effect shown in Fig. 8.7a. We also note that the entanglement has still survived, although much weaker and broader due to the dephasing process. Finally, in Fig. 8.16d, we see that interference fringes are observed in the plane where the two wave packets reconnect (the plot is across the line indicated in white, 55 nm from the entrance, in panels (b) and (c) of the figure). The number of fringes is limited by the waveguide size, but it is clear that one is seeing the equivalent of the two-slit experiment in this output plane. No magnetic field has been included in the simulation, so it is not a proper A-B oscillation simulation, but it is very instructive for the existence of the entanglement as well as the two-slit-like results that arose when the two packets reconnected at the output plane.

8.7.2 Photons

Although we are primarily interested in semiconductors in this book, it is fruitful to discuss briefly the role of entanglement with photons, primarily because this is often viewed with a Wigner function. But it is also of interest in quantum information, and this is likely to implement with circuits in semiconductors in the not-too-distant future.

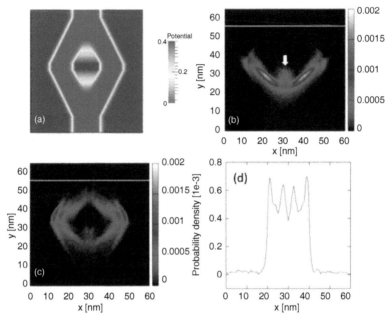

Figure 8.16 Propagation of a localized Wigner wave packet through an Aharonov–Bohm style ring. (a) The waveguides are defined by the potential. (b) At 45 fs, the input packet has clearly divided into two packets, which are entangled (white arrow). (c) At 65 fs, the packets have reached the turning points and developed fringes. (d) When the two packets reconnect (105 fs), they lead to interference fringes. The amplitude is plotted along the white line shown in (b) and (c). Reprinted from Ref. [87], with permission from author.

The coupling between a two-level atom and a resonant (quantized) electromagnetic cavity is described by a relatively old model, termed the Jaynes–Cumming model [90]. Generally, the field can interact with the atomic state and lead to Rabi oscillations of the atomic state population. However, something different happens in this particular model, in that the Rabi oscillations collapse and then reappear periodically. The temporal time over which this collapse and reappearance occur has been termed the revival time [91]. When the field in the cavity is in a coherent state, such as discussed earlier in this chapter, and the atom is prepared in an excited state, it is found that the atomic and field states become rapidly entangled, and then subsequently disentangle at one-half the revival time [92]. That is, the maximum entanglement occurs when the Rabi oscillations are strongest and the minimum entanglement occurs when the Rabi oscillations are minimal. More recently, it has been observed that the photon parity operator also shows similar behavior with one important difference. The peaks of the amplitude of the parity operator seem to occur when the Rabi oscillations are at their minimum. So the amplitude and decay of this parity operator seem to be shifted by a period from the Rabi oscillations [93]. In general, the atomic state Rabi oscillations occur with an average frequency $\Omega = 2\lambda\sqrt{n+1}$, where λ is an effective coupling constant between the atom and the cavity, and n is the occupation number of the single mode of the cavity. This occupation number is given as

$$n = a^\dagger a, \tag{8.86}$$

where $a^\dagger$ and a are the creation and annihilation operators, respectively, for the photons in the cavity mode. As the mode occupation builds up, the Rabi frequency increases and the amplitude of these oscillations increases. The photon parity operator is then defined as

$$\Pi_F = (-1)^n = e^{i\pi n}. \tag{8.87}$$

The normal expectation value of this operator may be found to be [93]

$$\left\langle \Pi_F(t) \right\rangle = e^{-n} \sum_{s=0}^{\infty} (-1)^s \frac{n^s}{s!} \cos\left(2\lambda t\sqrt{s+1}\right), \tag{8.88}$$

where the time dependence arises from the time dependence of the effective Rabi frequency. The interesting fact is that for $t = 0$, the cosine is unity, and the expectation value of the parity operator goes asymptotically to zero (the time is scaled with the same coupling factor λ). Hence, when the density peaks and the Rabi oscillations are strongest, the parity operator vanishes, and vice versa. In general, for the initial conditions discussed, the Wigner function of the field splits into a pair of counter-rotating components in phase space [94]. These two components reach a maximum separation at one-half the revival time and then recombine at the revival time. Hence, the peak of the parity operator corresponds to the maximum separation of the two parts of the Wigner function. The Wigner function itself can be written as a shifted expectation value of the parity operator, given as [95]

$$W(\alpha) = \frac{2}{\pi} \left\langle D(\alpha) \Pi_F D^\dagger(\alpha) \right\rangle, \tag{8.89}$$

where D is a displacement operator and a is a complex number that represents a point in phase–space; e.g., the real and imaginary parts correspond to position and momentum. The displacement operators work on the two wavefunctions in the density opertor, leading to the displaced versions that appear in Eq. (8.2). In Fig. 8.17, we plot the Wigner function of the parity operator at a time when this operator is near its maximum amplitude and at a positive peak (of the oscillating operator). The axes x and y correspond to the real and imaginary parts of α, and thus to position and momentum of the function. While the two major peaks are evident, there are clear interference fringes between them. These fringes themselves oscillate at the Rabi frequency when the parity amplitude is large.

Another experiment has shown the ability to actually measure a significant fraction of the Wigner function [96]. Here two photons are created by spontaneous parametric down conversion of a single high-energy photon. These photons are entangled as the pair of particles created in the EPR experiment discussed above. These two photons are guided into parallel paths and eventually into a beam splitter. Prior to arrival at the beam splitter, one photon undergoes a position translation by an amount 2δ, which amounts to the coordinate difference that appears in the wavefunctions of the Wigner transformation. At the same time, the second photon undergoes a phase shift μ, which corresponds to a momentum shift. After the beam splitter, the combined photons go to a pair of detectors. By varying the position and phase shifts, the authors are able to map out the Wigner function by monitoring the signals at the detectors. Again the negative going parts of the Wigner functions are indicators of entanglement in this photon pair. Other nonlinear optics experiments can be used to create non-classical states of the system. One example is the squeezed photon state, which produces non-classical behavior, and the Wigner function can be used to exhibit this non-classical behavior [97]. Others use photons to create what are called Schrödinger cat states, which again are examined with the use of the Wigner function [98].

8.7.3 Condensed Matter Systems

The current approach to quantum computing that seems to be the most fruitful is through the use of superconducting flux qubits. As a result, there is considerable interest in decoherence of these qubits, and how this is affected by the presence of the ensemble of electrons that also exist in the superconducting material. It is possible to cast this problem into one in which the coherence/entanglement can be followed by studying the evolution of the Wigner function of the reduced density matrix of the qubits [99]. Such studies have shown that, under some conditions, the presence of the electrons can even lengthen the coherence time.

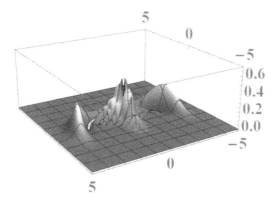

Figure 8.17 The Wigner function for the expectation of the parity operator at the scaled time for which this quantity is a maximum (near one-half the revival time [93]). The x- and y-axes correspond to the real and imaginary parts of α, as described in the text. This figure is courtesy of Richard Birrittella and Prof. Chris Gerry, and is used with their permission.

The Josephson junction is also interesting in its own right, as a nonlinear coupling or control device. For example, if two electromagnetic resonators are coupled in series with a Josephson junction, interesting quantum dynamics can be observed [100]. If the voltage that is applied to the Josephson junction leads to a Josephson frequency across the junction, which matches the sum of the frequencies of the two resonators, the Cooper pairs can flow through the entire circuit. This leads to the creation of photons in the two resonators that produce a non-classical, far from equilibrium state. Here the dynamics can be fruitfully explored by the study of a Wigner function describing the joint dynamics of the two coupled oscillators. The Wigner function for either of the two oscillators can then be found by integrating out the coordinates of the other oscillator. Needless to say, these individual oscillator Wigner functions appear to be Gaussians depending on the number operator for that oscillator, much like the Gaussians for the individual particles in Fig. 8.15. Study of the motion of the Wigner function leads to the dynamics of the coupled resonator–junction system and can even be used to study the quantum fluctuations in the system.

Problems

1. Suppose we have a Hamiltonian that includes a potential energy term of the form ax^4. How will this affect the equation of motion (8.5)?

2. Compare the different dynamics that result from Hamiltonians that are written as $2x^2p^2$, $x^2p^2 + p^2x^2$, or $2(xp)^2$. By dynamics, it is meant to find the time rate of change of the operators x and p in the normal quantum mechanical manner. What can you conclude about the proper symmetrization of operators.

3. For the three operator products in Problem 2, compute the commutator relations of each pair (three commutator products).

4. For a classical distribution function that has the (unnormalized) form

$$\exp\left(\frac{\hbar^2(\mathbf{k}-\mathbf{k}_d)^2}{2m^*k_BT}\right),$$

normalize the distribution. Then compute the first three moments $\mathbf{p}^0$, $\mathbf{p}$, and $\mathbf{p}^2$. Now consider the derivative terms that arise from the potential in Problem 1 and compute the three momentum averages of these derivative terms.

5. Let us consider a simple approach to the RTD in the ballistic transport limit. The structure for this has a 5 nm thick

undoped GaAs quantum well, with AlGaAs barriers of 3 nm thickness and 0.3 eV height. Using straightforward quantum mechanics, compute the transmission coefficient for one-dimensional tunneling through this structure. Then use this tunneling coefficient in the Landauer formula

$$J = \frac{e}{\pi\hbar} \int dE \, T(E + eV_a / 2)\left[f_L(E) - F_R(E + eV_a) \right],$$

where T is the tunneling coefficient calculated earlier, V_a is the applied bias, and the f are the Fermi–Dirac distributions on the left and right contacts, to compute the J–V_a curve for this device.

References

1. E. P. Wigner, *Phys. Rev.*, **40**, 749 (1932).

2. J. E. Moyal, *Proc. Cambridge Philos. Soc.*, **45**, 99 (1949).

3. I. B. Levinson, *Zh. Eksp. Teor. Fiz.*, **57**, 660 (1969); *Sov. Phys. JETP*, **30**, 362 (1970).

4. M. Nedjalkov, S. Selberherr, D. K. Ferry, D. Vasileska, P. Dollfus, D. Querlioz, I. Dimov, and P. Schwaha, *Ann. Phys.*, **328**, 220 (2013).

5. N. C. Dias and J. N. Prata, *Ann. Phys.*, **313**, 110 (2004).

6. U. Ravaioli, M.A. Osman, W. Pötz, N. Kluksdahl, and D. K. Ferry, *Physica*, **134B**, 36 (1985).

7. W. R. Frensley, *Phys. Rev. B*, **36**, 1570 (1987).

8. N. C. Kluksdahl, A. M. Kriman, D. K. Ferry, and C. Ringhofer, *Phys. Rev. B*, **39**, 7720 (1989).

9. T. B. Smith, *J. Phys. A*, **11**, 2179 (1978).

10. A. Janusis, A. Streklas, and K. Vlachos, *Physica*, **107A**, 587 (1981).

11. A. Royer, *Phys. Rev. A*, **43**, 44 (1991).

12. E. A. Remler, *Ann. Phys.*, **95**, 455 (1975).

13. A. M. Kriman, N. C. Kluksdahl, and D. K. Ferry, *Phys. Rev. B*, **36**, 5953 (1987).

14. H. Weyl, *The Theory of Groups and Quantum Mechanics* (Dover, New York, 1931).

15. A. Messiah, *Quantum Mechanics*, Vol. 1 (Wiley Interscience, New York, 1961), p. 442.

16. H. L. Grubin, D. K. Ferry, G. J. Iafrate, and J. R. Barker, in *VLSI Microelectronics: Microstructure Science*, Vol. 3, edited by N. Einspruch (Academic Press, New York, 1982), pp. 197–300.

17. G. J. Iafrate, H. L. Grubin, and D. K. Ferry, *Phys. Lett.*, **87A**, 145 (1982).

18. H.-W. Lee, *Phys. Rep.*, **259**, 147 (1995).

19. D. K. Ferry, *Quantum Mechanics*, 2nd Ed. (Institute of Physics, Bristol, UK, 2001).

20. A. L. Fetter, and J. D. Walecka, *Quantum Theory of Many-Particle Systems* (McGraw-Hill, New York, 1971).

21. H. Mehta, *J. Math. Phys.*, **5**, 677 (1964).

22. J. G. Kirkwood, *Phys. Rev.*, **44**, 31 (1933).

23. R. J. Glauber, *Phys. Rev.*, **131**, 2766 (1963).

24. E. C. G. Sundarshan, *Phys. Rev. Lett.*, **10**, 277 (1963).

25. M. C. Gutzwiller, *Chaos in Classical and Quantum Mechanics* (Springer-Verlag, New York, 1990).

26. K. Husimi, *Proc. Phys.-Math. Soc. Jpn.*, **22**, 264 (1940).

27. J. von Neumann, *Mathematical Foundations of Quantum Mechanics*, translated by R. T. Beyer (Princeton University Press, Princeton, 1955).

28. K. Takahashi, *J. Phys. Soc. Jpn.*, **55**, 762 (1986).

29. L. E. Ballentine, *Quantum Mechanics: A Modern Development.* (World Scientific, Singapore, 1998).

30. G. A. Luna-Acosta, K. Na, L. E. Reichl, and A. Krokhin, *Phys. Rev. B*, **53**, 3271 (1996).

31. A. Bäcker, S. Fürstberger, and R. Schubert, *Phys. Rev. E*, **70**, 036204 (2004).

32. B. Weingartner, S. Rotter, and J. Burgdörfer, *Phys. Rev. B*, **72**, 115342 (2005).

33. R. Brunner, R. Meisels, F. Kuchar, R. Akis, R., D. K. Ferry, and J. P. Bird, *Phys. Rev. Lett.*, **98**, 204101 (2007).

34. G. J. Iafrate, H. L. Grubin, and D. K. Ferry, *J. Physique Coll*, **C7**, 307 (1981).

35. I. B. Levinson, *Sov. Phys. JETP*, **30**, 362 (1970).

36. T. C. L. G. Sollner, W. D. Goodhue, P. E. Tannenwald, C. D. Parker, and D. D. Peck, *Appl. Phys. Lett.*, **43**, 588 (1983).

37. U. Ravaioli, M. A. Osman, W. Pötz, N.C. Kluksdahl, and D. K. Ferry, *Physica B+C*, **134B**, 36 (1985).

38. W. R. Frensley, *Phys. Rev. B*, **36**, 1570 (1987).

39. N. C. Kluksdahl, A. M. Kriman, D. K. Ferry, and C. Ringhofer, *Phys. Rev. B*, **39**, 7720 (1989).

40. P. Carruthers and F. Zachariason, *Rev. Mod. Phys.*, **55**, 245 (1983).

41. Y. I. Shokin, *The Method of Differential Approximation* (Springer-Verlag, Berlin, 1983).

42. R. Courant, K. Friedrichs, and H. Lewy, *Math. Ann.*, **100**, 32 (1928).

43. N. C. Kluksdahl, A. M. Kriman, and D. K. Ferry, *IEEE Electron Dev. Lett.*, **9**, 457 (1988).

44. H. L. Grubin, R. C. Buggeln, and J. P. Kreskovsky, *Superlattices Microstruct.*, **27**, 533 (2000).

45. H. L. Grubin and R. C. Buggeln, *Physica B*, **314**, 117 (2002).

46. X. Oriols, J. J. Garcia-Garcia, F. Martin, J. Suné, T. González, J. Mateos, and D. Pardo, *Appl. Phys. Lett.*, **72**, 806 (1998).

47. F. Rossi, C. Jacoboni, and M. Nedjalkov, *Semicond. Sci. Technol.*, **9**, 934 (1994).

48. M. Nedjalkov, I. Dimov, F. Rossi, and C. Jacoboni, *Math. Comput. Modell.*, **23**, 159 (1996).

49. P. Bordone, M. Pascoli, R. Brunetii, A. Bertoni, C. Jacoboni, and A. Abramo, *Phys. Rev. B*, **59**, 3060.

50. J. Garcia-Garcia, F. Martin, X. Oriols, and J. Suñé, *Appl. Phys. Lett.*, **73**, 3539 (1998).

51. K. L. Jensen and F. A. Buot, *IEEE Trans. Electron Dev.*, **38**, 2337 (1991).

52. J. R. Barker, *Physica*, **134B**, 22 (1985).

53. A. Bialas and A. Krzywicki, *Phys. Lett. B*, **354**, 134 (1995).

54. F. Rossi, P. Poli, and C. Jacoboni, *Semicond. Sci. Technol.*, **7**, 1017 (1992).

55. M. Pascoli, P. Bordone, R. Brunetti, and C. Jacoboni, *Phys. Rev. B*, **58**, 3503 (1998).

56. A. Bertoni, P. Bordone, R. Brunetti, and C. Jacoboni, *J. Phys.: Condens. Matter*, **11**, 5999 (1999).

57. P. Bordone, A. Bertoni, R. Brunetti, and C. Jacoboni, *VLSI Design*, **13**, 211 (2001).

58. C. Jacoboni, A. Bertoni, P. Bordone, and R. Brunetti, *Math. Comput. Simul.*, **55**, 67 (2001).

59. R. Brunetti, C. Jacoboni, and F. Rossi, *Phys. Rev. B*, **39**, 10781 (1989).

60. L. Shifren and D. K. Ferry, *Phys. Lett. A*, **285**, 217 (2001).

61. C. Dewdney and B. J. Hiley, *Found. Phys.*, **12**, 27 (1982).

62. L. Shifren, Dissertation, Arizona State University, December 2002, *unpublished.*

63. L. Shifren and D. K. Ferry, *Physica B*, **314**, 72 (2002).

64. L. Shifren, C. Ringhofer, and D. K. Ferry, *IEEE Trans. Electron Dev.*, **50**, 769 (2003).

65. M. Nedjalkov, H. Kosina, R. Kosik, and S. Selberherr, *Microelectron. Eng.*, **63**, 199 (2002).

66. L. Shifren and D. K. Ferry, *Phys. Lett. A*, **306**, 332 (2003).

67. M. A. Osman and D. K. Ferry, *Phys. Rev. B*, **36**, 6018 (1987).

68. L. Reggiani, P. Lugli, and A. P. Jauho, *Phys. Rev. B*, **36**, 6602 (1987).

69. P. Bordone, D. Vasileska, and D. K. Ferry, *Phys. Rev. B*, **53**, 386 (1996).

70. D. Querlioz, J. Saint-Martin, V.-N. Do, A. Bournel, and P. Dollfus, *IEEE Trans. Nanotechnol.*, **5**, 737 (2006).

71. D. Querlioz, P. Dollfus, V.-N. Do, A. Bournel, and V. L. Nguyen, *J. Comp. Electron.*, **5**, 443 (2006).

72. D. Querlioz, J. Saint-Martin, K. Huet, A. bournel, V. Aubry-Fortuna, C. Chassat, S. Galdin-Retailleau, and P. Dollfuss, *IEEE Trans. Electron. Dev.*, **54**, 2232 (2007).

73. D. Querlioz, H.-N. Nguyen, J. Saint-Martin, A. Bournel, S. Galdin-Retailleau, and P. Dollfus, *J. Comp. Electron.*, **8**, 324 (2009).

74. D. Querlioz, J. Saint-Martin, and P. Dollfus, *J. Comp. Electron.*, **9**, 224 (2010).

75. D. Querlioz, J. Saint-Martin, A. Bournel, and P. Dollfus, *Phys. Rev. B*, **78**, 165306 (2008).

76. S. Koba, R. Aoyagi, and H. Tsuchiya, *J. Appl. Phys.*, **108**, 064504 (2010).

77. M. Nedjalkov, H. Kosina, and S. Selberherr, in *Large Scale Scientific Computing*, Vol. 2907, edited by I. Lirkov et al. (Springer-Verlag, Heidelberg, 2004), pp. 178–184.

78. M. Nedjalkov, H. Kosina, E. Ungersboeck, and S. Selberherr, *Semicond. Sci. Technol.*, **19**, S226 (2004).

79. M. Nedjalkov, H. Kosina, S. Selberherr, C. Ringhofer, and D. K. Ferry, *Phys. Rev. B*, **70**, 115319 (2004).

80. J. M. Sellier, M. Nedjalkov, I. Dimov, and S. Selberherr, *J. Appl. Phys.*, **114**, 174902 (2013).

81. J. M. Sellier, M. Nedjalkov, I. Dimov, and S. Selberherr, *Math. Comput. Simul.*, **107**, 108 (2015).

82. J. M. Sellier, M. Nedjalkov, and I. Dimov, *Phys. Rep.*, **577**, 1 (2015).

83. Y. Yamada, H. Tsuchiya, and M. Ogawa, *IEEE Trans. Electron Dev.*, **56**, 1396 (2009).

84. A. Einstein, B. Podolsky, and N. Rosen, *Phys. Rev.*, **47**, 777 (1935).

85. E. Schrödinger, *Naturwissenschaften*, **23**, 807 (1935); translated *Proc. Am. Philosoph. Soc.*, **124**, 323 (1980).

86. D. Bohm, *Quantum Theory* (Prentice-Hall, Englewood Cliffs, NJ, 1951), Secs. 15–19.

87. P. Ellinghaus, Tech. Univ. Vienna, *unpublished.*

88. P. Ellinghaus, Dissertation, Tech. Univ. Vienna, February 2016.

89. M. V. Berry and N. L. Balazs, *J. Phys. A*, **12**, 625 (1979).

90. E. T. Jaynes and F. W. Cummings, *Proc. IEEE*, **55**, 89 (1963).

91. N. B. Norozhny, I. I. Sanchez-Mondragon, and J. H. Eberly, *Phys. Rev. A*, **23**, 236 (1981).

92. J. Gea-Banacloche, *Phys. Rev. Lett.*, **65**, 3385 (1990).

93. R. Birrittella, K. Chang, and C. C. Gerry, *Optics Commun.*, **354**, 286 (2015).

94. J. Eiselt and H. Risken, *Phys. Rev. A*, **43**, 346 (1991).

95. K. E. Cahill and R. J. Glauber, *Phys. Rev.*, **177**, 1882 (1969).

96. T. Douce, A. Eckstein, S. P. Wallborn, A. Z. Khoury, S. Ducci, A. Keller, T. Coudreau, and P. Milman, *Scientific Repts.*, **3**, 3530 (2013).

97. H.-L. Zhang, H.-C. Yuan, L.-Y. Hu, and X.-X. Xu, *Optics Commun.*, **356**, 223 (2015).

98. K. P. Seshadreesan, J. P. Dowling, and G. S. Agarwal, *Physica Scripta*, **90**, 074029 (2015).

99. M. Reboiro, O. Civitrese, and D. Tielas, *Phys. Scr.*, **90**, 074028 (2015).

100. A. D. Armor, B. Kubala, and J. Ankerhold, *Phys. Rev. A*, **91**, 184508 (2015).

Chapter 9

Real-Time Green's Functions I

As we have pointed out repeatedly in this book, the role of quantum transport is to move well beyond the understanding that traditionally has been garnered from treatments of the Boltzmann transport equation with classical or even semi-classical physics. In the earlier chapters, we developed the zero-temperature and thermal Green's functions, but these remained near equilibrium. The general feature of semi-classical transport in the presence of high electric fields, which is the case in nearly all semiconductor electron devices, is that the distribution function deviates significantly from the equilibrium form. This is not encountered in these earlier Green's functions, and one stays true to the Fermi–Dirac distribution in their applications. On the other hand, we encountered significant modifications of the density matrix and/or the Wigner distribution in the last two chapters. To bring these equivalent modifications into the world of Green's functions, we need to move into the real-time Green's functions. These latter forms are often called the nonequilibrium Green's functions (NEGF), but many, if not most, of the applications to which they are applied do not really require this extended approach. For example, we will see later that when the transport is ballistic, then everything is determined by the retarded and advanced Green's functions in the (equilibrium) contacts, which allows us to

An Introduction to Quantum Transport in Semiconductors
David K. Ferry
Copyright © 2018 Pan Stanford Publishing Pte. Ltd.
ISBN 978-981-4745-86-4 (Hardcover), 978-1-315-20622-6 (eBook)
www.panstanford.com

use the Landauer formula. The claim to use NEGF for these simpler applications is often one of just desiring to use the buzzword.

In our earlier introduction to Green's functions, we utilized the retarded and advanced Green's functions. These two functions bring us into the world of off-shelf interactions, as they describe the evolution of the traditional, classical delta function $\delta(E - \hbar^2 k^2/2m^*)$, for a single parabolic band, into broadened spectral density $A(E, k)$. This latter quantity is described by the imaginary parts of either the advanced or retarded Green's function. We will find, in the following, that these two Green's functions will carry forward into the real-time Green's function in the same generic form that they took earlier, although the details within this form can be very different. What is new to the use of the real-time Green's functions is the necessity to be able to solve for the nonequilibrium distribution of the carriers under the applied forces. As remarked, in the semi-classical world, this is achieved through solutions to the Boltzmann transport equation. Now we need to find new Green's functions, which will be described by quantum evolution equations, and which will give us the quantum distribution functions.

The additional Green's functions, which are proper correlation functions, seem to have first been used by Martin and Schwinger [1]. Many of us were then introduced to them through the book by Kadanoff and Bahm [2]. Many useful advances were then introduced by Keldysh [3]. The retarded and advanced Green's functions, discussed in Chapter 3, were defined by expectation values of the commutator product of the wavefunction and its adjoint, with the two at different positions and times. These new Green's functions do not involve the commutator products, but do involve the wavefunction and its adjoint. They may be defined as [2]

$$G^<(\mathbf{x}, \mathbf{x}'; t, t') = i\left\langle \widehat{\psi}^\dagger(\mathbf{x}', t')\widehat{\psi}(\mathbf{x}, t) \right\rangle$$

$$G^>(\mathbf{x}, \mathbf{x}'; t, t') = -i\left\langle \widehat{\psi}(\mathbf{x}, t)\widehat{\psi}^\dagger(\mathbf{x}', t') \right\rangle.$$

(9.1)

These two functions are usually referred to as the "less than" Green's function and the "greater than" Green's function, respectively. We will see that these Green's functions are related to the distribution function *and* the spectral density.

9.1 Some Considerations on Correlation Functions

It is important to note that these Green's functions may also be used in the case of thermal equilibrium, such as discussed with the Matsubara Green's functions in Chapter 5. In this case, they satisfy the same boundary conditions in imaginary time that were discussed there. Because of this, we may develop some very useful properties of these two functions. In the case of Hamiltonians that have translational and rotational invariance, these Green's functions are functions only of the differences in position and time, and not their absolute values. Hence, we may introduce the new variables

$$\mathbf{x} = |\mathbf{x} - \mathbf{x}'|, \quad t = |t - t'|. \tag{9.2}$$

We may now Fourier transform the two Green's functions to find

$$G^<(\mathbf{k}, \omega) = i \int d^3\mathbf{x} \int dt e^{-i\mathbf{k}\bullet\mathbf{x} + i\omega t} G^<(\mathbf{x}, t)$$
$$G^>(\mathbf{k}, \omega) = -i \int d^3\mathbf{x} \int dt e^{-i\mathbf{k}\bullet\mathbf{x} + i\omega t} G^<(\mathbf{x}, t). \tag{9.3}$$

Now a similar argument as that leading to Eq. (5.18) in thermal Green's functions can lead, in equilibrium, to the relation

$$G^<(\mathbf{x}, t) = -e^{\beta\mu} G^>(\mathbf{x}, t - i\hbar\beta)$$
$$G^<(\mathbf{k}, \omega) = e^{-\beta(\hbar\omega - \mu)} G^>(\mathbf{k}, \omega), \tag{9.4}$$

where μ is the electrochemical potential, or more usually the Fermi energy. These two conditions are known as boundary conditions. Importantly, these two Green's functions are also related to the spectral density in a similar manner to the retarded and advanced Green's function. That is, we relate the two correlation functions to the spectral density as

$$A(\mathbf{k}, \omega) = G^<(\mathbf{k}, \omega) + G^>(\mathbf{k}, \omega). \tag{9.5}$$

Using the second of equations (9.4) in the above form of the spectral density gives us the relations

$$G^<(\mathbf{k}, \omega) = f(\omega) A(\mathbf{k}, \omega)$$
$$G^>(\mathbf{k}, \omega) = [1 - f(\omega)] A(\mathbf{k}, \omega), \tag{9.6}$$

where

$$f(\omega) = \frac{1}{1 + e^{\beta(\hbar\omega - \mu)}} \tag{9.7}$$

is the Fermi–Dirac distribution function. Thus, we note that the energy is represented by the frequency that arises from the Fourier transformation of the difference in time coordinate. We expect that $G^<$ characterizes the electrons and $G^>$ characterizes holes, or empty states in the Fermi sea. Strictly speaking, these two relationships are obeyed only in thermal equilibrium. They are not valid in the nonequilibrium case. Nevertheless, they are often invoked as an *ansatz* [4] in the nonequilibrium situation, in which the Fermi–Dirac distribution is replaced with the actual far-from-equilibrium distribution. This is important, as it means that we can find the spectral density from the retarded and advanced functions. The correlation functions can then be found, and removing the spectral density results in a form for the far-from-equilibrium distribution function, in full analogy of solving the semi-classical Boltzmann equation. As a result, one often tries to cast the equation of motion for the correlation functions into a form comparable to the Boltzmann equation.

While the two correlation functions and the retarded and advanced functions are sufficient for any applications, there are a pair of useful functions that can also be defined from these four. These are the time-ordered and anti-time-ordered Green's functions

$$G_t(\mathbf{x}, \mathbf{x}'; t, t') = \vartheta(t - t')G^>(\mathbf{x}, \mathbf{x}'; t, t') + \vartheta(t' - t)G^<(\mathbf{x}, \mathbf{x}'; t, t')$$
$$G_{\bar{t}}(\mathbf{x}, \mathbf{x}'; t, t') = \vartheta(t' - t)G^>(\mathbf{x}, \mathbf{x}'; t, t') + \vartheta(t - t')G^<(\mathbf{x}, \mathbf{x}'; t, t'). \tag{9.8}$$

There are obviously a number of relationships between the six functions, and we can summarize these as

$$G_r = G_t - G^< = G^> - G_{\bar{t}} = \vartheta(t - t')(G^> - G^<)$$
$$G_a = G_t - G^> = G^< - G_{\bar{t}} = -\vartheta(t - t')(G^> - G^<). \tag{9.9}$$

An important aspect of the above, particularly with respect to Eq. (9.1), is that the expectation values are no longer taken with respect to a thermal equilibrium averaging or an averaging over the ground state, as these have no meaning in the nonequilibrium system. Rather, the bracket means an average over the available states of the nonequilibrium system, in which these states are weighted by the nonequilibrium distribution function.

While we used the single variables for the situation in which the system is described by a spatially homogeneous and rotationally invariant Hamiltonian, this is not the case in most applications in nonequilibrium systems. In these cases, we need to revert to the full complement of two spatial vector variables and two time variables. It is often useful in these situations to transform to the center-of-mass (or Wigner) coordinates through

$$\mathbf{R} = \frac{1}{2}(\mathbf{x} + \mathbf{x}'), \quad T = \frac{1}{2}(t + t'),$$
$$\mathbf{s} = (\mathbf{x} - \mathbf{x}'), \quad \tau = (t - t').$$

(9.10)

Then, the Fourier transform (9.3) may be applied to the $\mathbf{s}$ and τ variables as before. This allows us to arrive at a connection between the less-than function and the Wigner distribution from the last chapter, which may be written as

$$f_W(\mathbf{R}, \mathbf{k}, T) = \lim_{\tau \to 0} \int d^3 s \, G^<(\mathbf{R}, \mathbf{s}, T, \tau) e^{-i\mathbf{k} \cdot \mathbf{s}}$$
$$= \int \frac{d\omega}{2\pi i} G^<(\mathbf{R}, \mathbf{k}, T, \omega).$$

(9.11)

Here, the frequency that arises from the difference in time coordinate continues to play the role of the energy in the system. The fact that this energy is integrated out to obtain the Wigner function has certain implications. The foremost is that the momentum must now be related to the energy through an on-shell delta function, just as in the classical case. This does not mean that off-shell effects cannot arise, but these must be treated carefully as part of the scattering process, as discussed in the last chapter. The result of Eq. (9.11) is that the critical kinetic variable in the Wigner function is the momentum, and not the energy, although the two are certainly related. In a sense, the integration has removed the spectral density, as the Wigner function is the distribution function for the system. In that sense, solving the Wigner equation of motion is equivalent to solving a similar equation for the less-than Green's function. We will come to this latter equation in due time.

As we discussed in Chapter 4, proceeding to solve the quantum transport problem for complicated Hamiltonians, such as the case of the many-body interactions or electron–phonon interactions, is not particularly simple and a perturbative approach is usually used. However, this approach is not without its own problems; the perturbative series is difficult to evaluate and may not converge. Generating the perturbation series usually relies on the S-matrix (see Section 4.1) of the unitary operator (4.16)

$$\exp\left[-\frac{i}{\hbar}\int_{t'}^{t}V(t'')dt''\right],$$

(9.12)

where $V(t)$ is the perturbing potential interaction, written in the interaction representation. In nearly all cases, it is necessary to expand any perturbation series in terms of Green's function written in some reference basis, usually the ground state at $t \to -\infty$. In the equilibrium situation of Chapter 4, we returned to equilibrium at $t \to +\infty$. In the nonequilibrium situation, which is the normal case in nearly all semiconductor devices, this latter limit is just not allowed. Then, one must seek a better approach, or live with a questionable approximation. One suggestion is to use a modified contour, which begins (and perhaps ends) with the thermal Green's function [2, 3]. This contour is shown in Fig. 9.1. In order to avoid the need to proceed to $t \to +\infty$, a new time path has been suggested, in which the contour returns back toward the equilibrium state [5]. Here, we have an imaginary time leg, in which the Green's function evolves from the thermal Green's function and then evolves forward in time before reversing backward in time. The turning point is generally taken to be the maximum of either t or t'. For example, with t on the upper side of the turn, as shown and assumed to be the larger value, the lower leg is the time-ordered branch, while the upper leg would be the anti-time-ordered branch [6]. But this designation assumes that both times are on the same leg. When they are on different legs, the correlation functions (9.1) are the appropriate Green's functions. Is this the proper contour along which to evaluate the Green's functions? This is not entirely clear. We need a contour because we need to maintain normalization of the Green's functions, which is necessary if we are to cancel the disconnected diagrams in the perturbation series, as was achieved in Chapter 4. This contour serves that purpose. But in general, just as in classical systems, the less-than function will evolve into a steady state, which is a balance between the driving forces and the dissipative forces. This steady state has almost no connection with the thermal equilibrium Green's function. There is no requirement that the distribution function be a Fermi–Dirac distribution, and there is even some doubt as to the efficacy of the division indicated in Eq. (9.6). But in the absence of anything better, the contour of Fig. 9.1 is usually invoked. Perhaps it is fortunate that the electron–phonon interaction in semiconductors is weak, and we seldom have to go beyond the first-order terms.

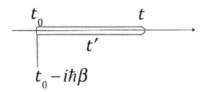

Figure 9.1   The standard contour for evaluating the real-time Green's functions in perturbation theory.

Keldysh introduced an innovation in the approach [3], in which he introduced a Green's function matrix. This matrix is written as

$$\mathbf{G} = \begin{bmatrix} G_t & G^> \\ G^< & -G_{\bar{t}} \end{bmatrix}. \tag{9.13}$$

We may think of this matrix in the following way. The rows and columns of the matrix refer to the two times in the Green's functions. Although it can be misleading to think of the rows and columns in terms of the portions of the trajectory in Fig. 9.1, we can relate them to where the times are on the trajectory. For example, $\mathbf{G}_{11}$ would be the Green's function one would get when both times are on the lower longitudinal leg of the trajectory. Similarly, $\mathbf{G}_{22}$ would be the Green's function one would get when both times are on the upper longitudinal leg of the trajectory. In both cases, we assume that $t > t'$. The other two Green's functions, the correlation functions, arise when one time is on the upper leg and one is on the lower leg. For the time progression, we replace the term "time ordering" with "contour ordering." The extension to a 3×3 in order to account for the extension in imaginary time to the thermal Green's function has been developed by Wagner [6, 7]. While the above form of the matrix is instructive, Keldysh actually rearranged it to reduce the number of elements that had to be considered. If we subtract G_t from each term in the matrix, and then multiply through with a minus sign, the new Keldysh matrix is

$$\mathbf{G} = \begin{bmatrix} 0 & G_a \\ G_r & G_K \end{bmatrix}, \tag{9.14}$$

where the Keldysh function is

$$G_K = G^> + G^<. \tag{9.15}$$

Here we have used the various interrelationships that are expressed in Eq. (9.9). A more complicated method of getting to the Keldysh matrix commonly used is to remove the minus sign from the anti-time-ordered term in Eq. (9.13) and then use a variety of rotations in the so-called Keldysh space to write $\mathbf{G}_K = \mathbf{L}\sigma_z \mathbf{G}\mathbf{L}^\dagger$, where the transformation matrices are given by

$$\mathbf{L} = \frac{1}{\sqrt{2}} \begin{bmatrix} 1 & -1 \\ 1 & 1 \end{bmatrix}, \quad \mathbf{L}^\dagger = \frac{1}{\sqrt{2}} \begin{bmatrix} 1 & 1 \\ -1 & 1 \end{bmatrix}, \tag{9.16}$$

and σ_z is a Pauli spin matrix. These transformations lead to

$$\mathbf{G}_K = \begin{bmatrix} G_r & G_K \\ 0 & G_a \end{bmatrix}, \tag{9.17}$$

which of course is just Eq. (9.14) with the rows interchanged.

We can now develop the equations of motion for the non-interacting forms of these Green's functions. That is, we seek the equations that the Green's functions will satisfy in the absence of any applied potentials and/or perturbing interactions. For this, we assume that the individual field operators are based on the wavefunctions that satisfy the basic Schrödinger equation. However, in the case of single-point potentials (local potentials that depend on only a single position, such as in the case of solutions to the Poisson's equation), this potential can be included in the Schrödinger equation. If we limit ourselves to only such potentials, then the equations of motion can be written as

$$\left(i\hbar \frac{\partial}{\partial t} - H_0(\mathbf{x}) - V(\mathbf{x}) \right) \mathbf{G}_{K0} = \hbar \mathbf{I}$$

$$\left(i\hbar \frac{\partial}{\partial t'} - H_0(\mathbf{x}') - V(\mathbf{x}') \right) \mathbf{G}_{K0} = \hbar \mathbf{I}, \tag{9.18}$$

where $\mathbf{I}$ is the 2×2 unit matrix.

As has been stated earlier, transport that reaches the steady state arises as a balance between the driving forces and the dissipative forces. The potentials above give the driving forces, as will a magnetic field if it is present. But to achieve a description of transport, we need to add the interaction terms to the Hamiltonian. As before, these will lead to self-energy terms Σ. With the new contour, it is possible to construct a Feynmann series expansion of the interaction terms from the contour-ordered unitary operator (9.12). This procedure seems to work well for the real-time Green's functions, as it did in the earlier chapters and seems to have been pursued almost universally in the field [8, 9]. The assumption is that the projection of the time axes back to t_0 allows the use of the pseudo-equilibrium to justify the use of an equivalent form of Wick's theorem. The various parts of the diagrams may then be regrouped into terms that represent the Green's functions themselves and terms that represent the interactions leading to the self-energy. The latter naturally can be expressed as a matrix, so that it is possible to write the equations of motion for the interacting Green's functions as

$$
\begin{aligned}
\left(i\hbar \frac{\partial}{\partial t} - H_0(\mathbf{x}) - V(\mathbf{x}) \right) \mathbf{G}_K &= \hbar \mathbf{I} + \Sigma \mathbf{G}_K \\
\left(i\hbar \frac{\partial}{\partial t'} - H_0(\mathbf{x}') - V(\mathbf{x}') \right) \mathbf{G}_K &= \hbar \mathbf{I} + \mathbf{G}_K \Sigma,
\end{aligned}
\tag{9.19}
$$

where

$$
\Sigma = \begin{bmatrix} \Sigma_r & \Sigma_K \\ 0 & \Sigma_a \end{bmatrix}.
\tag{9.20}
$$

The products that appear in Eq. (9.19) are actually integrals in the form

$$
AB(\mathbf{x}, \mathbf{x}', t, t') = \int d^3\mathbf{x}'' \int dt'' A(\mathbf{x}, \mathbf{x}'', t, t'') B(\mathbf{x}'', \mathbf{x}', t'', t') . \tag{9.21}
$$

The proper way to write these integrals is that in which the Keldysh Green's functions appear naturally, as will be discussed in the next section.

9.2 Langreth Theorem

While we have written down the self-energy and the Green's functions in the Keldysh form in Eq. (9.19), this result is not at all self-evident. For example, when we start with one of the Green's functions separately from the matrix and write the equations of motion for this, we do not get the identity matrix on the right-hand side, but a pair of delta functions, one for the two time arguments and one for the two space arguments (for example, see Eq. (3.19)) [10]. It turns out that the transformation matrix σ_z converts this to a proper identity matrix and then the remaining two terms in the transformation matrices leave it unchanged. Thus, the transformations described above can be applied to the non-interacting forms of the Green's functions as well.

For the product of the self-energy and the Green's function, however, the application of the non-unitary transformations in Keldysh space fails, as a general rule. The problem lies in defining just what the terms in the two matrices $\boldsymbol{\Sigma G}$ and $\boldsymbol{G\Sigma}$ should be. The logical solution to this dilemma is to compute directly the individual matrix elements in the Keldysh form. The problem that must be faced lies in the convolution integral over the internal time, and not in the spatial integration. Consider the product

$$C(t,t') = \int A(t,t'')B(t'',t')dt''. \tag{9.22}$$

The approach that is used to evaluate this integral has been given by Langreth [11]. Let us consider, for example, the situation if C represents the less-than function. For this case, t is on the outgoing (bottom) leg of the contour in Fig. 9.1, while t' is on the incoming leg (top) of the contour. The intermittent time t'' could be on either contour, and this would give different results depending on just where it lays. The solution was given by Langreth. There are two possibilities. First, let us deform the contour by taking it at t, and pulling it back to t_0, then returning out to t'. We call these two contours C_1 and C_2, respectively. Now, we still get different results depending on just which branch t'' resides. The solution is to split the integral into two parts, one of which arises for each leg where t'' resides, that is one for this time on C_1 and one for this time on C_2. This allows us to write Eq. (9.22) as

$$C^<(t,t') = \int_t^{-\infty} A(t,t'')B(t'',t')dt'' + \int_{-\infty}^t A(t,t'')B(t'',t')dt'' . \quad (9.23)$$

Clearly, in the first integral, which corresponds to C_1, B is the less-than function since its arguments fold around the C_2 contour. The function A is the anti-time-ordered function, but since $t > t''$, the Heaviside function inherent in this Green's function just gives us the less-than function. In the second integral, A is the time-ordered function and, under the same limits, gives us the greater-than function. These arguments allow us to rewrite Eq. (9.23) as

$$C^<(t,t') = \int_{-\infty}^t \theta(t - t'')(A^> - A^<)B^< dt'' = \int_{-\infty}^t A^r(t,t'')B^<(t'',t')dt'' .$$
$$(9.24)$$

In the second possibility, we deform the contour not from t, but from t'. This makes A always the less-than function, and then the two paths contribute to cause B to be the advanced function. These two possibilities must be combined to give a general result. Integrals such as this allow us to write the possible product formulas as

$$C^< = A_r B^< + A^< B_a, \quad C_r = A_r B_r$$
$$C^> = A_r B^> + A^> B_a, \quad C_a = A_a B_a. \quad (9.25)$$

Similar arguments can be used to form triple products should they arise. The key point here is that the use of the Langreth theorem tells us that the matrices that form the products ΣG and $G\Sigma$ should be written in the Keldysh form. Then the proper matrix products are achieved in each case. One interesting point, however, is that the equation for the Keldysh Green's function is actually two equations, which have been added together for simplicity. In practice, however, the equations for the less-than and greater-than functions should be separated and treated as two disjoint equations of motion.

9.3 Near-Equilibrium Approach

In the above, the totality of the additional interactions has been split into two parts. The first is the potential that is recognized as a single-site (only a single spatial variable) function, which is thereby local in its effect. As a result, the potential does not require the integral in Eq. (9.19). The second part is the self-energy interaction, which is a nonlocal, two-site term by its very nature. Normally, in the Fermi golden rule, the interaction between, e.g., the electron and the phonon is assumed to occur at a single site and is thus local in nature. As we discussed in the previous two chapters, however, this is not the general case for the self-energy. Those who purport to do quantum transport using a single-site representation of the self-energy are really doing little beyond the Boltzmann equation other than incorporating the possibility of spatial quantization and the formation of sub-bands. Here we will try to avoid this and follow the more general, but more complicated, formulation with the full nonlocal self-energy. However, we shall begin with the low electric field case, so that we are not that far from equilibrium. In this, we seek a quantum transport equation that has some equivalence to the normal Boltzmann equation. In essence, we will be seeking what is called the linear response of the system. For this, we assume that the system is spatially homogeneous and will utilize the Wigner coordinates (9.10). The approach to be followed was developed by Mahan and Hänsch [12–14].

9.3.1 Retarded Function

We begin with the retarded Green's function, which we separate out from the matrix equation given earlier, which leads to a pair of equations

$$
\left[i\hbar \frac{\partial}{\partial t} - H_0(\mathbf{x}) - V(\mathbf{x}) \right] G_r(\mathbf{x}, \mathbf{x}') = \delta(\mathbf{x} - \mathbf{x}') + \Sigma_r G_r
$$

$$
\left[i\hbar \frac{\partial}{\partial t'} - H_0(\mathbf{x}') - V(\mathbf{x}') \right] G_r(\mathbf{x}, \mathbf{x}') = \delta(\mathbf{x} - \mathbf{x}') + G_r \Sigma_r,
$$

(9.26)

where the last terms on the right-hand side are, of course, the convolution integral over an internal spatial coordinate. The retarded Green's function must satisfy both these equations. It is now convenient to introduce the coordinate transformation (9.10) and then add and subtract the resulting two equations. This leads us to the forms

$$\left[i\hbar\frac{\partial}{\partial\tau} + \frac{\hbar^2}{8m*}\nabla^2 + \frac{\hbar^2}{2m*}\frac{\partial^2}{\partial\mathbf{s}^2} + e\mathbf{E}\cdot\mathbf{R} \right]G_r(\mathbf{R},\mathbf{s},T,\tau)$$

$$= \delta(\tau)\delta(\mathbf{s}) + \frac{1}{2}\int dt'' \int d^3\mathbf{x}''(\Sigma_r G_r + G_r \Sigma_r)$$

$$\left[i\hbar\frac{\partial}{\partial T} + \frac{\hbar^2}{m*}\nabla\cdot\frac{\partial}{\partial\mathbf{s}} + e\mathbf{E}\cdot\mathbf{s} \right]G_r(\mathbf{R},\mathbf{s},T,\tau) = \int dt'' \int d^3\mathbf{x}''(\Sigma_r G_r - G_r \Sigma_r).$$

(9.27)

Here we have assumed that the potential is of no more than quadratic order, so that the major term is the electric field. In addition, the two products in the integrals on the right-hand side of the equation have not yet been transformed to the Wigner coordinates, as this is a complicated process, which will be dealt with later.

The next step is to Fourier transform on the relative coordinates in order to introduce the momentum and frequency, which leads to

$$\left[\hbar\Omega - E(\mathbf{k}) + \frac{\hbar^2}{8m*}\nabla^2 + e\mathbf{E}\cdot\mathbf{R} \right]G_r(\mathbf{R},\mathbf{k},T,\Omega)$$

$$= 1 + \frac{1}{2}\int d^3\mathbf{s}\int d\tau e^{i\mathbf{k}\cdot\mathbf{s}-i\Omega\tau}\int dt'' \int d^3\mathbf{x}''(\Sigma_r G_r + G_r \Sigma_r)$$

(9.28)

$$i\hbar\left[\frac{\partial}{\partial T} + \mathbf{v}_k\cdot\nabla + \frac{e\mathbf{E}}{\hbar}\cdot\frac{\partial}{\partial\mathbf{k}} \right]G_r(\mathbf{R},\mathbf{k},T,\Omega)$$

$$= \int d^3\mathbf{s}\int d\tau e^{i\mathbf{k}\cdot\mathbf{s}-i\Omega\tau}\int dt'' \int d^3\mathbf{x}''(\Sigma_r G_r - G_r \Sigma_r).$$

The second of these equations has the same streaming terms as the Boltzmann equation. The first equation presents some difficulties from the streaming terms, as it is more like a Schrödinger equation. These streaming terms give rise to a positional dependence in the otherwise homogeneous system. This would lead to a size dependence that is not at all in keeping with the physics of the homogeneous structure. This suggests that we are not in the most convenient gauge for the fields. It becomes more advantageous to introduce a further transformation of the energy and potential, which may be thought of as a gauge transformation as

$$\Omega + \frac{e\mathbf{E}\cdot\mathbf{R}}{\hbar} \to \omega, \quad \nabla \to \nabla + \frac{e\mathbf{E}}{\hbar}\frac{\partial}{\partial\omega}.$$

(9.29)

With these changes, the equations may be written as

$$\left[\hbar\omega - E(\mathbf{k}) + \frac{\hbar^2}{8m^*}\left(\nabla + \frac{e\mathbf{E}}{\hbar}\frac{\partial}{\partial\omega}\right)^2\right]G_r(\mathbf{R},\mathbf{k},T,\omega)$$

$$= 1 + \frac{1}{2}\int d^3s\int d\tau e^{i\mathbf{k}\cdot\mathbf{s}-i\Omega\tau}\int dt''\int d^3x''(\Sigma_r G_r + G_r\Sigma_r) \tag{9.30}$$

$$i\hbar\left[\frac{\partial}{\partial T} + \mathbf{v}_k\cdot\nabla + \frac{e\mathbf{E}}{\hbar}\cdot\left(\frac{\partial}{\partial\mathbf{k}} + \mathbf{v}_k\frac{\partial}{\partial\omega}\right)\right]G_r(\mathbf{R},\mathbf{k},T,\omega)$$

$$= \int d^3s\int d\tau e^{i\mathbf{k}\cdot\mathbf{s}-i\Omega\tau}\int dt''\int d^3x''(\Sigma_r G_r - G_r\Sigma_r).$$

The large omega in the integrals still needs to be transformed for these equations. It is clear that the exponential in the transformed collision terms will lead to phase interference effects, just as found in the previous two chapters. These effects will normally occur on distances smaller than the mean free path. The two equations (9.30) both contain terms linear in the electric field. But we have to remember that the most important factor in high-field transport does not arise from nonlinear terms in the electric field, but from the changes in the distribution function that arise as a result of the field. These equations are still exact to all orders in the field.

It is now time to address the scattering terms, which can be quite complicated. Hence, some form of simplification is necessary to evaluate them properly. Since this section is for linear response, we will introduce what has been called the gradient expansion [2]. What this entails is to expand the various functions around the center-of-mass coordinates, which assumes that these are larger than the difference coordinates. This approach limits the results to systems in which the potentials are slowly varying in space and time. Of course, this is not the case in mesoscopic systems, but then the latter are not really homogeneous, which is our assumption. To illustrate this procedure, we will explicitly treat only the first term in the collision integrals on the right-hand side of Eq. (9.30). We may rewrite this term as

$$I = \int d^3s e^{i\mathbf{k}\cdot\mathbf{s}}\int d\tau e^{-i\Omega\tau}\int dt''\int d^3x''\Sigma_r(\mathbf{x},\mathbf{x}'',t,t'')G_r(\mathbf{x}'',\mathbf{x}',t'',t').$$

$$\tag{9.31}$$

To proceed, we make the coordinate transformation to the center-of-mass coordinates, such as for the Wigner coordinates, but do not make the variable change as yet. Hence, we rewrite the integrand as (we illustrate this only for the spatial coordinates, but equivalent processing is done on the temporal coordinates as well)

$$\Sigma_r\left(\mathbf{x}-\mathbf{x}'',\frac{\mathbf{x}+\mathbf{x}''}{2}\right)G_r\left(\mathbf{x}''-\mathbf{x}',\frac{\mathbf{x}''+\mathbf{x}'}{2}\right). \tag{9.32}$$

Now we make the substitution $\mathbf{y} = \mathbf{x} - \mathbf{x}'$, so that the integrand is given by

$$\Sigma_r\left(\mathbf{y},\mathbf{x}-\frac{\mathbf{y}}{2}\right)G_r\left(\mathbf{x}-\mathbf{x}'-\mathbf{y},\frac{\mathbf{x}+\mathbf{x}'-\mathbf{y}}{2}\right). \tag{9.33}$$

We now introduce the explicit Wigner coordinates (9.10), so that the last expression becomes

$$\Sigma_r\left(\mathbf{y},\mathbf{R}+\frac{\mathbf{s}}{2}-\frac{\mathbf{y}}{2}\right)G_r\left(\mathbf{s}-\mathbf{y},\mathbf{R}-\frac{\mathbf{y}}{2}\right), \tag{9.34}$$

and finally, we let $\mathbf{w} = \mathbf{s} - \mathbf{y}$, so that we arrive at being able to write the integral as

$$I = \int\int d^3\mathbf{w}d^3\mathbf{y}e^{i\mathbf{k}\cdot(\mathbf{w}+\mathbf{y})}\int\int d\tau d\tau' e^{-i\Omega(\tau+\tau')}$$
$$\times\Sigma_r\left(\mathbf{y},\mathbf{R}+\frac{\mathbf{w}}{2},\tau',T+\frac{\tau}{2}\right)G_r\left(\mathbf{w},\mathbf{R}-\frac{\mathbf{y}}{2},\tau,T-\frac{\tau'}{2}\right). \tag{9.35}$$

The center-of-mass coordinates all appear with an offset, so that these may be separated by expanding the functions in a Taylor series, the so-called gradient expansion [2, 15]. When this is done, the Fourier transforms can be performed to that the integral becomes

$$I = \Sigma_r(\mathbf{k},\mathbf{R},\Omega,T)G_r(\mathbf{k},\mathbf{R},\Omega,T)+\frac{i}{2}\{\Sigma_r,G_r\}+... \tag{9.36}$$
$$\{\Sigma_r,G_r\}\equiv\frac{\partial\Sigma_r}{\partial\mathbf{R}}\cdot\frac{\partial G_r}{\partial\mathbf{k}}-\frac{\partial\Sigma_r}{\partial\mathbf{k}}\cdot\frac{\partial G_r}{\partial\mathbf{R}}+\frac{\partial\Sigma_r}{\partial\Omega}\cdot\frac{\partial G_r}{\partial T}-\frac{\partial\Sigma_r}{\partial T}\cdot\frac{\partial G_r}{\partial\Omega}.$$

The last expression is called the *bracket operator* and plays a role quite similar to the classical Poisson brackets. This is why the symbolism is used. The frequency derivatives are still with respect to the unshifted frequencies, so that this correction modifies the bracket operator to yield

$$\{\Sigma_r, G_r\} \equiv \frac{\partial \Sigma_r}{\partial \mathbf{R}} \cdot \frac{\partial G_r}{\partial \mathbf{k}} - \frac{\partial \Sigma_r}{\partial \mathbf{k}} \cdot \frac{\partial G_r}{\partial \mathbf{R}} + \frac{\partial \Sigma_r}{\partial \omega} \cdot \frac{\partial G_r}{\partial T} - \frac{\partial \Sigma_r}{\partial T} \cdot \frac{\partial G_r}{\partial \omega}$$
$$+ \frac{e\mathbf{E}}{\hbar} \left(\frac{\partial \Sigma_r}{\partial \omega} \cdot \frac{\partial G_r}{\partial \mathbf{k}} - \frac{\partial \Sigma_r}{\partial \mathbf{k}} \cdot \frac{\partial G_r}{\partial \omega} \right). \tag{9.37}$$

This gradient expansion has already produced a term that is linear in the electric field, which will be combined with an equivalent term on the left-hand side of the equation. Because we are only interested in the linear response (as mentioned, the distribution function will actually be of higher order in the field and create the nonlinear transport), we do not keep higher-order terms in the expansion.

With these results, we can write down an approximate, but closed pair of equations for the retarded Green's function. These are

$$\left[\hbar\omega - E(k) + \frac{\hbar^2}{8m^*} \left(\nabla + \frac{e\mathbf{E}}{\hbar} \frac{\partial}{\partial \omega} \right)^2 - \Sigma_r \right] G_r = 1$$

$$\left\{ \frac{\partial}{\partial T} + \mathbf{v}_k \cdot \nabla + \frac{e\mathbf{E}}{\hbar} \cdot \left[\left(1 - \frac{1}{\hbar} \frac{\partial \Sigma_r}{\partial \omega} \right) \frac{\partial}{\partial \mathbf{k}} + \left(\mathbf{v}_k + \frac{1}{\hbar} \frac{\partial \Sigma_r}{\partial \mathbf{k}} \right) \frac{\partial}{\partial \omega} \right] \right\} G_r \tag{9.38}$$

$$= -\frac{i}{\hbar} \{\Sigma_r, G_r\}.$$

Now the bracketed term refers solely to the leading, field-independent term in Eq. (9.37). These equations simplify considerably for near-equilibrium systems (which are the ones of interest here) that are spatially homogeneous and in steady state, so that the derivative with respect to $\mathbf{R}$ and T vanishes. Then the preceding equations can be written in the form

$$[\hbar\omega - E(k) - \Sigma_r]G_r = 1$$

$$e\mathbf{E} \cdot \left[\left(1 - \frac{1}{\hbar} \frac{\partial \Sigma_r}{\partial \omega} \right) \frac{\partial}{\partial \mathbf{k}} + \left(\mathbf{v}_k + \frac{1}{\hbar} \frac{\partial \Sigma_r}{\partial \mathbf{k}} \right) \frac{\partial}{\partial \omega} \right] G_r = 0. \tag{9.39}$$

The first of these equations can be simply solved, and the result can also be shown to satisfy the second of these equations for sufficiently small electric fields. This gives the retarded Green's function to be

$$G_r(\mathbf{k}, \mathbf{R}, \omega, T) = \frac{1}{\hbar\omega - E(k) - \Sigma_r(\mathbf{k}, \mathbf{R}, \omega, T)} + O(E^2).... \tag{9.40}$$

The retarded Green's function in this last equation has no first-order terms in the electric field [12, 16]. Thus, its form is unchanged from that of the equilibrium function, although of course the self-energy may have undergone some changes to its form. It may thus be expected that the spectral density will also have the equilibrium form exhibiting collisional broadening, but no intra-collisional field effect.

9.3.2 Less-Than Function

The long tedious procedure to obtain the retarded Green's function must now be replicated in order to obtain the less-than Green's function. This latter function will be related to the distribution function, and we will be able to obtain something akin to the Boltzmann equation. We begin by writing the equations for the less-than function as

$$
\left[\hbar\omega - E(k) + \frac{\hbar^2}{8m^*}\left(\nabla + \frac{eE}{\hbar}\frac{\partial}{\partial\omega} \right)^2 \right] G^< = \frac{1}{2}I^+
$$

$$
i\hbar\left[\frac{\partial}{\partial T} + \mathbf{v}_k\cdot\nabla + \frac{eE}{\hbar}\cdot\left(\frac{\partial}{\partial\mathbf{k}} + \mathbf{v}_k\frac{\partial}{\partial\omega} \right) \right] G^< = I^-,
$$

(9.41)

where the scattering functions are given by

$$
I^\pm = \int d^3\mathbf{s}\int d\tau e^{i(\mathbf{k}\cdot\mathbf{s}-\omega\tau+e\mathbf{E}\cdot\mathbf{R}\tau/\hbar)}\int d^3\mathbf{x}''\int dt''\left[\Sigma_r G^< + \Sigma^< G_a \pm G_r\Sigma^< \pm G^<\Sigma_a \right].
$$

(9.42)

When there are no collisions present, or when the system is in thermal equilibrium, the scattering terms on the right-hand side of the equations vanish due to detailed balance. The left-hand side of the second equation in Eq. (9.41) has the same form as the Boltzmann equation, except for the second term in the parentheses. This derivative with respect to the frequency (energy) suggests a constraint that may be placed on the connection between the classical and quantum regimes. If we resort to connecting the Wigner distribution to the classical distribution through Eq. (9.11), then these frequency derivatives vanish in going to the classical limit, and the second of Eq. (9.41) can be thought of as the quantum Boltzmann equation.

The gradient expansion can now be introduced in the scattering terms just as done above for the retarded Green's function. Our interest remains on the homogeneous steady-state solutions in semiconductors, so that we can ignore the variation with respect to **R** and *T*. Before proceeding, it is convenient to introduce some expansions of the previously obtained retarded functions, in which we expand these into their real and imaginary parts as

$$G_r = G_a^\dagger \equiv W - \frac{i}{2} A$$

$$\Sigma_r = \Sigma_a^\dagger \equiv S - i\Gamma. \tag{9.43}$$

Now carrying out the various transformations for the gradient expansion, we directly arrive at the new equations of motion for the less-than Green's function as

$$\left[\hbar\omega - E(k) - S + \frac{e\mathbf{E}}{\hbar} \cdot \left(\frac{\partial\Gamma}{\partial\mathbf{k}} \frac{\partial}{\partial\omega} - \frac{\partial\Gamma}{\partial\omega} \frac{\partial}{\partial\mathbf{k}} \right) \right] G^< = W\Sigma^< - \frac{e\mathbf{E}}{4\hbar} \cdot J(A)$$

$$e\mathbf{E} \cdot \left[\left(1 - \frac{\partial S}{\partial\omega} \right) \frac{\partial}{\partial\mathbf{k}} + \left(\mathbf{v}_k + \frac{1}{\hbar} \frac{\partial S}{\partial\mathbf{k}} \right) \frac{\partial}{\partial\omega} \right] G^< + 2\Gamma G^< = A\Sigma^< + \frac{e\mathbf{E}}{\hbar} \cdot J(W),$$

$$\tag{9.44}$$

where

$$J(u) = \frac{\partial\Sigma^<}{\partial\omega} \frac{\partial u}{\partial\mathbf{k}} - \frac{\partial\Sigma^<}{\partial\mathbf{k}} \frac{\partial u}{\partial\omega}. \tag{9.45}$$

While the spectral density may clearly be recognized in Eq. (9.43), the function Γ is related to the linewidth, or broadening, of the states involved.

We may now introduce another simplification, as we are in the near-equilibrium (low electric field) situation. Since the field multiplies the entire bracketed term in the second term of Eq. (9.44), we may replace the Green's functions within the brackets with the field-independent form since we seek only the linear response. In equilibrium, we may separate the less-than function into the spectral density and the distribution function, as in Eq. (9.6). Each term multiplying the electric field contains either the distribution function or its derivative with respect to frequency, since both $G^<$ and $\Sigma^<$ are proportional to the distribution function. In the absence of the field, the less-than function is just the equilibrium expression (9.6) and the two collision terms must vanish by detailed balance. Thus, it is the deviation of the distribution function from the pure Fermi–Dirac form that leads to the transport. Thus, those terms involving the derivative of the distribution must be of importance, and indeed the other terms will be seen to drop out of the equation. Thus, the second equation of (9.44) becomes

$$
ie\mathbf{E} \cdot \left[\left(\mathbf{v}_{\mathbf{k}} + \frac{1}{\hbar} \frac{\partial S}{\partial \mathbf{k}} \right) \Gamma + (\hbar\omega - E(k) - S) \frac{1}{\hbar} \frac{\partial \Gamma}{\partial \mathbf{k}} \right] A^2(\mathbf{k}, \omega) \frac{\partial f(\omega)}{\partial \omega} \tag{9.46}
$$

$$
= A\Sigma^< - 2\Gamma G^<.
$$

This result is the steady-state homogeneous form of the quantum Boltzmann equation. The right-hand side demonstrates the gain–loss mechanism, which although appearing somewhat different than the normal case may be easily cast into the more normal form. To linear terms in the electric field, this result is exact, and this result due to Mahan and Hänsch [12–14] represented a major step forward in the use of the real-time Green's functions.

It may be noted that the term in the second set of parentheses in Eq. (9.46) vanishes at the peak of the spectral density. It is reasonable to neglect this term, as well as the velocity correction term in the first set of parentheses. The velocity correction arises from the real part of the self-energy and is typically small. The second term in the square brackets is also a velocity correction but arises from the dispersion of the imaginary part of the self-energy. The neglect of these two terms is an approximation in which we are ignoring velocity corrections of the single-particle states, which is more or less equivalent to assuming that the self-energy provides a rigid shift of the energy band. This is not truly the case, but the approximation is not bad for the small fields of interest here. If we make these approximations, then Eq. (9.46) becomes

$$
A^2(\mathbf{k}, \omega) e\mathbf{E} \cdot \mathbf{v}_{\mathbf{k}} \Gamma \frac{\partial f(\omega)}{\partial \omega} = -i[A(\mathbf{k}, \omega)\Sigma^< - 2\Gamma G^<], \tag{9.47}
$$

and this may be rearranged to give the less-than Green's function as

$$
G^< = iA(\mathbf{k}, \omega) \left[\frac{\Sigma^<}{2i\Gamma} - \frac{A(\mathbf{k}, \omega)}{2} e\mathbf{E} \cdot \mathbf{v}_{\mathbf{k}} \frac{\partial f(\omega)}{\partial \omega} \right]. \tag{9.48}
$$

If we were to compare this with the linear response in the Boltzmann equation, it would be clear that the first term in the square brackets corresponds to the symmetrical near-equilibrium distribution function, while the second term corresponds to the streaming, current-carrying correction to the distribution.

To proceed in the quantum theme, it is now necessary to provide a form for the less-than self-energy. As expected, it is easy to show that it may be expressed in terms of the less-than Green's function, and this will provide an integral equation that must be solved (usually by iteration). Of course, this self-energy will also contain a phonon Green's function. Thus, we can write the self-energy for phonon scattering as a form of the Fermi golden rule as

$$\Sigma^<(\mathbf{k}, \omega) = \frac{2\pi}{\hbar} \sum_{\pm} \int \frac{d^3q}{(2\pi)^3} |M(\mathbf{k}, \mathbf{q})|^2 G^<(\mathbf{k} \pm \mathbf{q}, \omega \mp \omega_q). \quad (9.49)$$

In this expression, the term M is the matrix element for the electron–phonon interaction and is inserted as a representation of the phonon Green's function. The term ω_q is the phonon frequency and is usually taken to be a constant for the optical phonons. The plus/minus signs refer to the two processes of phonon emission (–) by which the electron loses energy by the emission of a phonon and phonon absorption (+) by which the electron gains energy. To proceed, we now make the *ansatz* that Eq. (9.48) can be rewritten in the form (which decouples the infinite series that would otherwise arise)

$$G^< = iA(\mathbf{k}, \omega)\left[f(\omega) - \Lambda(\mathbf{k}, \omega)e\mathbf{E} \cdot \mathbf{v}_{\mathbf{k}} \frac{\partial f(\omega)}{\partial \omega}\right], \quad (9.50)$$

and we are seeking to evaluate the function $\Lambda(\mathbf{k}, \omega)$. Using this form in Eq. (9.49) leads us to

$$\Sigma^<(\mathbf{k}, \omega) = if(\omega)\Gamma - ie\mathbf{E} \cdot \mathbf{v}_{\mathbf{k}} \frac{\partial f(\omega)}{\partial \omega}$$

$$\times \frac{2\pi}{\hbar} \sum_{\pm} \int \frac{d^3q}{8\pi^3} |M(\mathbf{k}, \mathbf{q})|^2 \frac{\mathbf{v}_{\mathbf{k}\pm\mathbf{q}}}{\mathbf{v}_{\mathbf{k}}} A(\mathbf{k} \pm \mathbf{q}, \omega \mp \omega_q)\Lambda(\mathbf{k} \pm \mathbf{q}, \omega \mp \omega_q),$$

$$(9.51)$$

where

$$\Gamma(\mathbf{k}, \omega) = \frac{2\pi}{\hbar} \sum_{\pm} \int \frac{d^3q}{8\pi^3} |M(\mathbf{k}, \mathbf{q})|^2 A(\mathbf{k} \pm \mathbf{q}, \omega \mp \omega_q). \quad (9.52)$$

This now gives us back our initial assertion on the form of the less-than Green's function if the kernel now satisfies the integral equation

$$
\Lambda(\mathbf{k}, \omega) = \frac{A(\mathbf{k}, \omega)}{2}
$$

$$
+ \frac{2\pi i}{\hbar \Gamma} \sum_{\pm} \int \frac{d^3\mathbf{q}}{8\pi^3} |M(\mathbf{k}, \mathbf{q})|^2 \frac{\mathbf{v}_{\mathbf{k} \pm \mathbf{q}}}{\mathbf{v}_{\mathbf{k}}} A(\mathbf{k} \pm \mathbf{q}, \omega \mp \omega_q) \Lambda(\mathbf{k} \pm \mathbf{q}, \omega \mp \omega_q).
$$

$$(9.53)$$

The kernel defined in Eq. (9.53) can be regarded as a *vertex correction*, just as those discussed earlier in Chapter 4. That is, the integral is solved to provide a modification of the actual scattering strength. In addition, the presence of the spectral density also tells us that the off-shell corrections are being provided here as well. We will also see this sequence of solutions of first the retarded function and then the less-than function in many other applications as we move through discussions of the real-time Green's functions.

9.4 A Single-Electron Tunneling Example

Let us now turn to a simple example in which we have a quantum dot coupled to a pair of reservoirs by tunneling barriers, as shown in Fig. 9.2. This creates a device that we have already treated in Section 6.6, and for which the modern high technology version was discussed in Section 1.4. The quantum dot is a strongly inhomogeneous device, and the carrier density within the dot can be far from its equilibrium value. The approach that will be followed here is from the initial work of Meir et al. [17] in which conductance oscillations arising from the charging and discharging of the dot were analyzed with real-time Green's functions. The details of the present approach largely follow that of Jauho et al. [18].

The reservoirs at each end of the device can be considered to have been decoupled from the dot in the remote past and are in thermal equilibrium. In each of these reservoirs, the distribution function is given by the respective Fermi energy. Similarly, the dot is considered to have been in thermal equilibrium in the remote past and is described by its own Fermi energy. Normally, one thinks of all the regions approaching a common equilibrium so that a single Fermi energy appears throughout. However, we now want to consider a situation in which the central quantum dot has not been allowed to equilibrate with the reservoirs. As one progresses from the distant past up to the present time, we consider a situation in which the coupling between the dot and the reservoirs are established and treated as perturbations. The actual couplings do not have to be small (but remain sufficiently small as to be described as tunneling). The time-dependent response can be similarly treated. In the distant past, the single-particle energies will have developed rigid time-dependent self-energy shifts, which, in the case of the non-interacting reservoirs, translate into extra phase factors for the propagators. A problem is that the calculations are made more difficult by the broken time translational invariance.

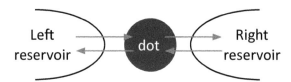

Figure 9.2 A conceptual structure for a quantum dot located between two metallic reservoirs. The arrows represent tunneling to/from the dot from/to the reservoirs.

The Hamiltonian for this system is broken into three pieces: $H = H_c + H_T + H_{dot}$, where H_c describes the reservoirs, H_T describes the coupling between the reservoirs and the dot via tunneling, and H_{dot} describes the central region quantum dot. In the reservoirs, the electrons are assumed to be non-interacting (as mentioned above) except for some overall self-consistent potential, which arises for example by a voltage difference between the two reservoirs. If we apply such a potential, which is also time varying, then the single-particle energies become time dependent as well, and may be writen as

$$E_{k\alpha}(t) = E^0_{k\alpha} + \Delta_\alpha(t), \tag{9.54}$$

where α is a label to denote the channel in the left or right reservoir, where these channels arise from the reduced dimensionality of the small reservoirs, in keeping with the Landauer formula described in Chapter 2. Here, however, we will narrow this set of channels and use the label to describe the left and/or right reservoir. The application of the bias potential does not change the occupation of either the channel or the reservoir in general, as this is determined by the thermal equilibrium distribution in the reservoir. Thus, we may write the Hamiltonian for the reservoirs as [17]

$$H_c = \sum_k \sum_{\alpha \subset L, R} E_{k\alpha}(t) c_{k\alpha}^\dagger c_{k\alpha},$$ (9.55)

where $c_{k\alpha}$ $(c_{k\alpha}^\dagger)$ are the annihilation (creation) operators for the density in the reservoirs, and the product appearing in this equation is the number of electrons in the particular state of the particular reservoir. The exact Green's function (we will use lower case letters for these non-interacting Green's functions) for the reservoirs in the uncoupled case may be written as

$$g_{k\alpha}^<(t,t') = i\left\langle c_{k\alpha}^\dagger(t') c_{k\alpha}(t) \right\rangle = if(E_{k\alpha}^0) e^{-\frac{i}{\hbar} \int_{t'}^{t} dt'' E_{k\alpha}(t'')}.$$ (9.56)

If the time limits were extended over all time, the exponential would give a delta function, but here represents the temporal version of the spectral density. A similar form can be written for the retarded Green's function as

$$g_{k\alpha}^r(t,t') = -i\theta(t-t')\left\langle \left\{ c_{k\alpha}(t), c_{k\alpha}^\dagger(t') \right\} \right\rangle$$
$$= -i\theta(t-t') e^{-\frac{i}{\hbar} \int_{t'}^{t} dt'' E_{k\alpha}(t'')}.$$ (9.57)

The quantum dot Hamiltonian depends on the specific geometry of the quantum region, as this gives rise to a particular set of levels. Two forms for this Hamiltonian will be considered. The first is just a generic version, equivalent to Eq. (9.55), but with a different set of operators for the electrons in the quantum dot. We write this form as

$$H_{dot} = \sum_m E_m(t) d_m^\dagger d_m ,$$ (9.58)

where the creation and annihilation operators create or destroy a state in the quantum dot. The second model that will be considered will have only a single resonant level in the quantum dot but will allow for the electrons in this dot to interact with phonons. We write this form of the Hamiltonian as

$$H'_{dot} = E_0 d^\dagger d + d^\dagger d \sum_q M(q)(a_q^\dagger + a_q).$$ (9.59)

The second term in the Hamiltonian represents the electron–phonon interaction, and the operators in the parentheses are phonon operators. Finally, the coupling between the reservoirs and the dot can be expressed by the tunneling interaction, which is written as a hopping term for the electrons, as

$$H_T = \sum_k \sum_{\alpha \subset L, R} \sum_n \left[V_{k\alpha,n}(t) c_{k\alpha}^\dagger d_n + h.c. \right].$$ (9.60)

As we mentioned above, the set of dot operators forms a complete orthonormal set of the available states within the dot, whether there are multiple levels or only a single level. The explicit term represents the transition of an electron from the dot to one of the reservoirs, and the indicated Hermitian conjugate would correspond to the transition from the reservoir to the dot.

9.4.1 Current

The current that flows into the dot from the left reservoir (which we take to be the source region and this is maintained at a ground potential) is easily related to the change in the density in this region due to transitions into the quantum dot. This current corresponds to electrons flowing *out of the dot* to the reservoir and leads to an increase in the number of carriers in the reservoir. This current may be expressed as

$$J_L(t) = e \left\langle \frac{\partial N_L}{\partial t} \right\rangle = \frac{ie}{\hbar} \langle [H, N_L] \rangle,$$ (9.61)

where

$$N_L = \sum_{k, \alpha = L} c_{k\alpha}^\dagger c_{k\alpha}.$$ (9.62)

The term H is just the total Hamiltonian, and e is taken as a positive quantity. Since both the reservoir term and the dot term commute with the Hamiltonian, one readily finds

$$J_L = \frac{ie}{\hbar} \sum_{k,n} \left[V_{kL,n} \left\langle c_{kL}^\dagger d_n \right\rangle - V_{kL,n}^* \left\langle d_n^\dagger c_{kL} \right\rangle \right],$$ (9.63)

where the second term has a reversed sign as it represents an electron moving to the dot from the left reservoir.

To proceed, it is useful to define two correlation functions that describe the coupling between the dot and one of the contacts. These correspond to the less-than and greater-than real-time Green's functions, which may be defined as

$$G_{k\alpha,n}^{<}(t,t') = i\langle c_{k\alpha}^{\dagger}(t')d_n(t)\rangle$$
$$G_{k\alpha,n}^{>}(t,t') = i\langle d_n^{\dagger}(t')c_{k\alpha}(t)\rangle.$$

(9.64)

Now these two Green's functions are related to one another by $G_{k\alpha,n}^{<} = -(G_{k\alpha,n}^{>})^*$; we can now rewrite Eq. (9.63) as

$$J_L = \frac{2e}{\hbar}\text{Re}\left\{\sum_{k,n}V_{kL,n}G_{k\alpha,n}^{<}(t,t)\right\}.$$

(9.65)

The task now is to evaluate this correlation function in order to find the current. The approach to this is standard, in that we develop an equation of motion for the Green's function, which couples this particular correlation function to the Green's functions in both the reservoir and the dot. This occurs via the interaction terms on the right-hand side of the equation, which must be separated via the Langreth theorem as previously. This equation of motion is then integrated and can be written as [18]

$$G_{k\alpha,n}^{<}(t,t') = \sum_{m}\int dt'' V_{k\alpha,m}^*(t'')\Big[G_{nm}^r(t,t'')g_{k\alpha}^{<}(t'',t')$$
$$+ G_{nm}^{<}(t,t'')g_{k\alpha}^a(t'',t')\Big],$$

(9.66)

where, for example, $G_{nm}^t(t,t') = -i\langle T\{d_m^{\dagger}(t')d_n(t)\}\rangle$, and this can be used with the interrelationships of the Green's functions in Eq. (9.9) to develop the full set of functions. In Eq. (9.66), the indices n and m both refer to the dot states, and the dot Green's functions correspond to mixing between the densities in the different dot levels. We can now use this result in the current equation, making use of the separation properties, as

$$J_L(t) = \frac{2e}{\hbar} \text{Im} \left\{ \sum_{k,n,m} V_{kL,n} \int_{-\infty}^{t} dt'' V_{kL,m}^* e^{\frac{i}{\hbar}\int_{t'}^{t''} d\eta E_{kL}(\eta)} \right.$$

$$\left. \times \left[G_{nm}^r(t,t'') f_L(E_{kL}) + G_{nm}^<(t,t'') \right] \right\}.$$

(9.67)

We recognize that the summation over k can be converted to an integration over the energy and the density of states. In addition, we can define the left-hand level broadening (the imaginary part of the left-hand self-energy for that reservoir) through the terms

$$\Gamma_{mn}^L(E,t',t'') = 2\pi \rho_L(E) V_{kL,n}(t') V_{kL,m}^*(t'') e^{\frac{i}{\hbar}\int_{t'}^{t''} d\eta E_{kL}(\eta)}.$$

(9.68)

Finally, we arrive at a simplified version of the current, which is

$$J_L(t)$$

$$= \frac{2e}{\hbar} \text{Im} \left\{ \sum_{n,m} \int_{-\infty}^{t} dt'' \int \frac{dE}{2\pi} \Gamma_{mn}^L(t'',t) \left[G_{nm}^r(t,t'') f_L(E_{kL}) + G_{nm}^<(t,t'') \right] e^{iE(t-t'')/\hbar} \right\}.$$

(9.69)

This is one of the main results of Ref. [13]. The current flowing into the quantum dot is a function of the local properties of the reservoir through its distribution function and spectral density, the latter of which leads to the level broadening. The first term in the square bracket contains the occupation of the reservoir states and corresponds to the hopping of carriers from the reservoir into the dot, while the last term corresponds to the equivalent out hopping of the carriers. Of course, an equivalent expression can be written for the current into the dot from the right-hand reservoir, and this will be necessary to treat the entire system.

9.4.2 Proportional Coupling in the Leads

By proportional coupling, we mean to invoke the general assumption that the linewidth functions of the two reservoirs differ only by a multiplicative constant. That is, the coupling through the two ends of the quantum dot may be written as

$$\Gamma_{mn}^L = \lambda \Gamma_{mn}^R.$$

(9.70)

In general, this will be satisfied if the two level shifts in the reservoirs are equal. While this is not an absolute requirement, it is likely to be almost true in most cases. The number of electrons tunneling to/from the dot is certainly much smaller than the population of the reservoirs, particularly if the latter are metals. The change in the population of the dot region can be simply expressed as

$$-e\frac{dN_{dot}(t)}{dt} = J_L(t) + J_R(t),$$
(9.71)

where both reservoir currents are inward flowing, and this last equation is just a statement of current continuity and Gauss' law. We can write the actual current as $J = xJ_L - (1-x)J_R$, which accounts for the fact that the particle current in the two reservoirs (or through the two tunneling regions) may differ somewhat in the time-varying case. This will lead to a fluctuation in the population of the dot. To proceed, we can introduce the current split as $J_L = xJ_L + (1-x)J_L$, from which we can use Eq. (9.71) to obtain

$$J_L(t) = xJ_L(t) + (1-x)\left[-J_R(t) - e\frac{dN_{dot}(t)}{dt}\right]$$
$$= \left[xJ_L(t) + (1-x)J_R(t)\right] - e(1-x)\frac{dN_{dot}(t)}{dt}.$$
(9.72)

We can now introduce Eq. (9.69) and an equivalent version for the right-hand current, so that

$$J_L(t) = -(1-x)e\frac{dN_{dot}(t)}{dt} + \frac{e}{\pi\hbar}\text{Im}\left\{\sum_{n,m}\int dt''\int dE e^{iE(t-t'')/\hbar}\Gamma^R_{mn}(t'',t)\right.$$

$$\times\left[\lambda xG^r_{mn}(t,t'')f_L(E) - (1-x)G^r_{mn}(t,t'')f_R(E)\right.$$

$$\left.\left. + \lambda xG^<_{mn}(t,t'') - (1-x)G^r_{mn}(t,t'')\right]\right\}.$$
(9.72)

The value of λ and of x are both arbitrary, so that we lose no generality by selecting a relation between two as $x = 1/(1+\lambda)$. This causes the terms on the third line above to vanish, which leads to the simpler form

$$J_L(t) = -\frac{\lambda}{1+\lambda}e\frac{dN_{dot}(t)}{dt} + \frac{e}{\pi\hbar}\frac{\lambda}{1+\lambda}\text{Im}\left\{\sum_{n,m}\int dt''\int dE e^{iE(t-t'')/\hbar}\right.$$

$$\left.\times\Gamma^R_{mn}(t'',t)G^r_{mn}(t,t'')\left[f_L(E) - (1-x)f_R(E)\right]\right\}.$$
(9.73)

The approach taken by these authors [18] is to decompose the couplings into energy- and time-dependent parts as $V_{k\alpha}(E, t) = u(t) V_\alpha(E_k)$, which implies that the interaction between the reservoirs and the quantum dot is turned on at $t = 0$. This now allows us to compute a long-time average, corresponding to the nonequilibrium steady state, of the current as

$$\langle J_{\mathrm{L}}(t) \rangle = \frac{e}{\pi\hbar} \int dE \left[f_{\mathrm{L}}(E) - f_{\mathrm{R}}(E) \right] \mathrm{Im} \left\{ \sum_{n,m} \frac{\Gamma^{\mathrm{R}}_{mn}(E)\Gamma^{\mathrm{L}}_{mn}(E)}{\Gamma^{\mathrm{R}}_{mn}(E)+\Gamma^{\mathrm{L}}_{mn}(E)} \right.$$

$$\left. \times \langle u(t)A_{mn}(E,t) \rangle \right\} \qquad (9.74)$$

where the first term has averaged to zero, which means that the number of electrons on the dot is in steady state, and

$$A_{mn}(E,t) = \int_{-\infty}^{t} dt'' e^{iE(t-t'')/\hbar + \frac{i}{\hbar}\int_{t''}^{t} d\eta \Delta(\eta)} u(t'') G^{r}_{mn}(t,t''). \qquad (9.75)$$

Here, Δ is the energy shift represented by the real part of the self-energy in the reservoirs. Because of the symmetry that now appears in the linewidth functions, it is no longer necessary to distinguish between the two reservoirs and the amplitude of the two currents is now equal. This last equation has the form of a Landauer relationship, although the summations over the different energy levels in the dots are still buried within the linewidths themselves. A more proper form will be attained later.

9.4.3 Non-interacting Resonant-Level Model

Let us now turn to the other case, described by the quantum dot Hamiltonian (9.59), where we have the electron–phonon interaction within the dot itself. For this purpose, we consider only a single level in the dot, so that the phonons will insert themselves into the tunneling process rather than scattering between levels in the dot. As above, we will continue to assume that the time dependence of the tunneling process can be divided into the product of two different terms. Here, however, we will also assume that this time dependence may be slightly different in the two reservoirs, as $V_{k\alpha}(t) = u_\alpha(t)V_\alpha(E_k)$ where the subscript n has gone away since we have only a single dot level. Now the time-dependent function is not necessarily a

Heaviside function, but could be any slowly varying time-dependent function. Further, we will ignore the real part of the self-energies, the level shifts, and consider the two imaginary parts of the self-energy to be energy independent. With these approximations, we can now write the retarded self-energy of the quantum dot as

$$\Sigma^r(t,t') = -i \sum_{\alpha \subset L, R} u_\alpha^*(t) u_\alpha(t') \int \frac{dE}{2\pi} e^{-iE(t-t')/\hbar} \theta(t-t') \Gamma_\alpha$$

$$\equiv -\frac{i}{2}\left[\Gamma_L(t) + \Gamma_R(t)\right]\delta(t-t'). \tag{9.76}$$

This result can now be used in the equation of motion for the retarded Green's function, which gives

$$G^{r,a}(t,t') = g^{r,a}(t,t') e^{\mp \frac{i}{2\hbar} \int_{t'}^{t} [\Gamma_L(t'') + \Gamma_R(t'')] dt''}. \tag{9.77}$$

Here the Green's function on the right-hand side is given by Eq. (9.57) without the subscripts, as this Green's function describes the single central dot state rather than the reservoir states. Within the spirit of these same approximations, we also can now write the less-than Green's function for the quantum dot as

$$G^<(t,t') = \frac{i}{\hbar} \int dt_1 \int dt_2 G^r(t,t_1) \Sigma^<(t_1,t_2) G^a(t_2,t')$$

$$= \frac{i}{\hbar} \int dt_1 \int dt_2 G^r(t,t_1) \left[\sum_{\alpha \subset L, R} \int \frac{dE}{2\pi} e^{-iE(t_1-t_2)/\hbar} \right.$$

$$\left. \times f_\alpha(E) \Gamma^\alpha(E,t_1,t_2) \right] G^a(t_2,t'). \tag{9.78}$$

This may now be used in our current equation (9.69) to give us

$$J_L = \frac{e}{\hbar}\left[\Gamma^L(t) N(t) + \int dE f_L(E) \int_{-\infty}^{t} dt' \Gamma^L(t',t) \text{Im}\{e^{iE(t-t')/\hbar} G^r(t,t')\} \right], \tag{9.80}$$

where

$$N(t) = \sum_{\alpha \subset L, R} \Gamma^\alpha \int \frac{dE}{2\pi} f_\alpha(E) |A_\alpha(E,t)|^2 \tag{9.81}$$

is the occupation of the quantum dot, and we have used

$$A_\alpha(E,t) = \int dt' u_\alpha(t') G^r(t,t') e^{iE(t-t')/\hbar - \frac{i}{\hbar} \int_t^{t'} \Delta_\alpha(t'') dt''} . \qquad (9.82)$$

If we now assume that the two time functions are the same and follow the same procedure in averaging the current, as done in the previous section, we again find that the average current may now be written as

$$\langle J \rangle = \frac{e}{\pi h} \frac{\Gamma^L \Gamma^R}{\Gamma^L + \Gamma^R} \int dE \, \mathrm{Im} \left\{ f_L(E) \langle u(t) A_L(E,t) \rangle - f_R(E) \langle u(t) A_R(E,t) \rangle \right\} \qquad (9.83)$$

This result does not readily show the role of the phonons in the tunneling process, but with the phonon process included, we find the retarded Green's function as

$$G^r(t) = -i\theta(t) e^{-[it(E_0 - \Delta) - \Gamma/2]/\hbar - \Phi(t)} , \qquad (9.84)$$

where

$$\Delta = \sum_q \frac{|M_q|^2}{\hbar \omega_q} \qquad (9.85)$$

$$\Phi(t) = \sum_q \frac{|M_q|^2}{\hbar \omega_q} \left[N_q (1 - e^{i\omega_q t}) + (N_q + 1)(1 - e^{-i\omega_q t}) \right].$$

With these factors, the current now becomes

$$\langle J \rangle = \frac{e}{\pi h} \frac{\Gamma^L \Gamma^R}{\Gamma^L + \Gamma^R} \int dE \left[f_L(E) - f_R(E) \right] \int_{-\infty}^{\infty} dt e^{iEt/\hbar} \, \mathrm{Im} \left\{ G^r(t) \right\} . \qquad (9.86)$$

The straightforwardness of this result should clearly illustrate the interacting current model that is used here [18].

9.5 Landauer Formula and Mesoscopic Devices

In Section 9.4.3, a result (9.74) was obtained that had a very close relationship to the Landauer formula. In this section, we want to follow a slightly different approach to the conductance and the Landauer formula. For this, we will consider a mesoscopic system, still in the absence of many-body interactions and scattering. We embed the active region, the quantum dot from above or some other small system for which ballistic transport can be assumed, between two contact regions [19]. The general idea is that there will be some coupling between the active region and the contacts, which can be described by a so-called self-energy Γ. In some sense, this approach follows the prototypical tunneling device formulated in the previous section. As before, we take a set of fermion operators in the contacts, which are given by $c_{k\alpha}$, $c_{k\alpha}^{\dagger}$, where α takes the values L and R for the left and right contact, respectively, and d_n, $d_n^{\dagger}$ in the "active" region, which correspond to creation (and annihilation) of a particle in a particular energy level within the active region. For this approach to be valid, the basic "active region" must be closed in the quantum-mechanical sense. That is, the connections to the environment must be tunneling barriers, which preserve the totality of the states within the active region. In fact, this approach can go beyond this limit, and the current-carrying states do not have to correspond to eigenvalues of the active region Hamiltonian. Now the overall Hamiltonian for this system remains that of the previous section with the terms given by Eq. (9.55) for the reservoirs, Eq. (9.58) for the active region, and Eq. (9.60) for the tunneling interaction between the active region and the reservoirs.

With this Hamiltonian, it is now possible to compute the current flow through the device and the resulting conductance by direct application of our Green's function techniques. Hence, the current density may be written as Eq. (9.63) for the left reservoir current, and an equivalent form for the right reservoir current. The left current can also be written as

$$ J = J_{\mathrm{L}} = \frac{e}{h} \sum_{k,n} \int_{-\infty}^{\infty} d\omega \left[V_{kL,n} G_{n,kL}^{<}(\omega) - V_{kL,n}^{\dagger} G_{kL,n}^{<}(\omega) \right]. \qquad (9.87) $$

The two Green's functions that appear in the last line of Eq. (9.87) can be defined by

$$ G_{k\alpha,n}^{<}(\omega) = \sum_{m} V_{k\alpha,m} \left[g_{k\alpha}^{t}(\omega) G_{mn}^{<}(\omega) - g_{k\alpha}^{<}(\omega) G_{mn}^{\bar{t}}(\omega) \right] $$

$$ G_{n,k\alpha}^{<}(\omega) = \sum_{m} V_{k\alpha,m}^{\dagger} \left[g_{k\alpha}^{<}(\omega) G_{nm}^{t}(\omega) - g_{k\alpha}^{\bar{t}}(\omega) G_{nm}^{<}(\omega) \right]. \qquad (9.88) $$

As in the previous section, the small gs are the equilibrium values of the various Green's functions in the reservoirs. Each of these two Green's functions above connects a transition from a contact to the active region, or vice versa. For example, the two correlation functions on the right-hand side can be written as

$$G_{nm}^< = i\langle d_m^\dagger d_n \rangle, \quad G_{nm}^> = -i\langle d_n d_m^\dagger \rangle$$

$$g_{k\alpha}^< = 2\pi i f_\alpha(\omega)\delta(\omega - E_{k\alpha}/\hbar) \tag{9.89}$$

$$g_{k\alpha}^> = -2\pi i\left[1 - f_\alpha(\omega)\right]\delta(\omega - E_{k\alpha}/\hbar),$$

and the various time functions now can be found from Eq. (9.9). With these various relationships, Eq. (9.87) can be rewritten as

$$J_{\mathrm{L}} = \frac{e}{h}\sum_{k,n}\int_{-\infty}^{\infty}d\omega\Bigg\{V_{kL,n}\sum_m V_{kL,m}^\dagger\left[g_{kL}^< G_{nm}^t(\omega) - g_{kL}^t G_{nm}^<(\omega)\right]$$

$$-V_{kL,n}^\dagger\sum_m V_{kL,m}\left[g_{kL}^t G_{mn}^<(\omega) - g_{kL}^< G_{mn}^t(\omega)\right]\Bigg\}. \tag{9.90}$$

This is our basic result, but with a little further manipulation, it can be put into a more meaningful form. First, we replace the sum over wavevector (in the contacts) with a sum over the energy, which means introducing the density of states, as

$$J_{\mathrm{L}} = \frac{ie}{\hbar}\sum_{n,m}\int dE\rho(E)V_{kL,n}V_{kL,m}^\dagger\left[f(E_{kL})(G_{nm}^r - G_{nm}^a) - G_{nm}^<\right]. \tag{9.91}$$

We now introduce the couplings that arise in the self-energies through the relations

$$\Gamma_{\mathrm{L}} = \sum_{n,m}\Gamma_{nm}^{\mathrm{L}} = 2\pi\sum_{n,m}\rho(E)V_{kL,n}V_{kL,m}^\dagger \tag{9.92}$$

for the left reservoir, with a similar version for the right reservoir. With this definition, the left-derived current (9.91) may be written as

$$J_{\mathrm{L}} = \frac{ie}{h}\int dE\mathrm{Tr}\left\{\Gamma_{nm}^{\mathrm{L}}f(E_{\mathrm{L}})(G_{nm}^r - G_{nm}^a) - \Gamma_{nm}^{\mathrm{L}}G_{nm}^<\right\}. \tag{9.93}$$

A similar result arises naturally for the current in the right reservoir, remembering that both these currents are *into* the quantum dot region. Then the average current for both leads together may be written as

$$J_{\mathrm{L}} = \frac{ie}{2h} \int dE \mathrm{Tr}\left\{ \left[\Gamma^{\mathrm{L}}_{nm} f(E_{\mathrm{L}}) - \Gamma^{\mathrm{R}}_{nm} f(E_{\mathrm{R}}) \right] (G^{\mathrm{r}}_{nm} - G^{\mathrm{a}}_{nm}) - (\Gamma^{\mathrm{L}}_{nm} - \Gamma^{\mathrm{R}}_{nm}) G^{<}_{nm} \right\}.$$
(9.94)

As we have pointed out above, we assume that the transport is ballistic, so that we can simplify the correlation functions in this region as

$$G^{<} = i f_{\mathrm{L}} \,\mathrm{Im}\{A_{\mathrm{L}}\} + i f_{\mathrm{R}} \,\mathrm{Im}\{A_{\mathrm{R}}\} = i f_{\mathrm{L}} G^{\mathrm{r}} \Gamma^{\mathrm{L}} G^{\mathrm{a}} + i f_{\mathrm{L}} G^{\mathrm{r}} \Gamma^{\mathrm{R}} G^{\mathrm{a}}$$
$$G^{\mathrm{r}} - G^{\mathrm{a}} = -2 \,\mathrm{Im}\{A_{\mathrm{L}} + A_{\mathrm{R}}\} = 2 G^{\mathrm{r}} (\Gamma^{\mathrm{L}} + \Gamma^{\mathrm{R}}) G^{\mathrm{a}}.$$
(9.95)

Finally, we can obtain the total current through the device as

$$
\begin{aligned}
J_{\mathrm{L}} = \frac{e}{h} \int dE \mathrm{Tr} & \left\{ \Gamma^{\mathrm{L}}_{nm} f(E_{\mathrm{L}}) G^{\mathrm{r}}_{nm} (\Gamma^{\mathrm{L}}_{nm} - \Gamma^{\mathrm{R}}_{nm}) G^{\mathrm{a}}_{nm} \right. \\
& - \Gamma^{\mathrm{R}}_{nm} f(E_{\mathrm{R}}) G^{\mathrm{r}}_{nm} (\Gamma^{\mathrm{L}}_{nm} - \Gamma^{\mathrm{R}}_{nm}) G^{\mathrm{a}}_{nm} \\
& \left. - (\Gamma^{\mathrm{L}}_{nm} - \Gamma^{\mathrm{R}}_{nm}) \left[f(E_{\mathrm{L}}) G^{\mathrm{r}}_{nm} \Gamma^{\mathrm{L}}_{nm} G^{\mathrm{a}}_{nm} + f(E_{\mathrm{R}}) G^{\mathrm{r}}_{nm} \Gamma^{\mathrm{R}}_{nm} G^{\mathrm{a}}_{nm} \right] \right\} \\
= \frac{e}{h} & \int dE \left[f(E_{\mathrm{L}}) - f(E_{\mathrm{R}}) \right] \mathrm{Tr}\left\{ \Gamma^{\mathrm{L}}_{nm} G^{\mathrm{r}}_{nm} \Gamma^{\mathrm{R}}_{nm} G^{\mathrm{a}}_{nm} \right\}.
\end{aligned}
$$
(9.96)

The last trace term is often written as $\mathrm{Tr}\{tt^{\dagger}\}$ where t is the mode-to-mode transmission matrix discussed in Chapter 2.

Now, as discussed above, this current density equation results from ballistic transport and the only contribution to the self-energy comes from coupling between the contacts and the active region, which is assumed to be tunneling in nature. If there were proper scattering processes considered, then the approximations for the correlation functions would fail. Thus, the approach is limited to really ballistic devices but cannot be applied to devices in which there is real dissipation occurring. There is an important result here. It may be noted that the final result depends only on the retarded and advanced Green's functions of the quantum dot region, as well as the two self-energies that define the coupling to the reservoirs. None of the important real-time Green's functions, such as the less-than and greater-than functions appear in this expression. Hence, it may be concluded that this result could as easily have been obtained using nothing but the thermal Green's functions—this is not really a meaningful application of the real-time Green's functions.

There have been many groups who have pursued this approach, or an earlier variant of it, for studying various mesoscopic systems.

This approach was used in the computation of the earliest studies of transport through a molecule attached to a pair of metallic leads, in a Green's function formalism [20]. Here the coupling between the molecule and the leads is treated as a rather mystical tunneling coefficient [21, 22]. Although most of these approaches use Green's functions, they can as easily be formulated with the scattering matrix approach of Chapter 2 [23–25]. The key aspect of any of these approaches is that one must know the atomic structure in the molecule as well as in the metallic leads. In the present treatment, as well as that of the last section, the atomic structure was found by assuming normal electron bands, as all the transport was in semiconductors. In the approach of these latter papers, however, the molecules are characterized by localized wavefunctions, while the metallic leads have iterant electron character. As a result, it is quite difficult to treat both the molecules and the leads with equal accuracy. The common approach is to use something like a tight-binding basis, but this really assumes localized orbitals, and it is difficult, for example, to accurately recover the work function of the metallic leads [25, 26]. If the work function cannot be accurately computed by the atomic structure calculation, then there is no chance that the real potential barrier between the lead and the molecule will be determined in a realistic manner. Hence, the coupling between the two is no more than an adjustable parameter, and the fact that the current is still not computed correctly, even with this adjustable parameter, is disappointing. Nevertheless, it is imperative in these latter approaches to have a viable method of computing the atomic structure throughout the mesoscopic system.

9.6 Green–Kubo Formula

The Kubo formula was developed originally as a linear-response approach to the applied fields, and the latter were represented by the vector potential [27], although there is slightly earlier work in this area by Green [28]. The use of the Kubo formula was a significant change from the normal treatment of the dominant streaming terms of the Boltzmann equation, or of the equivalent quantum transport equations, to the relaxation and/or scattering terms. With the Kubo formula, one concentrates on the relaxation processes through the correlation functions that describe the transport. Here we talk about how the real-time Green's functions fit into the Kubo formula, with the combination termed the Green–Kubo formula. Now if we note that the quantum-mechanical current is described by

$$J(\mathbf{r}) = \frac{e\hbar}{2im*}\left[\psi^\dagger(\mathbf{r})\frac{\partial\psi(\mathbf{r})}{\partial\mathbf{r}} - \frac{\partial\psi^\dagger(\mathbf{r})}{\partial\mathbf{r}}\psi(\mathbf{r})\right], \tag{9.97}$$

it is not too difficult to develop a Green's function form of this quantity to use in the Kubo formula. To begin with, we note that this latter equation can be written in terms of the Green's functions as

$$\langle J(\mathbf{r},t)\rangle = -\frac{e\hbar}{2m*}\lim_{\substack{\mathbf{r}\to\mathbf{r}'\\t\to t'}}\left(\frac{\partial}{\partial\mathbf{r}} - \frac{\partial}{\partial\mathbf{r}'}\right)G^<(\mathbf{r},\mathbf{r}',t,t') + i\frac{ne^2}{m*\omega}\mathbf{E}(\mathbf{r},t). \tag{9.98}$$

The last term represents the displacement current and will be ignored (in this term, **E** is the applied field and will also appear in the current–current correlation function itself). The actual value of the current found from the Kubo formula is given by the current–current correlation function, as

$$\langle J(\mathbf{r},t)\rangle = \frac{1}{\hbar}\int_0^t dt' \int d^3\mathbf{r}' \langle[J(\mathbf{r},t'), J(\mathbf{r}',t-t')]\rangle\cdot\mathbf{A}(\mathbf{r}',t'). \tag{9.99}$$

Here the term $\mathbf{A}(\mathbf{r}, t)$ within the integral is the vector potential, which describes the driving field as well as any magnetic field that may be present. Forming the two currents with the aid of Eqs. (9.97) and (9.98) and using a different choice of variable for the internal integration, we then find the resultant current in terms of the real-time Green's functions as

$$\langle J(\mathbf{r},t)\rangle = -\frac{ie^2\hbar}{4m*^2}\lim_{\substack{\mathbf{r}\to\mathbf{r}'\\t\to t'}}\left(\frac{\partial}{\partial\mathbf{r}} - \frac{\partial}{\partial\mathbf{r}'}\right)\cdot\int dt_s \int d^3\mathbf{s}\lim_{\substack{\mathbf{s}\to\mathbf{s}'\\t_s\to t'_s}}\left(\frac{\partial}{\partial\mathbf{s}} - \frac{\partial}{\partial\mathbf{s}'}\right)$$
$$\times\left[G^r(\mathbf{r},\mathbf{s}',t'_s)G^<(\mathbf{s},\mathbf{r}',t_s,t') + G^<(\mathbf{r},\mathbf{s}',t'_s)G^a(\mathbf{s},\mathbf{r}',t_s,t')\right]\mathbf{A}(\mathbf{s},t_s) \tag{9.100}$$

Actually, the Green's function product should be a two-particle Green's function involving four field operators. However, this has been expanded into the lowest-order Green's functions. It should be remembered, though, that this is an approximation, and higher-order Green's function products may be needed to treat certain physical processes. The result (9.100) can be Fourier transformed to give the a.c. conductivity for a homogeneous system (required to take the spatial Fourier transforms), at least on the average scale of the response functions, giving

$$
\sigma(\mathbf{k}, \omega) = -\frac{e^2 \hbar}{m^{*2} \omega} \int \frac{d^3 \mathbf{k}'}{(2\pi)^3} \int \frac{d\omega'}{2\pi} (\mathbf{k}' + \frac{\mathbf{k}}{2}) \cdot \mathbf{k}
$$

$$
\times \left[G^r(\mathbf{k}' + \frac{\mathbf{k}}{2}, \omega') G^<(\mathbf{k}' - \frac{\mathbf{k}}{2}, \omega' - \omega) \right. \tag{9.101}
$$

$$
\left. + G^<(\mathbf{k}' + \frac{\mathbf{k}}{2}, \omega') G^a(\mathbf{k}' - \frac{\mathbf{k}}{2}, \omega' - \omega) \right].
$$

After making a few change of variables, this form can be rewritten as

$$
\sigma(\mathbf{k}, \omega) = -\frac{e^2 \hbar}{m^{*2} \omega} \int \frac{d^3 \mathbf{k}'}{(2\pi)^3} \int \frac{d\omega'}{2\pi} \mathbf{k}_1 \cdot \mathbf{k}
$$

$$
\times \left[G^r(\mathbf{k}_1, \omega') G^<(\mathbf{k}_1 - \mathbf{k}, \omega' - \omega) + G^<(\mathbf{k}_1, \omega') G^a(\mathbf{k}_1 - \mathbf{k}, \omega' - \omega) \right].
$$

$$
\tag{9.102}
$$

This particular form differs somewhat from that used earlier, in that the real-time functions appear here. It should also be noted that the higher-order two-particle Green's functions have not been included, so this formulation is that of the lowest-order Green's functions, assuming that Wick's theorem is perfectly valid. We will remark about this again later. The form of Eq. (9.102) has not been utilized very much in transport calculations based on the real-time Green's functions. Nevertheless, it is important to note that the conductance here is an integral (actually, a double integral) over the current–current correlation function.

It is now useful to rearrange the terms into those more normally found in the equilibrium and zero-temperature forms of the Green's functions. For this, we make the *ansatz* (9.6), which is known to be correct in the equilibrium situation. We will see in the next chapter that it is also correct in some other approaches. With this ansatz, we can rewrite Eq. (9.102) as

$$
\sigma(\mathbf{k}, \omega) = -\frac{e^2 \hbar}{m^{*2} \omega} \int \frac{d^3 \mathbf{k}'}{(2\pi)^3} \int \frac{d\omega'}{2\pi} \mathbf{k}_1 \cdot \mathbf{k} \left[G^r(\mathbf{k}_1, \omega') G^a(\mathbf{k}_1 - \mathbf{k}, \omega' - \omega) f(\omega') \right.
$$

$$
- G^r(\mathbf{k}_1, \omega') G^r(\mathbf{k}_1 - \mathbf{k}, \omega' - \omega) f(\omega' - \omega)
$$

$$
\left. - G^r(\mathbf{k}_1, \omega') G^a(\mathbf{k}_1 - \mathbf{k}, \omega' - \omega) [f(\omega') - f(\omega' - \omega)] \right] \tag{9.103}
$$

The first two products in the square brackets will cancel one another. This can be seen by changing the frequency variables as $\omega'' = \omega'' - \omega$, and then using the fact that $\sigma(\mathbf{k}, \omega) = \sigma^*(\mathbf{k}, -\omega)$ and $G_r = G_a^*$. Thus, we are left with only the last term in the square brackets. For low frequencies (and we will go immediately to the long-time limit of $\omega = 0$), the distribution function can be expanded about ω', so that

$$\sigma(\mathbf{k},\omega) = -\frac{e^2\hbar}{m^{*2}\omega}\int \frac{d^3k'}{(2\pi)^3}\int \frac{d\omega'}{2\pi}\mathbf{k}_1\cdot\mathbf{k}G^r(\mathbf{k}_1,\omega')G^a(\mathbf{k}_1-\mathbf{k},\omega'-\omega)\frac{\partial f(\omega')}{\partial\omega'}.$$

(9.104)

Finally, at low frequencies and for homogeneous material, we arrive at the form

$$\sigma(\mathbf{k}) = -\frac{e^2\hbar}{m^{*2}}\int \frac{d^3k'}{(2\pi)^3}\int \frac{d\omega'}{2\pi}\mathbf{k}_1\cdot\mathbf{k}\left|G^r(\mathbf{k}_1,\omega')\right|^2\frac{\partial f(\omega')}{\partial\omega'}.$$

(9.105)

In the case of very low temperatures, one arrives back at the form used extensively in previous chapters. That is, we replace the derivative of the distribution function with the negative of a delta function at the Fermi energy. The sum over the momentum counts the number of states that contribute to the conductivity and results in the density at the Fermi energy at low temperature. In mesoscopic systems, where only a single transverse state may contribute (in a quantum wire, for example), the Landauer formula can easily be recovered when one recognizes that the squared magnitude of the Green's function represents the transmission of a particular mode. Even if there is no transverse variation, the integration over the longitudinal component of the wavevector will produce the difference in the Fermi energies at the two ends of the samples, as in the previous section. The reduction of the less-than function has allowed a separation of the density–density correlation function from the current–current correlation function. This latter is represented by the polarization that appears as the magnitude squared of the retarded Green's function.

The approach (9.93) has been extensively utilized by the Purdue group to model mesoscopic systems with the equivalent Landauer formula for nonequilibrium Green's functions [29, 30]. For mesoscopic waveguides in the linear-response regime, even with dissipation present, they have shown that this form can be extended to the use of a Wigner function, which can then be used to define a local thermodynamic potential and that reasonable results are obtained so long as these potentials are defined over a volume comparable in size to the thermal de Broglie wavelength. They have been particularly successful in probing inelastic tunneling in resonant-tunneling diodes through the emission of optical phonons. In general, however, the expression (9.105) is an approximation, in that the magnitude term in the retarded Green's function is really a lowest-order representation of the actual polarization that appears in the conductivity bubble, as discussed in Chapter 4. As was addressed in this earlier chapter, the extended version becomes

$$\sigma = -\frac{e^2\hbar}{m^{*2}} \int \frac{d^3\mathbf{k}}{(2\pi)^3} \int \frac{d\omega'}{2\pi} \frac{\partial f(\omega')}{\partial \omega'} \Pi(\mathbf{k},\omega') \tag{9.106}$$

where the polarization function satisfies the Bethe–Salpeter equation

$$\Pi(\mathbf{k},\omega') = G^r(\mathbf{k},\omega')G^a(\mathbf{k},\omega') \int \frac{d^3\mathbf{k}'}{(2\pi)^3} \left\{ \mathbf{k}\cdot\mathbf{k}'\delta(\mathbf{k}-\mathbf{k}') \right.$$
$$\left. + \frac{\mathbf{k}\cdot\mathbf{k}'}{k'^2}T(\mathbf{k}-\mathbf{k}')\Pi(\mathbf{k}',\omega') \right\}. \tag{9.107}$$

In this last expression, we have summed the conductivity over the Fourier variables in position to get the spatially averaged conductivity (equivalent to $\mathbf{r} = 0$, but with the origin at any point in the sample), and made a change of variables on the momentum variables for convenience. The polarization is the general product of the retarded and advanced functions that was dealt earlier, at least on a formal basis. While these results appear to be quite simple in form, it is important to point out that the form of the distribution function that appears in the equations still must be determined by the balance between the driving forces and the dissipative forces: In essence, the less-than Green's function must be determined in the nonequilibrium system.

9.7 Transport in a Silicon Inversion Layer

Carriers at the interface, for example, between a silicon dioxide layer and bulk silicon reside in a quasi-two-dimensional world as discussed in Section 2.4. The carrier momentum normal to the interface is quantized. This arises from the high potential barrier of the oxide and the depletion barrier within the bulk silicon. This potential well creates quantization of the momentum in the direction normal to the interface. The carriers then reside in one of a series of sub-bands, with free motion only in the directions parallel to the interface. The general transport in this system has been studied by a great variety of approaches, and the area was extensively reviewed more than a decade ago [31]. In general, the motion parallel to the interface can be separated from that normal to the interface, if the interface is a principal crystal axis (it is normal for the Si surface to be the [001] plane). In the normal direction, the sub-band minima are defined by the confinement energies, which result from the z-directed Schrödinger equation

$$H_{0\perp}\psi_n(z) = \varepsilon_n\psi_n(z). \tag{9.108}$$

The total energy is composed of this confinement energy and that of the free propagating plane waves in the inversion layer parallel to the interface, as

$$E_n = \frac{\hbar^2 k^2}{2m^*} + \varepsilon_n = E(k) + \varepsilon_n. \tag{9.109}$$

Here, $\mathbf{k}$ is now a two-dimensional momentum vector lying in the plane of the interface between the Si and the SiO_2. These quantum sub-bands that arise from the confining potentials have to be calculated self-consistently within the overall transport problem [32]. The self-energies that will be calculated later depend on the actual sub-band wavefunctions and energy levels. These self-energies, in turn, affect the actual population of the sub-bands and hence the self-consistent solution to the Schrödinger equation (9.108) and the Poisson's equation for the exact confining potential within the semiconductor. Usually, the latter two equations are solved within the Hartree approximation, but a better approach is to include the role of the exchange–correlation corrections to the sub-band energies. This calculation itself is a formidable numerical challenge if the confining potential is complicated in any way. Most often, some sort of approximation is made to this, but the results that are quoted below are obtained from a fully self-consistent solution, but without including the exchange–correlation corrections.

At low temperatures, the dominant scattering mechanisms in this structure are scattering by impurities, both within the bulk Si and in the oxide, and scattering off the potential fluctuations that arise from surface (interface) roughness. The latter is also important at higher temperatures (room temperature, for example) when the oxide field is high ($\sim 10^6$ V/cm), but it is dominated by normal phonon scattering at lower oxide fields. Here we will treat the low-temperature case.

Scattering by the ionized acceptors (in an n-channel device) that create the depletion charge also must include the image of these charges from reflection through the dielectric discontinuity at the silicon-oxide interface. The square of the matrix element for the scattering from the single impurity located at depth z_i and lateral position (a two-dimensional position vector) $\mathbf{r}_i$ is given by

$$\left|\langle \mathbf{k}, n | U(\mathbf{q}) | \mathbf{k}+\mathbf{q}, m\rangle\right|^2_{\text{depl}} = n_{\text{depl}} \left(\frac{e^2}{2\bar{\varepsilon}q}\right)^2 A_{nm}^2 \int_0^\infty O_{nm}^2(q, z_i)dz_i$$

$$(9.110)$$

for scattering between the n and m sub-bands, where $\bar{\varepsilon} = (\varepsilon_s + \varepsilon_{\text{ox}})/2$ is the average dielectric function. Here, $\mathbf{q}$ is the Fourier-transformed momentum exchange associated with the scattering process. However, there is no momentum conservation in the direction normal to the interface, since the quantization removes the z-momentum as a good quantum number. Hence, the overlap integral between initial and final sub-bands does not prohibit inter-sub-band scattering; rather, this integral is just the form factor arising from the finite extent of the electron gas in the quantization direction as

$$A_{nm}(q) = \int_0^\infty \psi_n^*(z)e^{-qz}\psi_m(z)dz .$$

$$(9.111)$$

The other factor in Eq. (9.110) describes the contribution of both the charge and its image. This latter expression is given by

$$O_{nm}(q, z_i) = \left(\frac{\varepsilon_s + \varepsilon_{\text{ox}}}{2\varepsilon_s}\right)e^{qz_i} + \left(\frac{\varepsilon_s - \varepsilon_{\text{ox}}}{2\varepsilon_s}\right)e^{-qz_i}$$

$$+ \left(\frac{\varepsilon_s + \varepsilon_{\text{ox}}}{2\varepsilon_s}\right)\left[\frac{a_{nm}^{(+)}(q, z_i)}{A_{nm}(q)}e^{-qz_i} + \frac{a_{nm}^{(-)}(q, z_i)}{A_{nm}(q)}e^{qz_i}\right] \quad (9.112)$$

where

$$a_{nm}^{(\pm)}(q,z_i) = \int_0^{z_i} \psi_n^*(z) e^{\pm qz} \psi_m(z) dz . \tag{9.113}$$

The quantity n_{depl} is the total depletion charge converted to a sheet density.

At the interface between the Si and the oxide, there are usually a significant number of defect states into which charge can be trapped. These states may arise from the disorder of the interface, from dangling bonds, or just from precipitated impurities. The effect of these charges is that of a sheet of scattering centers is located not exactly at the interface, but a small distance z_d (<0) into the oxide. This leads to a scattering strength of

$$\left| \langle \mathbf{k}, n | U_{\text{int}}(\mathbf{q}) | \mathbf{k}+\mathbf{q}, m \rangle \right|_{\text{int}}^2 = n_{\text{int}} \left(\frac{e^2}{2\bar{\varepsilon}q} \right)^2 A_{nm}^2 e^{2qz_d} . \tag{9.114}$$

By quite similar arguments, we also have to worry about charge, which may be trapped in the oxide itself. For a uniform distribution of charge through the oxide thickness d_{ox}, the scattering is given by

$$\left| \langle \mathbf{k}, n | U_{\text{ox}}(\mathbf{q}) | \mathbf{k}+\mathbf{q}, m \rangle \right|_{\text{ox}}^2 = n_{\text{ox}} \left(\frac{e^2}{2\bar{\varepsilon}q} \right)^2 A_{nm}^2 \frac{1-e^{-2qz_d}}{2q} . \tag{9.115}$$

In contrast to the previous cases of impurity scattering, the charge density n_{ox} is now a bulk quantity, which accounts for the extra factor of q in the denominator.

Surface-roughness scattering is associated with disorder at the interface or edge of the quantum-confining structure, and the general semi-classical approach considers how this random interface affects the transport by modulating the energy eigenvalue of each sub-band. Early treatments considered that the scattering strength, given in terms of the correlation function of the surface roughness, was described by a Gaussian correlation function for the roughness [31]. However, it was subsequently found from direct cross-sectional TEM imaging of the interface itself, that the roughness was much better described in terms of an exponential correlation [33]. These conclusions were corroborated by measurements of the surface itself (after oxidation and subsequent removal of the oxide by selective etches) with atomic force microscopy [34, 35]. In this case, the roughness correlation function is given by

$$S_{\exp}(q) = \frac{\pi \Delta^2 L^2}{(1 + q^2 L^2 / 2)^{3/2}},$$

(9.116)

where Δ and L are the mean height and correlation length of the roughness, respectively. In general then, the matrix element for scattering by the surface roughness is given as

$$\left| \langle \mathbf{k}, n | U_{sr}(\mathbf{q}) | \mathbf{k} + \mathbf{q}, m \rangle \right|_{sr}^2 = e^2 S(q) E_{\text{eff}}^2 A_{nm}^2(q),$$

(9.117)

where the effective electric field in the semiconductor is given as

$$E_{\text{eff}} = \frac{e}{\varepsilon}(n_{\text{depl}} + \eta n_{\text{inv}})$$

(9.118)

and n_{inv} is the two-dimensional inversion charge density in the semiconductor. There is some debate about the value of η with values of 1/2 expected in simple cases from Gauss' law [36] and 11/32 quoted for a variational calculation for the wavefunctions and energy levels in a triangular potential [37]. In general, the potential and matrix element in Eq. (9.117) is unscreened at the moment. Both Ando [38] and Saitoh [39] have pointed out that the roughness induces charge fluctuations, which means that it should be screened by a polarization function such as that of Eq. (4.56) (in two dimensions, of course). This means that the effective potential that appears in Eq. (9.118) can be written as

$$U_{\text{eff}}^2 = \frac{e^4 S^2(q) A_{nm}^2(q)(n_{\text{depl}} + \eta n_{\text{inv}})^2}{2\bar{\varepsilon} \left[1 + \frac{q_0}{q} F(q) \Pi(q) \right]^2},$$

(9.119)

where q_0 is an effective two-dimensional screening wavevector and $F(q)$ is a slowly varying function of order unity at low inversion densities and becomes smaller at higher densities. With these preliminary discussion of the scattering processes that are unique to this system, we are ready to move forward to the Green's functions.

9.7.1 Retarded Green's Function

As in the previous sections, nearly all one-electron properties of the interacting electron system can be determined from the electron propagator itself, which is the interacting retarded Green's function. The remaining factor that must be known is the distribution function, which will be treated with the less-than Green's function in the next section. Here we concentrate on the retarded function. In the interacting system, the retarded Green's function is related to the one-electron (non-interacting) retarded function by Dyson's equation and the introduction of the self-energy. Thus, we need first to find the form for the non-interacting retarded Green's function, then to construct the self-energy function, and finally to formulate the interacting retarded Green's function.

The equation of motion for the non-interacting retarded function G_{r0} is found from Eq. (9.18) by also setting both the potential terms $V(r)$ to zero. Then we can rewrite this equation as

$$\left(i\hbar\frac{\partial}{\partial t} - H_0(\mathbf{r})\right)G_{r0}(\mathbf{r},\mathbf{r}',t,t') = \hbar\delta(\mathbf{r}-\mathbf{r}')\delta(t-t'),\qquad(9.120)$$

where the equilibrium Hamiltonian $H_0(\mathbf{r})$ *includes* the confining potential of the quantum well at the interface. Because of the latter confinement, and because in general the system is separable, the non-interacting retarded Green's function will be expanded in a series over contributions from each of the sub-bands that form in the confinement potential. What were three-dimensional vectors $(\mathbf{r}, \mathbf{k})$ are now two-dimension vectors, with the added z-components (z, k_z), where $\mathbf{k}$ is defined by the confinement energy ε_n of the nth sub-band according to Eq. (9.109). We can then write this function in the Fourier transform form as

$$G_{r0}(\mathbf{k},\omega,z,z') = \sum_n \psi_n^\dagger(z')\psi_n(z)\frac{\hbar}{\hbar\omega - E(k) - \varepsilon_n + i\hbar\eta}\qquad(9.121)$$

where the last factor is written in analogy with Eq. (3.21). The unperturbed density of states for the two dimensions parallel to the oxide interface is given as [40]

$$\rho_{2D}(\hbar\omega) = \frac{g_v g_s m^*}{2\pi\hbar^2}\sum_n \theta(\hbar\omega - \varepsilon_n),\qquad(9.122)$$

where g_v and g_s are the valley and spin degeneracy factors, respectively. Normally, we take $g_s = 2$ in this work, as we have not spent time on spin-dependent transport.

Now the interacting retarded Green's function can be written almost immediately from Eq. (9.121) by introducing the self-energy in analogy with Eq. (4.48) as

$$G_r(\mathbf{k}, \omega, z, z') = \sum_n \psi_n^\dagger(z')\psi_n(z)g_{r,n}(\mathbf{k}, \omega)$$

$$= \sum_n \psi_n^\dagger(z')\psi_n(z)\frac{\hbar}{\hbar\omega - E(k) - \varepsilon_n - \Sigma_{r,n}(\mathbf{k}, \omega)}.$$

$$(9.123)$$

This diagonal approximation for the interacting retarded Green's function has been questioned, since a non-diagonal one might, for example, include a second summation over the various sub-band Green's function of different index within the first sum (and the two wavefunctions would then come from different sub-bands). However, the interaction between the various sub-bands is provided by the scattering processes themselves. We return to this point below. Thus, the appearance of diagonality in this equation is just an appearance; the result is fully non-diagonal once the self-energy is properly formulated. The retarded self-energy that appears in this last equation may be calculated, by the same sub-band expansion, to be

$$\Sigma_{r,n}(\mathbf{k}, \omega) = \int dz \int dz' \psi_n^\dagger(z')\Sigma_r(\mathbf{k}, \omega, z, z')\psi_n(z). \qquad (9.124)$$

Since impurity and surface-roughness scattering are independent scattering mechanisms, both of which may be assumed to be weak, they can be treated separately. This is different than the procedure used in Chapter 4, where impurity scattering was sufficient to induce disorder (diffusive transport) so that impurity scattering and electron–electron scattering had to be calculated by treatment of their interactions by multiple scattering. Here it has been assumed that the electron–electron interaction is screened only by itself and, in turn, provides simple screening to the Coulomb and surface-roughness interactions. Both of the latter two interactions are assumed to be sufficiently weak that interaction between these two scattering processes may be neglected. Hence, the self-energy can be calculated separately for each and then added together. In this approach, the self-energy can be written as

$$\Sigma_{r,n}(\mathbf{k}, \omega) = \sum_m \sum_i \sum_q |U_i(\mathbf{q})|^2 g_{r,m}(\mathbf{k} - \mathbf{q}, \omega). \qquad (9.125)$$

Here the index i sums over the various scattering processes such as impurities and surface roughness discussed above. It is clear at this point that the interacting retarded Green's function must be found self-consistently, not only because of the coupling among the Green's functions themselves but also because the various sub-bands are explicitly linked within the self-energy. The self-energy provides the renormalization of the sub-band energies as well as the scattering rates themselves.

At this point, we return to the discussion of the diagonality of Eq. (9.123). Let us consider a non-diagonal form that often appears in the literature. The interacting many-body propagator $G_{r,nm}(\mathbf{k}, \omega)$ can be related to the non-interacting version via the Dyson equation [41] as

$$[G_{r,nm}(\mathbf{k}, \omega)]^{-1} = [G_{r0,nm}(\mathbf{k}, \omega)]^{-1} - \Sigma_{nm}(\mathbf{k}, \omega). \qquad (9.126)$$

Let us now sum over the m index; we have

$$\sum_m [G_{r,nm}(\mathbf{k}, \omega)]^{-1} \equiv [G_{r,n}(\mathbf{k}, \omega)]^{-1} = [G_{r0,nn}(\mathbf{k}, \omega)]^{-1} - \sum_m \Sigma_{nm}(\mathbf{k}, \omega)$$

$$= [G_{r0,nn}(\mathbf{k}, \omega)]^{-1} - \Sigma_n(\mathbf{k}, \omega). \qquad (9.127)$$

We can easily now invert the equation and arrive at Eq. (9.123). Hence, the diagonal description used here is exact to the same order as any matrix formulation of the retarded Green's function.

In Fig. 9.3, the density of states for the lowest sub-band that forms in the Si–SiO$_2$ interface is shown for a combination of Coulomb and surface-roughness scattering [42]. It is clear that the sharp step in the density of states is broadened considerably, and this effect increases as the density (and hence the scattering) increases. At high densities, where the effective field is quite large, the band tailing is actually dominated by the surface-roughness scattering.

9.7.2 Less-Than Green's Function

We now want to turn our attention to the less-than Green's function, which will provide the quantum distribution function. The treatment here will be for a small value of the electric field, where the equations can be linearized in some sense. The high-electric-field case is left to the next chapter. Mahan [14] has given a full treatment of the linear-response approach to both the retarded and less-than Green's functions, which we treated earlier in this chapter. There, he used a gradient expansion in the self-energies to try to arrive at a closed set of equations. The approach here will be that of the multi-sub-bands present in an inversion layer, and the gradient expansion will not be used. Nevertheless, a closed Fredholm integral equation will be found for the scattering kernel in the distribution function (to be explained below), which is fully equivalent to that of this earlier treatment, although the latter was done for phonon scattering.

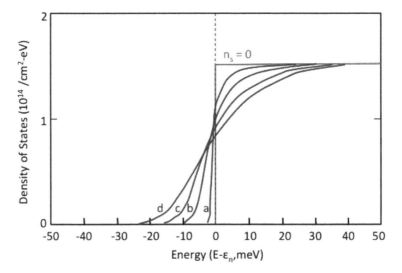

Figure 9.3 The broadened density of states function in the presence of impurity scattering and surface-roughness scattering for the lowest sub-band in Si ($n = 0$). The blue curve is that normally expected from Eq. (9.122) in the absence of interactions. The curve labels a, b, c, d are for 1, 2, 3, 4×10^{12} cm^{-2} inversion density. The values of Δ and L were taken to be 0.18 nm and 4 nm, respectively, for the surface-roughness scattering and a depletion charge of 10^{16} was taken. Adapted from Ref. [42].

The starting point is once again the equations of motion for the less-than Green's function, which was given in Eq. (9.19), albeit without the proper use of the Langreth theorem. These equations are

$$\left[i\hbar \frac{\partial}{\partial t} - H_0(\mathbf{r}) - e\mathbf{E}\cdot\mathbf{r} - V_c(\mathbf{r}) \right] G^<(\mathbf{r},\mathbf{r}',t,t') = \Sigma^r G^< + \Sigma^< G^a$$

$$\left[-i\hbar \frac{\partial}{\partial t'} - H_0(\mathbf{r}') - e\mathbf{E}\cdot\mathbf{r}' - V_c(\mathbf{r}') \right] G^<(\mathbf{r},\mathbf{r}',t,t') = G^r \Sigma^< + G^< \Sigma^a,$$

$$(9.128)$$

where the confinement potential has been separated out of the Hamiltonian proper. As before, this confinement potential affects only the z-component of the different Green's functions and leads to the sub-band expansion for these functions. Hence, the remaining variables are two-dimensional variables for motion along the interface. Now, we introduce the equivalent Wigner coordinates for these two-dimensional variables, which leads to the two-dimensional equations of motion given by Eq. (9.41) with the Green's functions replaced by the sub-band functions in analogy with Eq. (9.123). This is then Fourier transformed on the difference time and difference position, giving

$$\left[\hbar\omega - E(k) - \frac{\hbar^2}{8m*}\nabla^2 - e\mathbf{E}\cdot\mathbf{R} - \frac{\Sigma_n^r + \Sigma_n^a}{2} \right] g_n^<(\mathbf{R},\mathbf{k},T,\omega) = \frac{g_n^r + g_n^r}{2}\Sigma_n^<$$

$$\left[i\hbar\frac{\partial}{\partial T} - \frac{\hbar^2}{m*}\mathbf{k}\cdot\nabla - e\mathbf{E}\cdot\frac{\partial}{\partial \mathbf{k}} - (\Sigma_n^r - \Sigma_n^a) \right] g_n^<(\mathbf{R},\mathbf{k},T,\omega) = (g_n^r - g_n^r)\Sigma_n^<.$$

$$(9.129)$$

As previously, we treat a homogeneous channel, such as might occur in a large MOS capacitor. Hence, we expect that the derivatives with respect to $\mathbf{R}$ and T will vanish. However, care must be taken with this on account of the term in $e\mathbf{E}\cdot\mathbf{R}$. In essence, this term corresponds to a gauge shift as one moves along the channel and must be incorporated into the energy $\hbar\omega$ in order to be consistent, just as was done in Eq. (9.29). This suggests the change of variable $\hbar\omega - e\mathbf{E}\cdot\mathbf{R} \to \hbar\omega'$. In our case (of an Si inversion layer), the correction to the energy is quite small. The maximum size of the field term is of the order of $eE\lambda$, where λ is the mean free path. In an inversion layer at 300 K, the mobility may be of the order of 500 cm^2/Vs, which leads to a mean free path $\lambda \sim 10$–15 nm. Thus, the energy correction for a field of 100 V/cm is only 10–15 neV. The sub-band energies are typically tens of meV, so that this correction to the energy is quite small and may safely be ignored. A more important correction is the change in the gradient operation, in which the gradient terms in Eq. (9.129) are modified according to the second of equations (9.29). The added term in this case can be quite large. With these changes, the second of Eq. (9.129) becomes

$$
\left[-e\mathbf{E} \cdot \mathbf{v}_k \frac{\partial}{\partial \omega} - e\mathbf{E} \cdot \frac{\partial}{\partial \mathbf{k}} - \left(\Sigma_n^r - \Sigma_n^a \right) \right] g_n^<(\mathbf{R}, \mathbf{k}, T, \omega) = \left(g_n^r - g_n^r \right) \Sigma_n^< .
$$

(9.130)

This latter equation is often referred to as the quantum Boltzmann equation. This needs to be solved for us to find the less-than Green's function.

At this point, it is again useful to introduce some separation in the real and imaginary parts of the retarded functions via Eq. (9.43), but with the appropriate sub-band indices. We can also introduce formally the appropriate less-than self-energy as

$$
\Sigma_n^<(\mathbf{k}, \omega) = \sum_{i,m} \int \frac{d^2 q}{4\pi^2} W_{nm}^{(i)}(\mathbf{k} - \mathbf{q}) g_m^<(\mathbf{q}, \omega),
$$

(9.131)

where, as mentioned, we have used just the real part of the retarded self-energy. The index i goes over the various contributions to the self-energy from different scattering processes. Finally, we introduce the equivalent expansion of the less-than Green's function used in Eq. (9.50), as

$$
g_n^<(\mathbf{k}, \omega) = iA_n(\mathbf{k}, \omega) \left[f(\omega) - e\mathbf{E} \cdot \mathbf{v}_k \Lambda(\mathbf{k}, \omega) \frac{\partial f(\omega)}{\partial \omega} \right].
$$

(9.132)

This expression is now introduced into Eq. (9.130), along with the notational changes on the retarded functions mentioned above. When this is done, the terms are of two types. Either they contain the distribution function itself or they contain the derivative of the distribution function with respect to the energy (ω). The first set of terms gives us another expression for the spectral density, but this is just the same result as obtained from Eq. (9.123) and the two-dimensional equivalent of Eq. (9.40). That is, there are no corrections to this expression that are first-order in the electric field. The second set of terms provides the Fredholm integral equations for the kernel function introduced in Eq. (9.132). This is

$$\Lambda_n(\mathbf{k},\omega) = \frac{1}{2\Gamma_n(\mathbf{k},\omega)}\left[1 + \sum_{i,m}\int\frac{d^2q}{(2\pi)^2}\frac{\mathbf{v_k}\cdot\mathbf{v_q}}{v_k^2}W_{nm}^{(i)}(\mathbf{k}-\mathbf{q},\omega)\Lambda_m(\mathbf{q},\omega)\right].$$

(9.133)

Once this integral is solved, we may readily find the conductivity as

$$\sigma_{2D} = -\frac{e^2}{\pi\hbar}\sum_n\int_0^\infty dE(k)\int\frac{d\omega}{2\pi}E(k)\Lambda_n(\mathbf{k},\omega)A_n(\mathbf{k},\omega)\frac{\partial f(\omega)}{\partial\omega}.$$

This Green's function based theory of the inversion layer has been applied to realistic Si inversion layers [43, 44]. In Fig. 9.4, we plot the mobility of the electrons at 300 K as a function of the inversion density in the device. One will note that there is a peak in the mobility. At lower values of the inversion density, the scattering is dominated by the ionized acceptors in the p-type material. Even though the inversion density is low, there remains a very high concentration of the ionized acceptors due to the need to have a large band bending to invert the surface. At higher densities, the normal phonon scattering begins to play a role, and finally surface-roughness scattering becomes important at the highest densities. Another factor that is important in small MOSFETs is the quantization effects on the capacitance. Classically, the inversion density peaks right at the oxide–semiconductor interface. When quantization sets in, however, this is not the case as the wavefunction must tend to zero at this interface. Hence, the peak in the inversion density is some distance from the interface, and this means that there is an inversion capacitance (per unit area) of order e_s/d, where d is the distance from the interface of the density peak. This can have a dramatic effect on the device performance, as this capacitance is in series with the normal gate capacitance and, therefore, reduces the amount of inversion charge per volt on the gate. In Fig. 9.5, we plot the inversion capacitance computed for the same structure and conditions as for Fig. 9.4.

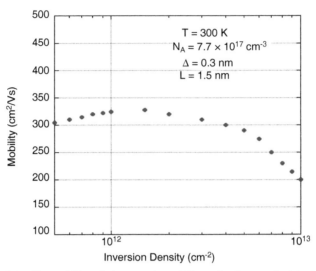

Figure 9.4 The mobility of electrons in an Si inversion layer, calculated from a real-time Green's function approach. Adapted from Ref. [43].

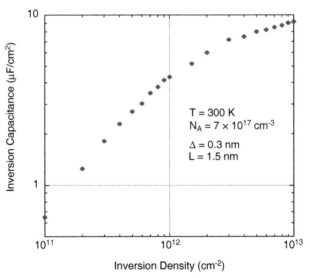

Figure 9.5 The inversion capacitance for an Si MOS structure simulated with the real-time Green's function technique. Adapted from Ref. [44].

Problems

1. Verify that the transformation matrices (9.16) produce the Keldysh matrix (9.17).
2. Carry through the various variable changes in Section 9.3.1 for the time variables and show that the results are as presented in Eq. (9.35).
3. Consider an apparent conducting solid, in which it has been determined that the self-energy has the form

$$\Sigma(\omega) = E(k)\left[0.1 - i4.0\left(\frac{\hbar\omega}{E(k)}\right)^{1/2} u_0(\omega)\right],$$

   where u_0 is the Heaviside step function. Plot Eq. (9.40) as a function of $\hbar\omega/E(k)$.
4. In resonant-tunneling diodes, the transmission is often written in terms of a simple single effective barrier rather than the double barrier form used in Eq. (9.86). In the case where the two barriers are identical, the total transmission rises to unity at the resonant level of the quantum well. What is the value of the transmission in Eq. (9.86) if $\Gamma_L = \Gamma_R$. Why isn't it unity? What is the value of the transmission if $\Gamma_L \gg \Gamma_R$. Why is Eq. (9.96) superior in this regard?

References

1. P. C. Martin and J. Schwinger, *Phys. Rev.*, **115**, 1342 (1959).
2. L. P. Kadanoff and G. Baym, *Quantum Statistical Mechanics* (Benjamin/ Cummings, Reading, MA, 1962).
3. L. V. Keldysh, *J. Exper. Theor. Phys.*, **47**, 1515 (1964) [translation *Sov. Phys. JETP*, **20**, 1018 (1965)].
4. P. Lipavsky, V. Spická, and B. Velicky, *Phys. Rev. B*, **34**, 6933 (1986).
5. A. Blandin, A. Nourtier, and D. W. Hone, *J. Physique*, **37**, 369 (1976).
6. M. Wagner, *Phys. Rev. B*, **44**, 6104 (1991); **45**, 11595 (1992).
7. Discussion in J. Rammer and H. Smith, *Rev. Mod. Phys.*, **58**, 323 (1986).
8. H. Haug and A. P. Jauho, *Quantum Kinetics in Transport and Optics of Semiconductors* (Springer, Berlin, 1997).

9. M. Bonitz, *Quantum Kinetic Theory* (Teubner, Stuttgart, 1998).

10. G. D. Mahan, in *Quantum Transport in Semiconductors*, edited by D. K. Ferry and C. Jacoboni (Plenum Press, New York, 1992), pp. 101–140.

11. D. C. Langreth, in *Linear and Nonlinear Electron Transport in Solids*, edited by J. T. Devreese and E. van Doren (Plenum Press, New York, 1976).

12. G. D. Mahan and W. Hänsch, *J. Phys. F*, **13**, L47 (1983).

13. W. Hänsch and G. D. Mahan, *Phys. Rev. B*, **28**, 1902 (1983).

14. G. D. Mahan, *Phys. Rep.*, **110**, 321 (1984), and references therein.

15. V. Spicka and P. Lipavsky, *Phys. Rev. B*, **52**, 14615 (1995).

16. T. Bornath, D. Kremp, W. D. Kraft, and M. Schlanges, *Phys. Rev. E*, **54**, 3274 (1996).

17. Y. Meir, N. S. Wingreen, and P. A. Lee, *Phys. Rev. Lett.*, **66**, 3048 (1991).

18. A. P. Jauho, N. S. Wingreen, and Y. Meir, *Phys. Rev. B*, **50**, 5528 (1994).

19. Y. Meir and N. S. Wingreen, *Phys. Rev. Lett.*, **68**, 2512 (1992).

20. V. Mujica, M. Kemp, and M. A. Ratner, *J. Chem. Phys.*, **101**, 6849 (1994), and references therein (we make no effort to be extensive in the list of references on this topic).

21. W. Tian, S. Datta, S. Hong, R. Reifenberger, J. I. Henderson, and C. I. Kubiak, *J. Chem. Phys.*, **109**, 2874 (1998).

22. L. Hall, J. R. Reimers, N. S. Hush, and K. Silverbrook, *J. Chem. Phys.*, **112**, 1510 (2000).

23. M. Di Ventra, S. T. Pantelides, and N. D. Lang, *Phys. Rev. Lett.*, **84**, 979 (2000).

24. R. Dahlke and U. Schollwöck, *Phys. Rev. B*, **69**, 085324 (2004).

25. G. Speyer, R. Akis, and D. K. Ferry, *J. Vac. Sci. Technol. B*, **24**, 1987 (2006).

26. G. Speyer, R. Akis, and D. K. Ferry, *J. Phys. Conf. Ser.*, **38**, 25 (2006).

27. R. Kubo, *J. Phys. Soc. Jpn.*, **12**, 570 (1957).

28. M. S. Green, *J. Chem. Phys.*, **22**, 398 (1954).

29. M. J. McLennan, Y. Lee, and S. Datta, *Phys. Rev. B*, **43**, 13846 (1991).

30. Y. Lee, M. J. McLennan, G. Klimeck, R. K. Lake, and S. Datta, *Superlattices Microstruct.*, **11**, 137 (1992).

31. T. Ando, A. B. Fowler, and F. Stern, *Rev. Mod. Phys.*, **54**, 437 (1982).

32. F. Stern and S. das Sarma, *Phys. Rev. B*, **30**, 840 (1984).

33. S. M. Goodnick, D. K. Ferry, C. W. Wilmsen, Z. Lillienthal, D. Fathy, and O. L. Krivanek, *Phys. Rev. B*, **32**, 8171 (1985).

34. T. Yoshinobu, A. Iwamoto, and H. Iwasaki, *Jpn. J. Appl. Phys.*, **33**, pt. 1, 383 (1994).

35. R. M. Feenstra, *Phys. Rev. Lett.*, **72**, 2749 (1994).

36. A. G. Sabnis and J. T. Clemens, *Proc. Intern. Electron Dev. Mtg.* (IEEE Press, New York, 1979), p. 18.

37. F. Fang and W. E. Howard, *Phys. Rev. Lett.*, **16**, 797 (1966).

38. T. Ando, *J. Phys. Soc. Jpn.*, **43**, 1616 (1977).

39. M. Saitoh, *J. Phys. Soc. Jpn.*, **42**, 201 (1977).

40. D. K. Ferry, *Transport in Semiconductor Mesoscopic Devices* (IOP Publishing, Bristol, UK, 2015), Sec. 2.2.

41. S. Das Sarma, in *Quantum Transport in Ultrasmall Devices*, edited by D. K. Ferry, H. L. Grubin, A.-P. Jauho, and C. Jacoboni (Plenum Press, New York, 1995), pp. 339–358.

42. S. M. Goodnick, R. G. Gann, D. K. Ferry, and C. W. Wilmsen, *Surf. Sci.*, **113**, 233 (1982).

43. D. Vasileska and D. K. Ferry, *IEEE Trans. Electron Dev.*, **44**, 577 (1997).

44. D. Vasileska, D. Schroder, and D. K. Ferry, *IEEE Trans. Electron Dev.*, **44**, 584 (1997).

Chapter 10

Real-Time Green's Functions II

The applications of the real-time Green's functions that have been discussed so far have been described in terms of being near to an equilibrium system. It is possible to move to an *open, far-from-equilibrium* system through the use of the real-time Green's functions just as well, although there are a few further problems that will arise. However, it has been pointed out that this interpretation may only be valid when the entropy production (by dissipation) is relatively small and the system remains near to equilibrium where, at least in the lowest order, Liouville's equation remains valid as an equation of motion for the appropriate statistical ensemble representing the carrier transport under the applied fields [1]. In *far-from-equilibrium* systems, however, there is usually strong dissipation, and the resulting statistical ensemble (even in steady state) is achieved as a balance between driving forces and dissipative forces; for example, it does not linearly evolve from the equilibrium state when the applied forces are "turned on." Enz [1] argues that there is no valid unitary operator (which describes the impact of perturbation theory) to describe the evolution through this symmetry-breaking transition to the dissipative steady state. As a consequence, there is no *general* formalism at this time that can describe the all-important far-from-equilibrium devices. On the contrary, though, the application of the real-time Green's functions to what are surely very strongly far-from-equilibrium systems—the excitation of electron–hole

An Introduction to Quantum Transport in Semiconductors
David K. Ferry
Copyright © 2018 Pan Stanford Publishing Pte. Ltd.
ISBN 978-981-4745-86-4 (Hardcover), 978-1-315-20622-6 (eBook)
www.panstanford.com

pairs in intense femtosecond laser pulses—has yielded results that suggest that their use in these systems is quite reliable for studying both the transition to the semi-classical Boltzmann theory and to explain experimentally observed details [2–4]. Indeed, even for a more normal study of condensed matter systems, there is a history of the real-time Green's functions being used to study the evolution from an arbitrary initial state, whether or not this initial state is a correlated state [5–8]. As a consequence of these latter studies, as well as initial attempts to actually begin to model real devices with these Green's functions, the situation is believed to be much better than this pessimistic view would warrant. Nevertheless, it is essential that one move carefully with these approaches to assure that the problematic approach is really meaningful [9].

There is also some concern in the literature about the interpretation of $f(\omega)$ arising in the *ansatz* (9.6). Is this the single-particle distribution function that is equivalent to the single-particle Boltzmann distribution? Or is it the distribution of quasi-particles arising in the many-body interacting state? Or both? In the case of metals, this is a meaningful question, as the conduction electrons are itinerant particles that are not much affected by the details of the periodic potentials. The sheer numbers of these electrons overcome many of the details such as this question raises. On the other hand, in the semiconductors with which we are interested, the so-called electrons (and holes) are already quasi-particles. They became so when we assigned them an *effective* mass, which is a dressing of the particles to account for the very real role the periodic potential plays in their existence. Hence, we may think of them as a localized wave packet whose wave properties lead to a particle-like behavior via defining this effective mass. And it is only through the discussion of this wave behavior and its connection to the quasi-particle do we arrive at the proper definition that should be used to obtain this effective mass for a particular band structure or material [10, 11]. Hence, further dressing of the mass via the development of a proper self-energy does not constitute a major change in the manner of thinking of these quasi-particles.

The last chapter has dealt with some applications of the real-time Green's functions to systems near to equilibrium, particularly so that the gradient expansion could be used relatively effectively. These were mainly homogeneous systems in which the steady-state response was an adequate result. In this chapter, we move beyond these limitations in several directions. We will first address one approach to a true high electric field transport situation, albeit one in which we retain the homogeneous condition. Then we will discuss the resonant-tunneling diode, and its treatment via the real-time Green's functions. This will parallel the treatment of this device in earlier chapters. But this will lead us to a discussion of the well-known NEMO/OMEN simulation code that evolved from this device now to arbitrary three-dimensional semiconductor devices [12]. Then we will devote the remainder of the chapter to the evolution of the system from an arbitrary initial state and the problems of high-frequency behavior. That is, we are going to devote a considerable amount of space to the problems of moving beyond simple d.c. transport [13].

10.1 Transport in Homogeneous High Electric Fields

In high electric fields, there can arise a significant interaction between the fields that drive the transport in the system and the dissipation within the system. The broadening that arises from the self-energy introduces the fact that there can be interference between the driving terms and the dissipative terms. This interference is of the same nature as the interference between the electron–electron interaction and the diffusive interaction with the impurities discussed in an earlier chapter. Here, the interference between the field and the scattering processes is termed the *intra-collisional field effect* [14], an effect we have treated in Chapters 7 and 8. However, the effect was known earlier; it gives rise to field-induced effects in interband absorption that lead to the Franz–Keldysh effect [15, 16]. Indeed, many of the properties of the Airy-function integrals that will be used here were worked out for a proper treatment of the Franz–Keldysh effect in interband absorption [17]. Now the idea for the Airy functions arises from studies of the quantization in an MOS inversion layer, where one approximation is that the potential well is triangular. This then leads to the solutions to Schrödinger's equation which are Airy functions. In the case, here, of a homogeneous electric field, the representation in one gauge is that the bands are slanted, and a given energy linearly increases in one direction with the potential $e\mathbf{E}\cdot\mathbf{R}$, a form extensively used in the last chapter. This leads us to suggest using the Airy transform on this coordinate, and this is the route we will follow in this section.

In the early years, it had proven difficult to develop a tractable quantum transport approach through the use of the real-time Green's functions. The needed approach must incorporate both the collisional broadening that arises from the scattering processes and the intra-collisional field effect. Moreover, this task was further complicated by the need to deal with the length and time scales relevant to modern mesoscopic semiconductor devices. The general approach usually followed that of the linear response used in previous chapters. Although the overall Green's function approach is rigorous in principle, most applications have been limited by the introduction of a center-of-mass transformation, in both space and time scales, and then by the use of a gradient expansion for slow variations around the center-of-mass coordinates, as done in the last chapter. These two processes, especially the gradient expansion, tend to limit the results to low fields and prevent application to the far-from-equilibrium system. One of the first works to go beyond this was by Jauho and Wilkins [18], who treated high-field transport in a resonant level (the impurity level is in the conduction band) impurity scattering system. In this work, the self-energies for electron–phonon as well as impurity scattering were formulated; in the end, however, they were forced to invoke the gradient expansion in order to achieve a result for the kinetic equation. Nevertheless, the power of the Green's function approach was evident.

Another problem that becomes important is the actual collision duration—the actual time required, say, for the emission or absorption of a phonon. Over the years, there has been some work that is really applicable to this. One approach using the Green's functions is that of Lipavsky et al. [19]. In this latter work, it was shown that different definitions of the instantaneous approximation (i.e., single-time, which is a neglect of the ω dependence) for the self-energy will lead to different effects of the field on the collision, which emphasizes that proper renormalization of the Green's functions must be maintained. This collision duration can be quite important in fast processes [20]. We will deal with this collision duration in Section 10.6.

The use of the center-of-mass Wigner transformations (9.10) introduces some non-physical variables into the description of transport. These, in fact, make the explicit assumption that the center-of-mass time $T = (t + t')/2$ has some physical significance (which it does in the limiting case $t = t'$). This is not the case, usually, and nothing makes the point clearer than the need to modify the basic separation of the less-than Green's function into the product of the spectral density and the distribution function as done in the normal ansatz (9.6). Approaches that have used the center-of-mass approach, with the gradient expansion, have been faced with the need to change the basic connection between $G^<$ and $f(\omega)$ found in Eq. (9.6) [21]. Consider the velocity autocorrelation function, which was one basis for the entire development of classical transport [22]. This is a function of the time difference $t - t'$, where t' is the initial time for the correlation function. Except for the limiting case mentioned earlier, the center-of-mass time does not enter into any physical process of interest. Thus, an approach that avoids this need is preferable for investigating the far-from-equilibrium transport.

As a consequence of the many problems mentioned here, a different approach has been developed to treat high-field transport with the real-time Green's functions. This approach arises from the observations that the proper wavefunctions in a high electric field are the Airy functions [17], as mentioned above. Here we transform the position variable along the field with a generalized Airy transform. This method is somewhat limited by the fact that it still assumes translational invariance in the transverse directions, through the use of the Fourier-transformed coordinates. The degree to which inhomogeneous systems can be treated by this approach is still not yet known, although it has been under investigation for some time. The use of the Airy transformation will allow us to diagonalize the non-interacting Green's function in the presence of the field alone and to achieve a simpler form of Dyson's equation [23].

The general approach is to utilize the Airy transformation of each portion of the coordinate system parallel to the electric field, which reduces the transverse coordinates effectively to a quasi-two-dimensional system. In this approach, quantization in the z-direction has continuous eigenvalues rather than the discrete ones that arise in a true quantum well. The general Airy transform is defined by

$$F(\mathbf{k}, s) = \int d^2\mathbf{r} \int \frac{dz}{2\pi L} e^{i\mathbf{k}\cdot\mathbf{r}} Ai\left(\frac{z-s}{L}\right) f(\mathbf{r}, z), \tag{10.1}$$

where $L = (\hbar^2/2m^*eE)^{1/3}$, E is the electric field, $Ai(x)$ is an Airy function of the first kind, and s is a spatial variable (the transform variable) related to the zero crossings of the Airy function. In a triangular quantum well, with infinite confinement on one side and a linear potential on the other, the discrete wavefunctions are defined by the set of zero crossings s_p. Here, however, with the open system, s is continuous. For the Green's function, of course, there will be two Airy transformations for the two different positions. In this transform space, a function that is diagonal in momentum (only a single $\mathbf{k}$, which is assumed here) and in s variables is translationally invariant in the transverse plane, but not in the z-direction. We also note that k and r are two-dimensional variables in the plane normal to z.

10.1.1 Retarded Function

The retarded Green's function, in the presence of the electric field, still satisfies the standard equation of motion. In these equations, the electric field is introduced in the scalar potential gauge as $e\mathbf{E}\cdot\mathbf{R}$. In the absence of the self-energy, these can be solved to give the solution to the non-interacting Green's function in the presence of the field, after Airy transforming as

$$G_r^0(\mathbf{k}, s, s', t, t') = -i\theta(t-t') e^{-iE(\mathbf{k}, s)(t-t')/\hbar} \delta(s-s'), \tag{10.2}$$

where

$$E(\mathbf{k}, s) = \frac{\hbar^2 k^2}{2m^*} + e|\mathbf{E}|s \tag{10.3}$$

in parabolic bands. This form has the distinct advantage that the Fourier transformation of the difference variable in time leads to

$$G_r^0(\mathbf{k}, s, s', t, t') = \frac{\hbar\delta(s-s')}{\hbar\omega - E(\mathbf{k}, s) + i\hbar\eta}, \tag{10.4}$$

which has the generic form expected for the retarded Green's function. There is a problem in using this form of the retarded Green's function because it assumes that the field was turned on in the infinite past (as indicated by the convergence factor η), but this ignores the transient part of the integration path used for the real-time functions. However, this is consistent with ignoring the "tail" segment connecting to the thermal equilibrium, which is consistent with seeking only the solution that is a balance between the driving force and the dissipative terms. Nevertheless, the approach here yields only the steady-state solution and cannot be used to evaluate the transient response of the carrier system.

The use of the above equilibrium form for the retarded non-interacting Green's function allows us to write the interacting Green's function as the integral solution to Eq. (9.19). In the Airy-transformed form, this equation then becomes

$$G_r(\mathbf{k}, s, s', \omega) = G_r^0(\mathbf{k}, s, s', \omega)\Big\{\delta(s - s')$$
$$+ \int ds'' \Sigma_r(\mathbf{k}, s, s'', \omega) G_r(\mathbf{k}, s'', s', \omega)\Big\}, \tag{10.5}$$

where the diagonality of the non-interacting function has been utilized to simplify the integral expression. Similarly, the second equation can be transformed to yield

$$G_r(\mathbf{k}, s, s', \omega) = \Big\{\delta(s - s') + \int ds'' G_r(\mathbf{k}, s'', s', \omega)$$
$$\times \Sigma_r(\mathbf{k}, s, s'', \omega)\Big\} G_r^0(\mathbf{k}, s, s', \omega). \tag{10.6}$$

To proceed, it is now necessary to develop an expression for the self-energy that appears in the integrals on the right side of the last two equations. Here we will treat the scattering due to optical phonons as the primary source for the self-energy. We have previously treated only scattering due to basically elastic processes, but in the presence of high fields, inelastic scattering due to the interaction of the carriers with the phonons is a dominant process. In treating this self-energy, a new consideration arises: The phonons are themselves characterized by a propagator describing their interaction. Normally, it is assumed that the phonons remain in thermal equilibrium (although much work in the laser excitation area has treated nonequilibrium phonons). Nevertheless, it is necessary to evaluate the paired propagators

$$\Sigma_r(\mathbf{r}, \mathbf{r}', t, t') = i[G(\mathbf{r}, \mathbf{r}', t, t')D(\mathbf{r}, \mathbf{r}', t, t')]_i. \tag{10.7}$$

It should be noted that the two Green's functions, $G(\mathbf{r}, \mathbf{r}', t, t')$ for the electron and $D(\mathbf{r}, \mathbf{r}', t, t')$ for the phonon, have the same set of spatial and temporal arguments. That is, they propagate in parallel. This means that they will create a convolution integral when Fourier transformed into the momentum and energy domain. But which Green's functions do we take to get the retarded product? By the same arguments used in the Langreth theorem of the previous chapter, one can easily show that the terms needed are [24]

$$\Sigma_r(\mathbf{r}, \mathbf{r}', t, t') = i\Big[G_r(\mathbf{r}, \mathbf{r}', t, t') D^{>}(\mathbf{r}, \mathbf{r}', t, t') + G^{<}(\mathbf{r}, \mathbf{r}', t, t') D_r(\mathbf{r}, \mathbf{r}', t, t') \Big].$$

(10.8)

The operator ordering in the last term is such that this term vanishes as the density goes to zero. As a result, for non-degenerate semiconductors (which is the case most often, especially at high electric fields), it is possible to ignore the second term as being small in comparison with the first term. Moreover, in most semiconductors, the scattering by phonons is weak (there is very little multiple scattering) and the self-energy can be calculated in the Born approximation; essentially, the Green's function in the first term is replaced by the non-interacting form, which is the free, field-assisted propagator (but see Ref. [25]). Since the retarded Green's function is still solved self-consistently within Dyson's equation, collisional broadening still will appear in the final form, and the field's presence in the non-interacting Green's function introduces the interference between the field and the scattering process.

The phonon Green's function for non-polar optical scattering can be expressed as (for the phonons in thermal equilibrium)

$$D_0^{>}(\mathbf{q}, \omega) = -i\pi \sum_{\pm} |M(\mathbf{q})|^2 \, \delta(\omega \pm \omega_0),$$

(10.9)

and it has been assumed that the phonons are non-dispersive. The matrix element $M(\mathbf{q})$ also includes the phonon occupation factors— the Bose–Einstein distributions for the phonons, while the sum runs over emission and absorption processes. With these various approximations, the retarded self-energy can be written as

$$\Sigma_r(\mathbf{k}, \omega) = \pi \int \frac{d^3\mathbf{q}}{(2\pi)^3} \sum_{\pm} |M(\mathbf{q})|^2 \, G_r^0(\mathbf{k} \pm \mathbf{q}, \omega \pm \omega_0).$$

(10.10)

It should be noted that this result *has not been Airy transformed.* The momentum variable is still a three-dimensional vector. To be useful, the momentum dependence in the field direction is transformed back to real space and then Airy transformed. More properly, the product form is first generated in real space and then the result is Airy and Fourier transformed using the delta function to break the convolution in Fourier space. In general, the matrix element for the non-polar optical modes is not momentum dependent, so there is no complication in the matrix element in these various transformations. Finally, after carrying out the integrations over momentum, the self-energy is found to be [23]

$$\Sigma_r(\mathbf{k}, s, \omega) = \frac{m^*}{\hbar^2}\sqrt{\frac{m^*\xi}{2}}\sum_{\pm}|M|^2\left[M(s, \omega\pm\omega_0)+iN(s, \omega\pm\omega_0)\right],$$

(10.11)

where

$$M(s, \omega) = Ai'^2(\zeta)+\zeta Ai^2(\zeta)$$
$$N(s, \omega) = Ai'(\zeta)Bi'(\zeta)+\zeta Ai(\zeta)Bi(\zeta)$$

(10.12)

and

$$\zeta = \frac{e|\mathbf{E}|s - \hbar\omega}{\xi}$$

$$\xi = \frac{3(e\hbar|\mathbf{E}|)^2}{2m^*}.$$

(10.13)

The real and imaginary parts of the self-energy are plotted in Figs. 10.1 and 10.2 for parameters suitable for silicon. The oscillatory behavior of the real part of the self-energy is quite interesting and indicates that the interaction of the field and the scattering process is creating an equivalent quasi-two-dimensional behavior in the electron gas, which is reinforced by looking at the imaginary part. The step-like structure that appears reinforces this interpretation.

Since the self-energy is diagonal in the Airy variable s, the integral in Dyson's equation collapses (due to an assumed delta function on the difference of the two arguments in the self-energy). There is no reason, then, not to also assume that the retarded Green's function is diagonal in the Airy transform variable. This leads to

$$G_r(\mathbf{k}, s, \omega) = \frac{\hbar}{\hbar\omega - E(\mathbf{k}, s) - \Sigma_r(\mathbf{k}, s, \omega)},$$

(10.14)

which is in the non-interacting form and actually has the form of the zero-temperature Green's function as well. Here, however, the energy term contains the Airy-transformed scalar potential, which also appears in the self-energy. This result differs from that of the previous chapters. The spectral density is then twice the imaginary part of the retarded Green's function

$$A(\mathbf{k}, s, \omega) = \frac{-2\mathrm{Im}\{\Sigma_r(\mathbf{k}, s, \omega)\}}{\left[\hbar\omega - E(\mathbf{k}, s) - \mathrm{Re}\{\Sigma_r(\mathbf{k}, s, \omega)\}\right]^2 + \left[\mathrm{Im}\{\Sigma_r(\mathbf{k}, s, \omega)\}\right]^2},$$

(10.15)

which, again, has the form found in the equilibrium (or zero-temperature) Green's functions. In Fig. 10.3, the spectral density is plotted for three different values of the electric field, again for the parameters appropriate to silicon. The shift and distortion introduced by the electric field are clearly evident in the figure. Note, however, that the normalization is maintained. Any integration over the spectral density will, therefore, show only small effects arising from the field distortion and shift.

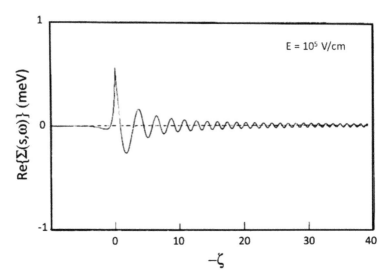

Figure 10.1 The real part of the retarded self-energy for an electric field of 10^5 V/cm in silicon at room temperature. The dashed curve corresponds to neglect of the intra-collisional field effect. The symbol ζ is defined in Eq. (10.13) in the text. Adapted from Ref. [23].

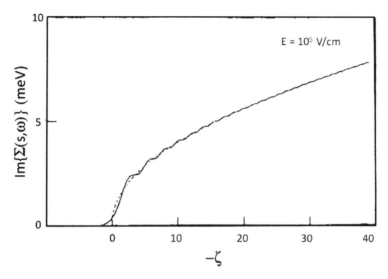

Figure 10.2 The imaginary part of the retarded self-energy for Silicon at 300 K and a field of 10^5 V/cm. The dashed curve corresponds to neglect of the intra-collisional field effect. The symbol ζ is defined in Eq. (10.13). Adapted from Ref. [23].

10.1.2   Less-Than Function

With the approximations in the above approach, it has been possible to get a good representation of the spectral density, as can be seen in Fig. 10.3. This result remains correct to all orders in the electric field since it is directly included without the need for any expansion in powers of the field (such as in the gradient expansion). Although the above approach is constrained to the case of weak scattering, this is the general case for semiconductors and their high mobilities. With the spectral function and the retarded Green's function now determined, it is possible to proceed to find the less-then function $G^<$. In the low-field case of the last chapter, it was found that we needed to solve an integral equation for the less-than function, especially in strong scattering where the self-energy results from higher-order interactions, which required treatment by the Bethe–Salpeter equation, as in Chapter 4. In the weak-scattering limit here, the approach will be somewhat easier, but an integral equation is still expected for the less-than function, particularly because there is an integral over the product of the self-energy and the less-than function in the equations of motion. This integral equation will actually be an integral equation for the distribution function, just as that which arises in Boltzmann transport.

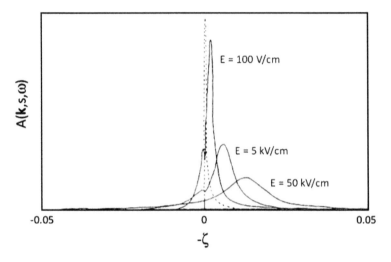

Figure 10.3 The spectral density for several values of the electric field in Si at 300 K. Here the carrier energy is comparable to the phonon energy. Adapted from Ref. [23].

The starting point, as in the case of the retarded equation, is the equation of motion for the Keldysh Green's function matrix. After introducing the Fourier and Airy transformations, two equations of motion result for the less-than function, which can be written

$$(\hbar\omega - E(k,s))G^<(\mathbf{k},s,s',\omega) = \int ds'' \Big[\Sigma_r(\mathbf{k},s,s'',\omega)G^<(\mathbf{k},s'',s',\omega)$$
$$+ \Sigma^<(\mathbf{k},s,s'',\omega)G_a(\mathbf{k},s'',s',\omega) \Big]$$

$$(\hbar\omega - E(k,s'))G^<(\mathbf{k},s,s',\omega) = \int ds'' \Big[G_r(\mathbf{k},s,s'',\omega)\Sigma^<(\mathbf{k},s'',s',\omega)$$
$$+ G^<(\mathbf{k},s,s'',\omega)\Sigma_a(\mathbf{k},s'',s',\omega) \Big].$$
$$(10.16)$$

At this point, it has already been established that the retarded Green's function, the advanced Green's function (which is merely its complex conjugate), and the corresponding self-energies are functions of a single Airy variable. This eliminates the integration on the right sides of the two equations above. We can then formulate the sum and differences of these two equations, which in the present case yield the same resulting equation. In essence, this result means that these two equations are self-adjoint, which is an important advantage of the Airy transform approach. The resulting equation is then

$$G^<(\mathbf{k}, s, s', \omega) = \frac{\left[G_r(\mathbf{k}, s, \omega) - G_a(\mathbf{k}, s', \omega)\right]\Sigma^<(\mathbf{k}, s, s', \omega)}{E(k, s) - E(k, s') - \Sigma_a(\mathbf{k}, s', \omega) + \Sigma_r(\mathbf{k}, s, \omega)}$$

$$= \frac{1}{\hbar}G_r(\mathbf{k}, s, \omega)G_a(\mathbf{k}, s', \omega)\Sigma^<(\mathbf{k}, s, s', \omega). \quad (10.17)$$

Because the retarded and advanced Green's functions have imaginary parts that are sharply peaked, we expect the product of these two functions to be small unless the arguments correspond to one another. This leads us to assume that both the less-than self-energy and the less-than Green's function are diagonal in the Airy transform variable. With this approximation, it is then possible to recognize that the less-than Green's function satisfies the ansatz that has been used in the previous chapter, and discussed above. That is, we can rewrite the first line of Eq. (10.17) as

$$G^<(\mathbf{k}, s, \omega) = iA(\mathbf{k}, s, \omega)f(\mathbf{k}, s, \omega), \quad (10.18)$$

since the first line of Eq. (10.17) already contains the spectral density within the square brackets in the numerator. Thus, we may define the quantum distribution function as the solution to

$$f(\mathbf{k}, s, \omega) = \frac{i\Sigma^<(\mathbf{k}, s, \omega)}{2\text{Im}\{\Sigma_r(\mathbf{k}, s, \omega)\}}. \quad (10.19)$$

The less-than self-energy function can be developed as easily as the retarded self-energy for the case of the non-polar optical phonons. As previously, it will be assumed that this function is diagonal in the Airy variable, so that we can write it (as in the previous section) as

$$\Sigma^<(\mathbf{k}, s, \omega) = \frac{1}{3^{1/6}L}\sum_\pm \int d^2\mathbf{q}\,|M(\mathbf{q})|^2 \int ds'\,Ai\left(\frac{s - s'}{3^{1/3}L}\right)G^<(\mathbf{k} \mp \mathbf{q}, s, \omega \pm \omega_0)$$

$$(10.20)$$

Here the factor L has been defined previously. This last equation, of course, also depends on the less-than Green's function. There is an important simplification that arises in the case of non-polar optical phonon scattering. In this last scattering process, the scattering is *isotropic,* that is, it does not depend on the direction or magnitude of the scattering wavevector $\mathbf{q}$. This is not a general result, but it is one that applies particularly to the case of such isotropic scattering. Thus, a change in variables can be carried out in the integration over the scattering momentum, and the resulting integration removes all momentum dependence from the right side of the equation. Hence, for isotropic scattering, the self-energy is independent of the momentum. Examination of the previous result, for the retarded self-energy, shows that the latter also is independent of the momentum. Thus, both of the self-energies, and also the resulting distribution function, are independent of the phonon momentum in the case of isotropic scattering. Thus, the resulting distribution function is dependent on only the energy ($\hbar\omega$) and the Airy variable. Thus, the integral equation for the distribution function can be written as

$$f(s,\omega)$$
$$= \frac{1}{2(3)^{1/6}L\,\mathrm{Im}\{\Sigma_r(s,\omega)\}}\sum_{\pm}|M|^2\int ds'\,Ai\left(\frac{s-s'}{3^{1/3}L}\right)\rho(s',\omega\pm\omega_0)f(s',\omega\pm\omega_0)$$

$$(10.21)$$

where

$$\rho(s,\omega)=\int d^2\mathbf{q}\,A(\mathbf{q},s,\omega) \qquad\qquad (10.22)$$

is the effective density of states at the final energy (in Airy transform space) of the scattering process. We note that the denominator of the prefactor plays the same role as the scattering function in the Boltzmann transport equation. In fact, the resulting formulation maintains much of the structure that is present in the path integral solution to the Boltzmann transport equation [26]. The actual distribution function that results from solving this appears as a normal Boltzmann-like distribution, but one which has been modified slightly to show the quasi-two-dimensional structure that is apparent in Figs. 10.1 and 10.2. In addition, there is more tailing to higher energies than would normally be expected in the classical result [27].

10.2   Resonant-Tunneling Diode

As we have found in previous chapters, the resonant-tunneling diode is the quintessential quantum device. We have treated it with the density matrix and the Wigner function. Now we shall approach it with the Green's function. Here, the electron–phonon interaction will be treated in the self-consistent first Born approximation [28]. That is, only a single scattering process is included, but it is retained to all orders in the perturbation series. The phonons will be treated by the usual approximation that they consist of a bath of independent harmonic oscillators that interact with the electrons locally in position. This is not applicable to the polar mode of the optical phonons (which is a relatively long-range Coulombic interaction), but it does apply to the normal deformable-ion approximation used in the optical modes. Finally, the phonon modes themselves will be traced out of the problem, leaving only the electron coordinates and the electron–phonon interaction. In this approach, the non-interacting electron Hamiltonian is described by the real-space terms

$$H_0 = \frac{(\mathbf{p} - e\mathbf{A})^2}{2m^*} + V(\mathbf{r}),$$ (10.23)

where $V(\mathbf{r})$ includes the linear potential drop, due to the applied bias, plus the conduction-band discontinuities that create the tunneling barriers. While the presence of the vector potential allows for the inclusion of a magnetic field, this term will be ignored here. Dephasing of the electrons is assumed to occur through their interaction with a bath of lattice vibrations, described by the creation and annihilation operators $a^\dagger$ and a, respectively, as previously. The electrons interact with this bath through an interaction described here in the site representation as

$$H_{eL}(\mathbf{r}) = \sum_i U\delta(\mathbf{r} - \mathbf{r}_i)(a^\dagger + a).$$ (10.24)

The sum of the two operators merely gives the amplitude of the lattice wave for that particular mode, and the interaction is assumed to occur at particular positions in space, which leads to the sums. If one assumes that a continuum of phonon modes exists, then the sum over the sites i becomes an integration over the modes.

The self-energies for the electron–phonon interaction are evaluated in the self-consistent Born approximation, in which, from the Langreth theorem, one can write

$$\Sigma^<(\mathbf{r},\mathbf{r}',\omega) = \int d\omega' G^<(\mathbf{r},\mathbf{r}',\omega-\omega')D^<(\mathbf{r},\mathbf{r}',\omega'),\qquad(10.25)$$

and similarly for the greater-than self-energy. The phonon Green's function is given by the normal Fermi golden rule in terms of the matrix elements between the electron states for the perturbing Hamiltonian (10.24). Rather than treat the full problem, it is assumed that the phonons scatter locally in space. In this sense, the retarded self-energy is written in terms of a simple energy shift and an imaginary part as

$$\Sigma_r(\mathbf{r},\mathbf{r}',\omega) = \left[\sigma(\mathbf{r},\omega) - \frac{i}{2\hbar\tau(\mathbf{r},\omega)}\right]\delta(\mathbf{r}-\mathbf{r}').\qquad(10.26)$$

The retarded Green's function is similarly found from the equation of motion

$$(\hbar\omega - H_0(\mathbf{r}) - \Sigma_r(\mathbf{r},\mathbf{r}',\omega))G^r(\mathbf{r},\mathbf{r}',\omega) = \hbar\delta(\mathbf{r}-\mathbf{r}').\qquad(10.27)$$

Now we begin to diverge from the normal approach with the ansatz. Instead, we assume that there exists a nonequilibrium distribution function for the electrons, which can be written as

$$f(\mathbf{r},\omega) = \frac{n(\mathbf{r},\omega)}{N_0(\mathbf{r},\omega)},\qquad(10.28)$$

and that there are electron and hole densities, which, along with the normalization factor N_0, can be written as

$$n(\mathbf{r},\omega) = \frac{i}{2\pi}G^<(\mathbf{r},\mathbf{r},\omega)$$

$$p(\mathbf{r},\omega) = -\frac{i}{2\pi}G^>(\mathbf{r},\mathbf{r},\omega)\qquad(10.29)$$

$$N_0(\mathbf{r},\omega) = n(\mathbf{r},\omega) + p(\mathbf{r},\omega) = -\frac{i}{2\pi}G_r(\mathbf{r},\mathbf{r},\omega).$$

To proceed, it is necessary to define the less-than and greater-than self-energies. As above, these are assumed to be local functions, from which one may approximate as

$$\Sigma^<(\mathbf{r},\mathbf{r}',\omega)=-i\frac{\hbar}{\tau_{p,n}(\mathbf{r},\omega)}\delta(\mathbf{r}-\mathbf{r}'),\qquad(10.30)$$

with the greater-than function being the complex conjugate of the right-hand side. In essence, these approximations return this calculation to the semi-classical realm, as the off-shell corrections are being ignored and the main quantum effect is the tunneling itself. Dyson's equation for the correlation function may now be written as

$$G^<(\mathbf{r},\mathbf{r}',\omega)=\int d\mathbf{r}_1\int d\mathbf{r}_2 G_r(\mathbf{r},\mathbf{r}_1,\omega)\,\Sigma^<(\mathbf{r}_1,\mathbf{r}_2,\omega)G_a(\mathbf{r}_2,\mathbf{r}',\omega),$$

$$(10.31)$$

which becomes simplified when the local approximation (10.30) is introduced. With this, the electron and hole distributions can be written as

$$f(\mathbf{r},\omega)=\frac{\hbar}{2\pi N_0(\mathbf{r},\omega)}\int d\mathbf{r}'\frac{|G_r(\mathbf{r},\mathbf{r}',\omega)|^2}{\tau_p(\mathbf{r}',\omega)}$$

$$p(\mathbf{r},\omega)=\frac{\hbar}{2\pi}\int d\mathbf{r}'\frac{|G_r(\mathbf{r},\mathbf{r}',\omega)|^2}{\tau_n(\mathbf{r}',\omega)}$$

$$N_0(\mathbf{r},\omega)=\frac{\hbar}{2\pi}\int d\mathbf{r}'\frac{|G_r(\mathbf{r},\mathbf{r}',\omega)|^2}{\tau_\phi(\mathbf{r}',\omega)} \qquad(10.32)$$

$$\frac{1}{\tau_\phi}=\frac{1}{\tau_p}+\frac{1}{\tau_n}.$$

The last five equations form a consistent process once an expression is obtained for the two initial scattering times that appear in Eq. (10.32). This is found by reinserting these equations into the expression for the less-than Green's function (10.31), which leads to

$$\frac{1}{\tau_p(\mathbf{r},\omega)}=\frac{\hbar}{2\pi}\int d\omega' F(\mathbf{r},\omega')N_0(\mathbf{r},\omega+\omega')f(\mathbf{r},\omega+\omega')$$

$$(10.33)$$

$$\frac{1}{\tau_n(\mathbf{r},\omega)}=\frac{\hbar}{2\pi}\int d\omega' F(\mathbf{r},\omega')N_0(\mathbf{r},\omega+\omega')[1-f(\mathbf{r},\omega+\omega')],$$

where

$$F(\mathbf{r},\omega')=U^2\rho(\mathbf{r},|\omega|)\cdot\begin{cases}N_q, & \omega>0,\\ N_q+1, & \omega<0,\end{cases}\qquad(10.34)$$

where ρ is the density of oscillator modes and N_q is the Bose–Einstein distribution. The separation of the frequency into positive and negative values distinguishes between phonon absorption and emission processes. Finally, one arrives at a closed integral equation for the electron distribution function

$$f(\mathbf{r},\omega) = \frac{1}{N_0(\mathbf{r},\omega)} \int d\mathbf{r}' \, |G_r(\mathbf{r},\mathbf{r}',\omega)|^2 \int d\omega' \, J(\mathbf{r}',\omega)$$
$$\times N_0(\mathbf{r}',\omega-\omega') f(\mathbf{r}',\omega-\omega'). \tag{10.35}$$

Two sets of boundary conditions are required in order to simulate an actual device. The first is on the retarded Green's function. Lake and Datta [28] assume perfectly absorbing boundary conditions. The retarded Green's function itself is solved on a finite lattice and is then extended analytically to infinity in both directions [29]. Inelastic scattering is included throughout the boundary contacts, so the entire boundary regions act like parts of the contact. The second set of boundary conditions is for the distribution itself. For this, it is assumed that the distribution in the contacts is a proper Fermi–Dirac distribution with the appropriate chemical potential for that contact. Finally, by manipulating the above equations, one can arrive at an appropriate current density [30] (as a function of both energy and position)

$$J(\mathbf{r},\omega) = \frac{eh}{2\pi} \int d\mathbf{r}' \frac{|G_r(\mathbf{r},\mathbf{r}',\omega)|^2}{\tau_\phi(\mathbf{r},\omega)} \left[\frac{f_{FD}(\omega-E_F/\hbar)}{\tau_\phi(\mathbf{r}',\omega)} - \frac{1}{\tau_p(\mathbf{r}',\omega)} \right]. \tag{10.36}$$

This result shows how the contact distribution propagates into the system. In Fig. 10.4, we illustrate some results of Lake and Datta [28] for the energy-dependent current distribution throughout a resonant-tunneling diode, in which inelastic scattering is important. The simulation was at 77 K, with 220 meV barriers, with the barriers and well each 5 nm thick. The base material is GaAs. In the figure, E_F is the Fermi energy in the emitter, E_c is the conduction-band edge in the emitter, and E_R is the energy of the resonance in the quantum well. The situation shown is such that the resonance is below the conduction-band edge, where the tunneling current should have been cutoff. Instead, the conduction electrons from the emitter reach the resonance through the emission of an optical phonon. This is phonon-assisted tunneling. The electrons are then able to tunnel through the second barrier to the drain end of the device.

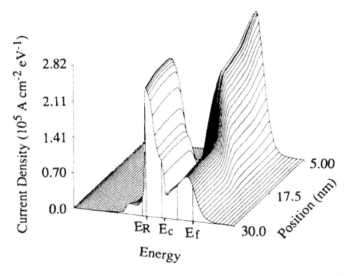

Figure 10.4 The energy- and position-dependent current density for a resonant-tunneling diode, showing the presence of inelastic tunneling involving the emission of an optical phonon. Reprinted with permission from Ref. [28], Copyright 1992, American Physical Society.

10.3 Nano-electronic Modeling

One of the most advanced simulation packages available began life with the simulation of one-dimensional resonant-tunneling diodes, just as in the previous section. This is the NEMO (nano-electronic modeling) code, which was originally developed at Texas Instruments. While the prototypical nanodevice is the small resonant-tunneling diode, small diameter nanowires, carbon nanotubes, and molecules stretched between metallic (or semiconducting) leads are other examples of quasi-one-dimensional structures of interest. Of course, the code has been extended in recent years to both two- and three-dimensional structures. Here the analytic band, effective mass approach used previously (such as with the scattering matrix, Lippmann–Schwinger equation approach to obtain the transmission) is no longer applicable. It is not the calculation of the transport that is no longer applicable, but the energy structure of the system under consideration that is likely to be modified by the small transverse dimensions. There have been several methodologies developed to study this, most of which use localized atomic orbitals (or basis functions) to build up the energy structure in the medium. This approach to computing the band structure is a real-space methodology and thus is amenable to systems that are physically small in spatial extent. Plane wave methods represent the other extreme, where the system may be spatially large but have a limited momentum space. Nevertheless, there are limits to the use of local orbitals, and thus, NEMO and many other approaches divide the system into semi-classical and quantum regions, using the full Green's functions only in the latter regions.

One of the earliest approaches, which utilized the empirical tight-binding basis set of atomic functions, was that of the Rome group [31]. This effort utilized the so-called exact exchange energy correction to the local density approximation to correct the band structure and has been used to study semiconductor nanostructures [32], band mixing in these structures [33], molecules stretched between metallic conductors [34], and carbon nanotubes [35]. A somewhat more famous energy structure program is SIESTA, which also uses localized atomic orbitals (but numerically simulated ones) [36, 37]. This program has been extended to computation of the conductance via Green's functions in TranSIESTA [38, 39]. These are all quite good approaches in this genre of simulation tools. However, to illustrate the approach, we will focus our attention to the NEMO approach.

There is an important difference between NEMO and these other codes. The NEMO atomic structure is found from semi-empirical tight-binding structure [40]. The tight-binding orbitals are characterized by overlap integrals (matrix elements in the Hamiltonian) that are adjusted to reach a best fit to experimental data. On the other hand, SIESTA uses parameterized orbitals and these parameters are adjusted by a full density functional approach [41, 42], in which the overall computation is iterated self-consistently to produce the energies. A similar approach is used in the FIREBALL code [43, 44], which is a first principles tight-binding approach in which the overlap integrals are evaluated exactly, rather than empirically.

These latter approaches are more physically based but have the well-known problem of producing a bandgap that is usually considerably smaller than seen experimentally. The empirical approach starts with at least getting the principal gaps correct, but the parameters are less physically based.

Some of the earliest work on the simulation of quantum structures was carried out at Purdue University under the direction of Supriyo Datta [28, 45]. This was among the earliest Green's function work to incorporate inelastic processes in a realistic manner. This basis was quickly expanded when several of the former Purdue students wound up at Texas Instruments developing the NEMO simulation code. As mentioned, this approach depends on detailed tight-binding atomic structure and Green's functions for evaluation of the transport and was developed and first applied for resonant-tunneling diodes [46, 47]. Here it was very important to assure that the tight-binding band structures gave accepted values for the effective mass in large structures [48], that it gave acceptable values for the known warpage in the valence bands [49], and yielded a set of parameters that could be utilized between various semiconductors [50]. A by-product of this careful development was the meeting of additional requirements in the presence of electromagnetic fields, which needed gauge invariance [51]. In any such code, it is important that it be applicable to realistic devices; e.g., devices that are likely to be experimentally realizable. Beyond this, it must be extendable to higher dimensions and to other device structures. It must handle spatially varying potentials, the inclusion of detailed scattering processes, and must be able to handle some definition of the contacts or boundary conditions that fit to experimentally realizable devices. Of course, most of the codes mentioned above will satisfy these constraints when the self-energy is included within the Green's function formulation.

To begin with this approach, we write the general Hamiltonian for the device under consideration in the form

$$H = H_{D,0} + H_{L,0} + H_{R,0} + H_{LD,0} + H_{RD,0}$$
$$+ H_{\text{POP}} + H_{\text{ac}} + H_{\text{ir}} + H_{\text{al}} + H_{\text{imp}}. \tag{10.37}$$

The first five terms are for the central device, the left and right contacts, and the interactions between the device and the left and right contacts, respectively. The last five terms represent the dissipative interactions with the polar optical phonons, the acoustic phonons, interface roughness, alloy scattering, and impurity scattering. These last five terms will be treated via perturbation theory and provide the self-energy correction to the total Hamiltonian. The two contact terms will be treated exactly but will also lead to a self-energy correction to the reduced Hamiltonian for the central device itself. This latter self-energy term will represent the injection (and ejection) of particles to (from) the device region. The device itself is broken into slices, just as in Chapter 2, for the scattering matrix approach. This means that the total Green's function will be built by a recursive technique just as the recursive Green's function was built in Section 3.3. Atomically, each individual layer in the device contains a cation sub-layer and an anion sub-layer. The position of the cation is designated by a vector that has a layer part and a transverse part indicating the cation location within the layer, as

$$\mathbf{R}_L = La\mathbf{a}_z + \mathbf{R}_{L,\perp}. \tag{10.38}$$

The corresponding anion location is displaced by $(a/2)(111)$, where a is the dimension of the primitive cell and, in Eq. (10.38), is the lattice constant. The localized anion and cation orbital wavefunctions are indicated by their locations and can be used to construct transverse plane wave states as

$$|c, L, \mathbf{k}\rangle = \frac{1}{\sqrt{N}} \sum_{\mathbf{R}_{L,\perp}} e^{i\mathbf{k}\bullet\mathbf{R}_{L,\perp}} |c, L, \mathbf{R}_{L,\perp}\rangle$$

$$|a, L, \mathbf{k}\rangle = \frac{1}{\sqrt{N}} \sum_{\mathbf{R}_{L,\perp}} e^{i\mathbf{k}\bullet(\mathbf{R}_{L,\perp}+\mathbf{d})} |a, L, \mathbf{R}_{L,\perp}\rangle, \quad \mathbf{d} = \frac{a}{2}(111). \tag{10.39}$$

The set of anion and cation states are a complete set of tight-binding orbitals at each anion and cation site, respectively. Thus, for a set sp^3s^*, for example, each set in Eq. (10.39) contains five separate orbitals that differ by the index a or c. Most of the simulations, at least for the resonant-tunneling diodes, have used just this set, but some have incorporated five d levels as well [52, 53]. In this formulation, $\mathbf{k}$ is a purely transverse vector and lies within the particular slice. We can now use these plane wave states to create the appropriate field operator

$$\hat{\psi}(\mathbf{r}) = \sum_{\mathbf{k},L} \left[\sum_c \langle \mathbf{r} \,|\, c, L, \mathbf{k} \,|\rangle \xi_{c,L,\mathbf{k}} + \sum_a \langle \mathbf{r} \,|\, c, L, \mathbf{k} \,|\rangle \xi_{a,L,\mathbf{k}} \right], \quad (10.40)$$

where ξ is the annihilation operator for the particular plane wave state. A particularly important point here is that the Schrödinger discretization is only in the z-direction (along the length of the device). The transverse dimensions, whether only a second dimension or both x and y transverse dimensions, are a quasi-periodic array for which the transverse band structure is given by the tight-binding orbital overlap integrals. This is a traditional layer approach to systems in which the z-direction has broken symmetry (broken periodicity), such as surfaces or interfaces. If the device is small in the transverse dimensions, it is well known that one must develop an assumed periodic structure in the transverse dimensions in order to develop these plane wave states. Now the overlap integrals, both in the slice and between slices, can be written as

$$\langle a, L, \mathbf{k} \,|\, H \,|\, \alpha', L', \mathbf{k} \rangle = D_{\alpha,\alpha',L}\delta_{LL'} - T_{\alpha,\alpha',L,L'}\delta_{L',L\pm j\neq 0}, \quad \alpha = a, c \, .$$
$$(10.41)$$

The second term represents the layer-to-layer coupling between the orbitals, and each element of T is an $m \times m$ block, where m is the number of atomic orbitals used. The span j tells us whether we are limiting ourselves to nearest neighbors ($j = 1$) or are going to more far-reaching interactions. The first term represents the interactions within a layer, and the diagonal $m \times m$ blocks contain on the diagonal the orbital eigenenergies and the local potential, while the off-diagonal elements in each block represent the anion–cation interactions. The presence of off-diagonal blocks would be indicative of beyond nearest-neighbor interactions.

It is possible to illustrate the nature of these various matrices by considering the simplest possible case first. In this case, we assume a single tight-binding orbital per site, with the cation and anion orbitals are lumped into a single equivalent "band," described by an effective mass at each site (or on each slice). Then the discretized Schrödinger equation becomes

$$H_0 = -\frac{\hbar^2}{2}\frac{d}{dz}\left(\frac{1}{m^*(z)}\frac{d}{dz}\right) + V_k(z) + \frac{\hbar^2 k^2}{2m_L} \qquad (10.42)$$

with

$$V_k(z) = V(z) + \frac{\hbar^2}{2m_L}\left(\frac{m_L}{m^*(z)} - 1\right). \tag{10.43}$$

Here, m_L is the in-plane effective mass while m^* is the effective mass in the z-direction. The appearance of the mass inside the derivative reflects the required conservation of probability current through the hetero-interfaces of the barriers. The matrix elements are now simply

$$D_L = \frac{\hbar^2}{2a^2}\left(\frac{1}{m^+} + \frac{1}{m^-}\right), \quad m^\pm = \frac{1}{2}(m_{L\pm1} + m_L)$$

$$T_{LL'} = \frac{\hbar^2}{(m_L + m_{L'})a^2}. \tag{10.44}$$

Now when we do the sp^3s^* orbitals, each anion and cation has five orbitals—four normal orbitals in the outer shell and the excited s-orbital. Hence, the Hamiltonian is block diagonal with 5×5 blocks. Each block is characterized by the cation (anion) eigenenergies on the diagonal and interactions among the cation (anion) orbitals, such as second-neighbor interactions or the spin–orbit interactions on the off-diagonals. The next blocks from the diagonal describe the interactions between the cation and the anion orbitals on nearest neighbors, and contain the Bloch sums over the lattice vectors in the transverse layer, which arise from the plane waves introduced above. The T-matrix is not block diagonal, but has 5×5 blocks describing interactions between cation (anion) orbitals in one layer with the anion (cation) orbitals in the adjacent layer.

We can understand the general approach by referring back to Section 6.5, where we illustrated the potential profile for a resonant-tunneling diode, with bias applied. The structure consists of two large reservoirs, left and right, and a short device. The left reservoir is described by layers $-\infty$ to 0. The central device is characterized by layers 1 to N, and the right reservoir is characterized by layers $N+1$ to ∞. Several self-energy terms are possible due to the interactions described in Eq. (10.37). For example, the self-energies Σ_r^{LD} and Σ_r^{RD} account for loss from the interaction between the central device and the contacts. The term $\Sigma^<$ accounts for injection from the contacts to the device. We can now write the less-than and greater-than correlation functions as

$$G^<_{\alpha,\alpha',L,L'}(\mathbf{k},t,t') = \frac{i}{\hbar}\left\langle \xi^\dagger_{\alpha',L'}(\mathbf{k},t')\xi_{\alpha,L}(\mathbf{k},t)\right\rangle$$

$$G^>_{\alpha,\alpha',L,L'}(\mathbf{k},t,t') = -\frac{i}{\hbar}\left\langle \xi_{\alpha,L}(\mathbf{k},t')\xi^\dagger_{\alpha',L'}(\mathbf{k},t')\right\rangle.$$

(10.45)

Similarly, we can write the retarded and advanced Green's functions as

$$G^r_{\alpha,\alpha',L,L'}(\mathbf{k},t,t') = \theta(t-t')\left[G^>_{\alpha,\alpha',L,L'}(\mathbf{k},t,t') - G^<_{\alpha,\alpha',L,L'}(\mathbf{k},t,t')\right]$$

$$G^a_{\alpha,\alpha',L,L'}(\mathbf{k},t,t') = \theta(t'-t)\left[G^>_{\alpha,\alpha',L,L'}(\mathbf{k},t,t') - G^<_{\alpha,\alpha',L,L'}(\mathbf{k},t,t')\right],$$

(10.46)

and the spectral density is given by the normal imaginary part of the bracketed terms in the equations. As we are concerned mainly with the long time limit, or steady-state response, we can Fourier transform on the time difference and work with the transformed Green's functions. Then, we have the normal relationships among the Green's functions such as

$$G^r_{\alpha,\alpha',L,L'}(\mathbf{k},\omega) = \int \frac{d\omega'}{2\pi}\frac{A_{\alpha,\alpha',L,L'}(\mathbf{k},\omega')}{\omega-\omega'} - \frac{i}{2}A_{\alpha,\alpha',L,L'}(\mathbf{k},\omega).$$

(10.47)

The first term is just the normal principal part of the integration, but arises in this form from the Kramers–Kronig relations.

Once the less-than correlation function is determined, then the density at slice L can be found from it as

$$n_L = \frac{2i\hbar}{Sa}\sum_{\mathbf{k},\alpha}\int \frac{d\omega}{2\pi}G^<_{\alpha,\alpha,L,L}(\mathbf{k},\omega),$$

(10.48)

where S is the cross-sectional area of the slice. Similarly, the current density through this slice is given by

$$J_L = \frac{2e}{S}\sum_{\mathbf{k},\alpha,\alpha'}\int \frac{d\omega}{2\pi}\sum_{L_1<L}\sum_{L_2>L}\left[T_{\alpha,\alpha',L,L_2}G^<_{\alpha',\alpha,L_2,L_1} - T_{\alpha,\alpha',L_2,L_1}G^<_{\alpha',\alpha,L,L_1}\right].$$

(10.49)

As mentioned above, the device is partitioned into left and right contact regions, which extend up to the active device region. Even with spatially varying potentials, these regions can be treated as reservoirs for the active device region. This approach allows for the injection of carriers from mixed, non-asymptotic states in the emitter (or collector), which is considered the left (or right) contact region. This will also allow the inclusion of scattering in the contact regions. This is most easily done by inclusion of an imaginary potential in the Hamiltonian for the contact regions, a necessary step to account for bound states, which may exist in these regions. The left and right contact regions are then collapsed into the appropriate self-energies, introduced above, which account for these regions in the active device regions.

To illustrate the approach with the minimum of extraneous notation, we will use the single band model introduced in Eq. (10.42). Hence, we will keep only the layer indices, and then we can write down Dyson's equation as

$$
\begin{aligned}
G^<_{L,L'} &= g^<_{L,L'} + g^r_{L,1}(-T_{1,0})G^<_{0,L'} + g^<_{L,1}(-T_{1,0})G^a_{0,L'} \\
&\quad + g^r_{1,N}(-T_{N,N+1})G^<_{N+1,L'} + g^<_{1,N}(-T_{N,N+1})G^a_{N+1,L'} \\
G^>_{L,L'} &= g^>_{L,L'} + g^r_{L,1}(-T_{1,0})G^>_{0,L'} + g^>_{L,1}(-T_{1,0})G^a_{0,L'} \\
&\quad + g^r_{1,N}(-T_{N,N+1})G^>_{N+1L'} + g^>_{1,N}(-T_{N,N+1})G^a_{N+1,L'},
\end{aligned}
\tag{10.50}
$$

where the lower case Green's functions are the equilibrium values obtained with the T-matrices set to zero. The first line in each expression represents the coupling to the left contact region, while the second line represents the coupling to the right contact region. Each of these Green's functions remains a function of both the transverse momentum $\mathbf{k}$ and the energy $E = \hbar\omega$. The Green's functions, which extend across the boundaries between the active device and the contacts, also satisfy their own Dyson's equations through

$$
\begin{aligned}
G^<_{0,L'} &= g^r_{0,0}(-T_{0,1})G^<_{1,L'} + g^<_{0,0}(-T_{0,1})G^a_{1,L'} \\
G^a_{0,L'} &= g^a_{0,0}(-T_{1,0})G^a_{1,L'} \\
G^>_{0,L'} &= g^r_{0,0}(-T_{0,1})G^>_{1,L'} + g^>_{0,0}(-T_{0,1})G^a_{1,L'}
\end{aligned}
\tag{10.51}
$$

for the left contact and

$$
G^<_{N+1,L'} = g^r_{N+1,N+1}(-T_{N+1,N})G^<_{N,L'} + g^<_{N+1,N+1}(-T_{N+1,N})G^a_{N,L'}
$$

$$G^a_{N+1,L'} = g^a_{N+1,N+1}(-T_{N+1,N})G^a_{N,L'}$$

$$G^>_{N+1,L'} = g^r_{N+1,N+1}(-T_{N+1,N})G^>_{N,L'} + g^>_{N+1,N+1}(-T_{N+1,N})G^a_{N,L'}$$

$$(10.52)$$

for the right contact. Substituting these last equations into the set (10.50) leads to the results, after some algebra,

$$G^<_{L,L'} = g^r_{1,1}\,\Sigma^{rB}_{1,1}\,G^<_{1,L'} + g^r_{1,1}\,\Sigma^{<B}_{1,1}\,G^a_{1,L'} + g^<_{1,1}\,\Sigma^{aB}_{1,1}\,G^a_{1,L'}$$

$$+g^r_{1,N}\,\Sigma^{rB}_{N,N}\,G^<_{N,L'} + g^r_{1,N}\,\Sigma^{<B}_{N,N}\,G^a_{N,L'} + g^r_{1,N}\,\Sigma^{aB}_{N,N}\,G^a_{N,L'},$$

$$(10.53)$$

where we have introduced the self-energies

$$\Sigma^{rB}_{1,1} = T_{1,0}g^r_{0,0}T_{0,1}, \qquad\qquad \Sigma^{aB}_{1,1} = T_{1,0}g^a_{0,0}T_{0,1},$$

$$\Sigma^{<B}_{1,1} = T_{1,0}g^<_{0,0}T_{0,1}, \qquad\qquad \Sigma^{<B}_{N,N} = T_{N,N+1}g^<_{N+1,N+1}T_{N+1,N},$$

$$\Sigma^{aB}_{N,N} = T_{N,N+1}g^a_{N+1,N+1}T_{N+1,N},\ \Sigma^{rB}_{N,N} = T_{N,N+1}g^r_{N+1,N+1}T_{N+1,N}.$$

$$(10.54)$$

By definition, the actual contact regions are considered to be in equilibrium. Hence, we can define the Green's functions for this area as

$$g^<_{0,0} = -f_{FD,L}(g^r_{0,0} - g^a_{0,0})$$

$$g^<_{N+1,N+1} = -f_{FD,R}(g^r_{N+1,N+1} - g^a_{N+1,N+1}).$$

$$(10.54)$$

Here, the subscripts L and R on the distribution function refer to the left and right contacts, respectively. It is also convenient to introduce the quantities

$$\Gamma^B_{1,1} = i\left(\Sigma^{rB}_{1,1} - \Sigma^{aB}_{1,1}\right)$$

$$\Gamma^B_{N+N} = i\left(\Sigma^{rB}_{N,N} - \Sigma^{aB}_{N,N}\right),$$

$$(10.55)$$

with which we obtain

$$\Sigma^B_{1,1} = if_{FD,L}\Gamma^B_{1,1}, \quad \Sigma^B_{N,N} = if_{FD,R}\Gamma^B_{N,N}.$$

$$(10.56)$$

In the nearest-neighbor tight-binding model, all the boundary self-energy terms (superscript B above) are zero except for the two cases $L = L' = 1$, $L = L' = N$. That is, the boundary self-energies only couple to the slices in the device that are adjacent to the contact regions. To get to the equation of motion for the Green's functions at the various slices, we operate with the bare (equilibrium) Green's function differential operator as

$$(\hbar\omega - H_0^D)G_{L,L'}^< = (\hbar\omega - H_0^D)\Big\{ g_{L,1}^r \, \Sigma_{1,1}^{rB} G_{1,L'}^< + g_{L,1}^r \, \Sigma_{1,1}^{<B} G_{1,L'}^a$$
$$+ g_{L,1}^< \, \Sigma_{1,1}^{aB} \, G_{1,L'}^a + g_{L,N}^< \Sigma_{N,N}^{aB} \, G_{N,L'}^a$$
$$+ g_{L,N}^r \Sigma_{N,N}^{rB} \, G_{N,L'}^< + g_{L,N}^r \Sigma_{N,N}^{<B} \, G_{N,L'}^a \Big\}.$$

$$(10.57)$$

We can now use the usual properties of the Green's functions, in the absence of scattering within the device region,

$$(\hbar\omega - H_0^D)g_{L,L}^r = 1$$
$$(\hbar\omega - H_0^D)g_{L,L'}^< = 0.$$

$$(10.58)$$

Using this result, we can simplify Eq. (10.57) to

$$(\hbar\omega - H_0^D)G_{L,L'}^< = \delta_{L,1}\Big[\Sigma_{1,1}^{rB} \, G_{1,L'}^< + \Sigma_{1,1}^{<B} \, G_{1,L'}^a \Big]$$
$$+ \delta_{L,N}\Big[\Sigma_{N,N}^{rB} \, G_{N,L'}^< + \Sigma_{N,N}^{<B} \, G_{N,L'}^a \Big].$$

$$(10.59)$$

As an example of the above approach, we consider the simple 3 × 3 block for a small device. Here, both L and L' run from 1 to 3. Then the matrices for the various Green's functions look like

$$[G^r] = \begin{bmatrix} \left(\hbar\omega - H_1 - \Sigma_{1,1}^{rB}\right) & T_{1,2} & 0 \\ T_{1,2} & (\hbar\omega - H_2) & T_{2,3} \\ 0 & T_{2,3} & \left(\hbar\omega - H_3 - \Sigma_{3,3}^{rB}\right) \end{bmatrix}^{-1}$$

$$(10.60)$$

and

$$[G^<] = [G^r] \begin{bmatrix} \Sigma_{1,1}^{rB} & 0 & 0 \\ 0 & 0 & 0 \\ 0 & 0 & \Sigma_{3,3}^{rB} \end{bmatrix} [G^r].$$

$$(10.61)$$

The boundary self-energies have been defined previously, and we need only evaluate these in terms of the boundary Green's functions. Here the length of the boundaries will affect the computational time, and in order to illustrate this, we take only three slices for each boundary. Then, if the potential is constant in all three slices of the boundary, we may write the left contact retarded function as

$$
g^r_{0,0} = \left(\left[\begin{array}{ccc} \left(\hbar\omega - H_{-2} - \Sigma^{rB}_{-2,-2} \right) & T_{-2,-1} & 0 \\ T_{-2,-1} & \left(\hbar\omega - H_{-1} \right) & T_{-1,0} \\ 0 & T_{-1,0} & \left(\hbar\omega - H_0 - \Sigma^{rB}_{0,0} \right) \end{array} \right]^{-1} \right)_{0,0}.
$$

(10.62)

Here the self-energy at the left end of this block is given by

$$
\Sigma^{rB}_{-2,-2} = T_{-2,-3} g^r_{-3,-3} T_{-3,-2} = T_{-2,-3} \left(\hbar\omega - H_{-3} + T_{-4,-3} \chi Z \chi^{-1} \right)^{-1} T_{-3,-2},
$$

(10.63)

where χ is a matrix of Bloch states propagating toward the device from the semi-classical region, and Z is a diagonal matrix of the wave propagation factors. For the simple model, we consider here that the Ts are simple real scalars, as T_L and T_R for the left and right contacts. Then,

$$
g^r_{-3,-3} = \frac{1}{T_L} e^{i\gamma_L a}, \quad \Sigma^{rB}_{-2,-2} = -2 T_L e^{i\gamma_L a}.
$$

(10.64)

The dispersion in the transverse directions comes from the eigenvalues of the slice Hamiltonian directly. If we call this transverse energy D, then the total energy arises from

$$
\hbar\omega - D_{-2} = -2 T_L \cos(\gamma_L a).
$$

(10.65)

Now the retarded Green's function (10.62) becomes

$$
g^r_{0,0} = \left(\left[\begin{array}{ccc} -T_L e^{-i\gamma_L a} & T_{-2,-1} & 0 \\ T_{-2,-1} & \left(\hbar\omega - D_{-1} \right) & T_{-1,0} \\ 0 & T_{-1,0} & \left(\hbar\omega - D_0 \right) \end{array} \right]^{-1} \right)_{0,0}.
$$

(10.66)

This is now easily solved to give the needed Green's functions for the solution for the device.

The current was given earlier in Eq. (10.49), where it was shown that we need to compute the less-than correlation function in order to find the current through this device structure. The coupling from the emitter to the device leads to the less-than function across this boundary being given by

$$G_{e,d}^< = \sum_{e,e'} \left[g_{e,e'}^r (-T_{e',d'}) G_{d',d}^< + g_{e,e'}^< (-T_{e',d'}) G_{d',d}^a \right] \qquad (10.67)$$

where the subscripts e, e', d, d' correspond to values of L, L' in the left contact (emitter) and the right contact (drain). Then for the nearest-neighbor coupling only, where the self-energy boundary term appears only in the first layer, we can write the current as

$$J_0 = \frac{2e}{S} \sum_k \int \frac{d\omega}{2\pi} Tr\left\{ \Gamma_{1,1}^B \left[i\Gamma_{1,1}^B G_{1,1}^< + i f_{FD,L} A_{1,1} \right] \right\}. \qquad (10.68)$$

If we now write

$$G_{1,1}^< = G_{1,1}^r \Sigma_{1,1}^{<B} G_{1,1}^a = i f_{FD,L} G_{1,1}^r \Gamma_{1,1}^B G_{1,1'}^a \qquad (10.69)$$

the current now becomes

$$J_0 = \frac{2e}{S} \sum_k \int \frac{d\omega}{2\pi} f_{FD,L} Tr\left\{ \Gamma_{1,1}^B \left[G_{1,1}^r \Gamma_{1,1}^B G_{1,1}^a + A_{1,1} \right] \right\}. \qquad (10.70)$$

The term in the square brackets is just the tunneling coefficient through the active device, whereas the scattering operator within the trace operation plays the role of the velocity within the device. The Landauer formula requires that we take account of the back-propagating states, and this leads to the modification of Eq. (10.70) to incorporate this, and

$$J_0 = \frac{2e}{S} \sum_k \int \frac{d\omega}{2\pi} Tr\left\{ \Gamma_{1,1}^B \left[G_{1,1}^r \Gamma_{1,1}^B G_{1,1}^a + A_{1,1} \right] \right\} \left(f_{FD,L} - f_{FD,R} \right). \qquad (10.71)$$

We can also use this to find the carrier density in each slice by finding the less-than Green's function for each slice. This gives

$$i G_{L,L}^< = f_{FD,L} G_{L,1}^r \Gamma_{1,1}^B G_{1,L}^a + f_{FD,R} \left[A_{L,L} - G_{L,1}^r \Gamma_{1,1}^B G_{1,L}^a \right]. \qquad (10.72)$$

This last expression is now integrated over the momentum and energy to get the total density on the slice, and this is used to drive Poisson's equation for the local potential and field. This need to do the integration over energy of the less-than Green's function can be quite difficult, and needs to be done at each and every slice in order to drive Poisson's equation. In higher dimensional devices, this can be tedious, especially as it must be done every update cycle of Poisson's equation. A great deal of accuracy is normally required, but routines now exist to do this integration via a complex integration off the energy axis in order to avoid the zeroes that exist there.

Let us now turn to how the recursion is carried out, as this set of equations described above is evaluated using the recursive Green's function approach, detailed in Section 3.3. The recursive Green's function approach is most useful in the nearest-neighbor tight-binding approach. In the absence of scattering, the term $g^r_{L,L'}$ of course, has two contributions, coming from the left and right contacts. One takes into account *exactly* from the left while coupling to the next slice to the right. Similarly, the second takes into account *exactly* from the right, while coupling to the next slice to the left. Hence, one starts with the retarded function in the right lead and propagates to the left as

$$g^r_{L,L;R} = \left[\hbar\omega - D_L - T_{L,L+1} g^r_{L+1,L+1;R} T_{L+1,L} \right]^{-1}.$$
(10.73)

Here D is the eigenvalue matrix for the transverse modes, as discussed above. Once we reach the left contact, this right-derived Green's function is to the boundary term to achieve

$$g^r_{1,1;R} = \left[\hbar\omega - D_1 - T_{1,0} g^r_{0,0;L} T_{0,1} + T_{1,2} g^r_{2,2;R} T_{2,1} \right]^{-1}.$$
(10.74)

If we needed only the current, we could stop here, as we have enough information in the ballistic case. However, to close the Poisson loop and compute the potential (and field) self-consistently, we also need to find the density. So we walk back to the right as

$$G^r_{L,L} = g^r_{L,L;R} + g^r_{L,L;R} T_{L,L-1} g^r_{L-1,L-1;R} T_{L-1,L} g^r_{L,L;R}$$
$$G^r_{L,1} = -g^r_{L,L;R} T_{L,L-1} G^r_{L-1,L}.$$
(10.75)

So this approach is an advanced form of the recursive Green's function, which builds in the nonequilibrium density approximations. However, there is no scattering in the system at present, so there has been no need to compute the complicated less-than correlation function accurately. Similarly, the only self-energy arises from the coupling to the contacts. However, we have established a powerful formalism that allows us to overcome these limitations. Yet at this point, the NEMO approach offers not much new except the inclusion of detailed band structure calculations within the Green's function approach. Hence, we now have to turn to the case of scattering to move beyond this limitation.

When we introduce scattering into the problem, this appears as new self-energy terms that modify the equations of motion. Within the device, these now become

$$\left(\hbar\omega - H_0 - \Sigma_r\right)G^r = 1$$

$$\left(\hbar\omega - H_0 - \Sigma_r\right)G^< = \left(\Sigma^< + \Sigma^{<B}\right)G^a. \tag{10.76}$$

Thus, we have to account for both the new retarded self-energy and a less-than self-energy in these equations of motion. These terms represent the dissipative scattering processes. Each new self-energy is a sum over all the various processes that contribute to the scattering, but when introduced in this manner, these terms do not break current continuity. However, these terms do not appear in a simple manner. For example, let us consider elastic scattering, the most straightforward set of processes in a classical system. Here the relaxation term can be written as

$$\Gamma = i\left(\Sigma^> - \Sigma^<\right) = \int \frac{d^2\mathbf{k}}{(2\pi)^2} iM(G^> - G^<) = \int \frac{d^2\mathbf{k}}{(2\pi)^2} MA. \tag{10.77}$$

where M is the kernel of the self-energy, containing the squared matrix element for the scattering, and A is the spectral density. Since A contains the function Γ, this expression must be solved self-consistently, and therein lies the difficulty. Thus, one usually adopts some approximation to evaluate this function. One such, discussed previously, is to assume the Born approximation, in which the self-energy correction to A is ignored. This is valid only for weak scattering, which is the usual assumption in semi-classical scattering, particularly in semiconductors where the mobilities are high. NEMO goes beyond this, however, to account for off-shell processes that become allowed when the full form of A is used.

To illustrate this problem, we will use polar optical phonon scattering. This process is inelastic, but the extensions to other scattering processes that are elastic can be found in a straightforward manner from this example. In general, the Hamiltonian for the electron–phonon interaction may be written as

$$H_{eL} = \frac{1}{\sqrt{V}} \sum_{L,\mathbf{k}} \sum_{\mathbf{q}} U_q e^{iq_z L} \left[e^{iq_z a/2} \underbrace{\sum \xi^\dagger_{L,\mathbf{k}} \xi_{L,\mathbf{k}}}_{\text{anions}} + \underbrace{\sum \xi^\dagger_{L,\mathbf{k}} \xi_{L,\mathbf{k}}}_{\text{cations}} \right] (a^\dagger_{\mathbf{q}} + a_{\mathbf{q}}),$$

(10.78)

where U_q is the matrix element, ξ are electron operators, $a_{\mathbf{q}}$ are the phonon operators, and the atomic sums run over the atoms in the slice. This Hamiltonian leads to the less-than self-energy term

$$\Sigma^<_{\alpha,\alpha',L,L'} = \frac{1}{V} \sum_{\mathbf{q}} |U_{\mathbf{k}-\mathbf{q}}|^2 \, e^{iq_z a(L - L' + v_{\alpha\alpha'})}$$

$$\times \left[(N_q + 1) G^<_{\alpha,\alpha',L,L'}(\mathbf{q}, \omega - \omega_q) + N_q G^<_{\alpha,\alpha',L,L'}(\mathbf{q}, \omega + \omega_q) \right].$$

(10.79)

The first term in the square brackets corresponds to phonon emission, while the second term corresponds to phonon absorption. The magnitude squared matrix element for polar optical phonon scattering is given by

$$|U_q|^2 = \frac{e^2 \hbar \omega_q}{2} \left(\frac{1}{\varepsilon_s(\infty)} - \frac{1}{\varepsilon_s(0)} \right) \frac{q^2}{(q^2 + q_D^2)^2},$$

(10.80)

where the ε's are the high-frequency and low-frequency dielectric constants and q_D is the Debye screening wave number. The latter assumes that the carrier density is sufficiently low that the material is non-degenerate so that Debye screening is the appropriate form of the screening to use. In the following, the terms other than the momentum ratio will be denoted by the simpler β. With this form, Eq. (10.79) can be evaluated to yield the expression

$$\Sigma^<_{\alpha,\alpha',L,L'} = \frac{\beta}{\pi} \int \frac{d^2 \mathbf{q}}{4\pi^2} Y(L - L' + v_{\alpha\alpha'}, \mathbf{k}, \mathbf{q})$$

$$\times \left[(N_q + 1) G^<_{\alpha,\alpha',L,L'}(\mathbf{q}, \omega - \omega_q) + N_q G^<_{\alpha,\alpha',L,L'}(\mathbf{q}, \omega + \omega_q) \right]$$

(10.81)

where

$$Y(x, \mathbf{k}, \mathbf{q}) = \int\limits_0^w dp_z \frac{\cos(q_z a x)}{\sqrt{(q^2 + k^2 + q_D^2)^2 - 4k^2 q_\perp^2}}$$

$$- \int\limits_0^w dp_z \frac{\cos(q_z a x) q_D^2 (q^2 + k^2 + q_D^2)}{\left[(q^2 + k^2 + q_D^2)^2 - 4k^2 q_\perp^2\right]^{3/2}}. \qquad (10.82)$$

The parameter w is a cutoff that is imposed upon the integral to assure that the integration does not pass through a singularity of the integrand. The retarded self-energy can be found in a similar manner, but we have to face the convolution integral, and the need for self-consistency in each of these self-energies and their corresponding Green's function.

The full matrix solutions will require an enormous amount of memory to carry out to the full level. Thus, weak scattering is a viable assumption, except that here one goes beyond the simple Born approximation. The approach is illustrated with the retarded Green's function and self-energy, whereas the above used the less-than functions. The same approach can be applied to either, and we chose the retarded functions merely to expand the examples. The imaginary scattering function can be written as

$$\Gamma(\mathbf{k}, \omega) = i\left[\Sigma^>(\mathbf{k}, \omega) - \Sigma^<(\mathbf{k}, \omega)\right]$$

$$\sim i\Sigma^>(\mathbf{k}, \omega) \sim iM \otimes G^> \sim M \otimes A. \qquad (10.83)$$

$$\Sigma^r(\mathbf{k}, \omega) \sim M \otimes G^r$$

Here the symbol $\otimes$ denotes the integration over transverse momentum and sum over emission and absorption, with the Green's functions on the right offset by the phonon energy. Now the equations for G^r and $G^<$ are uncoupled and can be solved independently. We solve for the retarded function by use of a continued fraction expansion as

$$G^r = (\hbar\omega - H_0 - \Sigma^{rB} - \Sigma^r)^{-1} = (\hbar\omega - H_0 - \Sigma^{rB} - M \otimes G^r)^{-1}$$

$$= \left\{\hbar\omega - H_0 - \Sigma^{rB} - M \otimes \left[\hbar\omega - H_0 - \Sigma^{rB} - M \otimes \left(\hbar\omega - H_0 - \Sigma^{rB} - \Sigma^r\right)^{-1}\right]^{-1}\right\}$$

$$(10.84)$$

The partial fraction expansion has been carried to third order. It is usually sufficient to drop the retarded self-energy in this last term of the expansion, as convergence can be achieved with the subsequent summations. Having found the retarded function, the less-than function can now be found from Eq. (10.53), which itself is an iterated process. Finally, the current density is found from Eq. (10.71). As previously, the Green's functions are built up using the scattering at each layer, and the recursion between layers to propagate from one end of the device to the other and back again. One interesting aspect of the simulation with the presence of phonons is the appearance of an inelastic tunneling peak (as opposed to the resonant-tunneling peak that forms the current peak), which typically appears in the valley region and illustrates the contribution of the scattering processes to the valley current [54]. The phonon replica peak has been seen in experiments in InGaAs/InAlAs resonant-tunneling diodes [55].

Hence, it is found that NEMO gives reasonably good agreement with experiment if all the correct physical processes are included. In fact, the agreement is sufficiently good that it can be an analytical tool to provide some feedback to the device fabricators about the actual dimensions of their structure. However, it is not perfect, but then no other simulation is perfect either. It is, however, a very good approach if used wisely in the hands of the knowledgeable. The NEMO code has been used in recent years for simulations beyond the simple 1D model discussed above. It has been applied to alloy semiconductors [56], Si and GaAs nanowires [57–59], carbon nanotubes [60], and Ge/Si core-shell structures [61]. Various of these approaches have also been combined for alloy nanowires [62].

The NEMO code described here is basically for one-dimensional devices, but improvements have appeared since its introduction. The atomic band structure has been expanded in order to accommodate strain in the device [63]. Toward the end of the last century, it was extended to a full three-dimensional simulation involving millions of atoms. This first venture into the three-dimensional world focused on self-assembled quantum dots [64]. This was extended further into the three-dimensional world for nanowires a few years later, although this work did not include dissipative scattering [65]. The full capability, including full scattering dynamics, appeared shortly afterward, under the new acronym OMEN [66]. Shortly after this, it was applied to strained Si MOSFETs [67]. It has since been used in the careful analysis of the physics of the single-atom transistor discussed earlier in Section 1.4.

As mentioned earlier in this section, there are other approaches that accomplish the same results. The most noted is based on the density functional approach of SIESTA [35, 36]. Like NEMO, SIESTA uses localized basis functions, although the latter is a *first principles* computation of the band structure, rather than the empirical evaluation of the overlap integrals that are used in NEMO. Such first principles approaches are usually defined as density functional theory (DFT). The transport addition to SIESTA, which appears in TranSIESTA [38, 39], also uses the full real-time Green's function formulation. Another approach using density functional tight-binding bands for the study of transport in molecules has also seen extensive usage [68].

Approaches to using the real-time Green's functions without the band structure solvers, e.g., using the effective mass approximation with analytic bands, have also appeared. There are a great many of these approaches, so I mention only a couple of them. These Green's functions have been applied to the transport through carbon nanotubes quite effectively [69]. They have also been used to study discrete doping in small FinFETs, again with an effective mass analytic band structure [70].

10.4 Beyond the Steady State

We now want to turn to consideration of transient phenomena in quantum transport, and how these may be treated with the real-time Green's functions. In many places in this book so far, we have treated conductance through the Landauer formula, in which a transmission coefficient for the device is evaluated in terms of the Fermi–Dirac functions in the two contacts. Many have tried to extend this approach to high-frequency devices. The results are perhaps best summarized by Landauer himself [71]:

> The literature contains a number of attempts to extend the viewpoint, which calculates the conductance from transmissive behavior, to high frequencies. In the prejudiced opinion of this author many of these, perhaps all, are faulty.

Perhaps we have been more successful since this was written, but I suspect that this is not the case. Consequently, we turn to approaches that rely on more established quantum transport techniques. In this section, we will describe some of the new problems that must be faced in this quest [72], while we will discuss various timescales of interest in the next section.

One of the most obvious problems with time-dependent transport is that the current contains contributions from the displacement current. In general, the displacement current at a point **r** in the device must satisfy

$$I(\mathbf{r}) = \int_S \varepsilon(\mathbf{r}) \frac{\partial \mathbf{E}(\mathbf{r})}{\partial t} \cdot d\mathbf{S}, \qquad (10.85)$$

where $\varepsilon(\mathbf{r})$ is the spatially varying dielectric function and **S** (and $d\mathbf{S}$) represents a cross-sectional area of the device at the point. Because the electric field is quite likely inhomogeneous throughout the device, there is no requirement that this current be uniform. Rather, it is the total current that must be continuous throughout the length of the device. Because of this and because the total device must be space charge neutral, the total current flowing into the device through the contacts must be zero. Many people ignore the displacement current because it can be shown that, in a space charge neutral device, the total displacement current passing through the enclosing surface vanishes. However, the device is not required to be space charge neutral at every point in the device. In fact, a transistor would not work under such a constraint. Consequently, the displacement current can be quite important within the device.

Another important aspect arises from Maxwell's equations. Quite generally, one can manipulate these equations so that the scalar and vector potentials satisfy wave equations separately, as

$$\nabla^2 V - \frac{1}{\mu_0 \varepsilon_s} \frac{\partial^2 V}{\partial t^2} = -\frac{\rho}{\varepsilon_s}$$
$$\nabla^2 \mathbf{A} - \mu_0 \varepsilon_s \frac{\partial^2 \mathbf{A}}{\partial t^2} = -\mu_0 \mathbf{J}, \qquad (10.86)$$

provided that the Lorentz condition, or Lorentz gauge,

$$\nabla \cdot \mathbf{A} + \mu_0 \varepsilon_s \frac{\partial V}{\partial t} = 0 \qquad (10.87)$$

is satisfied. One usually encounters the first of Eq. (10.86) in devices without the time derivative, which leads to Poisson's equation. But this is achieved only in the Coulomb gauge, in which $\nabla \cdot \mathbf{A} = 0$, which forces the scalar potential to be static via Eq. (10.87). Most electron transport studies assume the "electrostatic" approximation where Poisson's equation can be safely used. It is a reasonable approximation for nanometer devices below frequencies of a few tens of GHz. However, for larger devices or higher frequencies, care must be taken to incorporate properly all of the electromagnetic fields. This latter scenario represents the transition from electronics to electromagnetism, which remains mainly unexplored in the quantum transport community. But in the semi-classical world, full electromagnetic waves have been coupled to advanced device simulations [73, 74]. The fact that these approaches have had limited success lies more in the circuit problem rather than in the approach. By circuit problem, it is meant that the microwave circuit has to transition the rather low impedance of a transmission line into the very high impedance of the transistor gate input. In the circuit, this requires an extensive matching network, which rather complicates the coupling of the electromagnetics to the device simulation. On the other hand, the electrostatic approximation for the Poisson's equation has been rather successful in the semi-classical world. One of the highest frequency transistors in the world is the strained-InAs quantum well HEMT, which has demonstrated the ability to reach an f_{max} of 1.2–1.4 THz [75]. This has led to the demonstration of the first integrated amplifier to demonstrate significant gain at 1 THz [76]. It is important to note that device simulation indicated that these results were fully obtainable, even when the experimental state of the art was at a much lower frequency [77].

The scattering matrix, discussed in Chapter 2, has been successfully applied to a variety of time-dependent quantum phenomena. For example, Büttiker et al. considered different many-body approximations [78–80] to compute the high-frequency conductance and noise of mesoscopic systems, in which there remained overall charge neutrality and total current conservation. This approach led to the prediction of the quantization of the a.c. conductance in mesoscopic devices [81], a result that has recently been verified [82]. Another relevant technique is the density matrix discussed in Chapter 7 (with many variations grouped under the name of quantum master equations, such as those in Section 7.4). These have their origins in realizing that the complete knowledge of the many-body wavefunction is not necessary to predict the dynamic behavior of quantum devices. For example, the electron energy follows Fermi–Dirac statistics, but we do not need to know the exact energy of each electron. For situations with statistical uncertainty, the most general expression for a quantum system of electrons is the density operator [83, 84]. Similarly, the Wigner function offers a good approach, as was indicated in Chapter 8. While our main approach in this chapter will pursue the real-time Green's functions, it will be seen that it is often not possible to separate them from a density matrix or Wigner function. Finally, we will also introduce the time-dependent density functional theory, which has become popular in recent years for studying the transient situation.

10.5 Timescales

One of the crucial considerations in transient studies is the range of the characteristic timescales that exist in the semiconductor. For example, if the timescale of the transient is much longer than any of the characteristic times for transport, then it can be considered a quasi-steady-state situation. This points to the general proposition that when the timescale of interest is placed in the hierarchy of pertinent transport timescales, then one has an idea of which regions of the physics must be carefully treated and which may be safely approximated. All of this refers to the fact that the representation of any physical law is generally written for a particular timescale. As a result, it is not uncommon for two physical laws to have contradictions when they refer to different timescales. An example of this is the reversibility of Newton's force equation for a single particle and the apparent irreversibility of the diffusion equation for an ensemble of particles. These latter two processes operate on dramatically different timescales. The diffusion equation, in particular, operates on timescales long compared to the mean time between collisions so that the scattering effects tend to be averaged.

The importance of the various timescales is obvious in the analysis of nonequilibrium systems, since these systems present considerably more difficulty than that of equilibrium systems, due to the necessity of evaluating the time dependence of the various measurable properties. Indeed, these properties must be determined from equations describing them as evolving averages over the entire ensemble of particles. The ensemble itself, often characterized by the distribution function, or in the present case by the less-than Green's function, is evolving in time as it is a non-stationary quantity. As one can see from the variety of equations of motion we have so far discussed, this evolution is also a nonlinear one and is a non-ergodic one. In fact, it is important to evaluate carefully the various collection of timescales that are important.

In a semiconductor, numerous collisions occur and it is these collisions that provide the mechanism of exchange of energy and momentum among the carriers and from the carriers to the lattice. One should also note that it may well be, particularly in semiconductors, that the phononic bath of the lattice itself could well be driven out of equilibrium. In general, four generic timescales can be identified [85, 86]:

$$\tau_c < \tau < \tau_R < \tau_H. \tag{10.88}$$

The shortest of these, in most cases, is the collision duration (τ_c)— the average time it takes for a collisional process to actually occur. Generally, this time is quite short, but we have seen that it can be important and may actually be observable on the timescale of fast femtosecond laser pulse experiments. One might think of this time as that required to actually establish the energy-conserving delta function that appears in the Fermi golden rule. It is also related to the time over which the off-shell contributions, or the intra-collisional field effect, become important in transport. We will see in the next section that this time can be of the order of a few femtoseconds in most cases. This time must be distinguished from τ, which is the mean time between collisions. For scattering by the polar optical modes in GaAs, at room temperature, for example, this mean time between collisions can be of the order of 0.2–0.5 ps.

When we treat moment equations, as discussed in previous chapters, or are interested in the mobility, then we need to discuss relaxation times, such as the momentum or energy relaxation times. These differ from τ because they may produce different effects while also being an average over the nonequilibrium ensemble of carriers. For example, anisotropic scattering, such as impurity scattering or the polar optical mode scattering, does not efficiently relax the momentum along the electric field. Hence, the relaxation time is considerably longer than the mean time between collisions. In the semi-classical world, this is often taken into account for impurity scattering by the ubiquitous extra $(1 - \cos\theta)$ inserted into mobility calculations. Each of these relaxation times has its own timescale, but they typically are slightly larger than the mean time between collisions. For the same GaAs mentioned above, the momentum relaxation time arising in the mobility is about 0.1–0.3 ps. On the other hand, for mesoscopic devices at low temperatures, the energy relaxation time can be many nanoseconds [87].

The last of these is the hydrodynamic time τ_H, which is sometimes described as the time required to establish a uniform steady-state nonequilibrium ensemble, but this is not a particularly good definition. In fact, there are quite a few hydrodynamic times that can arise in semiconductors. One example might be *diffusion time*, which could be described as the time the carriers need to move a diffusion length, which turns out to be essentially the recombination time for minority carriers. There is a tendency to talk about mesoscopic devices as being in the diffusive regime when they are non-ballistic in nature, but this is a leftover concept from metals. In systems that can be driven out of equilibrium, there is no Einstein coefficient and, therefore, no good relationship between a diffusion constant and the mobility [88]. The closest to this might be to define a time given by the system size L and the mean velocity (the thermal velocity or the Fermi velocity), e.g., L/v_F. But there are many more hydrodynamic times. One important one is often called the *phase coherence time*, which is the time in which the phase of the quantum wave remains coherent. This is often confused with the inelastic mean free time. All inelastic scattering processes lead to loss of phase coherence, but there are other processes that also contribute to this loss of coherence.

Another important timescale, which is hydrodynamic in nature, is the period of the free carrier plasma oscillations

$$\tau_p = \frac{2\pi}{\omega_p} = 2\pi\sqrt{\frac{m^*\varepsilon_s}{ne^2}} \,, \tag{10.89}$$

which, for GaAs with 10^{17} cm^{-3} electrons, is about 0.3 ps, that is on the same scale as the mean time between collisions. A time related to this is the dielectric relaxation time, which describes the timescale over which charge fluctuations normally die out in the semiconductor

$$\tau_D = \frac{\varepsilon_s}{\sigma} = \frac{1}{\omega_p^2 \tau_m}, \tag{10.90}$$

where τ_m is the momentum relaxation time. For our GaAs, this is about 10 fs. These last two indicate how the clear separation of timescales in Eq. (10.88) is not always fulfilled.

Whether we are talking about equilibrium or nonequilibrium, the memory of the initial state is essentially destroyed on timescales larger than the mean free time between collisions. Hence, no matter what the initial distribution is, the final evolving distribution after a few picoseconds (using the numbers cited above) is essentially a balance between the evolving driving fields and the evolving relaxation processes. Hence, after discussing the collision duration in the next section, we turn to fast laser excitation of semiconductors and its quantum treatments.

10.6 Collision Duration

Classically, the treatment of a collision duration may be tracked back at least as far as Lord Rayleigh [89]. Quantum mechanically, however, the problem is better raised as asking just how long it takes to establish the Fermi golden rule, by which the quantum transition is determined. To my knowledge, this was first discussed by Paige in studies of the transport in Ge and by Barker in high electric field quantum transport [14, 90]. However, these latter approaches used an estimate based on the energy dependence of the matrix elements rather than trying to analyze the time dependence within the standard formulation of first-order time-dependent perturbation theory. This latter seems to have been first done by Geltman [91], via a Green's function approach to answer the question of how long it takes to ionize an atom, and by Lane in analyzing the decay of a field-created state [92]. In the idea of ionizing an atom from an incident electromagnetic field, it is important to understand that the field first creates a correlation between the initial and final states, which can lead to Rabi oscillations between these two states [93]. This correlation is most easily treated by wave mechanics. The conjugate field then breaks up this correlation, leading to the completed process. This approach was applied later by Haug and by Rossi and Kuhn in the treatment of very fast laser excitation of semiconductors [94, 95]. (This fast laser response will be discussed in the next section.) Hence, the electromagnetic field first created the correlation (or polarization) between the bands, and then the completed electron–hole pair creation was the second step, just as described above. In the above process, both Geltmann [91, 93] and, later, Lipavsky et al. [19] found that the approach to the asymptotic form, given by the Fermi golden rule, for long times shows weak oscillations about this asymptotic result. In Geltman's case, these oscillations center on the asymptotic result, indicating that the probability of scattering (or emission) is actually enhanced for short periods of time, although these oscillations die out rather quickly. On the other hand, Lipavsky et al. found that these oscillations only cause decreases in the probability for short periods of time before they die out.

In the current approach [96] to computing the collision duration time, we will utilize the real-time Green's functions, but follow the approach above for the absorption of a photon. That is, the electron is assumed to interact with the optical phonon through the polarization field of the phonon. First, a correlation between the initial and final states is created, and then this correlation is broken as the collision is completed. In other words, we can say that the electron is in a well-defined state of the momentum before the scattering event, and that it will be in a well-defined state of the momentum after the scattering is completed. During the scattering, the electron state is not well defined. This condition is mathematically described in terms of a generalized less-than Green's function that can be seen as a polarization induced between the initial and final states by the perturbation potential. We define the "collision duration time" as the time over which this generalized less-than Green's function is different from zero. The presence of oscillations in the development of the Fermi golden rule depends on the collisional broadening of the initial state. Oscillations only occur when the state collisional broadening (damping) is sufficiently weak, and then appear in a manner consistent with the earlier results discussed above.

The general state of the system is described in terms of electron and phonon states. Here we consider the perturbation induced by the electrons on the phonon population to be negligible, which means that we can factorize the states of the system in terms of products of electrons and equilibrium phonon states. Since the field operators included in the definition of the electron Green's function act only on the electron states, we can eliminate the dependence on the phonon states, summing over them without affecting the form of the Green's function. The self-energy Σ is usually a function of two positions (and times) and is obtained from a diagrammatic expansion of the perturbation interaction. *Here, however, we make two crucial deviations from the normal approach.* First, we follow the approach of Kuhn–Rossi [95] and Haug [94] (who work with the electromagnetic field of the photon) by introducing the polarization field of the longitudinal optical (LO) phonons as

$$\Sigma(\mathbf{q}, t) = \lambda_q \left(a_q^{\dagger} e^{-i(\mathbf{q} \cdot \mathbf{r} - \omega t)} + a_q e^{i(\mathbf{q} \cdot \mathbf{r} - \omega t)} \right), \tag{10.91}$$

where λ_q is the coupling matrix element, and a_q and $a_q^{\dagger}$ are the creation and annihilation operators for the phonons. For the polar optical mode, this becomes

$$\lambda_q = \frac{1}{q} \left(\frac{\hbar e^2}{2 V \eta \omega_0} \right)^{1/2}, \quad 1/\eta = \omega_0^2 \left(\frac{1}{\varepsilon_s(\infty)} - \frac{1}{\varepsilon_s(0)} \right), \tag{10.92}$$

where ω_0 is the longitudinal optical phonon frequency, q is the phonon wave number, V is the volume, and $\varepsilon_s(0)$ and $\varepsilon_s(\infty)$ are the low-frequency and high-frequency dielectric permittivities of the semiconductor of interest. Alternatively, in the inter-valley optical phonon, the coupling constant becomes

$$\lambda_q = \left(\frac{\hbar D^2}{2 \rho V \omega_0} \right)^{1/2}, \tag{10.93}$$

where D is the optical deformation potential and ρ is the mass density.

As usual, the Green's function is written in the Wigner coordinates and Fourier transformed in the different position. Here we Fourier transform only on the difference position and treat the result as independent of the center-of-mass position (homogeneity of the material). Hence, we deal with the electron correlation function

$G^<(\mathbf{k}, t, t')$. To approach the quantum transition problem, we explicitly introduce the electron correlation function between the initial and final states as the polarization that arises due to the phonon. This is represented by the generalized Green's function

$$G^<(\mathbf{k}_1, \mathbf{k}_2, t_1, t_2) = i\left\langle \psi^\dagger_{\mathbf{k}_2}(t_2)\psi_{\mathbf{k}_1}(t_1)\right\rangle. \tag{10.94}$$

With these two Green's functions and the general equations of motion, we can write the equations for these two Green's functions (for the polar interaction) as [97]

$$G^<(\mathbf{k}, t_1, t_2) = \frac{i}{\hbar}\left(\frac{\hbar e^2}{2\eta\omega_0}\right)\int_{t_1}^{t_2} dt'_2 e^{i[\omega_k(t_2 - t'_2) - \omega_0 t'_2]}$$

$$\times \int \frac{d^3q}{(2\pi)^{3/2}}\frac{\sqrt{N_q + 1}}{q}G^<(\mathbf{k}, \mathbf{k} + \mathbf{q}, t_1, t'_2)$$

$$G^<(\mathbf{k}, \mathbf{k} + \mathbf{q}, t_1, t'_2) = -\frac{i}{\hbar}\left(\frac{\hbar e^2}{2\eta\omega_0}\right)\int_{t_1}^{t_2} dt'_1 e^{i[\omega_k(t'_1 - t_1) - \omega_0 t'_2]}$$

$$\times \int \frac{d^3q'}{(2\pi)^{3/2}}\frac{\sqrt{N_q + 1}}{q}G^<(\mathbf{k} + \mathbf{q}', \mathbf{k} + \mathbf{q}', t', t_2). \tag{10.95}$$

To simplify, we will take $t_1 = 0$, and $t_2 = t$. Then these two equations can be combined to yield the process by which a phonon is emitted from state $\mathbf{k} + \mathbf{q}$ to state $\mathbf{k}$ as

$$G^<(\mathbf{k}, t) = \left(\frac{e^2}{2\hbar\eta\omega_0}\right)\int \frac{d^3q}{(2\pi)^3}\frac{N_q + 1}{q^2}\int_0^t dt'_1 \int_0^t dt'_2 e^{i(\omega_k - \omega_0)(t'_1 - t'_2)}$$

$$\times G^<(\mathbf{k} + \mathbf{q}, t'_1, t'_2). \tag{10.96}$$

In order to reduce the two-time propagator on the right-hand side of Eq. (10.96), we introduce the generalized Kadanoff–Baym ansatz, which has been suggested for time-dependent problems [97], for the initial state as

$$G^<(\mathbf{k}, t, t') = i[G^r(\mathbf{k}, t, t')f(t') - f(t)G^a(\mathbf{k}, t, t')], \tag{10.97}$$

which is defined in terms of the retarded and advanced Green's functions and an assumed distribution function. This distribution function may be taken either as the density matrix or as the less-than

Green's function, both at the initial time. The retarded and advanced Green's functions may be written as

$$G^r(\mathbf{k},t,t') = -i\theta(t-t')e^{(-i\omega_k-\gamma)(t-t')}$$
$$G^a(\mathbf{k},t,t') = i\theta(t'-t)e^{(i\omega_k-\gamma)(t'-t)}.$$

(10.98)

where γ is the imaginary part of the self-energy of the initial state. Inserting these functions into Eq. (10.96) and carrying out the integration over t_1', we arrive at

$$G^<(\mathbf{k},t) = \left(\frac{e^2}{\hbar\eta\omega_0}\right)\int \frac{d^3q}{(2\pi)^3}\frac{N_q+1}{q^2}\frac{\gamma}{\gamma^2+\alpha^2}\int_0^t dt' f(t')L(t'),$$

(10.99)

where $\alpha = \omega_k + \omega_0 - \omega_{k+q}$ and

$$L(t') = 1 - e^{-\gamma(t-t')}\left[\cos(\alpha(t-t')) - \frac{\alpha}{\gamma}\sin(\alpha(t-t'))\right]. \quad (10.100)$$

The time derivative of the Green's function in Eq. (10.99) gives the probability that an electron in an initially occupied state [f = 1] will wind up in the state with wavevector $\mathbf{k}$, at time t as the consequence of the emission of a polar optical phonon, in the case for which the interaction began at t = 0. The integrations can be carried out using some results from Lewis [98], and this gives

$$P(\mathbf{k},t) = \frac{C}{\sqrt{\omega_k}}\left\{\frac{\pi}{4}\ln[R(\omega_k,\omega_0,\gamma)] - e^{-\gamma t}\int_{\omega_k+\omega_0}^{\infty} L(\alpha,\gamma)d\alpha\right\},$$

(10.101)

where

$$C = \frac{e^2}{2\pi^2\hbar\eta\omega_0}\sqrt{\frac{2m^*}{\hbar}}f(N_q+1)$$

$$R(\omega_k,\omega_0,\gamma) = \frac{\sqrt{\omega_k+\omega_0+i\gamma}+\sqrt{\omega_k}}{\sqrt{\omega_k+\omega_0+i\gamma}-\sqrt{\omega_k}}\cdot\frac{\sqrt{\omega_k+\omega_0-i\gamma}+\sqrt{\omega_k}}{\sqrt{\omega_k+\omega_0-i\gamma}-\sqrt{\omega_k}}$$

$$L(\alpha,\gamma) = \frac{\gamma}{\gamma^2+\alpha^2}\left[\sin(\alpha t)+\frac{\alpha}{\gamma}\cos(\alpha t)\right]\cot^{-1}\left(\sqrt{\frac{\alpha-\omega_0}{\omega_k}-1}\right).$$

(10.102)

The first term on the right-hand side of Eq. (10.101) represents the Fermi "golden rule." The second term represents the time-dependent buildup of the scattering process. When $t = 0$, the integral over the possible initial frequencies of the Lorentzian in α cancels the constant contribution of the Fermi golden rule. From the physical point of view, at $t = 0$, the scattering is inhibited by the fact that the electron cannot identify a single well-defined phonon energy for the scattering. Comparing the definition of α below (10.99) with the equivalent in Eq. (10.102), we see that the integration is now over all possible frequencies of the initial state, at least as these are set by the phonon energy. This arises as the energy conservation, which is only enforced in the long time limit.

While the above development has been for the polar optical phonon, which is an intra-valley process, an equivalent one can be developed for the inter-valley phonon process. The results are quite similar, although the details are different due to the different matrix elements for this process, as shown in Eq. (10.93). These different matrix elements will carry over to slightly different time-dependent processes for the buildup and decay of the correlation between the initial and final states.

The integral in Eq. (10.101) is not amenable to an analytical solution, so we evaluate it numerically. Here we consider a bulk GaAs system. The decay parameter γ is the imaginary part of the self-energy and is defined as the lifetime of the initial state. We use for this a constant value, taken to be the mean scattering time of 0.14 ps. This leads to the value of $\gamma = 4.6$ meV. The phonon energy is the normal 36 meV, which corresponds to $\omega_0 = (E/\hbar) = 5.5 \times 10^{13}$ s^{-1}. In Fig. 10.5, we plot the scattering probability, normalized to the Fermi golden rule result for an energy of 36 meV above the phonon energy (the final state has an energy equal to the phonon energy). There is an initial rise in the probability and then a damped oscillatory decay to the final value. These oscillations correspond to remnants of the Rabi oscillation between the initial and final states that are set up by the correlation between these states. While the initial rise is rapid, the remnants of this correlation die rather hard, lasting for tens of femtoseconds. These oscillations imply that the scattering is actually enhanced for brief periods during this transient. This is consistent with the findings of Geltman [91] in atomic emission and with those of Lane [92] for decay of a prepared state. From Fig. 10.5, we can define the collision duration time as that time in which the initial and final states are *strongly* correlated. That is, we do not use the much longer time over which the oscillations decay, but the shorter time required for the rise to $(1 - 1/e) \sim 0.63$ in value. In this way, the contributions from the two terms are comparable.

In Fig. 10.6, we plot the collision duration time, defined above, for the polar optical intra-valley phonon (emission) scattering and the L phonon inter-valley (emission) scattering process. For the latter, we take $\gamma = 66$ meV, corresponding to the much shorter scattering time for this process. The inter-valley phonon is also taken to have an energy of 36 meV. The energy scale is referenced to the threshold energy required for each process. In the intra-valley process, the threshold is just the phonon energy. On the other hand, the L valleys lie some 290 meV above the Γ valley, so that the threshold for the inter-valley process is 326 meV. The inter-valley response shows much the same behavior as seen in Fig. 10.5, although the timescale is different. One additional difference is that the oscillatory behavior is damped more effectively in the inter-valley process due to the larger self-energy of the initial state for this process.

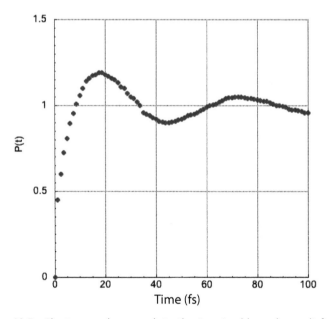

Figure 10.5 The temporal approach to the Fermi golden rule result for the emission of an optical phonon in the conduction band of GaAs. Details of the parameters are discussed in the text. Adapted from Ref. [96].

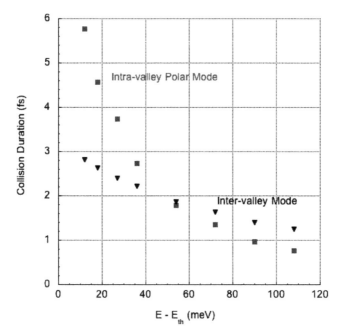

Figure 10.6 Comparison of the polar (intra-valley, squares) process to the inter-valley (diamonds) process. The energy threshold is discussed in the text. Adapted from Ref. [96].

10.7 Short-Pulse Laser Excitation

When one examines the interaction of light with a semiconductor on the short timescale, one must be concerned with the presence of coherent interactions. For these situations, knowledge of the photo-generated distribution functions of the electrons and holes is inadequate to describe the full range of the coherent interactions. One reason for this is that information about the interband polarization is required. This polarization is created by the coherent interaction of the light field with the wavefunctions in the conduction and valence band. In fact, the absorption process itself is initiated with the polarization of the electron–hole wavefunctions, which is then broken up by the final transition of the bound electron from the valence band to the conduction band. This is symbolized by the square magnitude of the optical matrix element that appears in the Fermi golden rule for the absorption process. To describe the polarization, a full quantum mechanical approach is required. There are a variety of such approaches, but we are interested here in the real-time Green's function method [4, 94, 99–101]. Since there are generally no electron–hole pairs prior to the intense, short laser pulse, this is an example of a dynamic evolution from a well-defined initial state.

Generally, we note from the foregoing sections that the Green's function approach usually involves single-particle propagators, and the many-body forms are built up via a variety of perturbation and integral equation approaches, such as the Bethe–Salpeter equations. Sometimes these can be simplified by factorization approaches such as the Kadanoff–Baym ansatz. Here we will follow the approach of Haug and Jauho [4] and treat the interband processes with the carriers interacting with the polar longitudinal optical phonons in a direct gap material such as GaAs. However, we treat the light field itself as a classical electromagnetic field with the interaction arising from the electric field of the optical plane wave. Alternatively, one often uses the vector potential, but the electric field is obtainable from this potential and the two are collinear in the plane wave. The spatial variation of the light field will be ignored as it is usually uniform over the size of a typical semiconductor sample. In this approach, the perturbing term in the Hamiltonian may be written as $\mathbf{r} \cdot \mathbf{E}$, and this interacts with the semiconductor to produce optical transitions between the valence and conduction bands. The time-dependent dipole interaction may be simply written as

$$H_{\mathrm{int}}(t) = -\frac{1}{2}\sum_{\mathbf{k}} |\mathbf{E}(t)| \big(d_{\mathbf{k}} a_{c,\mathbf{k}}^{\dagger} a_{v,\mathbf{k}} e^{-i\omega_{op}t} + h.c.\big), \qquad (10.103)$$

where the dipole matrix element in the direction of the electric field is

$$d_{\mathbf{k}} = e \int d^3 r u_{c,\mathbf{k}}^{*} \mathbf{a}_E \cdot \mathbf{r} u_{v,\mathbf{k}}. \qquad (10.104)$$

Here the us are the cell periodic parts of the Bloch function and $\mathbf{a}_E$ is a unit vector in the field direction. It has also been assumed that the time dependence of the electric field is given by

$$\mathbf{E}(t) = E_0 \mathbf{a}_E \cos(\omega_{op}t) \tag{10.105}$$

and the result (10.103) is sometimes called the rotating wave approximation. The energies of the electrons and holes are described by

$$E_c(k) = \frac{\hbar^2 k^2}{2m_c^*} = \hbar \omega_e$$

$$E_v(k) = -E_G - \frac{\hbar^2 k^2}{2m_v^*} = -E_G - \hbar \omega_h, \tag{10.106}$$

in order to keep with the normal variables of our Green's functions.

10.7.1 A Simpler First Approach

For the beginning, we will ignore the interactions of the laser-created electrons and holes and return to this aspect below. This simple approach will still lead us to the optical Bloch equations for the semiconductor. In particular, the spatially homogeneous real-time Green's functions are 2×2 matrices in the band indices, as

$$G_{\mu\nu}(\mathbf{k},t,t') = -i \left\langle T_C[a_{\mu,\mathbf{k}}(t) a_{\nu,\mathbf{k}}^\dagger(t')] \right\rangle. \tag{10.107}$$

Each of the two indices μ and ν take the values c and v for the conduction and valence bands, respectively. The four possible combinations give the four entries in the Green's function matrices. The operators themselves are ordered along the time contour as indicated by T_C, which is illustrated in Fig. 9.1. The self-energy corresponding to this real-time Green's function may be expressed as

$$\Sigma_{\mu\nu}(\mathbf{k},t,t') = \Sigma_{\mu\nu}^\delta(\mathbf{k},t,t') + \theta(t-t')\Sigma_{\mu\nu}^>(\mathbf{k},t,t')$$

$$+ \theta(t'-t)\Sigma_{\mu\nu}^<(\mathbf{k},t,t'). \tag{10.108}$$

The first term is a singular part appearing when the two times are the same and must be treated separately [6]. This term may be written as

$$\Sigma_{\mu\nu}^\delta(\mathbf{k},t,t') = \Sigma_{\mu\nu}^s(\mathbf{k},t)\delta(t-t'). \tag{10.109}$$

The two Dyson's equations for the less-than Green's function can now be written in its familiar form as

$$G_0^{-1}G^< = \Sigma^\delta G^< + \Sigma^r G^< + \Sigma^< G^a$$
$$G^< G_0^{-1} = G^< \Sigma^\delta + G^r \Sigma^< + G^< \Sigma^a. \tag{10.110}$$

Now we will subtract the second equation from the first and then set the two times to be equal, as

$$[G_0^{-1}, G^<]\Big|_{t=t'} = [\Sigma^\delta, G^<] + \Sigma^r G^< - G^< \Sigma^a + \Sigma^< G^a - G^r \Sigma^<. \tag{10.111}$$

If we ignore the interactions of the electrons and holes after they are created, the problem becomes much simpler. In essence, the only part of the self-energy that remains is the singular part that appears with the equal time situation, as

$$\Sigma_{\mu,v}^s(\mathbf{k},t) = d_\mathbf{k} E(t)(1-\delta_{\mu v}). \tag{10.112}$$

Using this result, the equation of motion for the less-than Green's function may written explicitly as

$$\left[i\hbar\frac{\partial}{\partial t} - \hbar(\omega_\mu - \omega_v)\right]G_{\mu,v}^<(\mathbf{k},t) = -d_\mathbf{k} E(t)\sum_\lambda \Big[(1-\delta_{\mu\lambda})G_{\lambda,v}^<(\mathbf{k},t)$$
$$-G_{\mu,\lambda}^<(\mathbf{k},t)(1-\delta_{\lambda v})\Big]. \tag{10.113}$$

In this simple case, we obtain a closed equation for the less-than Green's function, which gives the density matrix in this equal time situation. The diagonal elements of the density matrix define the electron and hole densities as

$$G_{\mu,\mu}^<(\mathbf{k},t) = i\left\langle a_{\mu,\mathbf{k}}^\dagger(t)a_{\mu,\mathbf{k}}(t)\right\rangle = in_{\mu,\mathbf{k}}(t) = i\rho_{\mu,\mu}(\mathbf{k},t). \tag{10.114}$$

The off-diagonal elements define the interband polarization. For example, $\rho_{c,v}(\mathbf{k},t)$ describes the annihilation of an electron in the valence band and the creation of an electron in the conduction band, producing the electron–hole pair. Because the photon momentum is so much smaller (about two orders of magnitude) than the electron or hole momentum, the transition is essentially vertical so that the electron and hole have the same wave numbers $\mathbf{k}$. The polarization may be written as

$$p(t) = \sum_\mathbf{k} d_\mathbf{k}\Big[\rho_{v,c}(\mathbf{k},t) + \rho_{c,v}(\mathbf{k},t)\Big], \tag{10.115}$$

so we now need to compute these last two elements of the density matrix. One may want to ask why we care about this polarization, since it is not the final act of creation of the electron–hole pair. Even though it is only a very short time process, it can be measured by nonlinear optical processes, known as time-resolved four-wave mixing [4]. And if it can be measured, it is a real parameter, even though it is neither electron nor hole.

Now we can proceed to evaluate the off-diagonal terms of the density matrix through the equations

$$\left[\frac{\partial}{\partial t}+i(\omega_e-\omega_h-E_G)\right]\rho_{c,v}(\mathbf{k},t)=-id_\mathbf{k}E(t)\left[\rho_{c,c}(\mathbf{k},t)-\rho_{v,v}(\mathbf{k},t)\right]$$

$$\frac{\partial\rho_{c,c}(\mathbf{k},t)}{\partial t}=-\frac{\partial\rho_{v,v}(\mathbf{k},t)}{\partial t}=-id_\mathbf{k}E(t)\left[\rho_{c,v}(\mathbf{k},t)-\rho_{v,c}(\mathbf{k},t)\right].$$

$$(10.116)$$

The complex conjugate of the first of Eq. (10.116) gives the equation for the last density matrix component, which already appears in the second of these equations. We want to get the exponential time variation into the problem, and this is done as follows:

$$e^{i\omega_{op}t}\left[\frac{\partial}{\partial t}+i\omega_r\right]\rho_{c,v}(\mathbf{k},t)$$

$$=\left[\frac{\partial}{\partial t}+i(\omega_r-\omega_{op})\right]\rho_{c,v}(\mathbf{k},t)e^{i\omega_{op}t}$$

$$=-\frac{i}{2}\omega_R\left[\rho_{c,c}(\mathbf{k},t)-\rho_{v,v}(\mathbf{k},t)\right]-\frac{\rho_{c,v}(\mathbf{k},t)}{\tau_2}e^{i\omega_{op}t}$$

$$\frac{\partial\rho_{c,c}(\mathbf{k},t)}{\partial t}=-\frac{i}{2}\omega_R\left[\rho_{c,v}(\mathbf{k},t)e^{i\omega_{op}t}-\rho_{v,c}(\mathbf{k},t)e^{-i\omega_{op}t}\right]-\frac{\rho_{c,c}(\mathbf{k},t)}{\tau_1}$$

$$(10.117)$$

Here we have introduced the reduced energy (frequency) through $\omega_r=\omega_e-\omega_h-E_G$, and $\omega_R=d_\mathbf{k}E_0$ is the Rabi frequency. In addition, a relaxation term has been added to the end of each of these two equations. These give a normal relaxation of the density (τ_1) and the polarization (τ_2). These are often called the longitudinal and transverse relaxation times, respectively.

10.7.2 Bloch Equations

The Bloch equations to be discussed in this section are not the traditional Bloch functions that describe the wavefunctions in a periodic lattice, but rather the three components of a Bloch vector more usually associated with two-level systems [102]. The three components of this vector, in the present application, are related to the elements of the density matrix discussed in the previous paragraphs. Here we can write these as

$$U_1(\mathbf{k}) = \rho_{c,v}(\mathbf{k},t)e^{i\omega_{op}t} + c.c$$

$$U_2(\mathbf{k}) = i\left[\rho_{c,v}(\mathbf{k},t)e^{i\omega_{op}t} - c.c\right] \tag{10.118}$$

$$U_3(\mathbf{k}) = \rho_{c,c}(\mathbf{k},t) - \rho_{v,v}(\mathbf{k},t).$$

The first two components relate to the real and imaginary parts of the polarization, after the optical frequency oscillations are removed, while the third component yields the inversion arising from the laser excitation. If we define the detuning as the difference between the photon energy and the energy of the excited electron and hole states as

$$\delta(\mathbf{k}) = \omega_r - \omega_{op}, \tag{10.119}$$

the equations of motion for the Bloch vector components can be written as

$$\frac{\partial U_1(\mathbf{k})}{\partial t} = -\frac{1}{\tau_2}U_1(\mathbf{k}) - \delta(\mathbf{k})U_2(\mathbf{k})$$

$$\frac{\partial U_2(\mathbf{k})}{\partial t} = -\frac{1}{\tau_2}U_2(\mathbf{k}) + \delta(\mathbf{k})U_1(\mathbf{k}) + \omega_R U_3(\mathbf{k}) \tag{10.120}$$

$$\frac{\partial U_3(\mathbf{k})}{\partial t} = -\frac{1}{\tau_1}[U_2(\mathbf{k}) + 1] - \omega_R U_2(\mathbf{k}).$$

This last set of equations is usually referred to as the optical Bloch equations as they are applied to the optics of two-level atoms [103]. We know that a vector $\mathbf{r}$ on the surface of the Bloch sphere will rotate according to

$$\frac{\partial \mathbf{r}}{\partial t} = \Omega \times \mathbf{r}, \tag{10.121}$$

where the rotation vector is defined from the above equations as

$$\Omega = -\omega_R \mathbf{a}_1 + \delta(\mathbf{k})\mathbf{a}_3 \tag{10.122}$$

(the $\mathbf{a}$'s are unit vectors for the axes defined by the Bloch equations). In the absence of damping, the Bloch vector has a magnitude of unity, the radius of the Bloch sphere, and in the ground state it is oriented along the $-\mathbf{a}_3$ direction. The presence of the light field and the detuning lead to a rotation of this vector. The rotations around the $\mathbf{a}_1$ axis are known as Rabi flops, while the detuning causes rotation around the $\mathbf{a}_3$ axis. (Finite rotations are in integral part of optical computing applications and various operations in quantum optics.) The dispersion in the electron and hole bands leads to the detuning due to the spread in energies of these carriers.

10.7.3 Interband Kinetic Equations

While the above treatment did not discuss the role of the phonons, we shall now take this into account. In the polar semiconductors, such as GaAs, the rapid cooling of the photo-excited electrons and holes arises from the emission of the polar optical phonons. As a result, this process can significantly affect the absorption spectra of the semiconductor. In the equal time limit discussed above, the equations for the elements of the Green's function tensor can be written as

$$\left[i\frac{\partial}{\partial t} + \omega_e - \omega_h - E_G \right] G^<_{\mu,v}(\mathbf{k},t)$$

$$= -d_k E(t) \sum_\lambda \left[(1-\delta_{\mu\lambda})G^<_{\lambda,v} - G^<_{\mu,\lambda}(1-\delta_{\lambda v}) \right]$$

$$+ \sum_\lambda \int_{-\infty}^t dt' \left[\Sigma^>_{\mu,\lambda}(t,t')G^<_{\lambda,v}(t',t) - \Sigma^<_{\mu,\lambda}(t,t')G^>_{\lambda,v}(t',t) \right.$$

$$\left. - G^>_{\mu,\lambda}(t,t')\Sigma^<_{\lambda,v}(t',t) + G^<_{\mu,\lambda}(t,t')\Sigma^>_{\lambda,v}(t',t) \right] \quad (10.123)$$

The first line describes the time development of free carriers without any interaction with the optical phonons. The last two lines account for the role played by the optical phonons on the overall process. As is normal, the frequency of the optical phonons is taken to be a constant, and the phonons are assumed to have a thermal bath, which means neglecting the occupation of the modes. This is a significant approximation, as the intense laser excitation can lead to a rather large phonon emission rate by the relaxing carriers, which in turn drives the phonon distribution out of equilibrium [104]. Nevertheless, in keeping with the discussion of the rest of this section, we will not treat the nonequilibrium phonons. Thus, the self-energies can be written as

$$\Sigma_{\mu,v}^<(\mathbf{k},t,t') = i\sum_q M_q^2 D^<(\mathbf{q},t,t')G_{\mu,v}^<(\mathbf{k}-\mathbf{q},t,t') \qquad (10.124)$$

and a complementary equation for the greater-than functions. Here, the phonon Green's function may be written as

$$D^<(\mathbf{q},t,t') = -i\sum_q \left[(N_q+1)e^{i\omega_0(t-t')} + N_q e^{-i\omega_0(t-t')}\right] \qquad (10.125)$$

with the squared matrix element being

$$M_q^2 = \frac{e^2\omega_0}{8\pi\varepsilon_0 q^2\Omega}\left(\frac{1}{\varepsilon_r(\infty)} - \frac{1}{\varepsilon_r(0)}\right), \qquad (10.126)$$

where the two relative dielectric constants are the optical and d.c. values for the polar material, and Ω is the volume of the crystal.

In order to proceed, it is useful to carry out a reduction of the correlation functions, through the use of the generalized Kadanoff–Baym ansatz [97]. In this case, the distribution function is replaced by the elements of the density matrix as

$$G_{\mu,v}^<(t,t') = -\sum_\lambda\left[G_{\mu,\lambda}^r(t,t')\rho_{\lambda,v}(t') - \rho_{\mu,\lambda}(t)G_{\lambda,v}^a(t,t')\right]$$

$$G_{\mu,v}^>(t,t') = \sum_\lambda\left\{G_{\mu,\lambda}^r(t,t')\left[\delta_{\lambda v} - \rho_{\lambda,v}(t')\right] - \left[\delta_{\mu\lambda} - \rho_{\mu,\lambda}(t)\right]G_{\lambda,v}^a(t,t')\right\}.$$

$$(10.127)$$

The general properties of the retarded and advanced Green's functions have been used to reach this final form. Finally, the retarded Green's function can be expressed as

$$G_{\mu,\mu}^r(\mathbf{k},t,t') = -i\theta(t-t')e^{(-i\omega_\mu - \gamma_\mu)(t-t')}, \qquad (10.128)$$

where γ_μ is the imaginary part of the retarded self-energy for the appropriate band, and

$$G_{c,v}^r(\mathbf{k},t,t') = \frac{i}{2}\theta(t-t')\frac{d_k E_0}{\delta(\mathbf{k})}\left(e^{i\delta(\mathbf{k})t} - e^{i\delta(\mathbf{k})t'}\right)e^{i\omega_h t' - i\omega_e t}. \qquad (10.129)$$

It has to be noted here that the development in terms of real-time Green's functions has largely degenerated into a set of equations for the density matrix elements in the mixed band configuration. This has arisen through the equal time approximations that have been liberally applied throughout the discussion of this section. This will be repeated in the next section and only illustrates the fact that time-varying solutions to the real-time Green's functions are much harder to arrive at, with the result that one most often results to other forms, such as the density matrix. One approach that has met with some success is to simply accept that the density matrices for the two bands are more easily treated via a modified semi-classical Monte Carlo procedure [3, 105]. In this latter approach, the optical Bloch equations are treated fully quantum mechanically, but the density matrix elements for the conduction and valence bands are treated via the Monte Carlo simulation approach. Thus, the polarization and coherence of the optical transition can be retained, while the less important scattering dynamics within the two bands are treated via a stochastic method.

10.8 Evolution from an Initial State

The equations for the real-time Green's functions become more intense when we are interested in the evolution from a given, correlated initial state. This becomes a more difficult problem in finding the transient behavior as well as the steady-state behavior. Many people have approached this, with varying degrees of success. Perhaps the earliest was Hall [5], who proceeded without much consideration of the contour of Fig. 9.1, but introduced several auxiliary functions that arose from particular summations of the Feynman diagrams. Perhaps more useful is the work of Danielewicz [6], that the evolution from a finite time, which involves a correlated initial state, is much more difficult than that from a time in the infinite past. In his approach, he added an imaginary time part of the contour, so that the correlated initial state could be prepared from an uncorrelated state by evolution in the imaginary time used for the Matsubara Green's functions treated in Chapter 5. This means that the Kadanoff and Baym equations for the real-time functions have extra terms arising from the imaginary time portion of the contour. Thus, the equation for the less-than function evaluated on the real-time parts of the contour becomes

$$\left(i\hbar\frac{\partial}{\partial t}+H_0\right)G^<(\mathbf{r},\mathbf{r}',t,t')=\int d^3\mathbf{r}_2\Sigma_{\mathrm{HF}}(\mathbf{r},\mathbf{r}_2,t,t)G^<(\mathbf{r}_2,\mathbf{r}',t,t')$$

$$+\int d^3\mathbf{r}_2\int_{t_0+i\tau_0}^{t}dt_2\Sigma^>(\mathbf{r},\mathbf{r}_2,t,t_2)G^<(\mathbf{r}_2,\mathbf{r}',t_2,t')$$

$$+\int d^3\mathbf{r}_2\int_{t}^{t'}dt_2\Sigma^>(\mathbf{r},\mathbf{r}_2,t,t_2)G^<(\mathbf{r}_2,\mathbf{r}',t_2,t')$$

$$+\int d^3\mathbf{r}_2\int_{t'}^{t_0-i\tau_0}dt_2\Sigma^<(\mathbf{r},\mathbf{r}_2,t,t_2)G^>(\mathbf{r}_2,\mathbf{r}',t_2,t')$$

$$(10.130)$$

Here, t_0 is the initial state, and the imaginary time propagation is into this state. In addition, the carrier–carrier interaction up to the Hartree–Fock approximation has been added in the first term on the right-hand side. Now, of course, it is possible to separate the Green's functions on this imaginary time contour, just as has been done for the Keldysh matrix itself. In particular, Wagner [106] has pursued this approach, which expands the 2×2 Keldysh matrix into a 3×3 matrix. However, neither of these approaches actually deals with a correlated initial state, but rather with a thermal equilibrium initial state.

One of the problems with a correlated initial state is that the buildup in the correlations among the carriers leads to effects that are not present in the thermal equilibrium state. In the correlated state, there must be additional terms in the kinetic equation that insure that the set of collision integrals reduces to zero as thermal equilibrium is approached [107]. Of course, in thermal equilibrium, detailed balance assures the vanishing of the collision integrals. It is fair to ask just to what does this initial correlation correspond? In Chapter 3, we dealt with the electron–electron interactions, where we encountered the two-particle Green's function in Eq. (3.61). At that point, we first assumed that the correlation was not important and reduced the two-particle Green's function to the product of two single-particle Green's function in Eq. (3.62) leading to the so-called Hartree approximation. We then went beyond this to incorporate the Hartree–Fock correction in Eq. (3.64). But this still remains an approximation to the full two-particle Green's function. To go further, we need to introduce the density–density correlation function as a correction with the two-particle Green's function as

$$G_2(\mathbf{r}_1, t_1; \mathbf{r}_2, t_2; \mathbf{r}_1', t_1'; \mathbf{r}_2', t_2') = G(\mathbf{r}_1, t_1; \mathbf{r}_1', t_1')G(\mathbf{r}_2, t_2; \mathbf{r}_2', t_2')$$
$$+ L(\mathbf{r}_1, t_1; \mathbf{r}_2, t_2; \mathbf{r}_1', t_1'; \mathbf{r}_2', t_2').$$

$$(10.131)$$

So the density–density correlation is the correction to the product of the two single-particle Green's functions and describes the interactions between the carriers sufficiently well that it cannot be reduced without further approximations. In order to simplify the notation in the following, let us introduce some short-hand notation as $1 = (\mathbf{r}_1, t_1)$, and similarly for the other variable sets that occur in the above equation. To proceed, we follow the approach of Refs. [7, 8, 108–110]. Then the dynamic equation for one of the real-time Green's functions becomes

$$\left[i\hbar \frac{\partial}{\partial t_1} - \frac{\hbar^2}{2m*} \nabla_1^2 - \Sigma_H(1) \right] G(1, 1') = \delta(1 - 1') + \int d3 V(1, 3) L(1, 3, 1', 3^+),$$

$$(10.132)$$

where the Hartree potential arises from the first terms in Eq. (10.131), and the short-hand notations used are

$$\int d3 = \int d^3\mathbf{r}_3 \int dt_3$$
$$3^+ = \lim_{t_3' \to t_3} (\mathbf{r}', t_3')$$

$$(10.133)$$

with the limit in the last line being such that t_3 is approached from the positive side. In Eq. (10.132), the density–density correlation function appears, and to find this, we must have an equation of motion for the two-particle Green's function in Eq. (10.131). Of course, as we know from the KKGBY hierarchy, this equation will, by necessity, involve the three-particle Green's function, as the equation becomes

$$\left[i\hbar \frac{\partial}{\partial t_1} - \frac{\hbar^2}{2m*} \nabla_1^2 - \Sigma_H(1) \right] G_2(1, 2, 1', 2')$$
$$= \delta(1 - 1')G(2, 2') - \delta(1 - 2')G(2, 1')$$
$$+ \int d3 V(1, 3) G_3(1, 2, 3, 1', 2', 3^+)$$

$$(10.134)$$

In these equations, $V(1,3)$ is the interaction amplitude. For the three-particle Green's function, an approximation referred to as the polarization approximation [109], will be used. This gives

$$G_3(1,2,3,1',2',3') = G(1,1')G(2,2')G(3,3') + G(1,1')L(2,3,2',3')$$
$$+ G(2,2')L(1,3,1',3') + G(3,3')L(1,2,1',2')$$
$$(10.135)$$

This equation will allow us to obtain a closed equation for the density–density correlation function, which may be rewritten as an integral equation

$$L(1,2,1',2') = L_0(1,2,1',2') - G_H(1,2')G(2,1')$$
$$+ \int d4 G_H(1,4)G(4,1') \int d3 V(4,3)L(2,3,2',3^+).$$
$$(10.136)$$

Here we have introduced the factor $(G_H^r)^{-1}$ as the left-hand side of Eq. (10.132) and the condition $(G_H^r)^{-1} L_0 = 0$ has been used. The term L_0 represents the contribution from the initial correlations. Hence, a condition on Eq. (10.136) is that when all times approach t_0, all terms except L_0 vanish from the right-hand side. Then the density–density correlation function evolves in time from this initial state according to this last equation. This propagation is given by

$$L_0(1,2,1',2') = \int d\mathbf{r}_1 d\mathbf{r}_2 d\mathbf{r}_1' d\mathbf{r}_2' G_H^r(1;\mathbf{r}_1,t_0) G_H^r(2;\mathbf{r}_2,t_0)$$
$$\times L_{00}(\mathbf{r}_1,\mathbf{r}_2,\mathbf{r}_1',\mathbf{r}_2',t_0) G_H^a(1';\mathbf{r}_1',t_0) G_H^a(2';\mathbf{r}_2',t_0).$$
$$(10.137)$$

Here the term L_{00} is the initial two-particle correlation function.

If we now insert Eq. (10.136) into Eq. (10.132) and restrict the development to the Born approximation, which is suitable for the weak scattering in most semiconductors, we arrive at

$$G_{HF}^{-1}(1,3)G(3,2) = \delta(1-2) + S_{init}(1,2) + \int d4[\Sigma_0(1,4) + \Sigma(1,4)]G(4,2),$$
$$(10.138)$$

where

$$G_{HF}^{-1}(1,2) = G_H^{-1}(1,2) + V(1,2)G^<(1,2) \qquad (10.139)$$

now incorporates the next order of correction to the Hartree potential, and

$$\Sigma_0(1,2) = \int d3d5V(1,3)G_H(1,2)V(2,5)L_0(3,5,3^+,5^+)$$

$$S_{init}(1,2) = \int d3V(1,3)L_0(1,3,2,3^+).$$

(10.140)

Both of these latter terms arise from the presence of the initial correlations, with the first of these obviously related to the correlated collisional effects. In addition, the self-energy term appearing in the square brackets is

$$\Sigma(1,2) = \int d3d5V(1,3)G_H(1,2)V(2,5)G_H(3,5^+)G(5,3^+).$$ (10.141)

The integration in Eq. (10.138) is performed along the Keldysh contour.

From the above equations, we can now find some of the individual Green's functions. For example, the retarded Green's function is found as

$$[G_{HF}^{-1} - \Sigma_0 - \Sigma^r]G^r(1,2) = \delta(1-2) + S_{init}^r(1,2).$$ (10.142)

This now leads to the Kadanoff–Baym equation

$$(G_{HF}^{-1} - \Sigma^r)G^< - G^<(G_{HF}^{-1} - \Sigma^a) = (\Sigma^< + \Sigma_0^<)G^a - G^r(\Sigma^< + \Sigma_0^<) + 2\mathrm{Re}\{S_{init}\}.$$
(10.143)

Using the generalized Kadanoff–Baym ansatz in the first line of Eq. (10.127), we can obtain the kinetic equation for the reduced density matrix [109]

$$\frac{\partial \rho(k,t)}{\partial t} = \frac{\partial}{\partial t}G^<(t,t) = M(k,t) + M_1(k,t) + M_2(k,t),$$ (10.144)

where the Wigner coordinates have been introduced prior to the Fourier transformations, and the individual terms are

$$M(k,t) = \frac{2}{\hbar^2}\mathrm{Re}\left\{ \int_{t_0}^{t} dt_1 \int \frac{dqdp}{(2\pi\hbar)^6}V^2(q)G^r(t,t_1,k-q)G^a(t_1,t,k)G^r(t,t_1,p+q)\right.$$

$$\times G^a(t_1,t,p)\big[\rho(t_1,k-q)\rho(t_1,p+q)[1-\rho(t_1,p)][1-\rho(t_1,k)]$$

$$\left. - \rho(t_1,k)\rho(t_1,p)[1-\rho(t_1,p+q)][1-\rho(t_1,k-q)]\big]\right\}$$

$$M_1(k,t) = \frac{2}{\hbar} \text{Im} \left\{ \int \frac{dqdp}{(2\pi\hbar)^6} V(q) G^r(t,t_0,k-q) G^a(t,t_0,k) G^r(t,t_0,p+q) \right.$$

$$\left. \times G^a(t,t_0,p) \left\langle \frac{p-k}{2} + q \middle| L_{00}(p+k,t_0) \middle| \frac{p-k}{2} \right\rangle \right\}$$

$$M_2(k,t) = \frac{2}{\hbar^2} \text{Re} \left\{ \int_{t_0}^{t} dt_1 \int \frac{dq}{(2\pi\hbar)^3} \Lambda_0(q,t,t_1) V^2(q) G^r(t,t_1,k-q) \right.$$

$$\left. \times G^a(t_1,t,k) \left[\rho(t_1,k-q) - \rho(t_1,k) \right] \right\}. \qquad (10.145)$$

Here the Wigner coordinate terms take the forms

$$L_{00}(x_1,x_2,x_3,x_4) = \left\langle x_1 - x_2 \middle| L_{00} \left(\frac{x_1+x_2}{2} - \frac{x_3+x_4}{2} \right) \middle| x_3 - x_4 \right\rangle$$

$$\Lambda_0(q,t,t') = \int dx e^{-iqx} \left\langle \frac{x}{2} \middle| L_0(0) \middle| \frac{x}{2} \right\rangle. \qquad (10.146)$$

The leading term in Eq. (10.145), M, is a collision integral that is a precursor to the one developed by Levinson, primarily for transport considering the density matrix or the Wigner function [111], which was discussed in Section 8.4. The term M_2 arises from Σ_0 and leads to corrections at the level of the third Born approximation, beyond what is necessary for weak scattering in semiconductors. Finally, the term M_1 gives corrections to the Levinson equations, which guarantee that the collision integrals vanish in equilibrium. As this is now an equation for the density matrix, moments of the kinetic equation can be developed just as in Section 7.5. But we can go a little further with the equations in Eq. (10.145). Going only to the second Born approximation, we can still write the retarded Green's function as

$$G^r(t,t',k) \sim -i\theta(t-t') e^{i(\hbar k^2/2m^*)(t'-t)} \qquad (10.147)$$

and then evaluate L_{00} as

$$\left\langle \frac{k-k'}{2} \middle| L_{00}(k+k') \middle| \frac{k-k'}{2} - q \right\rangle$$

$$= -\frac{P}{\Delta\varepsilon} V(q) \left\{ \rho_0(k)\rho_0(k')[1-\rho_0(k-q)][1-\rho_0(k'+q)] \right.$$

$$\left. -[1-\rho_0(k)][1-\rho_0(k')]\rho_0(k-q)\rho_0(k'+q) \right\} \qquad (10.148)$$

where P denotes the principal value,

$$\Delta\varepsilon = \frac{\hbar^2 k^2}{2m^*} + \frac{\hbar^2 k'^2}{2m^*} - \frac{\hbar^2 (k-q)^2}{2m^*} - \frac{\hbar^2 (k'+q)^2}{2m^*} \qquad (10.149)$$

and ρ_0 is the initial state Wigner function. Then the explicit collision integral M becomes

$$M(k,t) = \frac{2}{\hbar^2} \int_{t_0}^{t} dt_1 \int \frac{dqdk'}{(2\pi\hbar)^6} V^2(q) \cos\left(\frac{\Delta\varepsilon(t-t_1)}{\hbar}\right)$$

$$\times \{\rho(k-q,t_1)\rho(k'+q,t_1)[1-\rho(k',t_1)-\rho(k,t_1)]$$
$$-\rho(k',t_1)\rho(k,t_1)[1-\rho(k-q,t_1)-\rho(k'+q,t_1)]\}.$$
$$\qquad (10.150)$$

The additional term necessary arises from

$$M_1(k,t) = -\frac{2}{\hbar^2} \int_{t_0}^{t} dt_1 \int \frac{dqdk'}{(2\pi\hbar)^6} V(q)V_0(q) \cos\left(\frac{\Delta\varepsilon(t-t_1)}{\hbar}\right)$$

$$\times \{\rho_0(k-q)\rho_0(k'+q)[1-\rho_0(k')-\rho_0(k)]$$
$$-\rho_0(k')\rho_0(k)[1-\rho_0(k-q)-\rho_0(k'+q)]\}. \qquad (10.151)$$

From these, it is easy to see just how the two terms identically cancel in the equilibrium state.

Here, as in many such solutions to the real-time Green's functions in which the evolution from the initial state, whether correlated or not, the use of the Kadanoff–Baym ansatz in one form or another leads to an equation for the density matrix or the Wigner function. To recover the real-time Green's functions from this solution, one needs to *reconstruct* these functions. One method of doing so has been given as [9]

$$G^<(t,t') = -G^r(t,t')\rho(t') + \int_{t'}^{t} dt_1 \int_{t_0}^{t'} dt_2 G^r(t,t_1)\Sigma^<(t_1,t_2)G^a(t_2,t')$$

$$+ \int_{t'}^{t} dt_1 \int_{t_0}^{t'} dt_2 G^r(t,t_1)\Sigma^r(t_1,t_2)G^<(t_2,t')$$
$$\qquad (10.152)$$

for the time sequence $t > t' > t_0$, with a similar result for the case where t' is the later time. Such equations have leading terms that explicitly give the generalized Kadanoff–Bohm ansatz, which has been used above.

10.9 Time-Dependent Density Functional Theory

In Section 10.3, we discussed the use of the atomic lattice to construct the band structure for the actual device in which we were interested. This was used to then compute the transport with the aid of the Green's functions. This is not unique to quantum transport. There are several approaches (with available programs) to first compute the band structure for the device or small nanostructure in which one is interested, and then to compute the transport with an ensemble Monte Carlo approach. In the device, the transport modifies the density, and this density is used in Poisson's equation to update the local potential in the device. The change in potential means a modification of the transport and the cycle is repeated until self-consistency is obtained. Most of these band structures, like in Section 10.3, are usually done semi-empirically in which parameters are matched to available experimental data. However, it is possible to compute the band structure from first principles, where it is recognized that the many-body effects from the electrons will modify the band structure that is obtained. But the exact calculation of these many-body effects is quite difficult, and instead one or another various approximations to the exchange and correlation energy are usually made. These approximations are called density functionals, as the potential is usually a function of the density [112, 113]. This overall approach is thus called *density functional theory* (DFT). When the additional approximation is made that the density functional depends only on the local density (the local density gives a local potential, etc.), it is referred to as the *local density approximation* (LDA). These first principles approaches to the band structure have been pursued for a great many years. And DFT has had a relatively high degree of success in many condensed matter systems [114]. But in the case of semiconductors, the success has been tempered by what is called the bandgap problem.

Now in semiconductors, what can be called the ground state is the state at absolute zero, where all the outer shell electrons are comfortably confined to the valence band. As a result, DFT gives a good account of the valence bands as it is a ground-state calculation. On the other hand, the conduction bands represent excited states that are occupied when the temperature is raised sufficiently high, or impurities (donors) are introduced into the material. DFT has not historically given a good account of the conduction bands or of the bandgap, often failing by an order of magnitude (e.g., 0.1 eV gap versus the experimental 1.1 eV). The problem lies with the need for the full many-body interaction of the valence band in particular, as well as the modifications that arise with electrons in the excited states. In essence, there is a need to go beyond the Hartree–Fock approximations. The solution to this is with the so-called *GW* approximation for the self-energy of the electrons [115, 116]. Here the *G* stands for the Green's function formulation of the many-body interaction potential, while *W* stands for the Coulomb potential between the electrons. When this correction is made, it is generally much better at solving the bandgap problem with fairly high accuracy. Specifically, the many-body self-energy tends to lower the valence band by almost the correct amount (relative to the conduction bands). It is important to point out here that finding the potential from a density in a device with Poisson's equation is equivalent to only using the Hartree approximation. Adding in a correction arising from a density functional has often been done in devices, particularly for low-dimensional systems.

Now let us return to the idea of the density functional. This functional is determined by the local density according to one or another of the various functionals that have been adopted for the problem at hand [114]. It is obvious that, given a particular potential function, one can solve the one-electron Schrödinger equation for the wavefunctions. But to do so requires that we know that the potential is uniquely determined from the density. For the static case, this is the essence of the Hohenberg–Kohn approach [112]. But it was not clear that this would be true for a time-dependent system because of the nonlinearity, and possible retarded behavior, of the density arising from the many-body solutions to the Schrödinger equation. Yet this step is crucial, or else there would be no confidence that the density functional was meaningful. The existence of this mapping, from the density to the potential, was finally established for time-dependent systems by Runge and Gross [117]. What they also established was that there existed a cycle whereby the density functional would lead to the motion of charge that generates currents, and that this would lead to the time rate of change of the local density via the continuity equation. This would now give the new densities via the density functional and hence a new set of wavefunctions. This cycle then leads to the time-dependent behavior desired.

Although there are several excellent review articles on the time-dependent DFT [118, 119], we will illustrate the procedure by following a simpler and more direct description [120]. A typical system can be broken into three parts, which are the central device and the two contacts, just as described in previous chapters, and particularly in Section 10.3 and described pictorially in Fig. 9.2. This is the heart even of the Landauer formula discussed in the previous chapters. The entire device is initially assumed to be in thermal equilibrium, but as the case for most DFT, the temperature is set to zero. Current flows in the device only as a result of an applied bias, typically with the positive terminal at the right-hand contact (as depicted in Fig. 9.2). Initially, then, the electronic structure of the overall device is solved via the Kohn–Sham equations [113]. The Liouville–von Neumann equation is then used to propagate forward in time as

$$i\hbar \frac{\partial \rho(t)}{\partial t} = [H(t), \rho(t)] - \sum_{\alpha = L, R} \Lambda_\alpha(t), \tag{10.153}$$

where $\rho(t)$ is the Kohn–Sham density matrix, $H(t)$ is the Hamiltonian (without the applied potential), and Λ is the self-energy that arises from coupling to the leads, the applied potential, and scattering and dissipation within the device or through coupling to the leads. The density matrix is developed in an atomic basis, so that the wavefunctions are localized atomic orbitals, about which we will have more to say just below. It is the self-energy that is developed in a proper approach utilizing the real-time Green's functions, and

$$\Lambda_\alpha(t) = \frac{i}{\hbar}[\Sigma_\alpha, \rho(t)] + K_\alpha(t), \tag{10.154}$$

where Σ_α is the self-energy that arises from the interaction with the reservoirs and K_α is a term that represents the commutator of the Hamiltonian and the applied potential. The initial density matrix is evaluated by equilibrium Green's function theory (Chapters 3 and 4), and Eq. (10.153) is propagated by a stable numerical integration, such as the Runge–Kutta technique, although the self-energy and the applied potential are updated each time step in response to the new density matrix at that time step. Coming back to the TD-DFT, once the state of Eq. (10.153) is determined at each time step, the time-dependent current can be found from

$$I_\alpha(t) = -\text{Tr}\{\Lambda_\alpha(t)\} \tag{10.155}$$

In the end, the time development is found by solving the Green's function problem in a quasi-steady-state approach at each time step, and using the current and transport properties to give the new density for the next step.

There can be a problem with the use of local orbitals in systems with metallic leads, and this is the so-called work function problem. In general, in condensed matter systems, the material is assumed to be infinite in extent and the position of the Fermi energy is placed at the known position with respect to the vacuum energy. The difference between this vacuum energy and the Fermi energy is, of course, the work function. But in finite (small) systems, one has to be sure that the variation of the potential agrees with these infinite limits. For example, consider a metal–molecule–metal system, such as discussed in Ref. [120], and in a significant fraction of most TD-DFT papers. A sanity check is to remove the molecule in the DFT calculation and then ascertain the level of the energy in the gap between the two metals. This energy should sit above the Fermi level (the zero energy reference in most cases) by the known work function. If the metal DFT is developed with plane waves, which are a good representation of the iterant electrons, this is usually satisfied [121]. But this usually is not the case for local orbital DFT codes, such as SIESTA or FireBall. Indeed, we have found that the potential between the metal contacts can miss the vacuum potential by as much as an order of magnitude, just as in the bandgap problem. However, it can be attained by careful tuning of the local orbitals. The importance of this lies in the methodology of current flowing through the system. It is clear in these metal–molecule–metal systems that the dominant current is tunneling current [122]. Tunneling current depends exponentially on the amplitude of the tunnel barrier; hence, it is exceedingly important to get this barrier correct. This cannot be done unless the DFT approach is capable of exhibiting the large barriers characteristic of the work function when the molecule is removed.

One can see the efficacy of TD-DFT for problems such as the metal–molecule–metal system. However, is this level of detail needed for semiconductors? It is clear that the time dependence for simulations such as NEMO could easily be time dependent, since the transport leads to a density change, which is then used to compute the new potential at each step. Such time dependence is naturally incorporated in semi-classical simulations such as device modeling. But in these latter approaches, it is recognized that the band structure is dominated by the vastly larger number of valence electrons, and it has not been shown that the variations in the free carrier densities within the device lead to significant modifications of the band structure itself. Hence, it is not clear that such advanced approaches as TD-DFT are required for this scenario, but this may not be the case as we move to ever smaller devices.

Problems

1. Explain how the application of a homogeneous electric field throughout a semiconductor leads to a quasi-two-dimensional behavior, as in Section 10.1. Discuss how the choice of a particular gauge affects the problem.

2. Discuss how the current density for the RTD achieved in Eq. (10.36) differs from that found in Chapter 6. Why?

3. Assume we have a heterostructure between GaAs and AlAs, which occurs at $z = 0$. Let us consider discretizing the z-axis, with a grid spacing of 2 nm. Plot the value of the Hamiltonian in Eq. (10.42), neglecting any applied potential and the transverse kinetic energy, as a function of z near the interface. Use the common values for effective mass in these two materials.

4. Assume a value of $t = 10/\gamma$, and values of the kinetic energy of five times the phonon energy for the electron before emission of the phonon, and plot Eq. (10.100) as a function of the intermediate time.

5. We want to consider the conductivity of the channel in an Si MOSFET at low temperatures where the quantization becomes important. Show that the conductivity can be written as

$$\sigma_{2D} = \frac{2e^2}{h} \sum_{n} (E_F - E_n) \Lambda_n (E_F - E_n, E_F) \int_0^\infty \frac{dE_k}{2\pi} \frac{a_n(E_k, E_F)}{2\Gamma_m(E_k, E_F)}$$

where

$$\Lambda_n(E_k, \omega) = 1 + \sum_{m} \Lambda_n(\hbar\omega - E_m, \omega)$$

$$\times \iint \frac{d^2 k_1}{(2\pi)^2} \frac{\mathbf{k} \cdot \mathbf{k}_1}{k^2} T_{nm}(\mathbf{k} - \mathbf{k}_1) \frac{a_m(E_k, \omega)}{2\Gamma_m(E_k, \omega)}$$

is the ladder bubble from the Bethe–Salpeter equation (discussed in class), $T_{nm}(\mathbf{q})$ equals the summation of the squared matrix elements of all scattering mechanisms in the model, $a_n(\mathbf{k}, \omega) = -2\text{Im}[G^r_n(\mathbf{k}, \omega)]$ is the spectral density for sub-band n, and Γ_{nm} is the matrix element for a scattering process between sub-bands n and m. At low temperature, the dominant scattering mechanisms are impurity and surface-roughness scattering. Ref. Vasileska et al., *J. Vac. Sci. Technol. B*, **13**, 1841 (1995).

6. Let us assume that a magnetic field is added to the Hamiltonian in an otherwise homogeneous steady-state system. How would this change the spectral density (using real-time Green's functions)? Assume that there is elastic scattering present in the system, such as impurity scattering. Formulate the self-energy for impurity scattering and evaluate the less-than Green's function in the presence of a magnetic field, as discussed above.

References

1. C. P. Enz, *A Course on Many-Body Theory Applied to Solid-State Physics* (World Scientific Press, Singapore, 1992), p. 76.

2. H. Haug, in *Quantum Transport in Ultrasmall Devices*, edited by D. K. Ferry, H. L. Grubin, C. Jacoboni, and A.-P. Jauho (Plenum Press, New York, 1995).

3. T. Kuhn and F. Rossi, *Phys. Rev. B*, **46**, 7496 (1992).

4. H. Haug and A.-P. Jauho, *Quantum Kinetics in Transport and Optics of Semiconductors* (Springer, Berlin, 2004).

5. A. G. Hall, *J. Phys. A*, **8**, 214 (1975).

6. P. Danielewicz, *Ann. Phys.*, **152**, 239 (1984); **152**, 306 (1984).

7. V. G. Morozov and G. Röpke, *Ann. Phys.*, **278**, 127 (1999).

8. D. Semkat, D. Kremp, and M. Bonitz, *J. Math. Phys.*, **41**, 7458 (2000).

9. V. Spicka, B. Velicky, and A. Kalkova, *Phys. E*, **29**, 154 (2005); **29**, 175 (2005); **29**, 196 (2005).

10. W. Zawadzki, *Acta Phys. Pol.*, **123**, 132 (2013).

11. D. K. Ferry, *Semiconductors: Bonds and Bands* (Institute of Physics Publishing, Bristol, 2013), Sec. 2.6.

12. G. Klimeck and M. Luisier, *Comp. Sci. Eng.*, **10**, 28 (Mar./Apr. 2010).

13. X. Oriols and D. K. Ferry, *J. Comp. Electron.*, **12**, 317 (2013).

14. J. R. Barker, *J. Phys. C*, **6**, 2663 (1973).

15. W. Franz, *Z. Naturforsch.*, **13a**, 484 (1958).

16. L. V. Keldysh, *Zh. Eksperim. i. Teor. Phys.*, **34**, 1138 (1958) [Translation in *Sov. Phys. JETP*, **7**, 788 (1958)].

17. D. E. Aspnes, *Phys. Rev.*, **147**, 554 (1966).

18. A. P. Jauho and J. W. Wilkins, *Phys. Rev. B*, **29**, 1919 (1984).

19. P. Lipavsky, F. S. Khan, A. Kalvová, and J. W. Wilkins, *Phys. Rev. B*, **43**, 6650 (1991).

20. D. K. Ferry, A. M. Kriman, H. Hida, and S. Yamaguchi, *Phys. Rev. Lett.*, **67**, 633 (1991).

21. P. Lipavsky, F. S. Khan, and J. W. Wilkins, *Phys. Rev. B*, **44**, 3655 (1991).

22. D. K. Ferry, *Semiconductor Transport* (Taylor and Francis, London, 2000).

23. R. Bertoncini, A. M. Kriman, and D. K. Ferry, *Phys. Rev. B*, **41**, 1390 (1990).

24. J. Rammer and H. Smith, *Rev. Mod. Phys.*, **58**, 323 (1986).

25. J. R. Barker, *J. Physique*, **C7**, 245 (1981).

26. D. K. Ferry, *Semiconductors* (Macmillan, New York, 1991).

27. R. Bertoncini and A. P. Jauho, *Phys. Rev. B*, **44**, 3655 (1991).

28. R. Lake and S. Datta, *Phys. Rev. B*, **45**, 6670 (1992).

29. M. J. McLennan, Y. Lee, and S. Datta, *Phys. Rev. B*, **43**, 13846 (1991).

30. S. Datta, *J. Phys.: Condens. Matter*, **2**, 8023 (1990).

31. A. Di Carlo and P. Lugli, *Semicond. Sci. Technol.*, **10**, 1673 (1995).

32. A. Di Carlo, S. Pescetelli, M. Paciotti, P. Lugli, and M. Graf, *Solid State Commun.*, **98**, 803 (1996).

33. A. Di Carlo, S. Pescetelli, A. Kavorkin, M. Vladimirova, and P. Lugli, *Phys. Status Solidi B*, **204**, 275 (1997).

34. G. C. Solomon, A. Gagliardi, A. Pecchia, T. Frauenheim, A. Di Carlo, J. R. Reimers, and N. S. Hush, *J. Chem. Phys.*, **125**, 184702 (2006), and references therein.

35. L. Latessa, A. Pecchia, and A. Di Carlo, *IEEE Trans. Nanotechnol.*, **6**, 13 (2007).

36. P. Ordejón, E. Artacho, and J. M. Soler, *Phys. Rev. B*, **53**, 10441 (1996).

37. J. M. Soler, E. Artacho, J. D. Gale, A. Garcia, J. Junquera, P. Ordejón, and D. Sánchez-Portal, *J. Phys.: Condens. Matter*, **14**, 2745 (2002).

38. M. Brandbyge, J.-L. Mozos, P. Ordejón, J. Taylor, and K. Stokbro, *Phys. Rev. B*, **65**, 165401 (2002).

39. K. Stokbro, J. Taylor, M. Brandbyge, and P. Ordejón, *Ann. N. Y. Acad. Sci*, **1006**, 212 (2003).

40. J. C. Slater and G. F. Koster, *Phys. Rev.*, **94**, 1498 (1954).

41. P. Hohenberg and W. Kohn, *Phys. Rev.*, **136**, B864 (1964).

42. W. Kohn and L. Sham, *Phys. Rev.*, **140**, A1133 (1965).

43. O. F. Sankey and D. J. Niklewski, *Phys. Rev. B*, **40**, 3979 (1989).

44. A. A. Demkov, J. Ortega, O. F. Sankey, and M. P. Grumbach, *Phys. Rev. B*, **52**, 1618 (1995).

45. R. Lake, G. Klimeck, and S. Datta, *Phys. Rev. B*, **47**, 6427 (1993).

46. R. C. Brown, G. Klimeck, R. K. Lake, W. R. Frensley, and T. Moise, *J. Appl. Phys.*, **81**, 3207 (1997).

47. R. Lake, G. Klimeck, R. C. Bowen, and D. Javonivic, *J. Appl. Phys.*, **81**, 7845 (1997).

48. T. B. Boykin, G. Klimeck, R. C. Bowen, and R. Lake, *Phys. Rev. B*, **56**, 4102 (1997).

49. T. B. Boykin, L. J. Gamble, G. Klimeck, and R. C. Bowen, *Phys. Rev. B*, **59**, 7301 (1999).

50. G. Klimeck, R. C. Bowen, T. B. Boykin, and T. A. Cwik, *Superlattices Microstruct.*, **27**, 519 (2000).

51. T. B. Boykin, R. C. Bowen, and G. Klimeck, *Phys. Rev. B*, **63**, 245314 (2001).

52. S. Lee, F. Oyafuso, P. von Allmen, and G. Klimeck, *Phys. Rev. B*, **69**, 045316 (2004).

53. Y. Zheng, C. Rivas, R. Lake, K. Alam, T. B. Boykin, and G. Klimeck, *IEEE Trans. Electron Dev.*, **52**, 1097 (2005).

54. T. Sandu, G. Klimeck, and W. P. Kirk, *Phys. Rev. B*, **81**, 7845 (1997).

55. A. Celeste, L. A. Cury, J. C. Portal, M. Allovon, D. K. Maude, L. Eaves, M. Davies, M. Heath, and M. Maldonado, *Solid State Electron.*, **32**, 1191 (1989).

56. T. B. Boykin, N. Kharche, G. Klimeck, and M. Korkusinski, *J. Phys.: Condens. Matter*, **19**, 036203 (2007).

57. Y. Zheng, C. Rivas, R. Lake, K. Alam, T. B. Boykin, and G. Klimeck, *IEEE Trans. Electron Dev.*, **52**, 1097 (2005).

58. J. Wang, A. Rahman, A. Ghosh, G. Klimeck, and M. Lundstrom, *Appl. Phys. Lett.*, **86**, 093113 (2005).

59. M. Luisier, A. Schenk, W. Fichtner, and G. Klimeck, *Phys. Rev. B*, **74**, 205323 (2006).

60. G. Fiori, G. Iannaccone, and G. Klimeck, *IEEE Trans. Electron Dev.*, **53**, 1782 (2006).

61. G. Liang, J. Xiang, N. Kharche, G. Klimeck, C. M. Lieber, and M. Lundstrom, *Nano Lett.*, **7**, 642 (2007).

62. T. B. Boykin, M. Luisier, A. Schenk, N. Kharche, and G. Klimeck, *IEEE Trans. Nanotechnol.*, **6**, 43 (2007).

63. T. B. Boykin, G. Klimeck, R. C. Bowen, and F. Oyafuso, *Phys. Rev. B*, **66**, 125207 (2002).

64. G. Klimeck, S. S. Ahmed, H. Bae, N. Kharche, S. Clark, B. Haley, S. Lee, M. Naumov, H. Ryu, F. Saled, M. Prada, M. Korkusinski, and T. B. Boykin, *IEEE Trans. Electron Dev.*, **54**, 2079 (2007).

65. M. Luisier, A. Schenk, W. Fichtner, and G. Klimeck, *Phys. Rev. B*, **74**, 205323 (2006).

66. M. Luisier and G. Klimeck, *Phys. Rev. B*, **80**, 155430 (2009).

67. T. B. Boykin, M. Luisier, M. Salmani-Jelodar, and G. Klimeck, *Phys. Rev. B*, **81**, 125202 (2010).

68. J. R. Reimers, G. C. Solomon, A. Gagliardi, A. Bilic, N. S. Hush, T. Frauenheim, A. Di Carlo, and A. Pecchia, *J. Phys. Chem.*, **11**, 5692 (2007).

69. M. Pourfath, H. Kosina, and S. Selberherr, *Math. Comput. Simul.*, **79**, 1051 (2008).

70. R. Valin, A. Martinez, and J. R. Barker, *J. Appl. Phys.*, **117**, 164505 (2015).

71. R. Landauer, *Phys. Scripta*, **T42**, 110 (1992).

72. X. Oriols and D. K. Ferry, *J. Comp. Electron.*, **12**, 371 (2013).

73. J. S. Ayubi-Moak, S. M. Goodnick, S. J. Aboud, M. Saraniti, and S. El-Ghazaly, *J. Comp. Electron.*, **2**, 183 (2003).

74. N. Sule, K. J. Willis, S. C. Hagness, and I. Knezevic, *J. Comp. Electron.*, **12**, 563 (2013).

75. W. R. Deal, *Proceedings of the 39th International Conference on Infrared, Millimeter, and THz Waves* (IEEE Press, New York, 2014).

76. X. B. Mei, W. Yoshida, M. Lange, J. Lee, J. Zhou, P. H. Liu, K. Leong, A. Zamora, J. Padilla, S. Sarkozy, R. Lai, and W. R. Deal, *IEEE Electron Dev. Lett.*, **36**, 327 (2015).

77. R. Akis, J. S. Ayubi-Moak, N. Faralli, D. K. Ferry, S. M. Goodnick, and M. Saraniti, *IEEE Electron. Dev. Lett.*, **29**, 306 (2008).

78. M. Büttiker, A. Pêtre, and H. Thomas, *Phys. Rev. Lett.*, **70**, 4114 (1993).

79. M. Büttiker, H. Thomas, and A. Pêtre, *Z. Phys. B: Condens. Matter*, **B94**, 133 (1994).

80. M. Büttiker and S. Nigg, *Nanotechnology*, **18**, 044029 (2007).

81. M. Büttiker, H. Thomas, and A. Pêtre, *Phys. Lett. A*, **180**, 364 (1993).

82. J. Gabelli, G. Fève, J.-M. Berroir, B. Plaçais, A. Cavanna, B. Ettiene, Y. Jin, and D. C. Glattli, *Science*, **313**, 499 (2006).

83. M. di Ventra, *Electrical Transport in Nanoscale Systems* (Cambridge University Press, Cambridge, 2008).

84. I. Knezevic and D. K. Ferry, *Phys. Rev. A*, **69**, 012104 (2005).

85. G. V. Chester, *Rep. Prog. Phys.*, **26**, 411 (1963).

86. H. Mori, I. Oppenheim, and J. Ross, in *Studies in Statistical Mechanics*, Vol. 1, edited by J. de Boer and G. E. Uhlenbeck (North-Holland, Amsterdam, 1962).

87. D. K. Ferry, *Transport in Semiconductor Mesoscopic Devices* (IOP Publishing, Bristol, UK, 2015).

88. P. Price, in *Fluctuation Phenomena in Solids*, edited by R. E. Burgess (Academic Press, New York, 1965).

89. Lord Rayleigh, *Phil. Mag.*, **11**, 283 (1906).

90. E. G. S. Paige, *The Electrical Conductivity of Germanium, Progress in Semiconductors*, Vol. 8, edited by A. F. Gibson and R. E. Burgess (Wiley, New York, 1964).

91. S. Geltman, *J. Phys. B*, **10**, 831 (1977).

92. A. M. Lane, *Phys. Lett.*, **99A**, 359 (1983).

93. S. L. Haan and S. Geltman, *J. Phys. B*, **15**, 1229 (1982).

94. H. Haug, in *Optical Nonlinearities and Instabilities in Semiconductors*, edited by H. Haug (Academic, San Diego, 1988); *Phys. Status Solidi B*, **173**, 139 (1992).

95. F. Rossi and T. Kuhn, *Phys. Rev. Lett.*, **69**, 977 (1992); *Phys. Rev. B*, **46**, 7496 (1992).

96. P. Bordone, D. Vasileska, and D. K. Ferry, *Phys. Rev. B*, **53**, 3846 (1996).

97. P. Lipavsky, V. Spická, and B. Velicky, *Phys. Rev. B*, **34**, 6933 (1986).

98. R. R. Lewis, *Phys. Rev.*, **102**, 537 (1956).

99. H. Haug, *J. Luminesc.*, **30**, 171 (1985).

100. S. Schmitt-Rink, C. Ell, and H. Haug, *Phys. Rev. B*, **33**, 1183 (1986).

101. A. V. Kuznetsov, *Phys. Rev. B*, **44**, 8721 (1991).

102. E. Merzbacher, *Quantum Mechanics*, 2nd Ed. (John Wiley, New York, 1970), chap. 13.

103. P. Meystre and M. Sargent III, *Elements of Quantum Optics*, 2nd Ed. (Springer, Berlin, 1991), and references therein.

104. W. Pötz and P. Kocevar, *Phys. Rev. B*, **28**, 7040 (1980).

105. T. Kuhn and F. Rossi, *Phys. Rev. Lett.*, **69**, 977 (1992).

106. M. Wagner, *Phys. Rev. B*, **44**, 6104 (1991).

107. D. Lee, S. Fujita, and F. Wu, *Phys. Rev. A*, **2**, 854 (1970).

108. D. Semkat, D. Kremp, and M. Bonitz, *Phys. Rev. E*, **59**, 1557 (1999).

109. K. Morawetz, M. Bonitz, V. G. Morozov, G. Röpke, and D. Kremp, *Phys. Rev. E*, **63**, 020102 (2001).

110. D. Semkat, M. Bonitz, and D. Kremp, *Contrib. Plasma Phys.*, **43**, 321 (2003).

111. I. B. Levinson, *Fiz. Tverd. Tela Leningrad*, **6**, 2113 (1965) [*Sov. Phys. Sol. State*, **6**, 1665 (1965)]; *Zh. Eksp. Teor. Fiz.*, **57**, 660 (1969) [*Sov. Phys. JETP*, **30**, 362 (1970)].

112. P. Hohenberg and W. Kohn, *Phys. Rev.*, **136**, B864 (1964).

113. W. Kohn and L. J. Sham, *Phys. Rev.*, **140**, A1133 (1965).

114. J. Kohanoff, *Electronic Structure Calculations for Solids and Molecules* (Cambridge University Press, Cambridge, 2006).

115. L. Hedin, *Phys. Rev.*, **139**, A796 (1065).

116. L. Hedin and S. Lundqvist, in *Solid State Physics*, Vol. 23, edited by H. Ehrenreich, F. Seitz, and D. Turnbull (Academic, New York, 1969).

117. E. Runge and E. K. U. Gross, *Phys. Rev. Lett.*, **52**, 997 (1984).

118. G. Onida, L. Reining, and A. Rubio, *Rev. Mod. Phys.*, **74**, 601 (2002).

119. S. Kurth, *Fundamentals of Time-Dependent Density Functional Theory* (Springer, Berlin, 2012).

120. C. Oppenländer, B. Korff, and T. A. Niehaus, *J. Comp. Electr.*, **12**, 420 (2013).

121. G. Speyer, R. Akis, and D. K. Ferry, *J. Vac. Sci. Technol. B*, **24**, 1987 (2006); *J. Phys.: Conf. Ser.*, **38**, 25 (2006).

122. T. Morita and S. Lindsay, *J. Am. Chem. Soc.*, **129**, 7262 (2007) and references therein.

Chapter 11

Relativistic Quantum Transport

People in the semiconductor community are not usually worried significantly about relativistic transport. It is true that there are some processes that have their origins in relativistic effects, such as non-parabolic bands and spin–orbit interactions. But the consideration of this in transport in semiconductors is mainly after the fact and expressed in terms of simple analytic functions. This, however, came to a crashing halt when graphene was isolated from graphite layers in 2004 [1]. Graphene is a single atomic layer of carbon atoms, which are hexagonally coordinated. More importantly, it does not have a bandgap, and the bands are linear as a result of this fact. This means that the band structure in the neighborhood of the band crossing, which is termed the Dirac point, behaves like relativistic Dirac bands, but with a zero static mass. The carriers, be they electrons or holes, still are fermions, so they tend to be characterized with the phrase *massless Dirac fermions*. Of course, these carriers are not really massless, as they have a dynamic mass given the momentum and the linear bands, as we will discuss below. Then, a short time later, topological insulators appeared, and these materials have surface bands that are also of the Dirac form [2, 3].

So here we were with semiconductors whose properties are defined by the relativistic Dirac equation. It is no longer possible to ignore relativistic effects and live in the semi-classical world. So the purpose of this chapter is to give an introduction to relativity

An Introduction to Quantum Transport in Semiconductors
David K. Ferry
Copyright © 2018 Pan Stanford Publishing Pte. Ltd.
ISBN 978-981-4745-86-4 (Hardcover), 978-1-315-20622-6 (eBook)
www.panstanford.com

theory itself, and to see how it affects the quantum transport that we would like to understand in semiconductors. Of course, our interest is in the transport in these relatively new semiconductors that have the Dirac bands. In the next section, we will review the ideas of relativity theory through the Dirac equations. Then we will turn to a discussion of graphene and its band structure, with other ideas that further make it appear as a relativistic system. Following that, we will discuss the topological insulators due to their similarities to graphene. Then we turn our attention to quantum transport in these systems.

11.1 Relativity and the Dirac Equation

The need to develop a relativistic form of quantum mechanics was realized almost at the beginning, at least by Schrödinger who proposed an extension of his famous equation that meets the requirements of special relativity. Traditionally, in quantum mechanics, and in classical mechanics, we usually write the energy in terms of the momentum as

$$E = \frac{p^2}{2m^*}.$$
(11.1)

Of course, this relationship establishes the energy shell in momentum space as discussed in Section 1.1. Life off the shell arises from the quantization of the system, as discussed there. But we proceed slowly toward this goal by recognizing that the relativistic form of this equation is given by Einstein to be [4]

$$E^2 = p^2 c^2 + m_0^2 c^4.$$
(11.2)

If we now introduce the momentum operator and the normal form of the energy differential, we get for the free particle

$$-\hbar^2 \frac{\partial^2 \psi}{\partial t^2} = -\hbar^2 c^2 \nabla^2 \psi + m_0^2 c^4 \psi.$$
(11.3)

One can still have plane waves as solutions to this equation. If we adopt a traditional plane wave of the form $e^{i(\mathbf{k} \cdot \mathbf{r} - \omega t)}$, we merely get a new relationship between the frequency and the wavevector as

$$\hbar\omega = \pm(\hbar^2 c^2 k^2 + m_0^2 c^4)^{1/2}.$$
(11.4)

The positive root corresponds to our traditional free electron, but at this point, the negative root is ambiguous. Later, we see that Dirac identified this negative root with the positrons. However, in semiconductors, we shall identify this negative root with the holes in the system. Let us examine this a little closer and factor out the rest mass term, as

$$\hbar\omega = E = \pm m_0 c^2 \sqrt{1 + \frac{\hbar^2 k^2}{m_0^2 c^2}} \; . \tag{11.5}$$

If we identify the gap between the positive and negative roots of Eq. (11.4) as $E_G = 2m_0 c^2$, then this last equation can be rewritten as

$$E = \pm \frac{E_G}{2} \sqrt{1 + \frac{2\hbar^2 k^2}{m_0 E_G}} \; . \tag{11.6}$$

This is precisely one of the results obtained in k·p perturbation theory in semiconductor energy bands [5], or in the natural opening of gaps by the crystal potential in the free electron band theory [6], except that the zero of energy is now at the mid-gap position (normally in semiconductors, one takes the zero at either the top of the valence band or the bottom of the conduction band). Hence, the nonparabolic bands are often referred to as relativistic bands, as one can see from this similarity.

Equation (11.3) has come to be known with a different name, and that is the Klein–Gordon (or Klein–Gordon–Foch) equation, as these authors came up with it at about the same time. One problem with Eq. (11.3) is that it contains no spin variables, and so it has become associated with bosons. Nevertheless, it has some useful properties, in that the probability charge and probability current densities can be written as

$$Q = \frac{i\hbar}{2m_0 c^2} \left(\psi^* \frac{\partial \psi}{\partial t} - \psi \frac{\partial \psi^*}{\partial t} \right)$$

$$J = \frac{\hbar}{2im_0} (\psi^* \nabla \psi - \psi \nabla \psi^*). \tag{11.7}$$

We note that the latter equation is identical to its non-relativistic equivalent.

11.1.1 Dirac Bands

Dirac, however, approached the problem in a different manner [7]. To see this, we will follow the approach of Schiff [8]. Dirac wanted to have a Hamiltonian that would make his equations linear in the spatial dimensions. If we do this by taking the positive root in Eq. (11.4) and inserting into the Schrödinger equation, we do achieve this goal. Dirac approached this with the simplest Hamiltonian that is linear in the momentum and mass terms (nevertheless, it was a master stroke to realize this equation would work), which is

$$H = c\tilde{\alpha} \cdot \mathbf{p} + \beta m_0 c^2 , \tag{11.8}$$

where $\tilde{\alpha}$ is a three-dimensional vector. Realizing that the momentum is a vector operator, this leads to the wave equation

$$i\hbar \frac{\partial \psi}{\partial t} = -i\hbar c \tilde{\alpha} \cdot \nabla \psi + \beta m_0 c^2 \psi . \tag{11.9}$$

This form has come to be known as the Dirac equation. We still do not know the form of the new parameters $\tilde{\alpha}$ and β. As the former is a vector, that means we have four unknown parameters. We know first that the Hamiltonian cannot be a function of space or time, as that would give rise to position- and time-dependent energies and this would lead to extra forces that are not included in the wave equation. Second, we know that the time derivatives and spatial derivatives must appear only in $\mathbf{p}$ and E. This leads us to conclude that the new parameters cannot be functions of position or time or even energy or momentum. However, we do not yet know whether or not they are pure numbers, since we have no information on whether or not they commute, and quantum mechanics is rife with non-commuting operators and variables. We can step forward by recognizing that any solution to Eq. (11.9) should also satisfy Eq. (11.3), since any wave packet solutions to the latter equation should be similar to classical particles, which must have the standard connection between energy, momentum, and mass in relativity theory represented by Eq. (11.2).

We can reach a version of Eq. (11.2) by noticing that we can write Eq. (11.7) as

$$(E - c\tilde{\alpha} \cdot \mathbf{p} - \beta m_0 c^2)\psi = 0. \tag{11.10}$$

If we now multiply this equation on the left with $(E + c\tilde{\alpha} \cdot \mathbf{p} + \beta m_0 c^2)$, we arrive at

$$\left[E^2 - c^2(\tilde{\alpha}\cdot\mathbf{p})^2 - (\beta m_0 c^2)^2 - m_0 c^3(\tilde{\alpha}\cdot\mathbf{p}\beta + \beta\tilde{\alpha}\cdot\mathbf{p}) \right]\psi = 0 \;. \quad (11.11)$$

The term in the square brackets is not quite Eq. (11.2), although we would like it to be. First, let us expand the second term in the square brackets, as

$$(\tilde{\alpha}\cdot\mathbf{p})^2 = \alpha_x^2 p_x^2 + \alpha_y^2 p_y^2 + \alpha_z^2 p_z^2 + (\alpha_x\alpha_y + \alpha_y\alpha_x)p_x p_y \qquad (11.12)$$
$$+ (\alpha_y\alpha_z + \alpha_z\alpha_y)p_z p_y + (\alpha_z\alpha_x + \alpha_x\alpha_x)p_x p_z.$$

This would fit Eq. (11.2) if the anti-commutator terms vanished. Note that these are anti-commutators and not commutators, so the individual terms cannot be simple numbers. In addition, the values must have unity magnitude, as we require for Eq. (11.2) that

$$\alpha_x^2 = \alpha_y^2 = \alpha_y^2 = \beta^2 = 1$$
$$\{\alpha_x, \alpha_y\} = \{\alpha_x, \alpha_z\} = \{\alpha_z, \alpha_y\} = 0. \qquad (11.13)$$

Now let us expand the last term in the square brackets of Eq. (11.11), which leads to

$$\tilde{\alpha}\cdot\mathbf{p}\beta + \beta\tilde{\alpha}\cdot\mathbf{p} = (\alpha_x\beta + \beta\alpha_x)p_x + (\alpha_y\beta + \beta\alpha_y)p_y + (\alpha_z\beta + \beta\alpha_z)p_z \,,$$
$$(11.14)$$

which leads us to the additional anti-commutators

$$\{\alpha_x, \beta\} = \{\alpha_y, \beta\} = \{\alpha_z, \beta\} = 0. \qquad (11.15)$$

Hence, Eqs. (11.13) and (11.15) are constraints on our new parameters, which must be satisfied for the expansion (11.11) to be the same as Eq. (11.2).

Things that do not commute are often expressible as matrices and that will be true here as well. But we have some other considerations. As the Hamiltonian is Hermitian, the matrices must be square matrices in order to preserve the Hermitian properties. At most, one of the matrices can be diagonal, but the others cannot be as they do not commute with this one. Since β is a single matrix, it is natural to assume that this one can be diagonal, but since the eigenvalues are ± 1, we can formulate it as the lowest possible rank matrix as

$$\beta = \begin{bmatrix} 1 & 0 \\ 0 & -1 \end{bmatrix}. \qquad (11.16)$$

We know of a set of non-commuting matrices that are of the order 2, and these are the Pauli spin matrices given as

$$\sigma_x = \begin{bmatrix} 0 & 1 \\ 1 & 0 \end{bmatrix}, \quad \sigma_y = \begin{bmatrix} 0 & -i \\ i & 0 \end{bmatrix}, \quad \sigma_z = \begin{bmatrix} 1 & 0 \\ 0 & -1 \end{bmatrix}. \tag{11.17}$$

But note that σ_z is identical to β, so there is a problem with connecting these spinors to the α matrices. In fact, it is impossible to find four matrices that anti-commute at this order. If we go to third order, we would not have the correct symmetries, so the next lowest order is a rank 4 matrix. It turns out that the constraints can all be satisfied for rank 4 matrices, and we can find the four matrices that satisfy these constraints as

$$\beta = \begin{bmatrix} 1 & 0 & 0 & 0 \\ 0 & 1 & 0 & 0 \\ 0 & 0 & -1 & 0 \\ 0 & 0 & 0 & -1 \end{bmatrix}, \quad \alpha_x = \begin{bmatrix} 0 & 0 & 0 & 1 \\ 0 & 0 & 1 & 0 \\ 0 & 1 & 0 & 0 \\ 1 & 0 & 0 & 0 \end{bmatrix},$$

$$\alpha_y = \begin{bmatrix} 0 & 0 & 0 & -i \\ 0 & 0 & i & 0 \\ 0 & -i & 0 & 0 \\ i & 0 & 0 & 0 \end{bmatrix}, \quad \alpha_z = \begin{bmatrix} 0 & 0 & 1 & 0 \\ 0 & 0 & 0 & -1 \\ 1 & 0 & 0 & 0 \\ 0 & -1 & 0 & 0 \end{bmatrix}. \tag{11.18}$$

These can be written in a more compact manner in terms of the Pauli spinors as

$$\beta = \begin{bmatrix} I & 0 \\ 0 & -I \end{bmatrix}, \quad \tilde{\alpha} = \begin{bmatrix} 0 & \tilde{\sigma} \\ \tilde{\sigma} & 0 \end{bmatrix}. \tag{11.19}$$

Here I is the 2×2 unit matrix and 0 is a 2×2 zero matrix. The fact that we have the Pauli spinors in these matrices gives us a hint that the solutions to Eq. (11.10) involve spin. In fact, for the positive energy solutions, one of the two has spin "up" and the other spin "down." Similarly, the negative energy solutions are for spin positrons. In semiconductors, the latter are the holes. Finally, we note that the vector and scalar potentials can be added to the Dirac equation in the normal way, as Eq. (11.10) becomes

$$\left[E - e\varphi - c\tilde{\alpha} \cdot (\mathbf{p} - e\mathbf{A}) - \beta m_0 c^2 \right] \psi = 0. \tag{11.20}$$

11.1.2 Gamma Matrices

There are times when one wants to express the Dirac equation in a proper covariant form. This can be achieved by multiplying through by β, so as to express the equation in the form [9]

$$(\beta E + c\beta\tilde{\alpha}\bullet p + m_0 c^2)\psi = 0,$$ (11.21)

since $\beta^2 = 1$. To get the desired covariant form, the energy is defined as a term in a momentum 4-vector; e.g., a four-component vector, with the components

$$cp_x = p_1, \quad cp_y = p_2, \quad cp_z = p_3, \quad E = p_4.$$ (11.22)

Note that this notation is not universal, as many authors designate p_4 as p_0. This leads to a certain degree of confusion in the field, which I suppose the practitioners have learned to live with. In any case, while this is called the momentum 4-vector, the units are really energy, and this is why the c has been included in the true momentum terms. Then we can write Eq. (11.21) as

$$\left(\sum_\mu \gamma^\mu p_\mu - m_0 c^2\right)\psi = 0.$$ (11.23)

Here we have introduced the γ matrix, in which

$$\gamma^i = \beta\alpha_i, \quad i = 1, 2, 3$$
$$\gamma^4 = \beta.$$ (11.24)

In Eq. (11.23), when the index is a superscript, the matrix is a row matrix, while when the index is a subscript, the matrix is a column matrix. So the form indicated is the dot product of two 4-vectors. Usually, the notation is that any repeated index, such as in this first term, means that a summation must accompany the term. This allows a short-hand notation to omit explicitly writing the summation. Others also claim to use what are called natural units by taking $c = \hbar = 1$, but I shall refrain from this in the interest of clarity. If we wish to return to the form (11.9), then we have to recognize that the energy has now become an operator in relativity theory, and

$$p_4 = i\hbar\frac{\partial}{\partial t}.$$ (11.25)

We can uncover some properties of the γ matrix by using the anti-commutators of Eqs. (11.11) and (11.13). This allows us to write

$$(\gamma^4)^2 = \beta^2 = 1$$
$$(\gamma^4\gamma^i + \gamma^i\gamma^4) = \beta\beta\alpha_i + \beta\alpha_i\beta = \beta\beta\alpha_i - \beta\beta\alpha_i = 0 \qquad (11.26)$$
$$(\gamma^j\gamma^i + \gamma^i\gamma^j) = 0 \quad \text{for } i \neq j.$$

We should also point out that, like p_4 and p_0, γ^4 is often denoted as γ^0. These relations can be brought together into a single anti-commutator as

$$\{\gamma^\mu, \gamma^\nu\} = 2g^{\mu\nu}, \qquad (11.27)$$

which holds for all four components and g is the metric tensor

$$g = \begin{bmatrix} -1 & 0 & 0 & 0 \\ 0 & -1 & 0 & 0 \\ 0 & 0 & -1 & 0 \\ 0 & 0 & 0 & 1 \end{bmatrix}. \qquad (11.28)$$

Lastly, we note that the adjoint of the γ matrix is defined through the relations

$$(\gamma^4)^\dagger = \beta^\dagger = \beta = \gamma^4$$
$$(\gamma^i)^\dagger = (\beta\alpha_i)^\dagger = \alpha_i\beta = -\beta\alpha_i = -\gamma^i \qquad (11.29)$$

or

$$(\gamma^\mu)^\dagger = g^{\mu\mu}\gamma^\mu. \qquad (11.30)$$

These matrices are especially useful in writing the resulting equations for transport in the Dirac equations in a relativistically invariant form.

11.1.3   Wavefunctions

Since the matrices $\tilde{\alpha}$ and β are 4×4 matrices, we can expect that the wavefunction itself will be a 4-vector; that is, it will be a four-element column matrix. This is quite often reduced to a pair of two-element matrices, which we will call ψ_1 and ψ_2, and we can use the forms from (11.19). Then, Eq. (11.10) can be rewritten as

$$\left(E\begin{bmatrix} 1 & 0 \\ 0 & 1 \end{bmatrix} - c\begin{bmatrix} 0 & \tilde{\sigma}\cdot\mathbf{p} \\ \tilde{\sigma}\cdot\mathbf{p} & 0 \end{bmatrix} - \begin{bmatrix} 1 & 0 \\ 0 & -1 \end{bmatrix}m_0c^2 \right)\begin{bmatrix} \psi_1 \\ \psi_2 \end{bmatrix} = 0. \qquad (11.31)$$

Here each of the two components of the wavefunction is itself a two-component vector, which is sometimes called a spinor. We can shorten Eq. (11.31) to the more usual form as

$$\begin{bmatrix} E - m_0c^2 & -c\tilde{\sigma}\cdot\mathbf{p} \\ -c\tilde{\sigma}\cdot\mathbf{p} & E + m_0c^2 \end{bmatrix}\begin{bmatrix} \psi_1 \\ \psi_2 \end{bmatrix} = 0. \qquad (11.32)$$

This leads us to two coupled equations for the wavefunction, as

$$(E - m_0c^2)\psi_1 - c\tilde{\sigma}\cdot\mathbf{p}\psi_2 = 0$$
$$-c\tilde{\sigma}\cdot\mathbf{p}\psi_1 + (E + m_0c^2)\psi_2 = 0. \qquad (11.33)$$

A solution occurs only if the determinant of the square matrix vanishes, and this determinant should return us to Eq. (11.2). This will occur only if we have the relationship

$$(\tilde{\sigma}\cdot\mathbf{p})(\tilde{\sigma}\cdot\mathbf{p}) = p^2. \qquad (11.34)$$

Fortunately, it is easy to show that this is the case.

Let us first look for the positive energy solutions. Generally, the momentum term in Eq. (11.2) is smaller than the rest mass term, and this is certainly the case near the band extrema in semiconductors. If we look at the second equation of Eq. (11.33), the coefficient of the second spinor is much larger than that of the first spinor, so this tells us to expect that $\psi_2 \ll \psi_1$. Now if we take the magnitude of the energy (to assure that we take the positive energy root), then the second line of Eq. (11.33) gives us

$$\psi_2 = \frac{c\tilde{\sigma}\cdot\mathbf{p}}{E + m_0c^2}\psi_1, \qquad (11.35)$$

and indeed, this component is of order v/c smaller than ψ_1. Using this in the first equation of Eq. (11.33) leads to

$$\left[(E - m_0c^2) - c^2\frac{p^2}{E + m_0c^2} \right]\psi_1 = 0. \qquad (11.36)$$

If, as assumed, the kinetic energy is small compared to the rest mass term, then we can replace the denominator in the second term as just twice the rest mass term, and

$$\left[\frac{p^2}{2m_0} - (E - m_0 c^2) \right] \psi_1 = 0. \tag{11.37}$$

Thus, we remember that the zero of energy is taken midway between the two energy solutions, so that this last equation tells us that the kinetic energy (first term) is just the energy away from the minimum of the positive energy "band."

We must remember that this solution is still a two-component vector, which represents the two possible spin states for this solution. We can thus write this wavefunction, apart from a normalization constant, as

$$\psi_1^{(+)} = \begin{bmatrix} 1 \\ 0 \end{bmatrix}, \quad \psi_1^{(-)} = \begin{bmatrix} 0 \\ 1 \end{bmatrix}, \tag{11.38}$$

which introduces the spinors into the wavefunctions. With these solutions, we can find the wavefunctions for the smaller component as

$$\psi_2^{(+)} = \frac{\tilde{\sigma} \cdot \mathbf{p}}{2m_0 c} \psi_1^{(+)} = \frac{1}{2m_0 c} \begin{bmatrix} p_z & p_x - ip_y \\ p_x + ip_y & -p_z \end{bmatrix} \begin{bmatrix} 1 \\ 0 \end{bmatrix} = \frac{1}{2m_0 c} \begin{bmatrix} p_z \\ p_x + ip_y \end{bmatrix}$$

$$\psi_2^{(-)} = \frac{\tilde{\sigma} \cdot \mathbf{p}}{2m_0 c} \psi_1^{(-)} = \frac{1}{2m_0 c} \begin{bmatrix} p_z & p_x - ip_y \\ p_x + ip_y & -p_z \end{bmatrix} \begin{bmatrix} 0 \\ 1 \end{bmatrix} = \frac{1}{2m_0 c} \begin{bmatrix} p_x - ip_y \\ -p_z \end{bmatrix}.$$

$$\tag{11.39}$$

When we see the negative energy solutions, the roles of ψ_2 and ψ_1 are reversed, but the equations are the same except for a minus sign change in the final result for the smaller wavefunction.

Earlier, following Eq. (11.6), we talked about the similarity to energy bands found from the $\mathbf{k} \cdot \mathbf{p}$ theory. There, the perturbation theory led to admixing of the conduction bands and the valence bands. That is, the conduction band, which is primarily derived from the cation s-function, has a small admixture from the valence band anion p-functions. Here we see the same thing. The positive energy wavefunction has a small admixture from the negative energy wavefunctions. This is because the Hamiltonian (11.32) is not diagonal. Hence, the basis set we used to write this Hamiltonian is not the true eigenstates of the system. These are found by the above procedure, which gives the admixture of the positive and negative energy states.

11.1.4   Free Particle in a Field

As we already mentioned above in Eq. (11.20), the electromagnetic vector and scalar potentials can be introduced into the Dirac equation. Now we would like to make sure that Eq. (11.20) gives us results that reduce to the non-relativistic form and with the Klein–Gordon equation. Now if we introduce the vector and scalar equations into Eq. (11.2), we have

$$(E - e\phi)^2 - c^2(\mathbf{p} - e\mathbf{A})^2 - m_0^2 c^4 = 0. \tag{11.40}$$

It is this form that we desire to be consistent with Eq. (11.20). Here we proceed in exactly the same manner as in the approach to Eq. (11.11). We pre-multiply Eq. (11.20) with the expression

$$[E - e\phi + c\tilde{\alpha}\cdot(\mathbf{p} - e\mathbf{A}) + \beta m_0 c^2]. \tag{11.41}$$

This leads us to the expression

$$[(E - e\phi)^2 - c^2\tilde{\alpha}\cdot(\mathbf{p} - e\mathbf{A})\tilde{\alpha}\cdot(\mathbf{p} - e\mathbf{A}) - m_0^2 c^4$$
$$+ (E - e\phi)c\tilde{\alpha}\cdot(\mathbf{p} - e\mathbf{A}) - c\tilde{\alpha}\cdot(\mathbf{p} - e\mathbf{A})(E - e\phi)]\psi = 0. \tag{11.42}$$

Now the Pauli spin matrices have the property that

$$\tilde{\sigma}\cdot\mathbf{B}\tilde{\sigma}\cdot\mathbf{C} = \mathbf{B}\cdot\mathbf{C} + i\tilde{\sigma}\cdot(\mathbf{B}\times\mathbf{C}), \tag{11.43}$$

which leads to Eq. (11.34) in the absence of the vector potential. In the proper 4-vector form for Eq. (11.43), we should have $\tilde{\sigma}$ replaced by

$$\tilde{\sigma}' = \begin{bmatrix} \tilde{\sigma} & 0 \\ 0 & \tilde{\sigma} \end{bmatrix}. \tag{11.44}$$

Using these two equations, the expansion of the second term in Eq. (11.42) becomes

$$\tilde{\alpha}\cdot(\mathbf{p} - e\mathbf{A})\tilde{\alpha}\cdot(\mathbf{p} - e\mathbf{A}) = (\mathbf{p} - e\mathbf{A})^2 + ie\tilde{\sigma}'\cdot(\mathbf{p}\times\mathbf{A} + \mathbf{A}\times\mathbf{p})$$
$$= (\mathbf{p} - e\mathbf{A})^2 + ie\tilde{\sigma}'\cdot(-i\hbar\nabla\times\mathbf{A}) \tag{11.45}$$
$$= (\mathbf{p} - e\mathbf{A})^2 + e\hbar\tilde{\sigma}'\cdot\mathbf{B},$$

where **B** is the magnetic flux density. The last two terms in Eq. (11.42) can be written as

$$(E - e\phi)\tilde{\alpha}\cdot(\mathbf{p} - e\mathbf{A}) - \tilde{\alpha}\cdot(\mathbf{p} - e\mathbf{A})(E - e\phi)$$
$$= -e\tilde{\alpha}\cdot(E\mathbf{A} + \mathbf{A}E) - e\tilde{\alpha}\cdot(\phi\mathbf{p} - \mathbf{p}\phi) \tag{11.46}$$
$$= -ie\tilde{\alpha}\cdot\left(\frac{\partial\mathbf{A}}{\partial t} + \nabla\phi\right) = ie\tilde{\alpha}\cdot\mathbf{E},$$

where **E** is the electric field intensity. In arriving at this last expression, it must be recalled that these parameters are operators that work on a wavefunction.

With the above expressions for reducing the terms, we may now rewrite Eq. (11.42) in the form

$$[(E - e\phi)^2 - c^2(\mathbf{p} - e\mathbf{A})^2 - m_0^2 c^4 + e\hbar\tilde{\sigma}'\cdot\mathbf{B} + ie\tilde{\alpha}\cdot\mathbf{E}]\psi = 0. \quad (11.47)$$

To proceed, we write the energy as the rest mass energy (half the bandgap) and the kinetic energy so that

$$(E - e\phi)^2 - m_0^2 c^4 = (E - m_0 c^2 - e\phi)(E + m_0 c^2 - e\phi)$$
$$\approx (E_{\text{kin}} - e\phi)m_0 c^2. \quad (11.48)$$

The last term in Eq. (11.47), after we factor out $m_0 c^2$, is of order v^2/c^2 and can be ignored. Then, Eq. (11.47) becomes

$$\left[E_{\text{kin}} - \frac{1}{2m_0}\left(\mathbf{p} - \frac{e}{c}\mathbf{A}\right)^2 - e\phi + \frac{e\hbar}{2m_0}\tilde{\sigma}'\cdot\mathbf{B} \right]\psi = 0. \quad (11.49)$$

The pre-factor of the last term in square brackets is the Bohr magneton, and this term is recognized as the Zeeman splitting of the two spin states in each band, which is normally added by hand in the semi-classical transport case. In MKS units, the value of the Bohr magneton is 9.274×10^{-24} J/T. But we see that the proper approach of the Dirac equation leads to the normal terms in kinetic energy with the addition of the Zeeman term.

11.2 Graphene

Graphene is a single layer of graphite that has been isolated recently [1, 10]. Normally, the layers in graphite are very weakly bonded to one another, which is why graphite is found as the common pencil lead and is useful as a lubricant. The single layer of graphene, on the other hand, is exceedingly strong in the plane of the atoms and has been suggested for a great many applications. Here we wish to discuss the energy structure and then the transport. To begin, we refer to the crystal structure and reciprocal lattice shown in Fig. 11.1. Graphene is a single layer of C atoms, which are arranged in a hexagonal lattice. The unit cell contains two C atoms, which are non-equivalent. Thus, the basic unit cell is a diamond, which has a basis of two atoms. In Fig. 11.1, the unit vectors of the diamond cell are designated as $\mathbf{a}_1$ and $\mathbf{a}_2$, and the cell is defined by the red solid and dashed lines. The two inequivalent atoms are shown as the A (red) and B (green) atoms. The three nearest-neighbor vectors are also shown pointing from a B atom to the three closest A atoms. The reciprocal lattice is also a diamond, rotated by 90 degrees from that of the real space lattice, but the hexagon shown is usually used. There are two inequivalent points at two distinct corners of the hexagon, which are marked as K and K'. As we will see, the conduction and valence bands touch at these two points, so that they represent two valleys in either band. The two unit vectors of the reciprocal lattice are $\mathbf{b}_1$ and $\mathbf{b}_2$, with $\mathbf{b}_1$ normal to $\mathbf{a}_2$ and $\mathbf{b}_2$ normal to $\mathbf{a}_1$. The nearest-neighbor distance is $a = 0.142$ nm, and from this, one can write the lattice vectors and reciprocal lattice vectors as

$$\mathbf{a}_1 = \frac{a}{2}\left(3\mathbf{a}_x + \sqrt{3}\mathbf{a}_y\right), \quad \mathbf{b}_1 = \frac{2\pi}{3a}\left(\mathbf{a}_x + \sqrt{3}\mathbf{a}_y\right)$$
$$\mathbf{a}_2 = \frac{a}{2}\left(3\mathbf{a}_x - \sqrt{3}\mathbf{a}_y\right), \quad \mathbf{b}_2 = \frac{2\pi}{3a}\left(\mathbf{a}_x - \sqrt{3}\mathbf{a}_y\right),$$

(11.50)

where a is the lattice constant (nearest-neighbor distance). The two non-equivalent points K and K' are given by the reciprocal space vectors

$$K = \frac{2\pi}{3a}\left(1, \frac{1}{\sqrt{3}}\right), \quad K' = \frac{2\pi}{3a}\left(1, \frac{-1}{\sqrt{3}}\right).$$

(11.51)

From these parameters, we can construct the energy bands with a nearest-neighbor interaction. This was first done apparently by Wallace [11], and we basically follow his approach.

11.2.1 Band Structure

Because of the two atoms per unit cell, we must assume that the wavefunction has two basic components, one for the A atoms and one for the B atoms. Thus, we write the wavefunction in the following form

$$\psi(x, y) = \varphi_1(x, y) + \lambda\varphi_2(x, y) = \sum_A e^{i\mathbf{k}\bullet\mathbf{r}_A}\chi(\mathbf{r} - \mathbf{r}_A) + \sum_B e^{i\mathbf{k}\bullet\mathbf{r}_B}\chi(\mathbf{r} - \mathbf{r}_B).$$

(11.52)

where the $\chi's$ are the atomic wavefunctions, and the exponential factors are translation operators that move the atomic function to the indicated atoms. Here we have written both the position and momentum as two-dimensional vectors. Each of the two-component wavefunctions is a sum over the Bloch functions for each type of atom. Without fully specifying the Hamiltonian, we can write the Schrödinger equation as

$$H(\varphi_1 + \lambda\varphi_2) = E(\varphi_1 + \lambda\varphi_2). \tag{11.53}$$

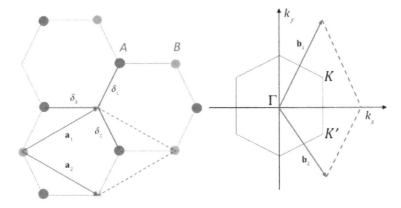

Figure 11.1 The crystal structure of graphene (left) and its reciprocal lattice (right). The *A* and *B* atoms are the two non-equivalent atoms in each unit cell (red diamond on left). This leads to two non-equivalent points *K* and *K'* in the Brillouin zone (right).

At this point, we pre-multiply Eq. (11.53) first with the complex conjugate of the first component of the wavefunction ($\varphi_1^{\dagger}$) and integrate, and then with the complex conjugate of the second component of the wavefunction ($\varphi_2^{\dagger}$) and integrate again. This leads to two equations, which can be written as

$$H_{11} + \lambda H_{12} = E$$
$$H_{21} + \lambda H_{22} = E. \tag{11.54}$$

with

$$H_{ij} = \int \varphi_i^{\dagger} H \varphi_j d^2\mathbf{r}. \tag{11.55}$$

As mentioned, we are only going to use nearest-neighbor interactions, so the diagonal terms become

$$H_{11} = \int \chi^*(\mathbf{r}-\mathbf{r}_A) H \chi(\mathbf{r}-\mathbf{r}_A) d^2\mathbf{r} = E_0$$

$$H_{22} = \int \chi^*(\mathbf{r}-\mathbf{r}_B) H \chi(\mathbf{r}-\mathbf{r}_B) d^2\mathbf{r} = E_0$$

(11.56)

since both atoms are C and thus have the same atomic wavefunctions. In graphene, the in-plane bonds that hold the atoms together are sp^2 hybrids, while the transport is provided by the p_z orbitals normal to the plane, and this is the atomic wavefunction of interest. For this reason, the local integral at the A atoms and at the B atoms should be exactly the same, and this is symbolized in Eq. (11.56) by assigning them the same net energy. However, in the empirical tight-binding approach, the values for E_0 are determined by experiment, in this case the use of X-rays. By the same process, the off-diagonal terms become

$$H_{12} = H_{21}^* = \int \varphi_1^\dagger H \varphi_2 d^2\mathbf{r} = \sum_{nn} e^{i\mathbf{k}\bullet(\mathbf{r}_B-\mathbf{r}_A)} \int \chi_A^* H \chi_B d^2\mathbf{r}$$

$$= \gamma_0 (e^{i\mathbf{k}\bullet\delta_1} + e^{i\mathbf{k}\bullet\delta_2} + e^{i\mathbf{k}\bullet\delta_3}).$$

(11.57)

The sum of the three exponentials shown in the parentheses is known as a *Bloch sum*. Each term is a displacement operator that moves the A atom basis function to the B atom where the integration is performed. The three nearest-neighbor vectors were shown in Fig. 11.1. We can write their coordinates, relative to a B atom (or an A atom) as shown in the figure, as

$$\delta_1 = \frac{a}{2}(1, \sqrt{3}), \quad \delta_2 = \frac{a}{2}(1, -\sqrt{3}), \quad \delta_3 = -a(1,0).$$

(11.58)

With a little algebra, the off-diagonal elements can now be written as

$$H_{12} = H_{21}^* = \gamma_0 \left[2e^{ik_x a/2} \cos\left(\frac{\sqrt{3}k_y a}{2}\right) + e^{-ik_x a} \right].$$

(11.59)

Now that the various matrix elements have been evaluated, the Hamiltonian matrix can be written from these values. This leads to the determinant

$$\begin{vmatrix} (E_0 - E) & \lambda H_{12} \\ H_{21} & \lambda(E_0 - E) \end{vmatrix} = 0.$$

(11.60)

This leads us to the result

$$E = E_0 \pm \gamma_0 \sqrt{1 + 4\cos^2\left(\frac{\sqrt{3}k_y a}{2}\right) + 4\cos\left(\frac{\sqrt{3}k_y a}{2}\right)\cos\left(\frac{3k_x a}{2}\right)} . \quad (11.61)$$

The result is shown in Fig. 11.2. The most obvious fact about these bands is that the conduction and valence bands touch at the six K and K' points around the hexagon reciprocal cell of Fig. 11.1. This means that there is no bandgap. Indeed, expansion of Eq. (11.61) for small values of momentum away from these six points shows that the bands are linear, just as for the Dirac equation in the previous section. This explains why we need to have a familiarity with the Dirac solutions in order to treat materials such as graphene. These bands have come to be known as massless Dirac bands, in that they give similar results to solutions to the Dirac equation with a zero rest mass. If this small momentum is written as ξ, then the energy structure has the form (with $E_0 = 0$)

$$E = \pm \frac{3\gamma_0 a}{2}\xi . \quad (11.62)$$

An important aspect of this is the velocity, which is given by

$$v = \frac{1}{\hbar}\frac{dE}{dk} = \frac{1}{\hbar}\frac{dE}{d\xi} = \pm\frac{3\gamma_0 a}{2\hbar} \equiv v_F . \quad (11.63)$$

As this velocity is constant throughout the linear bands, it has been termed the Fermi velocity. Experiments show that the width of the valence band is about 9 eV, so that $\gamma_0 \sim 3$ eV (the maximum value of the square root term in Eq. (11.61) is 3). Using this value and the result (11.63), we arrive at an effective Fermi velocity for these linear bands of about 9.7×10^7 cm/s. We should point out that, while the rest mass is zero, the dynamic mass of the electrons and holes is not zero, but increases linearly with ξ. This variation has also been seen experimentally, using cyclotron resonance to measure the mass [12], which we discuss below. The above approach to the band structure of graphene works quite well. At higher energies, the nominally circular nature of the energy "surface" near the Dirac points becomes trigonal, and this can have a big effect upon transport. For a more advanced approach, which also takes into account the sp^2 bands arising from the in-plane bonds, one needs to go to the approaches such as the empirical pseudo-potentials [13].

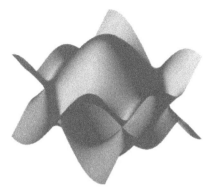

Figure 11.2 The conduction (upper) and valence (lower) bands in graphene. These bands touch at the K and K' points in the Brillouin zone, known as the Dirac points. These points are distributed around a hexagon as shown in Fig. 11.1.

11.2.2 Wavefunctions

In the preceding section, the band structure was computed with atomic wavefunctions in the tight-binding approach. Now we want to rewrite the Hamiltonian around the two Dirac points within the unit cell of the reciprocal lattice. The Hamiltonian can be written as

$$H = -\gamma_0 \sum_{nn,\sigma} (a_{\sigma,i}^{\dagger} b_{\sigma,j} + h.c.),\qquad(11.64)$$

where $a_{\sigma,i}^{\dagger}$ is the creation operator, which puts an electron with spin σ at site i on sublattice A, and $b_{\sigma,j}$ is the annihilation operator, which removes an electron with spin σ from site j on sublattice B. Here i and j are nearest neighbors in the lattice. We want to consider that the wavefunction at a particular lattice site can be written as a Bloch function

$$\psi_n(\mathbf{r}) = \frac{1}{\sqrt{N_C}} \sum_{\mathbf{k}} C(\mathbf{k}) e^{i\mathbf{k}\cdot\mathbf{R}_n},\qquad(11.65)$$

where N_C is the number of unit cells in the crystal. We wish to expand this around the two minima of the bands (the Dirac points). This means that we will have two contributions, one from the wavefunction around the K point and one from the contribution around the K' point. The wavefunction itself will be taken as a spinor

$$\psi = \begin{bmatrix} a_i \\ b_i \end{bmatrix} = \begin{bmatrix} a_{1,i}e^{-i\mathbf{K}\bullet\mathbf{R}_i} + a_{2,i}e^{-i\mathbf{K}'\bullet\mathbf{R}_i} \\ b_{1,i}e^{-i\mathbf{K}\bullet\mathbf{R}_i} + b_{2,i}e^{-i\mathbf{K}'\bullet\mathbf{R}_i} \end{bmatrix}, \tag{11.66}$$

and similarly for the adjoint operators. The first two terms constitute what we will call ψ_1, while the second two terms constitute what we call ψ_2. We now use these in the Dirac Hamiltonian (11.32), modified for the massless case and for the reduced speed of light, to give

$$\begin{bmatrix} 0 & v_F\tilde{\sigma}\bullet\mathbf{p} \\ v_F\tilde{\sigma}\bullet\mathbf{p} & 0 \end{bmatrix}\begin{bmatrix} \psi_1 \\ \psi_2 \end{bmatrix} = 0. \tag{11.67}$$

This produces two copies of the actual Hamiltonian equation, one of which holds around K and the other holds around K'. However, we note that the spin vector now has only two terms, as there is no wave momentum in the z-direction. Hence, the vector $\mathbf{p}$ is expanded around these two points and we use $\mathbf{k}$ as being referenced from the two points, that is, $\mathbf{k}$ replaces our ξ in Eq. (11.62) and is now a two-dimensional vector.

The wavefunction, in momentum space, for motion around the Dirac point K, has the form

$$\psi_{K,\pm}(\mathbf{k}) = \frac{1}{\sqrt{2}}\begin{pmatrix} e^{-i\theta_k/2} \\ \pm e^{i\theta_k/2} \end{pmatrix}, \tag{11.68}$$

where the $\pm$ signs refer to the same signs in Eq. (11.62). Hence, we have different wavefunctions for the positive velocity branch and the negative velocity branch. The phase factor θ_k is given as

$$\theta_k = \arctan\left(\frac{k_x}{k_y}\right), \tag{11.69}$$

and the components of the momentum have a particular orientation with respect to the Brillouin zone, as illustrated in Fig. 11.1. The wavefunction, for motion around the Dirac point K', has the form

$$\psi_{K',\pm}(\mathbf{k}) = \frac{1}{\sqrt{2}}\begin{pmatrix} e^{i\theta_k/2} \\ \pm e^{-i\theta_k/2} \end{pmatrix}. \tag{11.70}$$

Note that these two wavefunctions are related by time reversal symmetry, which is indicated by the existence of the reflection plane at $k_y = 0$. That is, time reversal symmetry is equivalent to reflection through this plane [10].

A relevant property for these wavefunctions is their helicity, and the proper quantum mechanical operator for this property is given as [10]

$$h = \frac{1}{2}\tilde{\sigma}\cdot\frac{\mathbf{p}}{p}.$$
(11.71)

It is clear that the two wavefunctions above are almost eigenstates of the helicity operator as

$$h\psi_K(\mathbf{r}) = \pm\frac{1}{2}h\psi_K(\mathbf{r}), \quad h\psi_{K'}(\mathbf{r}) = \mp\frac{1}{2}h\psi_{K'}(\mathbf{r}).$$
(11.72)

This tells us that the states close to the Dirac points are such that the spin operator has well-defined up or down states with regard to the momentum, or that they have a well-defined *chirality* or helicity. Now this has nothing to do with spin as we have not put spin into the wavefunctions. Rather the form of the spinor wavefunctions now refers to a spin-like quantity, which is referred to as the pseudo-spin of the wavefunction (the real spin still exists, but a magnetic field is usually required to uncover its presence). We can summarize this as noting that the electrons have a positive chirality, while the holes have a negative chirality for a given momentum state.

Another point with regard to Eqs. (11.68) and (11.70) is that the rotation of the direction of the momentum by 2π, which changes θ_k 2π, in fact only changes the wavefunction by a minus sign. One needs to rotate by 4π in order to recover the original form of the wavefunction. This is a well-known time reversal symmetry property of spin systems [14] and is another sign that the interpretation of the chirality in terms of pseudo-spin is the proper approach.

11.2.3 Effective Mass

When the electron moves in the potential that arises from the atoms (and from the varieties of electron–electron potentials), it tends to follow the energy bands and, thus, is not stationary in time. On the other hand, if we connect it to the Bloch wavefunction, then it sits with a wave momentum and corresponding energy at $\hbar k$ and can be thought of as being stationary at that momentum. We often associate these two concepts with the idea of a quasi-particle, which is an electron-like quantum excitation within the crystal, but with properties that are different than those of the free electron, even though the latter might well be the basis upon which the energy bands are constructed. At the end of the day, though, we need to give our quasi-particle an *effective* mass to complete the description and its response to external fields and potentials. In most textbooks, one finds a description of the effective mass, which is related to the energy through a second derivative of the latter. Unfortunately, this approach is only valid near the band extrema for gapped semiconductors. Moreover, an assumption of a constant effective mass is made in order to arrive at this formulation. This is not a valid assumption in any nonparabolic band, and certainly not in graphene, or any other Dirac-like material. Hence, the common second-derivative formula is not only misleading, it is absolutely wrong in the materials in which we are interested in this chapter.

So, then, what is the appropriate approach to be taken. To begin, the energy bands and the resulting motion arises from the Hamiltonian that may be written, in the simplest approach, as

$$H = \frac{p^2}{2m_0} + V(\mathbf{r}),$$ (11.73)

where the potential can be the pseudo-potential, or a modification of it to incorporate some form of the electron–electron energies. The quantity $\mathbf{p}$ is the conventional momentum operator from quantum mechanics, which is expressed as $p = -i\hbar\nabla$, in vector form. In order to arrive at an effective mass, we need to examine how this potential interacts with the Bloch functions and gives rise to a quasi-momentum, which we can also call the crystal momentum [15]. The approach we follow is based on ideas in Ref. [15], but more recently discussed by Zawadzki [16]. We will not go through the detailed derivation; that is available in Ref. 6. The major result, which is not surprising, is that the average velocity of the carrier is given by the first equality in Eq. (11.63). As the average velocity of the quasi-particle is given by the point derivative of the energy with respect to the quasi-momentum, this means that the average velocity of the quasi-particle must be directly related to the quasi-momentum, and this allows us to define this relationship via an effective mass:

$$\mathbf{P} = \hbar\mathbf{k} = m*\langle\mathbf{v}\rangle. \tag{11.74}$$

This *must* be regarded as the fundamental definition of the effective mass for the quasi-particle electron. It is different for every band, just as the Bloch function and the quasi-momentum are different for every band (and every momentum k). But this is dramatically different from what is commonly found in text books. Here we make the definition based on the existence of the quasi-momentum and the average velocity, which describe the motion of our quasi-particle arising from the Bloch function.

It is important to note that, if we combine the first equality in Eq. (11.63) with Eq. (11.74), we can write an expression for the mass in terms of the energy bands as

$$\frac{1}{m*} = \frac{1}{\hbar^2 k}\frac{\partial E(k)}{\partial k} \tag{11.75}$$

for a spherically symmetric band. When we apply this to graphene's energy structure given from Eqs. (11.62) and (11.63) as

$$E = \pm\hbar v_F k , \tag{11.76}$$

we obtain

$$m* = \frac{\hbar|k|}{v_F} . \tag{11.77}$$

This is exactly the value found for the cyclotron mass in Ref. [10], which has been confirmed by careful measurements in graphene [12]. In the latter paper, the authors plot the measured cyclotron mass as a function of the carrier density in graphene, whether electrons or holes. They observe that the mass increases as the square root of the density. This is completely consistent, as the density at low temperatures is given by

$$n = \frac{4}{(2\pi)^2}\int_0^{2\pi} d\theta \int_0^{k_F} k dk = \frac{k_F^2}{\pi} , \tag{11.78}$$

so that one expects the mass to vary as $\sqrt{\pi n}$, which is found in the experiments. Finally, it should be mentioned that the first derivative mass has long been associated with the cyclotron mass, as the cyclotron motion measures the cross-sectional area of the Fermi surface [17].

11.3 Topological Insulators

Topological insulators are materials in which a surface or interface provides a localized energy structure that has the Dirac-like bands of graphene [18], but generated with some additional properties. One prototypical material system is a heterostructure between HgTe and CdTe. In most semiconductors with a direct bandgap at the Γ point (center of the Brillouin zone), the bottom of the conduction band is composed of atomic S-orbitals from the cations, and often denoted as the Γ_6 or Γ_1 (the former is the so-called double group notation used when the spin–orbit interaction is included) band. On the other hand, the top of the valence band is usually composed of the anion P-orbitals, and denoted as the Γ_8 or Γ_{15} band. In HgTe, these two roles are reversed, so that the Γ_6 band lies below the Γ_8. Hence, HgTe is often referred to as having a negative bandgap. In the interface between these two materials, these bands must cross as they reverse their roles from HgTe to CdTe. Now this property has been known for quite some time as the HgTe/CdTe superlattice band structure was studied at least as early as 1979 [19, 20]. But for a topological insulator, one wants more to assure that the zero gap and its properties are topologically protected from disorder. In Fig. 11.3, we draw schematically how the interface bands extend from the bulk bands to provide the Dirac-like bands. One would like the bands to be such that, for example, one had spin up and the other spin down. In a topological insulator, the spin of one branch is locked at a right angle to their momentum (termed spin–momentum locking), so that carriers in the other branch have different spin and back-scattering is then forbidden. Time reversal symmetry was predicted to lead to these types of edge states in quantum wells of HgTe placed between layers of CdTe in 1987 [21], and it was experimentally observed in 2007 [22]. More recently, this type of surface band structure was predicted to occur in three-dimensional materials such as some Bi compounds [23], although the general basis for the topological insulator seems to have been put forward more generally in 2005 [24, 25]. The first experimental evidence for a three-dimensional topological insulator was in bismuth antimonide in 2008 [26].

The interest here, of course, is in transport in the Dirac-like bands, as they may occur in these materials. And it is clear that the Aharonov–Bohm effect can be observed in topological insulators

[27], as it has been observed in Bi_2Se_3 nanoribbons [28]. Bi_2Se_3 is a layered compound with a rhombohedral phase, but with covalent bonding in the layer and weak van der Waals bonding between the layers. In the nanoribbons, as it is on these surfaces, the protected state exists (the bulk is an insulator) [28]. Also seen in these nanoribbons is weak anti-localization [29]. Normally, one observes weak localization, which arises in disordered systems where a set of impurities can cause the electrons to move around two time-reversed paths (going around the ring of impurities in opposite directions) and lead to constructive interference upon back-scattering. This lowers the conductance through the sample and gives a resistance peak, which is easily destroyed by an applied magnetic field (the magnetic field breaks time reversal symmetry). However, when the spin–orbit interaction is important, as it is in the topological insulators, the spin is coupled to the momentum, as mentioned above. Hence the spins of the carriers in the two time-reversed paths are opposite to each other. As a result, the two paths interfere in such a manner that it results in a reduction in the resistance, and this is called weak anti-localization [30]. The magnetic field still leads to breaking the time-reversal symmetry and rapid decay of the signal.

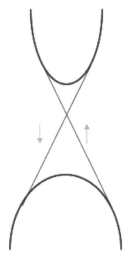

Figure 11.3 The curved bands describe the bulk bands of the topological insulator. The straight lines describe the surface bands that exist on the surface. The surface bands cross and have the Dirac-like form.

It will be recognized that the bands of the Dirac-like states in the topological insulator are precisely the same form as Eq. (11.76) for graphene. Thus, transport developed for Dirac bands in graphene will have strong similarities to the transport in topological insulators, and much of the discussion of the previous section carries over to these materials.

11.4 Klein Tunneling

In classical semiconductors, which possess a sizable bandgap, a potential barrier higher than the energy of the incident particle generally blocks transmission of the particle through the barrier. This is because the potential $V - E$, where E is the energy of the incident carrier, lowers the energy into the bandgap, where there are no propagating wave solutions. However, if the barrier has a finite thickness, then it is possible for the particle to tunnel through the barrier [31]. This is because with the wave interpretation, the part of the wave that penetrates the potential barrier does not fully decay before the back edge of the barrier is reached. Nevertheless, the transmission probability decays exponentially both with the height of the potential barrier and with its thickness. In the relativistic world of graphene, however, the barrier pushes the incident electron wave energy into the valence band, where there are allowed states. Nevertheless, the particle cannot transit the potential barrier (of infinite thickness) because the electron and hole states have different chirality, so the particle is still reflected. If the barrier has finite thickness, we come to the Klein paradox, which leads to a situation in which an incoming electron can penetrate a potential barrier if the height exceeds the rest energy mc^2 [32]. In graphene, the rest energy is zero, so that Klein tunneling can occur whatever the incident energy of the particle. When this happens, the transparency of the barrier depends only weakly on the barrier height and actually increases as the barrier height increases. The physics of the process is that the penetrating electron can couple to holes (positrons) under the barrier to affect the transmission, and matching between the two sets of wavefunctions leads to the high transparency [33].

The Klein paradox has been worked out for the chiral particles in graphene [34]. Thus, under a wide range of conditions, a potential barrier poses no obstacle to an electron or hole in graphene, and the concept of a Schottky barrier just does not work well, as discussed earlier in the chapter. Here, we follow the treatment of Ref. [34] to illustrate the conditions under which Klein tunneling appears. The basic premise is, of course, that the electrons and holes in graphene accurately follow the properties of the electrons and positrons that Klein analyzed. Let us consider a barrier of height V_0, which exists in the region $0 < x < L$. An electron at the Fermi energy approaches the barrier with wavevector k_F and with an angle ϕ defined by the longitudinal and transverse components of the Fermi wavevector through

$$k_x = k_F \cos\phi, \quad k_y = k_F \sin\phi. \tag{11.79}$$

Here, ϕ is the angle appearing in the exponential term of the wavefunctions (11.68) and (11.70). As in the classical tunneling problem, it is assumed that the barrier has infinitely sharp edges so that no disorder is introduced by these edges. The valley degeneracy gives the equivalent Dirac spinor of two wavefunctions corresponding to the pseudo-spin of the composite wavefunction. These two wavefunctions are written as [34]

$$\psi_1 = \begin{cases} (e^{ik_x x} + re^{-ik_x x})e^{ik_y y}, & x < 0, \\ (ae^{iq_x x} + be^{-iq_x x})e^{ik_y y}, & 0 < x < L, \\ te^{ik_x x + ik_y y}, & x > L, \end{cases}$$

$$\psi_2 = \begin{cases} s(e^{ik_x x + i\varphi} + re^{-ik_x x - i\varphi})e^{ik_y y}, & x < 0, \\ s'(ae^{iq_x x + i\theta} + be^{-iq_x x - i\theta})e^{ik_y y}, & 0 < x < L, \\ ste^{ik_x x + ik_y y + i\varphi}, & x > L. \end{cases} \tag{11.80}$$

Here, we use

$$q_x = \sqrt{\frac{(E - V)}{\hbar^2 v_F^2} - k_y^2}$$

$$\theta = \arctan\left(\frac{k_y}{q_x}\right) \tag{11.81}$$

$$s = \text{sgn}(E), \quad s' = \text{sgn}(E - V).$$

The angle θ is the refraction angle of the wave. Matching the various coefficients leads to the reflection coefficient

$$r = 2ie^{i\varphi}\sin(q_xL)\frac{\sin(\varphi)-ss'\sin(\theta)}{ss'[e^{-iq_xL}\cos(\varphi+\theta)+e^{iq_xL}\cos(\varphi-\theta)]-2i\sin(q_xL)}.$$

(11.82)

The transmission through the barrier is given by $T = |t|^2 = 1 - |r|^2$. In Fig. 11.4, the transmission is plotted as a function of incident angle for a barrier whose thickness is 100 nm [34]. The electron concentration outside the barrier is 5×10^{11} cm^{-2}, while the hole concentrations are 10^{12} cm^{-2} (red curve, where V_0 = 200 meV) and 3×10^{12} cm^{-2} (blue curve, where V_0 = 285 meV). Note that the transmission is unity for normal incidence and for a few other angles, which are energy dependent.

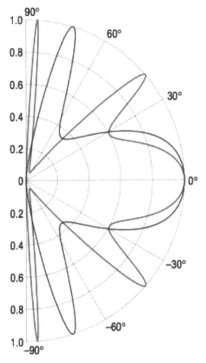

Figure 11.4 Transmission as a function of incident angle at a barrier in single-layer graphene. The red and blue curves are for a barrier of 200 and 285 meV, respectively, and the electron concentration outside the barrier is 5×10^{11} cm^{-2}. Reprinted by permission from Macmillan Publishers Ltd: *Nature Publishing Group*, Ref. [34], Copyright 2006.

11.5 Density Matrix and Wigner Function

As the reader may have already discovered from Chapters 7 and 8, the density matrix and the Wigner function are very intimately related with one another. The density matrix is defined in terms of the wavefunction at two different positions, with the diagonal terms being the local density in space. When the two positions are connected via the Wigner coordinate transformation (7.110) or (8.1), and then the function is Fourier transformed on the difference variable, we arrive at the Wigner function. So these two functions go through life in a connected fashion. We want to examine this further in this section in the spirit of the relativistic approach of this chapter. Before proceeding, we want to examine some of the notations that we have used in this chapter for this purpose, to connect to the general world of quantum relativity theory. We have already introduced the generalized momentum matrix as a four-term matrix, in which the first three terms are the normal components in our coordinate system, whereas p_4 is given as the energy E in Eq. (11.22). In addition, the gamma matrices also had four components, which were related to the $\tilde{\alpha}$ and β matrices, both of which were rank 4 matrices. On the other hand, the current and charge were left as a rank 3 vector and a scalar in Eq. (11.7), but their form will be different for the Dirac equation. Let us rewrite these in the form

$$\rho(\mathbf{r},t)=\psi^{\dagger}\psi$$

$$\mathbf{J}(\mathbf{r},t)=c\psi^{\dagger}\tilde{\alpha}\psi.$$

(11.83)

The meaning of the term $c\tilde{\alpha}$ becomes clear if we look at the time rate of change of the instantaneous position, using the Hamiltonian (11.8),

$$\frac{\partial \mathbf{r}}{\partial t}=-\frac{i}{\hbar}[\mathbf{r},H]=c\tilde{\alpha}.$$

(11.84)

Thus, this expression plays the role of an effective velocity for the wavefunction.

Now let us examine the second of equations (11.83), and evaluate it in terms of the gamma matrices, in order to bring the density and current into the covariant form. Let us write [9]

$$j_i=c\psi^{\dagger}\alpha_i\psi=c\psi^{\dagger}\beta\gamma^i\psi=c\psi^{\dagger}\gamma^4\gamma^i\psi,$$

(11.85)

where we have used Eq. (11.24). Let us define a new adjoint as

$$\bar{\psi} = \psi\gamma^4,\tag{11.86}$$

so that we can write Eq. (11.85) as

$$j_i = c\bar{\psi}\gamma^i\psi.\tag{11.87}$$

The rationale for this follows if we write the density as

$$\rho = \psi^\dagger\psi = \psi^\dagger(\gamma^4)^2\psi = \bar{\psi}\gamma^4\psi \equiv j_4.\tag{11.88}$$

This gives us a proper 4-vector for the current, which includes the density as its fourth member. And this form is properly covariant in a sense of leading to the continuity equation for the relativistic case.

The wavefunctions that appear in these equations are themselves 4-vectors. In the case of graphene, these represent, for a particular value of momentum, the electrons and holes in the two bands and the pseudo-spin of the wavefunctions. So here the spin we talk about is the pseudo-spin for the complex graphene wavefunctions. When we construct the Wigner function, we use j_4, which, in one dimension, leads us to [35]

$$\begin{aligned}
f_W(x, p) &= \frac{1}{2\pi\hbar}\int dy\, e^{-ipy}\left\langle x - \frac{y}{2}\left|j_4\right|x + \frac{y}{2}\right\rangle\\
&= \frac{1}{2\pi\hbar}\int dy\, e^{-ipy}\left\langle x - \frac{y}{2}\middle|\psi\right\rangle(\gamma^4)^2\left\langle\psi\middle|x + \frac{y}{2}\right\rangle\quad (11.89)\\
&= \frac{1}{2\pi\hbar}\int dy\, e^{-ipy}\psi^\dagger\left(x - \frac{y}{2}\right)\psi\left(x + \frac{y}{2}\right),
\end{aligned}$$

just as in Chapter 8. The difference here is that there are four wavefunctions for the four possible "particles," when electron–hole and pseudo-spin factors are considered. This would generally lead us to a 4×4 matrix with 16 distinct Wigner functions. However, when we are interested primarily in the transport of the electrons, or holes, without any correlation between them (such as the optical induced polarization during photon absorption in the previous chapter), the density matrix and the Wigner matrix become block diagonal with 2×2 matrices on the diagonal and zero blocks off the diagonal. This reduces us to eight elements in the overall matrices. So the impact of the Dirac Hamiltonian is to make the Wigner function a matrix-valued function of many elements. In graphene, where the rest mass

m_0 is zero, these two diagonal blocks give the same equation for the wavefunctions, as discussed following Eq. (11.67).

11.5.1 Equation of Motion

We now want to turn our attention to the equation of motion for the Wigner function with the Dirac Hamiltonian. We will approach this in a couple of ways, especially for the momentum term of the Hamiltonian. Our Hamiltonian is Eq. (11.67), although we want to add a scalar potential as in Eq. (11.41). Just as previously, this produces two identical equations, which are the expansions around K and K'. The wavefunctions are two-component spinors to account for the pseudo-spin in the wavefunctions, although the positive energy and negative energy solutions are separate blocks in the large Wigner matrix. We will ignore the mixing between the K and K', although the extension to these off-diagonal terms is straightforward. Traditionally, in quantum mechanics, the time derivative of any operator, such as the density matrix, is given by the Liouville equation

$$\frac{\partial \rho}{\partial t} = -\frac{i}{\hbar}[H, \rho].$$

(11.90)

However, the Wigner function possesses variables in phase–space, which are distinct from the operators that appear in normal quantum mechanics. The connection between the operators and these phase–space variables is through the characteristic function and the Weyl operator [36] discussed in Section 8.1. Moyal [37] generalized Eq. (11.90) through this approach to give what is called the Moyal* operation, expressed as

$$\frac{\partial f_{\mathrm{w}}}{\partial t} = -\frac{i}{\hbar}(H * f_{\mathrm{w}} - f_{\mathrm{w}} * H),$$

(11.91)

in which

$$* \equiv \exp\left[\frac{i\hbar}{2}\left(\frac{\overleftarrow{\partial}}{\partial \mathbf{r}}\frac{\overrightarrow{\partial}}{\partial \mathbf{p}} + \frac{\overleftarrow{\partial}}{\partial \mathbf{p}}\frac{\overrightarrow{\partial}}{\partial \mathbf{r}}\right)\right]$$

(11.92)

and the arrows denote the direction in which the quantum mechanical operators work. This is, of course, a generalization of the

Weyl operator. But using this, we can write the terms in Eq. (11.91) as [35]

$$H(\mathbf{x}, \mathbf{p}) * f_W(\mathbf{x}, \mathbf{p}) = H\left(\mathbf{x} + \frac{i\hbar}{2}\frac{\overrightarrow{\partial}}{\partial \mathbf{p}}, \mathbf{p} - \frac{i\hbar}{2}\frac{\overrightarrow{\partial}}{\partial \mathbf{x}}\right) f_W(\mathbf{x}, \mathbf{p})$$

$$\tag{11.93}$$

$$f_W(\mathbf{x}, \mathbf{p}) * H(\mathbf{x}, \mathbf{p}) = f_W(\mathbf{x}, \mathbf{p}) H\left(\mathbf{x} - \frac{i\hbar}{2}\frac{\overleftarrow{\partial}}{\partial \mathbf{p}}, \mathbf{p} + \frac{i\hbar}{2}\frac{\overleftarrow{\partial}}{\partial \mathbf{x}}\right).$$

Here the differentials operate on the phase–space variables in the Wigner function. Let us begin with just the Dirac Hamiltonian potential term, so that

$$(v_F \tilde{\alpha} \cdot \mathbf{p} - eV(\mathbf{x})) * f_W = \frac{1}{(2\pi\hbar)^3}\left[v_F \tilde{\alpha} \cdot \left(\mathbf{p} - \frac{i\hbar}{2}\frac{\overrightarrow{\partial}}{\partial \mathbf{x}}\right) - eV(\mathbf{x})\right]$$

$$\times \int d^3 y e^{-i\frac{\mathbf{y}}{\hbar}\cdot\left(\mathbf{p} + \frac{i\hbar}{2}\frac{\overleftarrow{\partial}}{\partial \mathbf{x}}\right)} \psi^\dagger\left(\mathbf{x} - \frac{\mathbf{y}}{2}\right)\psi\left(\mathbf{x} + \frac{\mathbf{y}}{2}\right)$$

$$= \frac{1}{(2\pi\hbar)^3}\int d^3 y e^{-i\mathbf{y}\cdot\mathbf{p}/\hbar}\left[v_F \tilde{\alpha} \cdot \left(\mathbf{p} - \frac{i\hbar}{2}\frac{\overrightarrow{\partial}}{\partial \mathbf{x}}\right) - eV\left(\mathbf{x} + \frac{\mathbf{y}}{2}\right)\right]$$

$$\times \psi^\dagger\left(\mathbf{x} - \frac{\mathbf{y}}{2}\right)\psi\left(\mathbf{x} + \frac{\mathbf{y}}{2}\right)$$

$$\tag{11.94}$$

and

$$f_W * (v_F \tilde{\alpha} \cdot \mathbf{p} - eV(\mathbf{x})) = \frac{1}{(2\pi\hbar)^3}\int d^3 y e^{-i\frac{\mathbf{y}}{\hbar}\cdot\left(\mathbf{p} - \frac{i\hbar}{2}\frac{\overrightarrow{\partial}}{\partial \mathbf{x}}\right)} \psi^\dagger\left(\mathbf{x} - \frac{\mathbf{y}}{2}\right)\psi\left(\mathbf{x} + \frac{\mathbf{y}}{2}\right)$$

$$\times \left[v_F \tilde{\alpha} \cdot \left(\mathbf{p} + \frac{i\hbar}{2}\frac{\overleftarrow{\partial}}{\partial \mathbf{x}}\right) - eV(\mathbf{x})\right]$$

$$= \frac{1}{(2\pi\hbar)^3}\int d^3 y e^{-i\mathbf{y}\cdot\mathbf{p}/\hbar}\left[v_F \tilde{\alpha} \cdot \left(\mathbf{p} + \frac{i\hbar}{2}\frac{\overrightarrow{\partial}}{\partial \mathbf{x}}\right) - eV\left(\mathbf{x} - \frac{\mathbf{y}}{2}\right)\right]$$

$$\times \psi^\dagger\left(\mathbf{x} - \frac{\mathbf{y}}{2}\right)\psi\left(\mathbf{x} + \frac{\mathbf{y}}{2}\right)$$

$$\tag{11.95}$$

which combine to yield

$$\frac{\partial f_W}{\partial t} + v_F \tilde{\alpha} \cdot \nabla f_W + \frac{1}{(2\pi\hbar)^3} \int d^3 y e^{-iy \cdot p/\hbar} \left[V\left(x - \frac{y}{2}\right) - V\left(x + \frac{y}{2}\right) \right]$$
$$\times \psi^\dagger\left(x - \frac{y}{2}\right)\psi\left(x + \frac{y}{2}\right) = 0.$$

(11.96)

As discussed above, the pre-factor of the gradient operation in the second term is an effective velocity. Apart from this distinction, the equation of motion is formally the same as that obtained in non-relativistic quantum mechanics. Of course, the right-hand side is not zero but is the collision term. At this point, the collision operator has not been worked out for the relativistic case for phonons, so it is probably just as useful to introduce the results of Section 8.4 for this purpose. The last term on the left can be rewritten in a more recognizable form by introducing a delta function and an integral over it, as

$$I = \frac{1}{(2\pi\hbar)^3} \int d^3 y \int d^3 y' e^{-ip\cdot y/\hbar} \psi^\dagger\left(x - \frac{y}{2}\right)\psi\left(x + \frac{y}{2}\right)$$
$$\times \delta(y - y')\left[V\left(x - \frac{y'}{2}\right) - V\left(x + \frac{y'}{2}\right) \right]$$
$$= \frac{1}{(2\pi\hbar)^3} \int d^3 y \int d^3 y' e^{-ip\cdot y/\hbar} \psi^\dagger\left(x - \frac{y}{2}\right)\psi\left(x + \frac{y}{2}\right)$$
$$\times \int d^3 p' e^{-ip'\cdot(y-y')/\hbar}\left[V\left(x - \frac{y'}{2}\right) - V\left(x + \frac{y'}{2}\right) \right]$$
$$= \int d^3 y' \int d^3 p' e^{ip'\cdot y'/\hbar} f_W(x, p + p')\left[V\left(x - \frac{y'}{2}\right) - V\left(x + \frac{y'}{2}\right) \right]$$
$$= \int d^3 p' M(x, p') f_W(x, p + p').$$

(11.97)

Now we recognize that the last term is a proper convolution of the Wigner function with the Weyl-transformed potential, with

$$M(x, p) = \int d^3 y e^{ip\cdot y/\hbar}\left[V\left(x - \frac{y}{2}\right) - V\left(x + \frac{y}{2}\right) \right]. \quad (11.98)$$

This is often seen with the exponential replaced by a sine function.

11.5.2 Another Approach

Another approach to obtaining the equation of motion is to first transform the Hamiltonian via the Moyal–Weyl transformation into phase-space. This generates a phase-space propagator, which is then integrated with the Wigner function itself to give the resulting equation of motion. Quite generally, we can write [37]

$$\frac{\partial f_W(\mathbf{r}, \mathbf{p})}{\partial t} = \int d^3x \int d^3p' S(\mathbf{r}, \mathbf{p}; \mathbf{x}, \mathbf{p}') f_w(\mathbf{x}, \mathbf{p}'), \tag{11.99}$$

where the kernel propagator is given by the Wigner–Weyl transform of the Hamiltonian as

$$S(\mathbf{r}, \mathbf{p}; \mathbf{x}, \mathbf{p}') = \frac{1}{\hbar} \int d^3y \int d^3q \, e^{i[\mathbf{y}\cdot(\mathbf{p}'-\mathbf{p})+\mathbf{q}\cdot(\mathbf{x}-\mathbf{r})]/\hbar}$$

$$\times \left[H\left(\mathbf{x} - \frac{\mathbf{y}}{2}, \mathbf{p}' + \frac{\mathbf{q}}{2}\right) - H\left(\mathbf{x} + \frac{\mathbf{y}}{2}, \mathbf{p}' - \frac{\mathbf{q}}{2}\right) \right]. \tag{11.100}$$

Let us now take the first term in the Dirac Hamiltonian for graphene, as

$$S_1(\mathbf{r}, \mathbf{p}; \mathbf{x}, \mathbf{p}') = \frac{1}{\hbar} \int d^3y \int d^3q \, e^{i[\mathbf{y}\cdot(\mathbf{p}'-\mathbf{p})+\mathbf{q}\cdot(\mathbf{x}-\mathbf{r})]/\hbar}$$

$$\times \left[v_F \tilde{\alpha} \cdot \left(\mathbf{p}' + \frac{\mathbf{q}}{2}\right) - v_F \tilde{\alpha} \cdot \left(\mathbf{p}' - \frac{\mathbf{q}}{2}\right) \right]$$

$$= \frac{1}{\hbar} v_F \tilde{\alpha} \cdot \int d^3y \int d^3q \, e^{i[\mathbf{y}\cdot(\mathbf{p}'-\mathbf{p})+\mathbf{q}\cdot(\mathbf{x}-\mathbf{r})]/\hbar} \mathbf{q}$$

$$= \frac{1}{\hbar} v_F \delta(\mathbf{p}' - \mathbf{p}) \tilde{\alpha} \cdot \int d^3q \, e^{i\mathbf{q}\cdot(\mathbf{x}-\mathbf{r})/\hbar} \mathbf{q}$$

$$= -v_F \delta(\mathbf{p}' - \mathbf{p}) \delta(\mathbf{x} - \mathbf{r}) \tilde{\alpha} \cdot \frac{\partial}{\partial \mathbf{r}} \tag{11.101}$$

Now let us turn to the potential term, which is (remember that it has a negative sign with it in the Hamiltonian)

$$S_2(\mathbf{r}, \mathbf{p}; \mathbf{x}, \mathbf{p}') = e \int d^3y \int d^3q \, e^{i[\mathbf{y}\cdot(\mathbf{p}'-\mathbf{p})+\mathbf{q}\cdot(\mathbf{x}-\mathbf{r})]/\hbar}$$

$$\times \left[V\left(\mathbf{x} + \frac{\mathbf{y}}{2}\right) - V\left(\mathbf{x} - \frac{\mathbf{y}}{2}\right) \right]$$

$$= \delta(\mathbf{x} - \mathbf{r}) \int d^3y \, e^{i\mathbf{y}\cdot(\mathbf{p}'-\mathbf{p})/\hbar} \left[V\left(\mathbf{x} + \frac{\mathbf{y}}{2}\right) - V\left(\mathbf{x} - \frac{\mathbf{y}}{2}\right) \right]. \tag{11.102}$$

If we insert these two terms into Eq. (11.99), we arrive at

$$\frac{\partial f_W(\mathbf{r},\mathbf{p})}{\partial t} = -v_F\tilde{\alpha}\cdot\frac{\partial f_W(\mathbf{r},\mathbf{p})}{\partial \mathbf{r}}$$

$$+e\int d^3p'\int d^3y e^{i\mathbf{y}\cdot(\mathbf{p}'-\mathbf{p})/\hbar}\left[V\left(\mathbf{r}+\frac{\mathbf{y}}{2}\right)-V\left(\mathbf{r}-\frac{\mathbf{y}}{2}\right)\right]f_W(\mathbf{r},\mathbf{p}').$$

$$(11.103)$$

For the last term, we follow the procedure of the last section and insert a delta function to uncouple the first exponential from the potential. This leads to

$$I = \int d^3p'\int d^3y\int d^3y' e^{i\mathbf{y}'\cdot(\mathbf{p}'-\mathbf{p})/\hbar}\left[V\left(\mathbf{r}-\frac{\mathbf{y}'}{2}\right)-V\left(\mathbf{r}+\frac{\mathbf{y}'}{2}\right)\right]$$

$$\times\ \delta(\mathbf{y}-\mathbf{y}')f_W(\mathbf{r},\mathbf{p}')$$

$$= \int d^3p'\int d^3y\int d^3y' e^{i\mathbf{y}'\cdot(\mathbf{p}'-\mathbf{p})/\hbar}\left[V\left(\mathbf{r}-\frac{\mathbf{y}}{2}\right)-V\left(\mathbf{r}+\frac{\mathbf{y}}{2}\right)\right]$$

$$\times\int d^3p'' e^{i\mathbf{p}''\cdot(\mathbf{y}-)/\hbar}f_W(\mathbf{r},\mathbf{p}')$$

$$= \int d^3p'\int d^3y\int d^3y' e^{i\mathbf{y}'\cdot(\mathbf{p}'-\mathbf{p}''-\mathbf{p})/\hbar}\left[V\left(\mathbf{r}-\frac{\mathbf{y}}{2}\right)-V\left(\mathbf{r}+\frac{\mathbf{y}}{2}\right)\right]$$

$$\times\int d^3p'' e^{i\mathbf{p}''\cdot\mathbf{y}/\hbar}f_W(\mathbf{r},\mathbf{p}')$$

$$= \int d^3p''\int d^3y e^{i\mathbf{p}''\cdot\mathbf{y}/\hbar}\left[V\left(\mathbf{r}-\frac{\mathbf{y}}{2}\right)-V\left(\mathbf{r}+\frac{\mathbf{y}}{2}\right)\right]f_W(\mathbf{r},\mathbf{p}+\mathbf{p}'').$$

$$(11.104)$$

This is, of course, the same as the last line of Eq. (11.97), as it should be.

11.5.3 Further Considerations

In Chapters 7 and 8, we developed the moments of the equations of motion for the density matrix and the Wigner function. These moments have a parallel to the classical moments of the Boltzmann transport equation and are often referred to as the hydrodynamic approach to transport. In addition to its use to study transport, the quantum hydrodynamic equations are a fast-developing field in applied mathematics [38]. As in the classical case, each level of the hierarchy is coupled to the next higher level, so that an uncoupling scheme is desired. Some methods were discussed in the previous chapters, but a general approach is the minimum entropy principle [39]. In the case of the Wigner function, the uncoupling is accomplished by an assumption that it will relax to a suitable steady-state solution, which depends only on the moments of interest. Such a steady-state solution is assumed to be the minimizer of a suitable entropy function under the constraints of the moments included in the solution set. This approach has been extended to graphene [40], where, because of the more complicated wavefunction discussed above, they require far more moment equations for the problem. For example, the normal two equations for density and current now become five equations, due to the four distinct probabilities in the problem. Adding the energy equation gives four more equations unless it is assumed that all four states, described by the four components of the wavefunction, have a common energy. Of course, to actually solve these equations for graphene, one needs to have the relativistic collision integrals, which to my knowledge have not been fully developed. Hence, it is probably still acceptable to use those discussed in these earlier chapters.

In Section 2.7, we discussed the use of Bohm trajectories to simulate the motion of particles in a quantum system, an approach that has been significantly developed [41]. But the presence of the Moyal* operator brings the possibility to use Moyal trajectories for the same purpose [42]. In the case of a general quadratic Hamiltonian and Bohmian trajectories, it has been proven that a coherent state wave packet will follow the classical trajectories, although in the case of interactions between the particles, this is only an approximation valid for short times. In the case of Moyal trajectories, the "particles" are in the proper phase–space representation and follow the phase–space evolution according to the Heisenberg approach. By construction, then, these evolutions are state independent [42]. In this latter work, it is shown that the Moyal trajectories deviate from the classical trajectories specifically due to the quantum corrections. Of course, the Dirac Hamiltonian is not quadratic, but this approach may still have some advantages in the relativistic world and should be investigated further.

11.6   Green's Functions in the Relativistic World

When we deal with the Dirac band structure, the basic equation for the bare equilibrium Green's function changes from that found with the Schrödinger equation. Instead, we must use the appropriate Dirac equation, and the bare Green's function derives from [43]

$$(E - c\tilde{\alpha} \cdot \mathbf{p} + m_0 c^2)G(\mathbf{r}, \mathbf{r}') = i\delta(\mathbf{r} - \mathbf{r}'), \qquad (11.105)$$

and, of course, for graphene, the rest mass is zero. As in Chapter 4, the interactions lead to a self-energy term, which takes this bare Green's function into the retarded and advanced functions. This certainly has an effect on simulations using Green's functions. Yet, there are two approaches to develop the transport through nanostructures, even if they are composed of a material like graphene, which possesses the Dirac band structure.

11.6.1   Analytical Bands

If we use the analytic bands that are given by the Dirac bands, the approach is similar to the use of the effective mass approximation with gapped semiconductors. However, here the Green's function in Eq. (11.105) becomes a matrix Green's function, which in the extreme case has 16 elements, just as for the Wigner function in the previous section. In the situation where the electrons and holes are uncoupled, this reduces to the same block diagonal matrix, in which the diagonal blocks are each 2×2 matrices.

Let us consider a graphene nanoribbon in which the two-dimensional graphene is mapped into a one-dimensional Dirac model with N-sites in the one-dimensional chain [44, 45]. The resulting matrix is a tri-block-diagonal matrix of dimension $2N \times 2N$. The diagonal blocks are given by, at site n,

$$h_n = \begin{bmatrix} V_n & -v_F i p_y \\ v_F i p_y & V_n \end{bmatrix} = \begin{bmatrix} V_n & -i v_F \hbar k_y \\ i v_F \hbar k_y & V_n \end{bmatrix}, \qquad (11.106)$$

and the nearest-neighbor hopping interactions are in the first off-diagonal blocks as

$$h_{n-1,n} = \frac{i\hbar v_F}{2l_0} \begin{bmatrix} 0 & 1 \\ 1 & 0 \end{bmatrix}$$

$$h_{n+1,n} = -\frac{i\hbar v_F}{2l_0} \begin{bmatrix} 0 & 1 \\ 1 & 0 \end{bmatrix}.$$

(11.107)

Here, l_0 is an effective one-dimensional cell size due to the discretization of the graphene lattice into the one-dimensional array of cells. For an infinitely long graphene nanoribbon, with an external bias of V_0 *for each cell*, the dispersion relation becomes [45]

$$E(k_x, k_y) = V_0 \pm \left(\frac{\hbar v_F}{l_0}\right) \sqrt{(k_y l_0)^2 + \sin^2(k_x l_0)},$$

(11.108)

and this leads to an induced bandgap, due to quantization in the transverse direction, of $E_g = 2\hbar v_F k_y$ at $k_x = 0$.

For a finite-length graphene nanoribbon, the ribbon is coupled to the contact regions at the source and drain by a self-energy term. These self-energy terms can be evaluated just as for the real-time Green's functions in Chapters 9 and 10, and the solutions iterated as discussed in these previous chapters. The retarded Green's function at a specific energy is given by

$$G(E) = (EI - H - \Sigma_S - \Sigma_D)^{-1},$$

(11.109)

where the two self-energies arise from the coupling between the "device" region and the contacts at the source (left) and drain (right). We discuss these self-energies further below. This Green's function now leads to the density on each lattice site in response to the local potential, and Poisson's equation can be solved from the density to give the new local potential at each site. Once the process has converged to the proper self-consistent potential and density, one can evaluate the current through the device with [46]

$$I_{DS} = \left\langle \int_{-\infty}^{\infty} J_{m,m+1}(E) dE \right\rangle,$$

(11.110)

where the angle brackets indicate an average over the slices. As this is a numerical procedure, there will be noise in the current determined at different slices, and this average reduces the noise and is based on current continuity. That is, current continuity means that the current should be the same at each node. The individual current at the node is given as

$$J_{m,m+1} = i\frac{e}{h}\sum_{k_y}[H_{m,m+1}(k_y)G^<_{m+1,m}(E,k_y) - H_{m+1,m}(k_y)G^<_{m,m+1}(E,k_y)],$$

(11.111)

where the indicated Green's function is the off-diagonal block in the Green's function found from the normal approach. For this approach, the Keldysh matrix representation still exists [47], although the individual Green's functions have the indicated structure discussed above.

11.6.2   Use of Atomic Sites in DFT

In contrast to the analytical band approach, we can use an atomic site basis and compute directly both the band structure and the Green's functions from this basis, just as was done in gapped semiconductors discussed in previous chapters. This approach builds in the atomic site band structure, whether density functional theory is used or the simpler tight-binding bands are used. We will illustrate the latter, in which each site is coupled to its neighbor via the tight-binding interaction discussed in Section 11.2.1 earlier. We consider a device region that is coupled to a left and right contact region, so that the overall Hamiltonian can be written as [48]

$$H = \begin{bmatrix} H_L & H_{LD} & 0 \\ H_{DL} & H_D & H_{DR} \\ 0 & H_{RD} & H_R \end{bmatrix}.$$

(11.112)

In these block terms, the elements are $-\gamma_0$, which is the nearest-neighbor coupling energy discussed in Section 11.2.1. Hence, the elements differ from zero only at these nearest-neighbor coupling sites. The device Hamiltonian H_D is a square matrix of dimension $N \times N$, where N is the number of atoms in the device itself. The other two diagonal terms represent the left and right contact regions and will be as large as the number of atoms in these regions, which could be infinite (we discuss this further below). The device Hamiltonian is tri-block-diagonal. The main diagonal block contains only the local potential on its diagonal, if one exists, and the terms $-\gamma_0$ in the first off-diagonal elements on either side of the main diagonal. These latter sites are the couplings between the nearest neighbors. For our example, we consider that the edges are zig-zag edges and take two rows of atoms per slice, which allows each atom to be coupled to its nearest neighbors in the slice. This simplifies the spacing of the atoms but complicates the coupling from one slice to another, which is defined by the off-diagonal blocks of the Hamiltonian. For example, suppose we have eight atoms in each slice. Then for forward coupling (see Fig. 11.5), atom 1 in the slice actually couples to atom 2 of the next slice, so this coupling term is off the diagonal of the H_{ij} block of the Hamiltonian matrix. Similarly, atom 4 of the slice couples to atom 3 of the next slice. Similarly, atom 5 of the slice couples to atom 6 of the next slice, so this coupling is on the diagonal and so on. Hence, the coupling block of the Hamiltonin looks like

$$H_{ij} = H_{ji}^{\dagger} = \begin{bmatrix} 0 & -\gamma_0 & 0 & 0 & 0 & 0 & 0 & 0 \\ 0 & 0 & 0 & 0 & 0 & 0 & 0 & 0 \\ 0 & 0 & 0 & 0 & 0 & 0 & 0 & 0 \\ 0 & 0 & -\gamma_0 & 0 & 0 & 0 & 0 & 0 \\ 0 & 0 & 0 & 0 & 0 & -\gamma_0 & 0 & 0 \\ 0 & 0 & 0 & 0 & 0 & 0 & 0 & 0 \\ 0 & 0 & 0 & 0 & 0 & 0 & 0 & 0 \\ 0 & 0 & 0 & 0 & 0 & & -\gamma_0 & 0 \end{bmatrix}. \tag{11.113}$$

If we do the entire matrix (11.112), we have to invert the infinite matrix. Instead, we follow common practice and treat the coupling from the device to the leads by a self-energy term. This means that we can write the device Green's function as

$$G_D(E) = (EI - H_D - \Sigma_L - \Sigma_R)^{-1}. \tag{11.114}$$

The two self-energies are given by

$$\Sigma_L = H_{DL} G_L H_{LD}$$
$$\Sigma_R = H_{DR} G_R H_{RD}. \tag{11.115}$$

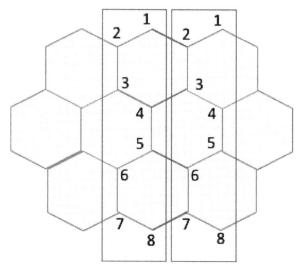

Figure 11.5 Illustration of the slice-to-slice coupling for eight atoms In a slice. Each slice is outlined by a box, and the atoms are numbered in each slice for reference. The slice-to-slice coupling is indicated by the violet bonds.

Here the two coupling Hamiltonians have the same form as H_{ij} above, with the exception that the coupling terms in general have an energy different from the normal nearest-neighbor coupling and can be taken as V_0. The left and right Green's functions are just those arising from the Hamiltonian of the left and right lead atoms, which can be limited in extent. Then, the self-energies satisfy the Dyson equation

$$\Sigma_L = V_0^\dagger (EI - H_L - \Sigma_L)^{-1} V_0$$
$$\Sigma_R = V_0 (EI - H_R - \Sigma_{\sqrt{L}})^{-1} V_0^\dagger. \tag{11.116}$$

With the use of the Landauer formula, we can now write the conductance through the structure as

$$G = \frac{2e^2}{h} \mathrm{Tr}\left\{ \Gamma_L G_D \Gamma_R G_D^\dagger \right\}, \tag{11.117}$$

where

$$\Gamma = -2\mathrm{Im}\{\Sigma\}. \tag{11.118}$$

It is obvious that we have used the approximation that there is no scattering within the body of the "device," and that the applied potential was vanishingly small. Hence, we have not had to build explicitly the less-than and greater-than Green's functions in order to determine the density (and potential at each site) self-consistently. There is, of course, no restriction to use this approach, other than that it is easier to grasp the methodology. The introduction of scattering and the use of potentials that are real (and large) merely complicate the situation, as described in the last chapter, but do not push us away from the tenets of this approach.

11.7 Adding the Magnetic Field

We added the magnetic field directly to the Dirac equation in Eq. (11.41), but here we want to see how this significantly modifies the traditional spacing of Landau levels in energy by $\hbar\omega_c$, where $\omega_c = eB/m^*$ is the cyclotron mass (11.77). To see how this occurs, let us expand Eq. (11.41) for the case of graphene as

$$v_F \tilde{\alpha}\cdot(\mathbf{p}-e\mathbf{A}) = v_F[\sigma_x(p_x-eA_x)+\sigma_y(p_y-eA_y)] \qquad (11.119)$$

$$= v_F \begin{bmatrix} 0 & (p_x-ip_y) \\ (p_x+ip_y) & 0 \end{bmatrix}$$

$$- ev_F \begin{bmatrix} 0 & (A_x-iA_y) \\ (A_x+iA_y) & 0 \end{bmatrix}.$$

Let us now assume that we are dealing with a graphene nanoribbon whose transport axis is in the x-direction, so we take the vector potential as $\mathbf{A} = By\mathbf{a}_x$, so that this can be rewritten as

$$v_F \tilde{\alpha}\cdot(\mathbf{p}-e\mathbf{A}) = v_F[\sigma_x(p_x-eA_x)+\sigma_y(p_y-eA_y)]$$

$$= v_F \begin{bmatrix} 0 & (p_x-eBy-ip_y) \\ (p_x-eBy+ip_y) & 0 \end{bmatrix}. \qquad (11.120)$$

We remember that the wavefunction is written as a spinor, as in Eq. (11.67). Hence, to find the upper component, we insert one of the equations into the other, and then we make the normal substitution that the wavefunction can be written as [49]

$$\psi(x,y) = e^{ikx}\varphi_B(y). \qquad (11.121)$$

Then we find that the energy can be written as

$$E^2 = v_F^2 \left[p_y^2 - m^* \hbar \omega_c + m^{*2} \omega_c^2 (y - y_0)^2 \right],$$ (11.122)

where we have the cyclotron frequency mentioned above and the positional shift is given as

$$y_0 = l_B^2 k, \quad l_B^2 = \frac{\hbar}{eB},$$ (11.123)

where the latter equation defines the magnetic length. Let us renormalize Eq. (11.122), for the moment by defining the reduced energy as

$$\tilde{E} = \frac{E^2}{2mv_F^2} + \frac{\hbar \omega_c}{2},$$ (11.124)

so that Eq. (11.122) can be written as

$$\tilde{E} = \frac{p_y^2}{2m^*} + \frac{1}{2} m^* \omega_c^2 (y - y_0)^2,$$ (11.125)

which we recognize as a harmonic oscillator. Thus, the magnetic field produces the same harmonic oscillator motion as in the non-relativistic case. The difference here is that this is in terms of the reduced energy, rather than the actual energy. These reduced eigenvalues are given as

$$\tilde{E}_n = \left(n + \frac{1}{2} \right) \hbar \omega_c,$$ (11.126)

where n is an integer as n = 0, 1, 2, ... If we now unfold to the real energy, we find that the energy levels in the Dirac bands are given by (we add the $\pm$ to indicate the two types of carriers)

$$E = \pm \sqrt{(2m^* v_F^2) \cdot n \hbar \omega_c}.$$ (11.127)

In the Dirac bands, the eigenenergies of the Landau levels now are separated by a factor that depends on the square root of the magnetic field, instead of the linear behavior in gapped semiconductors. Moreover, there is a Landau level at zero magnetic field that does not occur in the normal case. This is a much different behavior and leads to some interesting effects in materials such as graphene [10].

Problems

1. Show that the pair of equations (11.5) satisfy the continuity equation for probability.
2. Verify Eq. (11.34).
3. Verify Eq. (11.82).

References

1. K. S. Novoselov, A. K. Geim, S. V. Morozov, D. Jiang, Y. Zhang, S. V. Dubonos, I. V. Grigorieva, and A. A. Firsov, *Science*, **306**, 666 (2004).

2. Y. Hatsugai, *J. Phys. Soc. Jpn.*, **74**, 1374 (2005).

3. M. B. Hastings, *Europhys. Lett.*, **70**, 824 (2005).

4. E. Schrödinger, *Ann. Phys.*, **81**, 109 (1926).

5. E. O. Kane, *J. Phys. Chem. Sol.*, **1**, 249 (1957).

6. D. K. Ferry, *Semiconductors, Bonds and Bands* (IOP Publishing, Bristol, 2013).

7. P. A. M. Dirac, *Proc. Roy. Soc.*, **A117**, 610 (1928).

8. L. I. Schiff, *Quantum Mechanics*, 2nd Ed. (McGraw-Hill, New York, 1955), Sec. 43.

9. B. R. Desai, *Quantum Mechanics with Basic Field Theory* (Cambridge University Press, Cambridge, 2010).

10. A. H. Castro Neto, F. Guinea, N. M. R. Peres, K. S. Novoselov, and A. K. Geim, *Rev. Mod. Phys.*, **81**, 109 (2009).

11. P. R. Wallace, *Phys. Rev.*, **71**, 622 (1947).

12. K. S. Novoselov, A. K. Geim, S. V. Morozov, D. Jiang, M. I. Katsnelson, I. V. Gregorieva, S. V. Dubonos, and A. A. Firsov, *Nature*, **438**, 197 (2005).

13. M. V. Fischetti, J. Kim, S. Narayanan, Z.-Y. Ong, C. Sachs, D. K. Ferry, and S. J. Aboud, *J. Phys.: Condens. Matter*, **25**, 473202 (2013).

14. S. Weinberg, *The Quantum Theory of Fields*, Vol. 1 (Cambridge University Press, Cambridge, 2005).

15. R. A. Smith, *Wave Mechanics of Crystalline Solids* (Chapman and Hall, London, 1961).

16. W. Zawadzki, *Acta Phys. Pol. A*, **123**, 132 (2013).

17. C. Kittel, *Quantum Theory of Solids* (John Wiley, New York, 1963), p. 227.

18. J. Moore, *IEEE Spectrum*, **48**, 38 (July 2011).

19. J. N. Schulman and T. C. McGill, *J. Vac. Sci. Technol.*, **16**, 1513 (1979).

20. J. N. Schulman and T. C. McGill, *J. Vac. Sci. Technol.*, **17**, 1118 (1980).

21. O. A. Pankratov, S. V. Pakhomov, and B. A. Volkov, *Solid State Commun.*, **61**, 93 (1987).

22. M. König, S. Wiedmann, C. Brüne, A. Roth, H. Buhmann, L. W. Molenkamp, X.-L. Qi, and S.-C. Zhang, *Science*, **318**, 766 (2007).

23. L. Fu and C. L. Kane, *Phys. Rev. B*, **76**, 045302 (2007).

24. Y. Hatsugai, *J. Phys. Soc. Jpn.*, **74**, 1374 (2005).

25. M. B. Hastings, *Europhys. Lett.*, **70**, 824 (2005).

26. D. Hsieh, D. Qian, L. Wray, Y. Xia, Y. S. Hor, R. J. Cava, and M. Z. Hasan, *Nature*, **452**, 970 (2008).

27. J. H. Bardarson and J. E. Moore, *Rept. Prog. Phys.*, **76**, 056501 (2013).

28. H. Peng, K. Lai, D. Kong, S. Meister, Y. Chen, X.-L. Qi, S.-C. Zhang, Z.-X. Shen, and Y. Cui, *Nature Mater.*, **9**, 225 (2010).

29. D. K. Ferry, *Transport in Semiconductor Mesoscopic Devices* (IOP Publishing, Bristol, 2015).

30. S. Hikami, A. I. Larkin, and Y. Nagaoka, *Prog. Theor. Phys.*, **63**, 707 (1980).

31. D. K. Ferry, *Quantum Mechanics: An Introduction for Device Physicists and Electrical Engineers*, 2nd Ed. (Institute of Physics Publishing, Bristol, U.K., 2001).

32. O. Klein, *Z. Phys.*, **53**, 157 (1929).

33. P. Krekora, Q. Su, and R. Grobe, *Phys. Rev. Lett.*, **92**, 040406 (2004).

34. M. I. Katsnelson, K. S. Novoselov, and A. K. Geim, *Nature Phys.*, **2**, 620 (2006).

35. K. Ma, J.-H. Wang, and Y. Yuan, *Chin. Phys. C*, **35**, 11 (2011).

36. H. Weyl, *The Theory of Groups and Quantum Mechanics* (Dover, New York, 1931).

37. J. E. Moyal, *Proc. Cambridge Philos. Soc.*, **45**, 99 (1949).

38. A. Jüngel, *Transport Equations for Semiconductors* (Springer, Berlin, 2009).

39. P. Degond and C. Ringhofer, *J. Stat. Phys.*, **118**, 625 (2005).

40. N. Zamponi and L. Barletti, *Math. Meth. Appl. Sci.*, **34**, 807 (2011).

41. X. Oriols and J. Mompart, *Applied Bohmian Transport: From Nanoscale Systems to Cosmology* (Pan Stanford, Singapore, 2012).

42. N. Costa Dias and J. Nuno Prata, *J. Math. Phys.*, **48**, 012109 (2007).

43. V. N. Gribov and J. Nyiri, *Quantum Electrodynamics* (Cambridge University Press, Cambridge, 2001).

44. S.-K. Chin, D. Seah, K.-T. Lam, G. S. Samudra, and G. Liang, *IEEE Trans. Electron Dev.*, **57**, 3144 (2010).

45. S.-K. Chin, K.-T. Lam, D. Seah, and G. Liang, *Nanoscale Res. Lett.*, **7**, 114 (2012).

46. K.-T. Lam, S.-K. Chin, D. W. Seah, S. B. Kumar, and G. Liang, *Jpn. J. Appl. Phys.*, **49**, 04DJ10 (2010).

47. M. Bonitz, *Quantum Kinetic Theory* (Teubner, Stuttgart, 1998).

48. L. Huang, Y.-C. Lai, D. K. Ferry, R. Akis, and S. M. Goodnick, *J. Phys.: Condens. Matter*, **21**, 344203 (2009).

49. T. Stegmann and A. Lorke, *Ann. Phys. (Berlin)*, **527**, 723 (2015).

Index

For Product Safety Concerns and Information please contact our EU
representative GPSR@taylorandfrancis.com
Taylor & Francis Verlag GmbH, Kaufingerstraße 24, 80331 München, Germany

www.ingramcontent.com/pod-product-compliance
Ingram Content Group UK Ltd.
Pitfield, Milton Keynes, MK11 3LW, UK
UKHW051941210425
457613UK00029B/1